FOR REFERENCE

Do Not Take From This Room

Hazardous Materials Response Handbook

Second Edition

Hazardous Materials Response Handbook

Second Edition

Based on the 1992 editions of NFPA 471, *Recommended Practice for Responding to Hazardous Materials Incidents*; NFPA 472, *Standard for Professional Competence of Responders to Hazardous Materials Incidents*; and NFPA 473, *Standard for Competencies for EMS Personnel Responding to Hazardous Materials Incidents*.

Edited by

Gary Tokle

Senior Fire Service Training and Education Specialist

National Fire Protection Association
Quincy, Massachusetts

NFPA®

Hazardous materials response handbook

This second edition of the *Hazardous Materials Response Handbook* is an essential reference for everyone who responds to hazardous materials incidents. It covers the spectrum of hazardous materials preparedness, from recognizing risks to undergoing decontamination. State-of-the-art, international guidelines are included, together with the latest NFPA hazardous materials standards.

This handbook features the complete texts of NFPA's hazardous materials response documents, *Recommended Practice for Responding to Hazardous Materials Incidents* (NFPA 471), *Standard for Professional Competence of Responders to Hazardous Materials Incidents* (NFPA 472), and *Standard for Competencies for EMS Personnel Responding to Hazardous Materials Incidents* (NFPA 473), along with explanatory commentary, which is integrated between requirements.

All NFPA codes and standards are processed in accordance with NFPA's *Regulations Governing Committee Projects*. The commentary in this handbook is the opinion of the author, an expert in the field of hazardous materials. However, it has not been processed in accordance with the NFPA *Regulations Governing Committee Projects*, and therefore it shall not be considered to be, nor relied upon as, a Formal Interpretation of the meaning or intent of any specific provision or provisions of NFPA 471, NFPA 472, or NFPA 473. The language contained in NFPA 471, 472, and 473, rather than this commentary, represents the official position of the Committee on Hazardous Materials Response Personnel and of the NFPA.

This handbook is a must for all public, military, industrial, and volunteer fire departments that respond to hazardous materials incidents, as well as for those responsible for preparing and training the response community.

Project Manager: Kathleen M. Robinson
Project Editor: Kimberley Foster
Composition: Colleen Bachelder, Cathy Ray, Claire McRudin, and Nancy Maria
Illustrations: Smizer Design
Cover Design: Graphic Advertising Design
Production: Donald McGonagle and Stephen Dornbusch

Contents

Recommended Practice for Responding to Hazardous Materials Incidents, NFPA 471

Standard for Professional Competence of Responders to Hazardous Materials Incidents, NFPA 472

Standard for Competencies for EMS Personnel Responding to Hazardous Materials Incidents, NFPA 473

Supplements

Foreword

Although many people think that interest in hazardous materials is a recent development, certain entities have been concerned about this particular issue for quite a while. In fact, the National Fire Protection Association has been developing standards that deal with various hazardous materials for decades. Consider, for example, the various NFPA codes that cover flammable liquids, oxidizers, gases, pesticides, dangerous chemicals, chemical reactions, and explosives, as well as other NFPA documents that pertain to the identification, classification, and handling of some of these materials.

It is true, however, that there has been a great deal more interest in the subject of late, for several reasons. First and foremost, people have come to realize that the world's resources are both precious and limited, a fact that has been hammered home by a number of increasingly distressing environmental disasters. Notable among these was the 1986 chemical fire in Basel, Switzerland, which did prolonged damage to the Rhine River, and the 1989 oil spill from the Exxon Valdez, which left large stretches of the Alaskan coastline coated with oil.

Of course, the environment was not the only victim of hazardous materials incidents in the past decades. It is safe to say that no hazardous materials catastrophe made more of an impact on the public's consciousness than the toxic gas leak in Bhopal, India, which killed thousands of people in 1984. Another 500 people died in LP-Gas explosions in Mexico in 1983, and six fire fighters lost their lives in a 1988 explosion in Kansas City, Missouri.

Similar incidents are likely to occur, and they will invariably result in demands for more and better regulations. In the United States, the Environmental Protection Agency, a major regulator of hazardous substances, is enforcing its regulations with increasing efficiency.

Another driving force in the increasing interest in hazardous materials has been a growing concern for worker safety and health. The Occupational Safety and Health Administration (OSHA) has developed rules that help focus attention on the needs of employees and on the obligations of their employers. Labor unions and litigation have helped bring about appropriate observance of the many and varied OSHA regulations.

Concern for worker safety has also resulted in the realization that emergency responders need, and are entitled to, adequate protection against the hazardous materials with which they deal. In 1984, the International Society of Fire Service Instructors (ISFSI) and the International Fire Service Training Association (IFSTA) asked NFPA to develop standards for responders to hazardous materials incidents. NFPA sought public support for the project, and the response indicated that many thought that there was indeed a need for such standards.

When it began establishing the technical committee that would handle the project, NFPA anticipated—and received—an eager response not only from the fire service but from other key interest groups, such as regulators, labor, manufacturers, and educators, as well.

The late Chief Warren Isman of Fairfax County, Virginia, was appointed chairman of the committee, and Captain Gerry Grey of the San Francisco Fire Department was appointed vice-chair. Unfortunately, Chief Isman passed away while NFPA 471 and 472 were being revised. The hazardous materials response community owes a great deal to his dedication and commitment, and his efforts will long be remembered as having been instrumental in the development of NFPA 471, 472, and 473.

The first NFPA staff liaison to the committee was Martin Henry, assistant vice president in charge of NFPA's Public Fire Protection Division, who also edited the first edition of this book. Mr. Henry's expertise was very helpful to the committee as it wrestled with the challenge of developing comprehensive standards for emergency response to hazardous materials incidents.

As the current NFPA staff liaison, I would like to extend my thanks to all those who helped develop NFPA's hazardous materials standards. I would also like to thank all those who are currently members of the committee. They carry on the excellent work begun by the original committee to ensure adequate standards for emergency responders. The commitment of both the committee members and the nonmembers who have contributed has truly been outstanding and is clearly demonstrated in the quality of the documents they have produced.

In closing, I would like to repeat a statement Mr. Henry made in the first edition of this book: Allow me to extol the NFPA standards-making process. It is an amazing process to witness, and, as committee members struggle to come to consensus over often contentious issues, it is rewarding to review the excellent documents it has produced. I am always impressed by the results.

Acknowledgments

I would like to express my gratitude to the many people who helped me with this book. I can't begin to list them all individually, but I would particularly like to thank the following groups:

The members of the Hazardous Materials Response Personnel Committee.

Those who, although not members of the committee, have faithfully contributed their expertise to the committee.

All those who submitted public proposals and public comments for the committee's consideration.

The many people who sent me needed material or reviewed various portions of the book.

I owe a special thank you to Deputy Chief Mary Beth Michos of the Department of Fire and Rescue Services of Montgomery County, Maryland. Chief Michos chaired the Subcommittee for Emergency Medical Service Operations at Hazardous Materials Incidents, which developed NFPA 473, *Standard for Competencies for Emergency Medical Personnel Responding to Hazardous Materials Incidents*. In addition, Chief Michos wrote the commentary included in this handbook for NFPA 473.

I would also like to thank Manuel H. Ehrlich, Jr., who chairs the Chemical Manufacturers Association's Chemical Incident Responder Training Task Group, for his committee's assistance with the commentary for Chapter 6 of NFPA 472.

Finally, I would like to thank the NFPA staff members who patiently worked with me during my introduction to the process of editing a handbook, especially Kimberley Foster, Project Editor; Colleen Bachelder, Cathy Ray, Claire McRudin, and Nancy Maria, Composition; Donald McGonagle and Stephen Dornbusch, Production; and Kathleen Robinson, Project Manager.

Standard and Commentary for NFPA 471, Recommended Practice for Responding to Hazardous Materials Incidents

1992 Edition

The text and illustrations that make up the commentary on the various sections of NFPA 471 are printed in black. The text of the standard itself is printed in blue.

Paragraphs that begin with the letter "A" are extracted from Appendix A of the standard. Appendix A material is not mandatory. It is designed to help users apply the provisions of the standard. In this handbook, material from Appendix A is integrated with the text, so that it follows the paragraph it explains. An asterisk (*) following a paragraph number indicates that explanatory material from Appendix A will follow.

1

Administration

1-1* Scope.

This practice applies to all organizations that have responsibilities when responding to hazardous materials incidents and recommends standard operating guidelines for responding to such incidents. It specifically covers planning procedures, policies, and application of procedures for incident levels, personal protective equipment, decontamination, safety, and communications.

A-1-1 Many of the recommendations in this document are based on U.S. federal laws and regulations that were in effect at the time of adoption. Users should carefully review laws and regulations that may have been added or amended or that may be required by other authorities. Users outside the jurisdiction of the U.S. should determine what requirements may be in force at the time of application of this document.

NFPA's definition of a recommended practice is "a document containing only advisory provisions (using the word 'should' to indicate recommendations) in the body of the text."

Note that this document does not apply only to fire departments. The committee intends that the practices outlined in this recommended practice be suitable for use by, and be applicable to, all persons who might be called upon to respond to a hazardous materials incident. Fire service personnel come immediately to mind as front-line emergency responders, as do the police on frequent occasions. Others affected include government and industrial responders, as well as workers who daily engage in operations that involve hazardous materials in the broadest sense of the definition.

1-2 Purpose.

The purpose of this document is to outline the minimum requirements that should be considered when dealing with responses to hazardous materials incidents and to specify operating guidelines for responding to hazardous materials incidents. It is not the intent of this recommended practice to restrict any jurisdiction from using more stringent guidelines.

Hazardous materials incidents will differ widely in terms of intensity, duration, and significance, as will the responders' capabilities and preparedness. Every incident brings a new set of variables, and deciding on the appropriate course of action involves subjective judgments. Nonetheless, helpful guidelines suitable for all incidents are available, and the purpose of this recommended practice is to point them out to responders.

1-3 Application.

The recommendations contained in this document should be followed by organizations that respond to hazardous materials incidents and by incident commanders responsible for managing hazardous materials incidents.

The federally mandated local emergency planning committees (LEPCs) identify those organizations that may be summoned to assist at hazardous materials incidents. Incident commanders are, or should be, aware that they may have to confront such incidents, and they should prepare themselves accordingly. NFPA 472, Chapter 5, outlines the competencies required of an incident commander.

1-4 Definitions.

Authority Having Jurisdiction.* The "authority having jurisdiction" is the organization, office or individual responsible for "approving" equipment, an installation or a procedure.

A-1-4 Authority Having Jurisdiction. The phrase "authority having jurisdiction" is used in NFPA documents in a broad manner since jurisdictions and "approval" agencies vary as do their responsibilities. Where public safety is primary, the "authority having jurisdiction" may be a federal, state, local or other regional department or individual such as a fire chief, fire marshal, chief of a fire prevention bureau, labor department, health department, building official, electrical inspector, or others having statutory authority. For insurance purposes, an insurance inspection department, rating bureau, or other insurance company representative may be the "authority having jurisdiction." In many circumstances the property owner or his designated agent assumes the role of the "authority having jurisdiction"; at government installations, the commanding officer or departmental official may be the "authority having jurisdiction."

Confinement. Those procedures taken to keep a material in a defined or local area.

Containment. The actions taken to keep a material in its container (e.g., stop a release of the material or reduce the amount being released).

For clarity, a distinction is made between containment and confinement. Thus, a large fuel oil tank is designed to *contain* its stored product, while the dike surrounding the tank is designed to *confine* the fuel oil in the event of a spill. The use of plugging or patching techniques falls under containment, while improvising a dike to direct or limit a spill is confinement.

Contaminant. A hazardous material that physically remains on or in people, animals, the environment, or equipment, thereby creating a continuing risk of direct injury or a risk of exposure outside of the hot zone.

A contaminant possesses some property or characteristic that renders it threatening. Inherent in the definition of a contaminant are the concepts that the offending material is present where it should not be and does not belong, and that the offending material is somehow toxic or harmful.

Figure 1.1 Containment involves keeping a hazardous material in a suitable container.

Contamination.   The process of transferring a hazardous material from its source to people, animals, the environment, or equipment, which may act as a carrier.

This is one of the more important considerations for emergency responders, from a health and life safety standpoint. The importance of determining whether personnel or equipment has become contaminated cannot be stressed too strongly. Because personnel are often unaware that contamination has occurred, it is imperative that procedures be established ahead of time to ensure that proper monitoring and decontamination take place.

Control.   The defensive or offensive procedures, techniques, and methods used in the mitigation of a hazardous materials incident, including containment, extinguishment, and confinement.

This includes whatever measures are taken to address an incident. Control methods will vary depending on the type of incident and the emergency responders' level of training.

Control Zones. The designation of areas at a hazardous materials incident based upon safety and the degree of hazard. Many terms are used to describe the zones involved in a hazardous materials incident. For purposes of this document, these zones are defined as the hot, warm, and cold zones.

Control zones should be established at all hazardous materials incidents. The committee used the terms *cold*, *warm*, and *hot* to describe these zones because they are easily understood and clearly suggest the nature of the situation one would expect to encounter within the zones.

Decontamination (Contamination Reduction). The physical and/or chemical process of reducing and preventing the spread of contamination from persons and equipment used at a hazardous materials incident.

In the broad sense, decontamination includes the preventive measures taken to protect against contamination, which should be avoided to whatever extent is practical. In a narrower sense, decontamination involves the safe and effective physical removal of contaminants or the use of a chemical treatment that renders the contaminants less harmful. Intelligent planning, the use of proper protective equipment, and the deployment of control zones are major factors to be considered in dealing with the problem. The extent of decontamination needed is based largely on the degree of harm associated with the contaminant.

Decontamination is also referred to as *contamination reduction*.

Degradation. (a) A chemical action involving the molecular breakdown of a protective clothing material or equipment due to contact with a chemical. (b) The molecular breakdown of the spilled or released material to render it less hazardous during control operations.

Degradation can also occur if protective clothing is not properly maintained and stored. Physical circumstances can accelerate degradation. See Supplement 10 for additional information.

Where used in discussions of mitigation efforts, the term *degradation* refers to the breakdown of complex chemicals into simpler forms. We speak of products as being biodegradable, meaning that they can be broken down into basic elements by bacteria. Bacteria have been used successfully in breaking down underground spills of petroleum products.

Emergency. A sudden and unexpected event calling for immediate action.

Implicit in this definition is the notion that the event requires urgent action for control or remediation in order to minimize the danger to people, the environment, or property.

Figure 1.2 An emergency can happen at fixed facilities or on any transportation corridor that passes through the community.

Environmental Hazard.   A condition capable of posing an unreasonable risk to air, water, or soil quality and to plants or wildlife.

An environmental hazard usually involves a release or potential release of a hazardous material that will endanger the environment. It should be treated as being no less significant than more immediately visible occurrences, such as fires and explosions, since people are a part of the environment and public health and welfare are severely affected by environmental hazards. The Valdez oil spill in early 1989 is an example of an environmentally hazardous incident that has had long-lasting and widespread adverse effects.

Hazard/Hazardous.   Capable of posing an unreasonable risk to health, safety, or the environment; capable of causing harm.

Figure 1.3 On April 21, 1980, fire erupted at a chemical disposal facility in Elizabeth, New Jersey. Fire fighters encountered exploding barrels of unknown chemicals, toxic fumes, and searing heat.

Hazard Sector. That function within an overall incident management system that deals with the mitigation of a hazardous materials incident. It is directed by a sector officer and principally deals with the technical aspects of the incident.

In some areas, the hazard sector may be referred to as the hazard division, but the functions of both are basically the same. The hazard sector is established to provide an organizational structure that will allow responders to supervise and control the operations that are essential at a hazardous materials incident. These include the tactical objectives carried out in the hot zone, site access control, and decontamination.

Hazard Sector Officer. The person responsible for the management of the hazard sector.

In some areas, the hazard sector officer may be called the hazardous materials group supervisor. In either case, he or she is responsible for managing the three operations listed above. Rescue operations also come under the direction of this position. The hazard sector officer is not responsible for controlling activities outside the control zones.

Hazardous Material.* A substance (gas, liquid, or solid) capable of creating harm to people, the environment, and property. See specific regulatory definitions in Appendix A.

A-1-4 **Hazardous Material.** There are many definitions and descriptive names being used for the term hazardous material, each of which depends on the nature of the problem being addressed.

Unfortunately, there is no one list or definition that covers everything. The U.S. agencies involved, as well as state and local governments, have different purposes for regulating hazardous materials that, under certain circumstances, pose a risk to the public or the environment.

(a) *Hazardous Materials.* The U.S. Department of Transportation (DOT) uses the term *hazardous materials*, which covers eleven hazard classes, some of which have subcategories called divisions. DOT includes in its regulations hazardous substances and hazardous wastes as Class 9 Miscellaneous Hazardous Materials, both of which are regulated by the U.S. Environmental Protection Agency (EPA), if their inherent properties would not otherwise be covered.

(b) *Hazardous Substances.* EPA uses the term *hazardous substance* for the chemicals that, if released into the environment above a certain amount, must be reported, and, depending on the threat to the environment, federal involvement in handling the incident can be authorized. A list of the hazardous substances is published in 40 CFR Part 302, Table 302.4. The U.S. Occupational Safety and Health Administration (OSHA) uses the term hazardous substance in 29 CFR Part 1910.120, which resulted from Title I of SARA and covers emergency response. OSHA uses the term differently than EPA. Hazardous substances, as used by OSHA, cover every chemical regulated by both DOT and EPA.

(c) *Extremely Hazardous Substances.* EPA uses the term *extremely hazardous substance* for the chemical that must be reported to the appropriate authorities if released above the threshold reporting quantity. Each substance has a threshold reporting quantity. The list of extremely hazardous substances is identified in Title III of the Superfund Amendments and Reauthorization Act (SARA) of 1986 (40 CFR Part 355).

(d) *Toxic Chemicals.* EPA uses the term *toxic chemical* for chemicals whose total emissions or releases must be reported annually by owners and operators of certain facilities that manufacture, process, or otherwise use a listed toxic chemical. The list of toxic chemicals is identified in Title III of SARA.

(e) *Hazardous Wastes.* EPA uses the term *hazardous wastes* for chemicals that are regulated under the Resource, Conservation, and Recovery Act (40 CFR Part 261.33). Hazardous wastes in transportation are regulated by DOT (49 CFR Parts 170-179).

(f) *Hazardous Chemicals.* OSHA uses the term *hazardous chemical* to denote any chemical that would be a risk to employees if exposed in the workplace. Hazardous chemicals cover a broader group of chemicals than the other chemical lists.

(g) *Dangerous Goods.* In Canadian transportation, hazardous materials are called dangerous goods.

Figure 1.4 *Some materials present multiple hazards, falling in a number of classifications.*

Class 1 (Explosives)

Explosive means any substance or article, including a device, that is designed to function by explosion (i.e., an extremely rapid release of gas and heat) or that, by chemical reaction within itself, is able to function in a similar manner even if not designed to function by explosion. Explosives in Class 1 are divided into six divisions. Each division will have a letter designation.

Division 1.1 consists of explosives that have a mass explosion hazard. A mass explosion is one that affects almost the entire load instantaneously.

Examples of Division 1.1 explosives include black powder, dynamite, and TNT.

Division 1.2 consists of explosives that have a projection hazard but not a mass explosion hazard.

Examples of Division 1.2 explosives include aerial flares, detonating cord, and power device cartridges.

Division 1.3 consists of explosives that have a fire hazard and either a minor blast hazard or a minor projection hazard or both, but not a mass explosion hazard.

Examples of Division 1.3 explosives include liquid-fueled rocket motors and propellant explosives.

Division 1.4 consists of explosive devices that present a minor explosion hazard. No device in the division may contain more than 25 g (0.9 oz) of a detonating material. The explosive effects are largely confined to the package and no projection of fragments of appreciable size or range are expected. An external fire must not cause virtually instantaneous explosion of almost the entire contents of the package.

Examples of Division 1.4 explosives include line-throwing rockets, practice ammunition, and signal cartridges.

Division 1.5 consists of very insensitive explosives. This division is comprised of substances that have a mass explosion hazard but are so insensitive that there is very little probability of initiation or of transition from burning to detonation under normal conditions of transport.

Examples of Division 1.5 explosives include prilled ammonium nitrate fertilizer-fuel oil mixtures (blasting agents).

Division 1.6 consists of extremely insensitive articles that do not have a mass explosive hazard. This division is comprised of articles that contain only extremely insensitive detonating substances and that demonstrate a negligible probability of accidental initiation or propagation.

Class 2

Division 2.1 (Flammable Gas) means any material that is a gas at 20°C (68°F) or less and 101.3 kPa (14.7 psi) of pressure, a material that has a boiling point of 20°C (68°F) or less at 101.3 kPa (14.7 psi) and that:

(a) Is ignitable at 101.3 kPa (14.7 psi) when in a mixture of 13 percent or less by volume with air; or

(b) Has a flammable range at 101.3 kPa (14.7 psi) with air of at least 12 percent regardless of the lower limit.

Examples of Division 2.1 gases include inhibited butadienes, methyl chloride, and propane.

Division 2.2 (Nonflammable, Nonpoisonous Compressed Gas, Including Compressed Gas, Liquefied Gas, Pressurized Cryogenic Gas, and Compressed Gas in Solution) A nonflammable, nonpoisonous compressed gas means any material (or mixture) that exerts in the packaging an absolute pressure of 280 kPa (41 psia) at 20°C (68°F).

A cryogenic liquid means a refrigerated liquefied gas having a boiling point colder than –90°C (–130°F) at 101.3 kPa (14.7 psi) absolute.

Examples of Division 2.2 gases include anhydrous ammonia, cryogenic argon, carbon dioxide, and compressed nitrogen.

Division 2.3 (Poisonous Gas) means a material that is a gas at 20°C (68°F) or less and a pressure of 101.3 kPa (14.7 psi or 1 atm), a material that has a boiling point of 20°C (68°F) or less at 101.3 kPa (14.7 psi), and that:

(a) Is known to be so toxic to humans as to pose a hazard to health during transportation, or;

(b) In the absence of adequate data on human toxicity, is presumed to be toxic to humans because, when tested on laboratory animals, it has an LC_{50} value of not more than 5,000 ppm.

Examples of Division 2.3 gases include anhydrous hydrogen fluoride, arsine, chlorine, and methyl bromide.

Hazard zones are associated with Division 2.3 materials:

Hazard zone A — LC_{50} less than or equal to 200 ppm.

Hazard zone B — LC_{50} greater than 200 ppm and less than or equal to 1,000 ppm.

Hazard zone C — LC_{50} greater than 1,000 ppm and less than or equal to 3,000 ppm.

Hazard zone D — LC_{50} greater than 3,000 ppm and less than or equal to 5,000 ppm.

Class 3 (Flammable Liquid)

Flammable liquid means any liquid having a flash point of not more than 60.5°C (141°F).

Examples of Class 3 liquids include acetone, amyl acetate, gasoline, methyl alcohol, and toluene.

Hazard zones are associated with Class 3 materials:

Hazard zone A — LC_{50} less than or equal to 200 ppm.

Hazard zone B — LC_{50} greater than 200 ppm and less than or equal to 1,000 ppm.

Combustible Liquid

Combustible liquid means any liquid that does not meet the definition of any other hazard class and has a flash point above 60°C (140°F) and below 93°C (200°F). Flammable liquids with a flash point above 38°C (100°F) may be reclassified as a combustible liquid.

Examples of combustible liquids include mineral oil, peanut oil, and No. 6 fuel oil.

Class 4

Division 4.1   (Flammable Solid) means any of the following three types of materials:

(a) Wetted explosives — explosives wetted with sufficient water, alcohol, or plasticizers to suppress explosive properties.

(b) Self-reactive materials — materials that are liable to undergo, at normal or elevated temperatures, a strongly exothermic decomposition caused by excessively high transport temperatures or by contamination.

(c) Readily combustible solids — solids that may cause a fire through friction and any metal powders that can be ignited.

Examples of Division 4.1 materials include magnesium (pellets, turnings, or ribbons) and nitrocellulose.

Division 4.2   (Spontaneously Combustible Material) means any of the following materials:

(a) Pyrophoric material — a liquid or solid that, even in small quantities and without an external ignition source, can ignite within 5 minutes after coming in contact with air.

(b) Self-heating material — a material that, when in contact with air and without an energy supply, is liable to self-heat.

Examples of Division 4.2 materials include aluminum alkyls, charcoal briquettes, magnesium alkyls, and phosphorus.

Division 4.3   (Dangerous When Wet Material) means a material that, by contact with water, is liable to become spontaneously flammable or to give off flammable or toxic gas at a rate greater than 1 L per kg of the material, per hour.

Examples of Division 4.3 materials include calcium carbide, magnesium powder, potassium metal alloys, and sodium hydride.

Class 5

Division 5.1   (Oxidizer) means a material that may, generally by yielding oxygen, cause or enhance the combustion of other materials.

Examples of Division 5.1 materials include ammonium nitrate, bromine trifluoride, and calcium hypochlorite.

Division 5.2   (Organic Peroxide) means any organic compound containing oxygen (O) in the bivalent -O-O- structure that may be considered a derivative of hydrogen peroxide, where one or more of the hydrogen atoms have been replaced by organic radicals.

Division 5.2   (Organic Peroxide) materials are assigned to one of seven types:

Type A — organic peroxide that can detonate or deflagrate rapidly as packaged for transport. Transportation of type A organic peroxides is forbidden.

Type B — organic peroxide that neither detonates nor deflagrates rapidly, but that can undergo a thermal explosion.

Type C — organic peroxide that neither detonates nor deflagrates rapidly and cannot undergo a thermal explosion.

Type D — organic peroxide that detonates only partially or deflagrates slowly, with medium to no effect when heated under confinement.

Type E — organic peroxide that neither detonates nor deflagrates and shows low, or no, effect when heated under confinement.

Type F — organic peroxide that will not detonate, does not deflagrate, shows only a low, or no, effect if heated when confined, and has low or no explosive power.

Type G — organic peroxide that will not detonate, does not deflagrate, shows no effect if heated when confined, and has no explosive power, is thermally stable, and is desensitized.

Examples of Division 5.2 materials include dibenzoyl peroxide, methyl, ethyl ketone peroxide, and peroxyacetic acid.

Class 6

Division 6.1   (Poisonous Material) means a material, other than a gas, that is either known to be so toxic to humans as to afford a hazard to health during transportation, or in the absence of adequate data on human toxicity, is presumed to be toxic to humans, including irritating materials that cause irritation.

Examples of Division 6.1 materials include aniline, arsenic compounds, carbon tetrachloride, hydrocyanic acid, and tear gas.

Division 6.2   (Infectious Substance) means a viable microorganism, or its toxin, that causes or may cause disease in humans or animals. Infectious substance and etiologic agent are synonymous.

Examples of Division 6.2 materials include anthrax, botulism, rabies, and tetanus.

Class 7

Radioactive material means any material having a specific activity greater than 0.002 microcuries per gram (μCi/g).

Examples of Class 7 materials include cobalt, uranium hexafluoride, and "yellow cake."

Class 8

Corrosive material means a liquid or solid that causes visible destruction or irreversible alterations in human skin tissue at the site of contact, or a liquid that has a severe corrosion rate on steel or aluminum.

Examples of Class 8 materials include nitric acid, phosphorus trichloride, sodium hydroxide, and sulfuric acid.

Class 9

Miscellaneous hazardous material means a material that presents a hazard during transport, but that is not included in another hazard class, including:

(1) Any material that has an anesthetic, noxious, or other similar property that could cause extreme annoyance or discomfort to a flight crew member so as to prevent the correct performance of assigned duties; and
(2) Any material that is not included in any other hazard class, but is subject to the DOT requirements (a hazardous substance or a hazardous waste).

Examples of Class 9 materials include adipic acid, hazardous substances (e.g., PCBs), and molten sulfur.

ORM-D Material

An ORM-D material means a material that presents a limited hazard during transportation due to its form, quantity, and packaging.

Examples of ORM-D materials include consumer commodities and small arms ammunition.

Forbidden

Forbidden means prohibited from being offered or accepted for transportation. Prohibition does not apply if these materials are diluted, stabilized, or incorporated in devices.

Incident. An emergency involving the release or potential release of a hazardous material, with or without fire.

This is an event that requires some type of response from the emergency response community.

Incident Commander. The person responsible for all decisions relating to the management of the incident. The incident commander is in charge of the incident site. This is equivalent to the on-scene incident commander as defined by 29 CFR 1910.120.

The incident commander is responsible for the overall control of operations at a hazardous materials incident, although the actual command may change from one person to another as an incident develops and becomes more complex. If multiple agencies are likely to be involved in an incident, which is generally the case where hazardous materials are involved, their roles and responsibilities should be clarified ahead of time. Some incident command systems provide for a "unified command" when multiple agencies are involved. However, the way in which the system is structured is less important than (a) having a system for command and control, (b) ensuring that the parties involved are aware of their roles, and (c) conducting training exercises to work out any problems.

Incident Management System. An organized system of roles, responsibilities, and standard operating procedures used to manage and direct emergency operations. Such systems are sometimes referred to as incident command systems (ICS).

Figure 1.5 *Potential releases of hazardous materials must also be dealt with during the mitigation process.*

Mitigation.   Actions taken to prevent or reduce product loss, human injury or death, environmental damage, and property damage due to the release or potential release of hazardous materials.

All actions taken to address a hazardous materials incident fall under the broad meaning of the term mitigation. The mitigation process serves to resolve or remediate the situation. Final mitigation can be an enduring procedure.

Monitoring Equipment.   Instruments and devices used to identify and quantify contaminants.

Monitoring equipment serves several important functions at an incident. It can be used to determine what personal protective equipment will be needed, to establish control zones, to identify unknowns, and to determine the level of contamination or the effectiveness of decontamination.

National Contingency Plan.*   Policies and procedures of the federal agency members of the National Oil and Hazardous Materials Response Team. This document provides guidance for responses, remedial action, enforcement, and funding mechanisms for hazardous materials incident responses.

A-1-4 **National Contingency Plan.** See *Code of Federal Regulations*: 40 CFR, Part 300, Subchapters A through J.

The National Contingency Plan came about as a result of Presidential Executive Order 12580 of January 23, 1987. It reads, in part, as follows:

Superfund Implementation

By the authority vested in me as President of the United States of America by Section 115 of the Comprehensive Environmental Response, Compensation, and Liability Act of 1980, as amended (42 U.S.C. 9615 et seq.) ("the Act"), and by Section 301 of Title 3 of the United States Code, it is hereby ordered as follows:

Section 1. National Contingency Plan.

(a)(1) The National Contingency Plan ("the NCP") shall provide for a National Response Team ("the NRT") composed of representatives of appropriate Federal departments and agencies for national planning and coordination of preparedness and response actions, and regional response teams as the regional counterpart to the NRT for planning and coordination of regional preparedness and response actions.

(2) The following agencies (in addition to other appropriate agencies) shall provide representatives to the National and Regional Response Teams to carry out their responsibilities under the NCP: Department of State, Department of Defense, Department of Justice, Department of the Interior, Department of Agriculture, Department of Commerce, Department of Labor, Department of Health and Human Services, Department of Transportation, Department of Energy, Environmental Protection Agency, Federal Emergency Management Agency, United States Coast Guard, and the Nuclear Regulatory Commission.

(3) Except for periods of activation because of a response action, the representative of the Environmental Protection Agency ("EPA") shall be the chairman and the representative of the United States Coast Guard shall be the vice chairman of the NRT and these agencies' representative shall be co-chairs of the Regional Response Teams ("the RRTs"). When the NRT or an RRT is activated for a response action, the chairman shall be the EPA or United States Coast Guard representative, based on whether the release or threatened release occurs in the inland or coastal zone, unless otherwise agreed upon by the EPA and United States Coast Guard representatives.

(4) The RRTs may include representatives from State governments, local governments (as agreed upon by the States), and Indian tribal governments. Subject to the functions and authorities delegated to Executive departments and agencies in other sections of this Order, the NRT shall provide policy and program direction to the RRTs.

(b)(1) The responsibility for the revision of the NCP and all of the other functions vested in the President by Sections 105(a), (b), (c), and (g), 125, and 301(f) of the Act is delegated to the Administrator of the Environmental Protection Agency ("the Administrator").

(2) The function vested in the President by Section 118(p) of the Superfund Amendments and Reauthorization Act of 1986 (Public Law 99-499) ("SARA") is delegated to the Administrator.

(c) In accord with Section 107(f)(2)(A) of the Act and Section 311(f)(5) of the Federal Water Pollution Control Act, as amended [33 U.S.C. 1321(f)(5)], the following shall be among those designated in the NCP as Federal trustees for natural resources:

(1) Secretary of Defense;
(2) Secretary of the Interior;
(3) Secretary of Agriculture;
(4) Secretary of Commerce;
(5) Secretary of Energy.

(d) Revisions to the NCP shall be made in consultation with members of the NRT prior to publication for notice and comment. Revisions shall also be made in consultation with the Director of the Federal Emergency Management Agency and the Nuclear Regulatory Commission in order to avoid inconsistent or duplicative requirements in the emergency planning responsibilities of those agencies.

(e) All revisions to the NCP, whether in proposed or final form, shall be subject to review and approval by the Director of the Office of Management and Budget ("OMB").

Additional sections of the Superfund Executive Order cover the following subjects:

Response and related authorities	Financial responsibility
Cleanup schedules	Employee protection
Enforcement	Management of the Hazardous
Liability	Substance Superfund and claims
Litigation	Federal facilities and general provisions

Penetration. The movement of a material through a suit's closures, such as zippers, buttonholes, seams, flaps, or other design features of chemical-protective clothing, and through punctures, cuts, and tears.

Protection against penetration is vital. Proper storage, maintenance, and testing of protective clothing and equipment must be part of a response team's standard procedures. See Supplement 10.

Permeation. A chemical action involving the movement of chemicals, on a molecular level, through intact material.

Note the difference between penetration and permeation: The latter takes place at the molecular level through the undamaged material itself. Different fabrics have different levels of resistance to chemical permeation, and all will absorb chemicals over a period of time.

NFPA 1991[1], 1992[2], and 1993[3] provide criteria for manufacturer permeation testing and certification. The user should make sure the chemical-protective clothing he or she purchases meets the appropriate standard and has been certified as meeting the standard. In any event, extreme caution on the part of the user is advised. Available data cannot cover every situation that may be encountered. See Supplement 10.

Protective Clothing. Equipment designed to protect the wearer from heat and/or hazardous materials contacting the skin or eyes. Protective clothing is divided into four types:

(a) Structural fire fighting protective clothing;
(b) Liquid splash-protective clothing;
(c) Vapor-protective clothing; and
(d) High temperature-protective clothing.

Response. That portion of incident management in which personnel are involved in controlling (defensively or offensively) a hazardous materials incident. The activities in the response portion of a hazardous materials incident include analyzing the incident, planning the response, implementing the planned response, and evaluating progress.

The actions taken, whether offensive or defensive, center on implementing the steps that will lead to containment, confinement, or extinguishment. See the definition of "planned response" in NFPA 472 for additional information.

Sampling.   Sampling is the process of collecting a representative amount of gas, liquid, or solid for analytical purposes.

Sampling is essential for evaluating a hazardous materials incident. It must begin as early in the incident as possible, and it should continue periodically until the incident commander deems it no longer necessary. Initial sampling will help responders determine what courses of emergency action are necessary, especially as they relate to response personnel, public safety, and environmental impact.

Figure 1.6 Industry personnel may have instruments that can be used to identify the hazard. For example, a portable photoionizing trace gas analyzer is routinely used to detect a wide variety of organic and inorganic vapors. This equipment may be available to responding fire fighters.

Stabilization.   The point in an incident at which the adverse behavior of the hazardous material is controlled.

This can be considered an intermediate step in the mitigation process. An analogy may be made to a fire fighting operation in which the fire is declared "under control." Although there may be hours or days of work remaining to be done, the situation is considered to be "in hand." It will not worsen, even though the response forces may be a long way from termination procedures.

Waste Minimization. Treatment of hazardous spills by procedures or chemicals designed to reduce the hazardous nature of the material and/or to minimize the quantity of waste produced.

One of the immediate goals of all initial responders and response units should be to minimize the quantity of waste produced. Their actions should be directed at controlling—or, especially, at not increasing—the extent of the hazard. Subsequent responders with special equipment and expertise can undertake treatments or procedures that will reduce or neutralize the nature of the waste.

Figure 1.7 Ideally, hazardous materials should be handled according to regulations from the time they are generated until they are disposed of. Response to a hazardous materials incident must also be geared to the ultimate goal of safe disposal, with minimal contamination.

References Cited in Commentary

1. NFPA 1991, *Standard on Vapor-Protective Suits for Hazardous Chemical Emergencies*, 1990 edition.
2. NFPA 1992, *Standard on Liquid Splash-Protective Suits for Hazardous Chemical Emergencies*, 1990 edition.
3. NFPA 1993, *Standard on Support Function Protective Garments for Hazardous Chemical Operations*, 1990 edition.

2

Incident Response Planning

2-1 Emergency Preparedness.

Planning is an essential part of emergency preparedness. The development of both facility response plans and community emergency plans is required by numerous state and federal laws, including SARA, Title III, "The Emergency Planning and Community Right to Know Act of 1986." Planning guides and reference materials are listed in Appendix B-2.4.

To the extent possible, everyone in the response community who may become involved in a hazardous materials incident should be included in the incident planning process. The purpose of planning is to be able to handle an incident as effectively and efficiently as possible. Planning should be a continuing process, and periodically testing the plan is essential to its success. Consult Supplement 4 for an edited version of the *Hazardous Materials Emergency Planning Guide*.

Figure 2.1   In preparing a hazardous materials emergency response plan, study plans from other communities.

2-2 Planning Team.

A planning team is necessary for developing the hazardous materials emergency plan. Local, state, and federal planning guidelines should be reviewed and consulted by the planning team when preparing plans for hazardous materials incidents.

As the *Emergency Planning Guide* points out, a team approach encourages the entire community to participate in the planning process. Several communities can combine to form a hazardous materials advisory council (HMAC), which then becomes a resource for the individual planning team. Members of the team should be selected on the basis of certain considerations, according to the National Response Team (NRT). These include ability, commitment, authority, expertise, and compatibility. Members should represent all elements of the community.

Supplement 6 provides guidance for developing an emergency response plan.

2-3 Annual Review.

As a minimum, an annual review and update of the hazardous materials emergency plan is necessary.

It is important that someone be assigned responsibility for keeping the plan current. The NRT recommends that a plan be reviewed at least annually, if not every six months. Title III of SARA requires an annual review. It may be useful to have the planning team meet regularly to review the plan and ensure that it is up-to-date.

2-4 Training Exercise.

As a minimum, a training exercise should be conducted annually to determine the adequacy and effectiveness of the hazardous materials emergency plan.

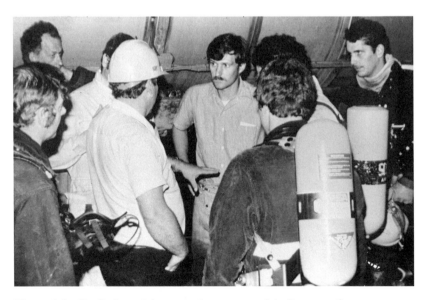

Figure 2.2   Realistic training exercises are good indicators of an emergency plan's effectiveness.

The training exercise can be a full-scale simulation or a tabletop exercise, and it should be followed by a review or critique. Training must be ongoing in order to keep the plan updated. The plan should indicate how it will be tested and how often.

The exercise will allow the different agencies and organizations involved the opportunity to work together to identify any weaknesses or deficiencies in the plan. It will also allow the responders to understand better their roles and responsibilities.

3

Response Levels

3-1* Planning Guide.

Table 3-1 is a planning guide intended to provide the user with assistance in determining incident levels for response and training. Potential applications to a jurisdiction's response activities may include development of standard operating procedures; implementation of a training program using the competency levels of NFPA 472, *Standard for Professional Competence of Responders to Hazardous Materials Incidents*; acquisition of necessary equipment; and development of community emergency response plans. When consulting this table, the user should refer to all of the incident condition criteria to determine the appropriate incident level.

A-3-1 These incidents can be considered as requiring either offensive operations or defensive operations.

Offensive operations include actions taken by a hazardous materials responder, in appropriate chemical-protective clothing, to handle an incident in such a manner that contact with the released material may result. This includes: patching or plugging to slow or stop a leak; containing a material in its own package or container; and cleanup operations that may require overpacking or transfer of a product to another container.

Defensive operations include actions taken during an incident where there is no intentional contact with the material involved. This includes: elimination of ignition sources, vapor suppression, and diking or diverting to keep a release in a confined area. It requires notification and possible evacuation, but does not involve plugging, patching, or cleanup of spilled or leaking materials.

Jurisdictions have the responsibility to develop standard operating procedures that equate levels of response to levels of training indicated in NFPA 472, *Standard for Professional Competence of Responders to Hazardous Materials Incidents.* Depending on the capabilities and training of personnel, first responder operational level may equate to incident level one and the technician level may equate to incident level two, for example.

Response personnel should operate only at that incident level that matches their knowledge, training, and equipment. If conditions indicate a need for a higher response level, then additional personnel, appropriate training, and equipment should be summoned.

Several response organizations have established a tiered response capability. For example, if a reported incident is classified as a Level 1 condition, the responding units will be able to handle it. A Level 2 incident requires greater response capability, and a Level 3 incident needs even more sophisticated equipment and highly trained personnel.

It is not the intent of Table 3-1 that response capabilities be absolutely determined on the basis of the incident level; nor must every jurisdiction have tiered response capability. The table is intended to serve as a guide should the responder need help in determining the real or potential seriousness of an incident.

Some of the terms and conditions in the table warrant explanation:

Placard Not Required. Refers to U.S. Department of Transportation regulations. However, the absence of a placard should not be taken as an assurance that the contents are harmless.

NFPA 0 or 1 in All Categories. A reference to NFPA 704, *Standard System for the Identification of the Fire Hazards of Materials.*[1] This standard deals with a labeling system that advises on three hazard conditions: health, flammability, and reactivity. There are five degrees of intensity, ranging from 0 through 4. A 0 or 1 in all three hazard conditions indicates relatively low hazard.

Class 9 (previously identified as ORM A, B, and C) and ORM-D. ORM means Other Regulated Materials.

Table 3-1 Planning Guide for Determining Incident Levels, Response, and Training

Incident Level	One	Two	Three
Product Identification	Placard not required, NFPA 0 or 1 all categories, all Class 9 and ORM-D.	DOT placarded, NFPA 2 for any categories, PCBs without fire, EPA regulated waste.	Class 2, Division 2.3—poisonous gases, Class 1, Division 1.1 and 1.2—explosives, organic peroxide, flammable solid, materials dangerous when wet, chlorine, fluorine, anhydrous ammonia, radioactive materials, NFPA 3 & 4 for any categories including special hazards, PCBs & fire, DOT inhalation hazard, EPA extremely hazardous substances, and cryogenics.
Container Size	Small (e.g., pail, drums, cylinders except one-ton, packages, bags)	Medium (e.g., one-ton cylinder, portable containers, nurse tanks, multiple small packages).	Large (e.g., tank cars, tank trucks, stationary tanks, hopper cars/trucks, multiple medium containers).
Fire/Explosion Potential	Low.	Medium.	High.
Leak Severity	No release or small release contained or confined with readily available resources.	Release may not be controllable without special resources.	Release may not be controllable even with special resources.
Life Safety	No life threatening situation from materials involved.	Localized area, limited evacuation area.	Large area, mass evacuation area.
Environmental Impact (Potential)	Minimal.	Moderate.	Severe.
Container Integrity	Not damaged.	Damaged but able to contain the contents to allow handling or transfer of product.	Damaged to such an extent that catastrophic rupture is possible.

Class 9 Miscellaneous includes *miscellaneous hazardous materials,* which refers to materials that present a hazard during transport, but that are not included in another hazard class, including:

(a) Any material that has an anesthetic, noxious, or similar property that could cause a flight crew such annoyance or discomfort as to prevent them from correctly performing their assigned duties; and
(b) Any material that is not included in any other hazard class but is subject to the DOT requirements (a hazardous substance or a hazardous waste).

Examples of Class 9 materials include adipic acid, hazardous substances such as PCBs, and molten sulfur.

ORM-D Material. A material that presents a limited hazard during transportation due to its form, quantity, and packaging. Examples of ORM-D materials include consumer commodities and small arms ammunition.

PCBs without Fire. Polychlorinated biphenyls (PCBs) present serious health threats to skin and the liver. Even without the added hazard of fire, they are sufficiently harmful to responders to warrant a Level 2 condition.

EPA Regulated Waste. These can be found in 40 CFR 261 (Code of Federal Regulations).

Class 2, Division 2.3 — Poisonous Gases (prior to 1991 classified as Poison A). Examples are arsine, hydrocyanic acid, and phosgene. These are extremely dangerous poisons.

Class 1, Division 1.1 and 1.2 — Explosives (prior to 1991 classified as Explosives A or B). Examples of Division 1.1 are dynamite and black powder. Examples of Division 1.2 are propellent explosives and rocket motors.

Organic Peroxide. Many organic peroxides are highly flammable, and most will decompose readily when heated. In some cases, the decomposition can be violent.

Flammable Solid. Examples are pyroxylin plastics, magnesium, and aluminum powder.

Materials Dangerous When Wet. Included in this category are sodium and potassium metals and calcium.

Chlorine. A greenish yellow gas that is highly toxic and irritating.

Fluorine. An extremely reactive and intensely poisonous yellow gas.

Anhydrous Ammonia. A very toxic and corrosive gas.

Radioactive Materials. Materials that spontaneously emit ionizing radiation having a specific activity greater than 0.002 microcuries per gram.

DOT Inhalation Hazard. Inhalation hazards are measured in terms of TLV/ TWA (threshold limit value/time-weighted average). Shipping papers must indicate inhalation hazard and containers must be marked "Inhalation Hazard." Vehicles will be placarded "Poison" or "Poison Gas," in addition to the primary hazard requirements.

EPA Extremely Hazardous Substances. EPA has published a list of 366 such substances.

Cryogenics. Cryogenics are extremely cold liquefied gases (–200°F) that can cause severe damage to skin or other body parts.

Container Size. The larger the container, the greater the potential for risk, hence the increase in the level of incident condition.

Fire/Explosion Potential. The assumption in each case is that the incident is not simply a fire, but that some hazardous material is involved. Where there is no fire, Level 1 may be appropriate, depending on other prevailing conditions. If a container is involved in fire, Level 3 may be more appropriate. It is conceivable that a fire involving a container can be handled safely by a responding fire department without the assistance of any hazardous materials response

Figure 3.1 Hazardous wastes may be found stored in containers that have corroded or are unsecured. Poorly contained wastes may mix, forming new compounds that are more unstable and toxic than their original components.

HAZARDOUS MATERIALS RESPONSE HANDBOOK

personnel. Nonetheless, appropriate authorities would have to be notified and alerted to the situation. It is also vitally important to keep in mind that containers involved in fire can overpressurize and fail in a catastrophic manner. Every precaution must be taken when approaching containers that are exposed to fire.

Leak Severity. The selection of levels obviously depends on the extent of the leak and the likelihood that it can be controlled.

Life Safety. The number of people potentially exposed is a major determining factor in selecting the appropriate level.

Environmental Impact. The terms used are general. Judgment is required, and experts should be consulted. The environmental impacts of an incident may not be known at the start of the incident, and they are frequently more severe than anticipated.

Container Integrity. Extreme care must be taken with damaged containers before allowing them to be transferred. Once again, experts in this field should be consulted. If doubt exists, the incident should be considered as Level 3.

Reference Cited in Commentary

1. NFPA 704, *Standard System for the Identification of the Fire Hazards of Materials,* 1990 edition.

4

Site Safety

4-1 Emergency Incident Operations.

4-1.1 Emergency incident operations should be conducted in compliance with Chapter 6 of NFPA 1500, *Standard on Fire Department Occupational Safety and Health Program,* or 29 CFR 1910.120 or EPA.

All fire fighters can be categorized as at least first responders. Most will be qualified at the first responder operational level. See NFPA 472, *Standard for Professional Competence of Responders to Hazardous Materials Incidents.*

The first responder category will also apply to most police officers, public utility workers, emergency medical technicians, employees at hazardous waste facilities, and drivers of trucks carrying hazardous materials. In many of these cases, the first responder awareness level will apply. In any event, OSHA 1910.120 is applicable, and specified training will be required. Chapter 6 of NFPA 1500 deals with emergency operations and covers organization, safety requirements, and incidents involving special hazards. This last category certainly applies to hazardous materials incidents.

4-1.2* An incident management system should be implemented at all hazardous materials incidents. Operations should be directed by a designated incident commander and follow established written standard operating procedures.

A-4-1.2 Though in the following text 29 CFR 1910.120 is cited, it should be understood that some states will adopt these regulations under state OSHA plans and others will adopt these regulations through adoption of a similar regulation established by EPA and appropriate state agencies.

Various agencies involved in emergency response to hazardous materials incidents use different types of incident management systems. The important point is that there is a system and that everyone is familiar with the way it functions. NFPA 1561, *Standard on Fire Department Incident Management System*[1], provides excellent guidance for developing an effective incident management system, clearly addressing the components that must be included in order to operate in as organized and safe a manner at an incident as possible. It should be kept in mind that an effective incident management system is one of the most important aspects of operating safely during an emergency. See Supplement 7.

4-1.3 An emergency response plan describing the general safety procedures that are to be followed at an incident should be prepared in accordance with 29 CFR 1910.120. These procedures should be thoroughly reviewed and tested.

All elements of the response community are required to have an emergency response plan outlining standard operating procedures for hazardous materials incidents. The plan should cover a number of subjects, including an incident command procedure, pre-incident planning, site control, evacuation policies, decontamination procedures, medical response plans, and training.

4-2 Ignition Sources.

Ignition sources should be eliminated whenever possible at incidents involving releases, or probable releases, of ignitable materials. Whenever possible, electrical devices used within the hot zone should be certified as intrinsically safe by recognized organizations.

Most ignition sources, such as matches, smoking materials, and open flames, are well known and easily recognized. However, other ignition sources, such as pilot lights, incidental static discharges, and electrical sparks, often receive less attention. Whenever flammable vapor is present, all sources of ignition should be removed.

4-3 Control Zones.

Control zone names have not been consistently applied at incidents. The intent of this section is to show areas of responder control. The various zones or areas at a typical emergency response site are shown in Figure 4-3.

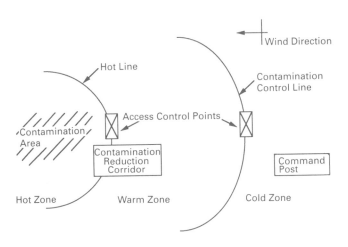

Figure 4-3   Diagram of control zones.

Control zones should be established at an incident as soon as possible to reduce contamination by controlling and directing the operations and movements of personnel at the incident. A site map that shows wind direction and topography will prove helpful. Boundaries for the various control zones are established using information gathered by monitoring the incident. Personnel should move only through the access control points in order to maintain control of the site and to prevent spread of contamination across zones.

Control zones should be monitored to ensure that they are in the proper locations as the incident progresses. It may be necessary to expand the zones or, in some cases, reduce them.

4-3.1* **Hot Zone.** Area immediately surrounding a hazardous materials incident, extending far enough to prevent adverse effects from hazardous materials releases to personnel outside the zone. This zone is also referred to as the exclusion zone or restricted zone in other documents.

A-4-3.1 Access into the hot zone is to be limited to those persons necessary to control the incident. A log is to be maintained at the access control point to record entry and exit time of all personnel in the hot zone.

Personnel may enter the hot zone, where contamination is likely to occur, for several reasons, including the need for sampling, cleanup, or spill control measures. The boundary lines should be clearly delineated by hazard tape, signs, or rope. It may be appropriate to have more than one access control point per zone so that there are separate entrance and exit points. All personnel within the hot zone should wear the level of protective equipment the incident commander has determined to be appropriate. Differing levels of protection may be appropriate in the same area, depending on the specific task being performed.

4-3.2 **Warm Zone.** The area where personnel and equipment decontamination and hot zone support takes place. It includes control points for the access corridor and thus assists in reducing the spread of contamination. This is also referred to as the decontamination, contamination reduction, or limited access zone in other documents.

The severity of contamination should decrease as responders move from the hot line toward the cold zone because of the effective decontamination procedures in the warm zone. Within the contamination reduction corridor, there should be one decontamination line for personnel and another for heavy equipment.

Personnel entering the warm zone from the cold zone should wear the level of protection required to operate in the warm zone. Personnel leaving the warm zone for the cold zone should remove the protective clothing they have worn in the warm zone.

The warm zone should be large enough to accommodate all the decontamination procedures that will take place in it.

4-3.3 Cold Zone. This area contains the command post and such other support functions as are deemed necessary to control the incident. This is also referred to as the clean zone or support zone in other documents.

Personnel in the cold zone may wear normal work clothes. The cold zone should be upwind of the hot zone and as far away from it as is practical. Support functions in the cold zone might include site security, medical support, reserve equipment, and a field laboratory.

4-4 Communications.

Effectively handling any emergency depends on establishing and implementing a coordinated communications program. Everyone operating at the scene must be involved, to varying degrees, in the communications loop. In addition, the communications network must extend beyond the immediate scene of operations, since dispatching centers are also part of the system.

Hazardous materials incidents are often more complicated than routine emergencies, a fact that increases the need for an effective communications system. Contact with many other agencies is often essential, and information available to those operating at the scene may be limited, especially in the early stages of an incident.

4-4.1 When personal protective clothing or remote operations inhibit communications, an effective means of communications, such as radios, should be established.

Radio equipment has improved substantially in recent years. Every consideration should be given to equipping workers with two-way radio equipment, particularly where protective clothing isolates the individual. There are now effective methods for communicating with personnel in totally encapsulating suits.

4-4.2 The frequencies employed in these radios should be "dedicated" and not used or shared with other local agencies.

Where multichannel radios are available to personnel, the incident commander can designate a particular channel for all on-scene communications.

4-4.3 Communication should be supplemented by a prearranged set of hand signals and hand-light signals to be used when primary communication methods fail. Hand-lights employed for this purpose should be in accordance with NFPA 70, *National Electrical Code®*, for use in hazardous environments.

Figure 4.1 Maintaining effective communications is a top priority at any hazardous materials incident.

Figure 4.2 Two-way radios link site commanders with entry personnel.

Where hand signals are to be employed, they should be simple to use and understand, and they should be limited to commands that are essential. Since an operation may well involve personnel from different disciplines and agencies, it is important that everyone involved in the operation understand and be able to use the appropriate hand or hand-light signals.

4-5 Monitoring Equipment.

Consult the commentary in NFPA 472 for additional information on monitoring equipment.

4-5.1 Monitoring equipment operates on several different principles and measures different aspects of hazardous materials releases. Examples of this equipment are:

(a) Oxygen meters

Air contains 21 percent oxygen by volume under normal conditions. Sixteen percent is needed to support human life. Atmospheres that contain less than 19.5 percent oxygen are generally considered to be deficient.

Several instruments can be used to determine the oxygen level, and many of these are small and easy to use. One type of meter measures the partial pressure of oxygen in air using an electrochemical sensor; the reading is then converted to the oxygen concentration.

(b) Combustible gas indicator (explosimeter)

This instrument measures the concentration of a combustible gas or vapor. The instrument burns the combustible gas, and the increased heat this causes is measured to provide information about the actual concentration of the gas or vapor.

(c) Carbon monoxide meter

Carbon monoxide is a colorless, odorless, and tasteless toxic gas that can prove fatal in very small concentrations. Several instruments can be used to measure its presence. Some use a catalytic combustion process, then measure the heat that is produced to determine how much carbon monoxide is present.

(d) pH meter

(e) Radiation detection instruments

Figure 4.3 *Monitoring equipment can be used to measure various types of hazardous materials present at an incident.*

There are four types of radiation: alpha particles, beta particles, gamma rays, and neutron particles. Radiation detection equipment includes survey meters and dosimeters. The dosimeter measures accumulated exposure to gamma radiation. Survey meters detect and measure beta and gamma radiation. Devices that measure neutron radiation are very expensive, but incidents resulting in neutron emissions are rare.

(f) Colorimetric detector tubes

These devices, widely used to evaluate airborne gases and vapors, are also known as detector tubes or gas indicators. They are easy to operate and can detect many contaminants.

The devices have a pump and a colorimetric indicator tube that contains a substance that reacts when contaminated air is drawn through the tube. Manufacturer's instructions and conversion tables help the responder determine the concentration of the contaminant.

(g) Organic vapor analyzer

These are direct reading devices capable of detecting all organic vapors. They can be certified as suitable for use in Class I, Division 1 atmospheres.

The manufacturer's instructions must be followed.

(h) Photoionization meter

These monitoring devices indicate the presence of many organic and some inorganic gases and vapors. Initially, they detect the total concentration of gases or vapors. Other probes can then be used to identify some of the compounds. These meters use ultraviolet radiation to ionize molecules, thereby producing a current proportionate to the number of ions present.

(i) Air sampling devices

Some air sampling devices collect a specific volume of air, while others pass a determined volume of air through a medium that removes the contaminant for sampling. The first type of metering technique is called *instantaneous sampling*; the second is known as *integrated sampling*.

(j) Other meters to measure specific products such as chlorine, hydrogen sulfide, or ethylene oxide

Metering instruments can be made to select a specified chemical or gas. For example, various chemicals absorb infrared energy at specified frequencies, and the instrument can be configured to limit the frequencies it will test to those characteristic of a specific gas or chemical. Where a single chemical is present, the device will indicate it. Where more than one chemical is present, other methods will also have to be used.

(k) pH paper or strips

These monitor the corrosivity of a substance by measuring its acidity or alkalinity.

(l) Organic vapor badge or film strip

This type of area sampling device, which measures personal exposure over time, samples the wearer's breathing zone and measures the dose to which the badge is exposed. Some of these devices are battery operated. Miniature dosimeters have been developed that a worker can wear on his or her lapel to measure exposure over a specified sampling period.

(m) Mercury badge

(n) Formaldehyde badge or strip.

4-5.2 All monitoring equipment should be operationally checked prior to use and periodically calibrated as per manufacturers' specifications.

Reference Cited in Commentary

1. NFPA 1561, *Standard on Fire Department Incident Management System*, 1990 edition.

5

Personal Protective Equipment

5-1 General.

It is essential that personal protective equipment meeting appropriate NFPA and OSHA standards be provided, maintained, and used. Protection against physical, chemical, and thermal hazards must be considered when selecting personal protective equipment.

Personal protective equipment includes structural fire fighting clothing, chemical-protective clothing, and high temperature-protective clothing. The hazardous materials incident responder need not be equipped with all three types, however. Many hazardous materials incidents do not involve fire, for example, so structural fire fighting clothing may not be needed. And first responders at the awareness or operational levels may include police officers, emergency medical service personnel, and fixed site workers, who are not expected to be equipped with structural fire fighting clothing.

Whatever personal protective equipment is used must meet the relevant standards. It must be used carefully, since there are hazards associated with its use. Among other things, personal protective equipment can provide the wearer with a false sense of security. It can also induce rapid fatigue and heat stress, limit vision, and impair communications. Selecting the appropriate protection is important, since overprotection can cause needless hardship.

See NFPA 472 for more information about personal protective clothing.

5-1.1 A written personal protective equipment program should be established in accordance with 29 CFR Part 1910.120. Elements of the program should include personal protective equipment selection and use; storage, maintenance, and inspection procedures; and training considerations. The selection of personal protective clothing should be based on the hazardous materials and/or conditions present and be appropriate for the hazards encountered.

Providing a complete range of protective equipment is not enough. Also needed are a training program that teaches the responder which items to use and how to use them correctly, and a maintenance, storage, and inspection program to ensure that the equipment is serviceable when it is needed.

It is critical that a clear personal protective equipment program be established. Personnel should never operate beyond the stage for which they have been trained and/or equipped. To do so puts them and the wider community in danger. Personal safety and self-protection are not the only concerns at work; an inappropriate action could have wide-ranging negative consequences.

5-1.2 Protective clothing and equipment used to perform fire suppression operations, beyond the incipient stage, should meet the requirements of Chapter 5 of NFPA 1500, *Standard on Fire Department Occupational Safety and Health Program*. Structural fire fighting protective clothing is not intended to provide chemical protection to the user.

The following excerpts from NFPA 1500 are worth citing:

"5-1.1 The fire department shall provide each member with the appropriate protective clothing and protective equipment to provide protection from the hazards of the work environment to which the member is or may be exposed. Such protective clothing and protective equipment shall be suitable for the tasks that the member is expected to perform.

5-1.2 Protective clothing and protective equipment shall be used whenever the member is exposed or potentially exposed to the hazards for which it is provided.

5-1.3 Members shall be fully trained in the care, use, inspection, maintenance, and limitations of the protective clothing and protective equipment assigned to them or available for their use.

5-1.4 Protective clothing and protective equipment shall be used and maintained in accordance with manufacturers' instructions. A maintenance and inspection program shall be established for protective clothing and protective equipment. Specific responsibilities shall be assigned for inspection and maintenance."[1]

Structural fire fighting clothing is not designed to offer the user chemical protection. Every fire department and every fire fighter is generally considered to have at least first responder awareness level capability and, even more likely, first responder operational capability; their structural fire fighting equipment is governed by OSHA regulation and by NFPA 1500. Yet that equipment should not be considered appropriate protection from exposure to chemical spills, even though it may provide protection in some cases. In other words, structural fire fighting equipment and chemical-protective clothing constitute two different categories of personal protective clothing, even though self-contained breathing apparatus may be common to each.

It is conceivable that a structural fire in a chemical facility would require the use of chemical-protective clothing rather than structural fire fighting clothing, even where the situation is clearly a structural fire. This is a judgment call, and the decision falls to the officer in charge of the incident.

Figure 5.1 Respiratory protection must be the primary concern in nonencapsulated protective clothing ensembles. Positive-pressure self-contained breathing apparatus will provide the highest level of respiratory protection.

5-2 Respiratory Protective Equipment.

5-2.1 Self-contained breathing apparatus (SCBA) should meet the requirements of NFPA 1981, *Standard on Open-Circuit Self-Contained Breathing Apparatus for Fire Fighters.*

Figure 5.2   Full facepiece, self-contained breathing apparatus (SCBA).

When an inhalation hazard is present or potentially present during an emergency response, positive-pressure self-contained breathing apparatus (SCBA) is the minimum type of respiratory protection that responders must wear. Once it has been determined through monitoring that a decreased level of protection is appropriate, the incident commander can allow the use of other types of respiratory protection.

Open-circuit SCBA provides air to the mask from a cylinder, and the wearer exhales directly to the atmosphere. When wearing closed-circuit SCBA, on the other hand, responders rebreath the exhaled, recycled gases, which have been chemically scrubbed and supplemented with oxygen from a supply source. The latter type of apparatus has a longer operating time.

5-2.2 Personal alert safety systems should meet the requirements of NFPA 1982, *Standard on Personal Alert Safety Systems (PASS) for Fire Fighters.*

NFPA 1500 requires fire service members involved in rescue, fire fighting, or other hazardous duty to use a PASS device, which sounds an alarm if the wearer does not move after 30 (±5) seconds. The standard also recommends that the device be worn on the protective clothing and used whether SCBA is being used or not.

5-2.3 Air Purifying Respirators. These devices are worn to filter particulates and contaminants from the air. They should only be worn in atmospheres where the type and quantity of the contaminants are known and sufficient oxygen is known to be present.

Air purifying respirators do not have an air source, so their use is very limited. They must rely on filters designed to purify the ambient air in which they are used. These respirators cannot be used in atmospheres that contain less than 19.5 percent oxygen or in atmospheres that contain contaminants immediately dangerous to life or health. When air purifying respirators are used, the oxygen level and the contaminated atmosphere must be monitored.

5-3 Chemical-Protective Clothing.

5-3.1 Chemical-protective clothing (CPC) is made from special materials and is designed to prevent the contact of chemicals with the body. Chemical-protective clothing is of two types: totally encapsulating and nonencapsulating.

Selecting the appropriate chemical-protective clothing is all-important. It is essential that the wearer be adequately trained in its use and limitations and that the manufacturer's instructions be followed. Totally encapsulating suits are generally one-piece suits designed to protect the wearer against gases and vapors. Nonencapsulating suits usually come in component parts and will not protect against gases and vapors, since they are not gastight.

Factors to consider when acquiring protective clothing include its cost, the ease with which it can be decontaminated, and the anticipated life of the garment.

See NFPA 472 and Supplement 10 for more information.

5-3.2 A variety of materials are used to make the fabric from which clothing is manufactured. Each material will provide protection against certain specified chemicals or
mixtures of chemicals. It may afford little or no protection against certain other chemicals. It is most important to note that there is no material that provides satisfactory
protection from all chemicals. Protective clothing material must be compatible with the
chemical substances involved, consistent with manufacturers' instructions.

Knowing the limitations of a particular protective garment is essential to
safe operation at a hazardous materials incident. This presupposes an awareness of the material from which the suit is made and of the manufacturer's
instructions regarding the garment's performance against specified chemicals
for specified exposure times.

5-3.3 Performance requirements must be considered in selecting the appropriate
chemical-protective material. These would include chemical resistance, permeation,
penetration, flexibility, abrasion, temperature resistance, shelf life, and sizing criteria.

The purpose of chemical-protective clothing is to protect the wearer from
exposure. In order to accomplish that end, performance requirements have
been established. There are many variables at work, since no single material protects against every chemical and different types of materials are more
effective than others in protecting against specific chemicals. Combinations
of chemicals are a complicating factor. In addition, there are some chemicals for which there is no effective, enduring protective material.

5-3.3.1 Chemical resistance is the ability of the material from which the protective
garment is made to prevent or reduce degradation and permeation of the fabric
by the attack chemical. Degradation is a chemical action involving the molecular
breakdown of the material due to contact with a chemical. The action may cause
the fabric to swell, shrink, blister, or discolor; become brittle, sticky, or soft; or deteriorate. These changes permit chemicals to penetrate the suit more rapidly or increase
the probability of permeation.

Manufacturers of chemical-protective clothing will furnish charts explaining the chemical resistance a product offers against degradation and permeation. However, manufacturers' information can be misleading, since their
conclusions may be based on old data, raw material performance, inconsistent test methods, and laboratory-type (rather than real-world) usage.

Mixtures of chemicals can significantly increase the rate at which protective clothing degrades. A chemical that permeates the clothing can allow other

chemicals in the mixture to enter the material and further degrade the clothing. To counter this problem, some protective clothing is now layered to offer additional protection against mixtures of chemicals.

5-3.3.2 Permeation is a chemical action involving the movement of chemicals, on a molecular level, through intact material. There is usually no indication that this process is occurring. Permeation is defined by two terms, permeation rate and breakthrough time. Permeation rate is the quantity of chemical that will move through an area of protective garment in a given period of time, usually expressed as micrograms of chemical per square centimeter per minute. Breakthrough time is the time required for the chemical to be measured on the inside surface of the fabric. The most desirable protective fabric is one that has the longest breakthrough time and a very low permeation rate. Breakthrough times and permeation rates are not available for all the common suit materials and the variety of chemicals that exist. Manufacturers' data and reference sources should be consulted. Generally, if a material degrades rapidly permeation will occur rapidly.

Additional variables that may affect permeation and breakthrough time are associated with actual-use conditions, such as temperature and humidity. Others include the thickness of the clothing and the concentration of the chemical it is intended to resist.

5-3.3.3 Penetration is the movement of material through a suit's closures, such as zippers, buttonholes, seams, flaps, or other design features. Torn or ripped suits will also allow penetration.

Zippers and closures on totally encapsulating suits are designed to be gastight, and the suit itself is designed for internal positive-pressure operation. Vapor should not penetrate. Nonencapsulating suits should protect against splashes, but they do not cover all parts of the body. The clothing should protect against splash penetration, however.

5-4 Thermal Protection.

5-4.1 Proximity Suits. These suits provide short duration and close proximity protection at radiant heat temperatures as high as 2,000°F (1093°C) and may withstand some exposure to water and steam. Respiratory protection must be provided with proximity suits.

Proximity suits are not designed for fire entry. They are two-piece or three-piece ensembles made of a heat-reflective material with layers of insulating linings. The wearer must also use special heat-reflective mittens and boots. Proximity suits are often worn over other protective clothing. NFPA 1500,[2] Section 5-4, provides guidelines on the use of clothing for proximity fire fighting. NFPA 1976, *Standard on Protective Clothing for Proximity Fire Fighting*,[3] provides criteria for the design of proximity clothing.

Figure 5.3 *Proximity suits are primarily used for close-proximity, short-duration exposures to both flame and radiant heat temperatures as high as 2,000°F (1093°C). They will withstand exposures to steam, liquids, and weak corrosive chemicals.*

5-4.2 Fire Entry Suits. This type of suit provides protection for brief entry into total flame environments at temperatures as high as 2,000°F (1093°C). This suit is not effective or meant to be used for rescue operations. Respiratory protection must be provided with fire entry suits.

This suit is composed of a coat, pants, boots, gloves, and a hood, each made up of many layers of flame-retardant materials. The outer layer is usually aluminized.

Figure 5.4 *Fire entry suits offer effective protection for short-duration entry into total flame environments, such as those one may find at petrochemical fires. They can withstand prolonged exposures to radiant heat levels as high as 2,000°F (1093°C).*

5-4.3 **Overprotection Garments.** These garments are worn in conjunction with chemical-protective encapsulating suits.

5-4.3.1 **Flash Cover Protective Suit.** Flash cover suits are neither proximity nor fire entry suits. They provide limited overprotection against flashback only. They are worn outside of other protective suits and are used only when the risks require them.

5-4.3.2 **Low Temperature Suits.** Low temperature suits provide some degree of protection of the encapsulating chemical-protective clothing from contact with low temperature gases and liquids. They are worn outside of the encapsulating chemical-protective clothing and are used only when the risk requires them.

5-5 Levels of Protection.

Personal protective equipment is divided into four categories based on the degree of protection afforded.

NOTE: An asterisk (*) after the description indicates optional, as applicable.

Those who have not been adequately trained to select and use personal protective equipment should not be permitted to wear such equipment at a hazardous materials incident. The training should be thorough and frequent so that the responder becomes intimately familiar with the equipment's limitations and handicaps. Responders should be trained to select, don, operate, test, clean, maintain, and care for the clothing.

During an incident, responders may have to change the level of protection they are wearing, from a high level of protection to a lower one or vice versa. Such a decision is made by the safety officer, based on his or her evaluation of the hazards to personnel present at the scene.

5-5.1 **Level A.** To be selected when the greatest level of skin, respiratory, and eye protection is required. The following constitute Level A equipment; it may be used as appropriate.

5-5.1.1 Pressure-demand, full facepiece, self-contained breathing apparatus (SCBA), or pressure-demand supplied air respirator with escape SCBA, approved by the National Institute of Occupational Safety and Health (NIOSH).

5-5.1.2 Vapor-protective suits: totally encapsulating chemical-protective suits (TECP suits) constructed of protective clothing materials; covering the wearer's torso, head, arms, and legs; having boots and gloves that may be an integral part of the suit, or separate and tightly attached; and completely enclosing the wearer by itself or in combination with the wearer's respiratory equipment, gloves, and boots. All components of a TECP suit, such as relief valves, seams, and closure assemblies, should provide equivalent chemical resistance protection. Vapor-protective suits should meet the requirements in NFPA 1991, *Standard on Vapor-Protective Suits for Hazardous Chemical Emergencies.*

5-5.1.3 Coveralls.*

5-5.1.4 Long underwear.*

5-5.1.5 Gloves, outer, chemical-resistant.

5-5.1.6 Gloves, inner, chemical-resistant.

5-5.1.7 Boots, chemical-resistant, steel toe and shank.

5-5.1.8 Hard hat (under suit).*

5-5.1.9 Disposable protective suit, gloves, and boots (depending on suit construction, may be worn over totally encapsulating suit).

5-5.1.10 Two-way radios (worn inside encapsulating suit).

5-5.2 **Level B.** The highest level of respiratory protection is necessary but a lesser level of skin protection is needed. The following constitutes Level B equipment; it may be used as appropriate.

5-5.2.1 Pressure-demand, full facepiece, self-contained breathing apparatus (SCBA), or pressure-demand supplied air respirator with escape SCBA, NIOSH approved.

5-5.2.2 Hooded chemical-resistant clothing that meet the requirements of NFPA 1992, *Standard on Liquid Splash-Protective Suits for Chemical Emergencies* (overalls and long-sleeved jacket, coveralls, one- or two-piece chemical-splash suit, disposable chemical-resistant overalls).

5-5.2.3 Coveralls.*

5-5.2.4 Gloves, outer, chemical-resistant.

5-5.2.5 Gloves, inner, chemical-resistant.

5-5.2.6 Boots, outer, chemical-resistant, steel toe and shank.

5-5.2.7 Boot covers, outer, chemical-resistant (disposable).*

5-5.2.8 Hard hat.

5-5.2.9 Two-way radios (worn inside encapsulating suit).

5-5.2.10 Face shield.*

5-5.3*   Level C. The concentration(s) and type(s) of airborne substance(s) is known and the criteria for using air purifying respirators are met. The following constitute Level C equipment; it may be used as appropriate.

A-5-5.3 Refer to OSHA 29 CFR 1910.134.

5-5.3.1 Full-face or half-mask, air purifying respirators, self-contained positive pressure breathing apparatus (NIOSH approved).

5-5.3.2 Hooded chemical-resistant clothing that meets the requirements of NFPA 1993, *Standard on Support Function Protective Garments for Hazardous Chemical Operations* (overalls, two-piece chemical-splash suit, disposable chemical-resistant overalls).

5-5.3.3 Coveralls.*

5-5.3.4 Gloves, outer, chemical-resistant.

5-5.3.5 Gloves, inner, chemical-resistant.

5-5.3.6 Boots, outer, chemical-resistant, steel toe and shank.

5-5.3.7 Boot covers, outer, chemical-resistant (disposable).*

5-5.3.8 Hard hat.

5-5.3.9 Escape mask.*

5-5.3.10 Two-way radios (worn under outside protective clothing).

5-5.3.11 Face shield.*

5-5.4   Level D. A work uniform affording minimal protection, used for nuisance contamination only. The following constitute Level D equipment; it may be used as appropriate.

5-5.4.1 Coveralls.

5-5.4.2 Gloves.*

5-5.4.3   Boots/shoes, chemical-resistant, steel toe and shank.

5-5.4.4   Boots, outer, chemical-resistant (disposable).*

5-5.4.5   Safety glasses or chemical-splash goggles.

5-5.4.6   Hard hat.

5-5.4.7   Escape mask.*

5-5.4.8   Face shield.*

5-6 Types of Hazards.

The types of hazards for which levels A, B, C, and D protection are appropriate are described below.

5-6.1   Level A protection should be used when:

5-6.1.1   The hazardous material has been identified and requires the highest level of protection for skin, eyes, and the respiratory system based on either the measured (or potential for) high concentration of atmospheric vapors, gases, or particulates; or the site operations and work functions involve a high potential for splash, immersion, or exposure to unexpected vapors, gases, or particulates of material that are harmful to skin or capable of being absorbed through the intact skin;

5-6.1.2   Substances with a high degree of hazard to the skin are known or suspected to be present, and skin contact is possible; or

5-6.1.3   Operations must be conducted in confined, poorly ventilated areas, and the absence of conditions requiring Level A have not yet been determined.

Under all circumstances, the totally encapsulating protective clothing must be compatible with the particular substance involved in the incident.

Table 5.1 EPA/OSHA Protection Levels*

Level of Protection	Equipment	Protection Provided	Should Be Used When:	Limiting Criteria
A	RECOMMENDED: • Pressure-demand, full-facepiece SCBA or pressure-demand, supplied-air respirator with escape SCBA. • Fully encapsulating chemical-resistant suit. • Inner chemical-resistant gloves. • Chemical-resistant safety boots/shoes. • Two-way radio communications. OPTIONAL: • Cooling unit. • Coveralls. • Long cotton underwear. • Hard hat. • Disposable gloves and boot covers.	The highest available level of respiratory, skin, and eye protection.	• The chemical substance has been identified and requires the highest level of protection for skin, eyes, and the respiratory system based on either: —measured (or potential for) high concentration of atmospheric vapors, gases, or particulates or —site operations and work functions involving a high potential for splash, immersion, or exposure to unexpected vapors, gases, or particulates of materials that are harmful to skin or capable of being absorbed through the intact skin. • Substances with a high degree of hazard to the skin are known or suspected to be present, and skin contact is possible. • Operations must be conducted in confined, poorly ventilated areas until the absence of conditions requiring Level A protection is determined.	Fully encapsulating suit material must be compatible with the substances involved.

* Based on EPA protective ensembles.

Table 5.1 EPA/OSHA Protection Levels, continued

Level of Protection	Equipment	Protection Provided	Should Be Used When:	Limiting Criteria
B	RECOMMENDED: • Pressure-demand, full facepiece SCBA or pressure-demand supplied-air respirator with escape SCBA. • Chemical-resistant clothing (overalls and long-sleeved jacket; hooded, one- or two-piece chemical splash suit; disposable chemical-resistant one-piece suit). • Inner and outer chemical-resistant gloves. • Chemical-resistant safety boots/shoes. • Hard hat. • Two-way radio communications. OPTIONAL: • Coveralls. • Disposable boot covers. • Face shield. • Long cotton underwear.	The same level of respiratory protection but less skin protection than Level A. It is the minimum level recommended for initial site entries until the hazards have been further identified.	• The type and atmospheric concentration of substances have been identified and require a high level of respiratory protection, but less skin protection. This involves atmospheres: —with IDLH concentrations of specific substances that do not represent a severe skin hazard; or —that do not meet the criteria for use of air purifying respirators. • Atmosphere contains less than 19.5 percent oxygen. • Presence of incompletely identified vapors or gases is indicated by direct-reading organic vapor detection instrument, but vapors and gases are not suspected of containing high levels of chemicals harmful to skin or capable of being absorbed through the intact skin.	• Use only when the vapor or gases present are not suspected of containing high concentrations of chemicals that are harmful to skin or capable of being absorbed through the intact skin. • Use only when it is highly unlikely that the work being done will generate either high concentrations of vapors, gases or particulates or splashes of material that will affect exposed skin.

Table 5.1	EPA/OSHA Protection Levels, continued			
Level of Protection	Equipment	Protection Provided	Should Be Used When:	Limiting Criteria
C	RECOMMENDED: • Full facepiece, air purifying, canister-equipped respirator. • Chemical-resistant clothing (overalls and long-sleeved jacket; hooded, one- or two-piece chemical splash suit; disposable chemical-resistant one-piece suit). • Inner and outer chemical-resistant gloves. • Chemical-resistant safety boots/shoes. • Hard hat. • Two-way radio communications. OPTIONAL: • Coveralls. • Disposable boot covers. • Face shield. • Escape mask. • Long cotton underwear.	The same level of skin protection as Level B, but a lower level of respiratory protection.	• The atmospheric contaminants, liquid splashes, or other direct contact will not adversely affect any exposed skin. • The types of air contaminants have been identified, concentrations measured, and a canister is available that can remove the contaminant. • All criteria for the use of air purifying respirators are met.	• Atmospheric concentration of chemicals must not exceed IDLH levels. • The atmosphere must contain at least 19.5 percent oxygen.
D	RECOMMENDED: • Coveralls. • Safety boots/shoes. • Safety glasses or chemical splash goggles. • Hard hat. OPTIONAL: • Gloves. • Escape mask. • Face shield.	No respiratory protection. Minimal skin protection.	• The atmosphere contains no known hazard. • Work functions preclude splashes, immersion, or the potential for unexpected inhalation of or contact with hazardous levels of any chemicals.	• This level should not be worn in the Exclusion Zone. • The atmosphere must contain at least 19.5 percent oxygen.

5-6.2 Level B protection should be used when:

5-6.2.1* The type and atmospheric concentration of substances have been identified and require a high level of respiratory protection, but less skin protection;

A-5-6.2.1 This involves atmospheres with IDLH (immediately dangerous to life and health) concentrations of specific substances that do not represent a severe skin hazard, or that do not meet the criteria for use of air purifying respirators.

5-6.2.2 The atmosphere contains less than 19.5 percent oxygen;

5-6.2.3 The presence of incompletely identified vapors or gases is indicated by a direct-reading organic vapor detection instrument, but the vapors and gases are known not to contain high levels of chemicals harmful to skin or capable of being absorbed through the intact skin; or

Level B protection cannot keep out vapors or gases that contain high concentrations of chemicals that may harm the skin or that may be absorbed through the skin. Level B should not be used when the work being performed is likely to generate high concentrations of gases or splashes of material that will affect exposed skin.

5-6.2.4 The presence of liquids or particulates is indicated, but they are known not to contain high levels of chemicals harmful to skin or capable of being absorbed through the intact skin.

5-6.3 Level C protection should be used when:

5-6.3.1 The atmospheric contaminants, liquid splashes, or other direct contact will not adversely affect or be absorbed through any exposed skin;

5-6.3.2 The types of air contaminants have been identified, concentrations measured, and an air purifying respirator is available that can remove the contaminants; and

5-6.3.3 All criteria for the use of air purifying respirators are met.

Level C protection is not appropriate where atmospheric concentrations of chemicals exceed IDLH levels, nor where the atmosphere contains less than 19.5 percent oxygen.

5-6.3.4* Atmospheric concentration of chemicals must not exceed IDLH levels. The atmosphere must contain at least 19.5 percent oxygen.

A-5-6.3.4 Refer to OSHA 29 CFR 1910.134.

5-6.4 Level D protection should be used when:

5-6.4.1 The atmosphere contains no known hazard; and

5-6.4.2* Work functions preclude splashes, immersion, or the potential for unexpected inhalation of or contact with hazardous levels of any chemicals.

A-5-6.4.2 Combinations of personal protective equipment other than those described for Levels A, B, C, and D protection may be more appropriate and may be used to provide the proper level of protection.

Level D protection is not appropriate for personnel operating in the warm zone. The atmosphere in which Level D protection is used must contain at least 19.5 percent oxygen.

References Cited in Commentary

1. NFPA 1500, *Standard on Fire Department Occupational Safety and Health Program*, 1992 edition.
2. Ibid.
3. NFPA 1976, *Standard on Protective Clothing for Proximity Fire Fighting*, 1992 edition.

6

Incident Mitigation

6-1 Control.

This chapter will address those actions necessary to ensure confinement and containment (the first line of defense) in a manner that will minimize risk to both life and the environment in the early, critical stages of a spill or leak. Both natural and synthetic methods can be employed to limit the releases of hazardous materials so that effective recovery and treatment can be accomplished with minimum additional risk to the environment or to life.

Ludwig Benner, Jr., offers a popular definition of hazardous materials: "Something that jumps out of its container at you when something goes wrong and hurts or harms the thing it touches."[1] An important element of that definition is that the harmful "something" is normally controlled or contained. Only when it is outside its normal controlling element does a hazardous materials incident occur.

It is only reasonable, then, that mitigating an incident must involve controlling the material that is presenting the problem. The types of control are divided into confinement and containment, and the methods of mitigation are either physical or chemical. This approach presents some order to the process and simplifies it for better understanding.

6-2 Types of Hazardous Materials.

All hazardous materials can be subdivided into three general categories, based on the principal characteristic that makes them harmful or dangerous: chemical, biological, or radioactive. Of course, other safety hazards exist at all emergencies. These include dangers typically associated with the response itself and with the operations responders undertake up to the point at which they bring the incident under control.

6-2.1 Chemical Materials. Those materials that pose a hazard based upon their chemical and physical properties.

Examining the U.S. Department of Transportation list of hazard classes reveals that most of the classes would fall under the chemical hazard type of material. The effect of exposure to chemical hazards can be either acute or chronic.

6-2.2 Biological Materials. Those organisms that have a pathogenic effect to life and the environment and can exist in normal ambient environments.

Examples of biological hazards are those whose packaging requires an "Etiologic Agents" label. Such hazards include toxins or microorganisms that cause diseases such as cholera, tetanus, and botulism. Disease-causing organisms might be found in waste from hospitals, laboratories, and research institutions.

6-2.3 Radioactive Materials. Those materials that emit ionizing radiation.

U.S. DOT lists three classes of radioactive materials, with Class I being the least harmful. Packaging requirements for radioactive materials vary depending on the hazard potentials the material itself presents. The three types of harmful radiation emitted by radioactive materials are alpha, beta, and gamma.

6-3 Physical States of Hazardous Materials.

Hazardous materials may be classified into three states, namely gases, solids, and liquids. They can be stored or contained at a high or low pressure. All three states may be affected by the environment in which the incident occurs. The emergency responder must take into account conditions such as heat, cold, rain, or wind, which can have a significant effect on the methods used to accomplish a safe operation.

All matter exists in one of the three states listed. Properties associated with each state have a bearing on how a specific material appears or behaves in the environment. For example, a liquid with a boiling point below 100°F (37.8°C) tends to give off substantial vapor at ambient temperatures. A gas with a vapor density substantially heavier than air may collect at low points and migrate along the ground until it mixes with air.

6-4 Methods of Mitigation.

There are two basic methods for mitigation of hazardous materials incidents, physical and chemical. Table 6-4.1 lists many physical methods and Table 6-4.2 lists many chemical methods that may be acceptable for mitigation of hazardous materials incidents. Recommended practices should be implemented only by personnel appropriately prepared by training, education, or experience.

Some of the methods listed for mitigating an incident require a high degree of specialized training on the part of the responder and require the use of sophisticated technical equipment. On the other hand, some mitigation efforts might be carried out by personnel at the first responder operational level. For example, diking or blanketing a liquid spill of diesel fuel can often be accomplished easily. Transferring that same product from a damaged tank truck to another tank truck, however, would require more specialized training and equipment than the first responder would be expected to have. Other operations, such as vent and burn techniques, should only be attempted by highly specialized personnel. In every case, good judgment must be used.

6-4.1* Physical Methods. Physical methods of control involve any of several processes or procedures to reduce the area of the spill, leak, or other release mechanism. In all cases, methods used should be acceptable to the incident commander. The selection of personal protective clothing should be based on the hazardous materials and/or conditions present and should be appropriate for the hazards encountered. Refer to Table 6-4.1.

Table 6-4.1   Physical Methods of Mitigation⁵

Method	Chemical Gases LVP*	Chemical Gases HVP**	Chemical Liq.	Chemical Sol.	Biological Gases LVP	Biological Gases HVP	Biological Liq.	Biological Sol.	Radiological Gases LVP	Radiological Gases HVP	Radiological Liq.	Radiological Sol.
Absorption	yes	yes	yes	no	no	no	yes⁴	no	no	no	yes	no
Covering	no	no	yes	yes	no	no	yes	yes	no	no	yes³	yes³
Dikes, Dams, Diversions & Retention	yes	yes⁶	yes	yes	no	no	yes	yes	no	no	yes	yes
Dilution	yes	yes	yes	yes	no	no	no	no	yes	no	yes	yes
Overpack	yes	no	yes	yes	yes	no	yes	yes	yes	no	yes	yes
Plug/Patch	yes	yes	yes	yes	yes	yes	yes	yes	yes	yes	yes	yes
Transfer	yes	no	yes	yes	yes	no	yes	yes	no	no	yes	yes
Vapor Suppression (Blanketing)	no	no	yes	yes	no	no	yes	yes	no	no	no	no
Venting¹	yes	yes	yes	no	yes	no	no	no	yes²	no	no	no

* Low Vapor Pressure
** High Vapor Pressure

1. Venting of low pressure gases is recommended only when an understanding of the biological system is known. Venting is allowed when the bacteriological system is known to be nonpathogenic, or if methods can be employed to make the environment hostile to pathogenic bacteria.
2. Venting of low vapor pressure radiological gases is allowed when the gas(es) is/are known to be alpha or beta emitters with short half lives. Further, this venting is only to be allowed after careful consultation with a certified health physicist.
3. Covering should be done only after consultation with appropriate experts.
4. Absorption of liquids containing bacteria is permitted where the absorption bacteria or environment is hostile to the bacteria.
5. For substances involving more than one type, the most restrictive control measure should be used.
6. Water dispersion on certain vapors and gases only.

A-6-4.1 Procedures described in 6-4.1.1 through 6-4.1.11 should be completed only by personnel trained in those procedures.

6-4.1.1* **Absorption.** Absorption is the process in which materials hold liquids through the process of wetting. Absorption is accompanied by an increase in the volume of the sorbate/sorbent system through the process of swelling. Some of the materials typically used as absorbents are sawdust, clays, charcoal, and polyolefin-type fibers. These materials can be used for confinement, but it should be noted that the sorbed liquid can be desorbed under mechanical or thermal stress. When absorbents become contaminated, they retain the properties of the absorbed hazardous liquid, and they are, therefore, considered to be hazardous materials and must be treated and disposed of accordingly. See ASTM F716, *Method of Testing Sorbent Performance of Absorbents*, for further information.

Many commercially available products are suitable for use as absorbents. Different types of absorbents are designed for different types of spilled materials. Absorbents can help reduce vapor generation and can facilitate cleanup procedures.

A-6-4.1.1 Absorbents saturated with volatile liquid chemicals can create a more severe vapor hazard than the spill alone because of severely enlarged surface area for vapor release.

6-4.1.2 **Covering.** Refers to a temporary form of mitigation for radioactive, biological, and some chemical substances such as magnesium. It should be done after consultation with a certified health physicist (in the case of radioactive materials) or other appropriate experts.

6-4.1.3 **Dikes, Dams, Diversions, and Retention.** These refer to the use of physical barriers to prevent or reduce the quantity of liquid flowing into the environment. Dikes or dams usually refer to concrete, earth, and other barriers temporarily or permanently constructed to hold back the spill or leak. Diversion refers to the methods used to physically change the direction of flow of the liquid. Vapors from certain materials, such as liquefied petroleum gas (LPG), can be dispersed by means of a water spray.

These techniques are the most commonly employed methods of controlling releases because responders can always improvise and simple methods of confinement can be devised with a little ingenuity. On the other hand, substantial liquid spills of hazardous materials can pose insurmountable problems.

In addition to the techniques listed, trenches can be used to collect spilled liquids, and pumps can transfer the materials to containers or to a containment system. Earthen dikes or dams can be erected quickly under favorable conditions, and even sand bags can be used in the damming effort. Commercial booms are available and are widely used to control spills, especially spills on waterways. See Figure 6.1.

Figure 6.1 Boom being placed across a waterway.

6-4.1.4 Dilution. Refers to the application of water to water-miscible hazardous materials. The goal is to reduce the hazard to safe levels.

Responders should not use water indiscriminately or without knowing what effect it will have. The addition of water to a liquid spill can add to confinement problems. Many materials react with water, thus increasing the intensity of the incident. Even if a material is water-soluble, the amount of water necessary to achieve a safe level could well render dilution an impractical approach. Nonetheless, it is a viable option in many instances.

6-4.1.5 **Overpacking.** The most common form of overpacking is accomplished by the use of an oversized container. Overpack containers should be compatible with the hazards of the materials involved. If the material is to be shipped, DOT specification overpack containers must be used. (The spilled materials still should be treated or properly disposed of.)

If it is possible to do so, a leaking drum or container should be temporarily repaired to reduce spillage before the container is placed in an overpack container. Reducing a leak in a container can sometimes be accomplished simply by repositioning the container. Holes can be covered, and temporary patches applied. Overpack containers typically have a form-fitting gasket in a lid that can be tightly secured with a ring-type closure.

A leaking container can be put into an overpack drum or container by placing the overpack on its side and sliding the smaller container into it, by lowering the overpack over the leaking container and then tipping it upright, or by using mechanical equipment to raise and lower the leaking container into the overpack container.

It is important that the overpack container be labeled in accordance with U.S. DOT regulations for the particular product it is carrying.

Figure 6.2 One method of overpacking involves lowering the overpack over the leaking container, then tipping it upright.

Responders must try to make sure that containers weakened by deterioration do not fail. Responders must also try to avoid injury when lifting or moving containers, particularly large or heavy containers. Based on the hazards that are present, personnel should wear the appropriate chemical-protective clothing and respiratory protection.

6-4.1.6 Plug and Patch. Plugging and patching refers to the use of compatible plugs and patches to reduce or temporarily stop the flow of materials from small holes, rips, tears, or gashes in containers. The repaired container may not be reused without proper inspection and certification.

Plugging involves putting something into a hole to reduce both the size of the hole and the flow from the hole. Tapered wooden plugs are often used. No matter what material the plug is made of, however, it must be compatible with both the product and the container. For example, soft pine may not be appropriate for plugging a strong acid leak. Again, responders must wear the appropriate chemical-protective clothing and respiratory protection.

Patching involves placing something over a hole to keep the material inside the container from leaking out. Patches are generally secured with clamps or adhesives. Patches designed to repair leaks in pipes of various sizes are commercially available.

Limiting or restricting a leak is an important condition of the mitigation process, so it is essential that responders master this skill. At all times, however, the safety of the responder must be paramount.

See Figures 6.3(a) and (b) for examples of plugging and patching materials.

6-4.1.7 Transfer. Transfer refers to the process of moving a liquid, gas, or some forms of solids, either manually, by pump, or by pressure transfer, from a leaking or damaged container or tank. Care must be taken to ensure the pump, transfer hoses and fittings, and container selected are compatible with the hazardous material. When flammable liquids are transferred, proper concern for electrical continuity (such as bonding/grounding) must be observed.

Materials should be transferred from one tank truck to another by personnel who are skilled and practiced in the procedure. The incident commander will be in charge of the operation and is responsible for seeing that proper precautions are taken. However, responders must rely on the experience of industry personnel who are appropriately trained and equipped to

Figure 6.3(a)   Examples of plugging and patching materials.

Figure 6.3(b)   Plugging materials being installed.

perform the transfer operation. For additional information about gasoline tank truck emergencies, see "Gasoline Tank Truck Emergencies: Guidelines and Procedures"[2] by Hildebrand and Noll.

It is essential that all electrical equipment used to transfer flammable liquids be grounded or bonded and that it be approved for such usage. NFPA 70, *National Electrical Code*®[3] (*NEC*®), should be consulted for additional information on this subject.

See NFPA 472, Chapter 4, for more information.

6-4.1.8 Vacuuming. Many hazardous materials may be placed in a containment simply by vacuuming them up. This has the advantage of not causing an increase in volume. Care must be taken to ensure compatibility of materials. The exhaust air may be filtered, scrubbed, or treated as needed. The method of vacuuming will depend on the nature of the hazardous material.

6-4.1.9 Vapor Dispersion. Vapors from certain materials can be dispersed or moved by means of a water spray. With other products, such as liquefied petroleum gas (LPG), the gas concentration may be reduced below the lower flammable limit through rapid mixing of the gas with air, using the turbulence created by a fine water spray. Reducing the concentration of the material through the use of water spray may bring the material into its flammable range.

6-4.1.10* Vapor Suppression (Blanketing). Vapor suppression refers to the reduction or elimination of vapors emanating from a spilled or released material through the most efficient method or application of specially designed agents. A recommended vapor suppression agent is an aqueous foam blanket.

A-6-4.1.10 One technique available for handling a spill of a hazardous liquid is the application of foams to suppress the vapor emanating from the liquid. This technique is ideally suited for liquid spills that are contained, i.e., diked. It can also be used where the spill is not confined. In all cases this technique should only be undertaken by personnel who have been trained in the use of foam concentrate for vapor suppression. Training in the use of foam as a fire extinguishing agent is not sufficient to qualify an individual for applying foam application as a vapor-suppressing agent.

Vapor-suppressing foam concentrates vary in their effectiveness depending on a number of factors. These factors can include the type of foam, the 25 percent drainage time of the foam, the rate of application of the foam, and the depth of the foam blanket. These variables serve to emphasize the need for training of the person selecting this technique for applying foam as a vapor-suppressing medium.

Figure 6.4 A fine water spray can be used to disperse some hazardous vapors.

Foams are produced by mechanically mixing a dilute solution of the foam concentrate and water with air, producing an expanded foam. Foams have been developed basically as fire extinguishing agents. Data has also been developed on their capability to suppress vapor release from water-immiscible flammable or combustible hydrocarbon liquids.

Foam and specific foam concentrates for each category, along with the definitions of appropriate terms, can be obtained by consulting NFPA 11, *Standard for Low Expansion Foam and Combined Agent Systems*; NFPA 11A, *Medium- and High-Expansion Foam Systems*; NFPA 16, *Foam-Water Sprinklers*; and ASTM Standard Guide F1129, *Using Aqueous Foams to Control the Vapor Hazard from Immiscible Volatile Liquids.*

The use of fire fighting foam as a vapor suppressant involves some considerations that are different from those required for fire extinguishing agents. It should be noted that fire fighting foams are predominantly restricted to use on water-immiscible hydrocarbon liquids or polar compounds. They are not usually effective on inorganic acids or bases, nor on liquefied gases, including the hydrocarbon gases such as methane and propane, in controlling vapor release. One of the obvious issues of concern is the stability of the foam blanket as a function of time. This can be approached by looking at what is referred to as the foam quality.

Foam quality is generally measured in terms of foam expansion ratio and foam 25 percent drainage time. Foam expansion ratio refers to the volume of foam solution. AFFF is often used in nonaspirating equipment such as water fog nozzles. Nonaspirated AFFF solutions have significantly limited effectiveness in comparison with aspirated foam solutions in vapor suppression.

The 25 percent drainage time is that time which is required for 25 percent of the foam solution to drain from the foam. This is the property that is generally used to measure the stability of the foam. The slower the drainage of any expanded foam, regardless of expansion, the more effective and longer lasting is the foam blanket. This assumes weather conditions are ideal.

An important factor is the vapor pressure of the liquid that is being suppressed. Liquid vapor pressures can vary widely. The higher the vapor pressure, the slower the control time. The key to effective use of foam as a vapor-suppressing agent is to have a continuous foam blanket on the fuel surface. Films from AFFF/FFFP are no guarantee of effective vapor control.

It is important to recognize that there are some limitations in the use of foam fire extinguishing agents in vapor suppression. As indicated, these materials are basically designed for flammable hydrocarbon liquids. They have severe shortcomings for inorganic acids or bases or liquefied gases. They should not be used for vapor suppression of these categories of volatile hazardous chemicals without consultation with the manufacturer of the specific foam agent being considered.

Most fire fighting foam concentrates have a limited range of pH tolerance. pH is a measure of the acidity or alkalinity of inorganic acids or bases. pH tolerance is the level that the bubble wall of the foam can tolerate before collapsing catastrophically. A few surfactant foams and the polar compound-type foams have good pH tolerance. Most of the protein, fluoroprotein, AFFFs, and high-expansion foams are not suitable for inorganic acids or bases.

Liquefied gas spills may be controlled by the application of high-expansion foam blankets. Low-expansion foams are not effective for liquefied gas spill control. Because of the large temperature differential between the liquefied gas and the foam, the drainage from the foam initially exaggerates boiloff from the spill. The higher the expansion, the lesser the effect of the drainage. For liquefied gases and all water-reactive inorganic materials, the foam should exhibit the best chemical resistance and expansion ratios to ensure maximum water retention consistent with the condition of the spill site. Since each spilled material can have unique properties, the manufacturer of the foam concentrate should be consulted for directions.

The great differences in the chemistry of flammable hydrocarbon liquids and the water-reactive inorganic materials have resulted in the development of foam concentrates specifically applicable to the inorganic chemicals. Few fire fighting foams have capabilities of vapor suppression of the inorganic acids and bases. For effective control, special foam concentrates should be employed.

Some special foam concentrates are specific for either acids or alkalis but not both. Further, they are not applicable to all inorganic materials nor are they effective in fire suppression. In many cases, their effectiveness is limited, and intermittent foam make-up may be required to maintain the foam blanket. Others, usually containing a polymer modification, can cover a wide range of materials, both acids and bases, and may possess some degree of fire resistance. This is important for those inorganic materials that may also pose a fire hazard.

At present, there is no single foam concentrate that is truly effective against all categories of volatile hazardous chemicals. A few possess limited capabilities in most categories, but they are compromise materials sacrificing in one category to provide some capability in other categories. These may, however, be the best choice for first responders where an overall capability is essential.

In all cases, however, the foam manufacturer or the manufacturer's literature should be consulted to provide specific guidance for the chemical to be treated.

NOTE: Vapor suppression can also be considered a chemical method of mitigation.

While vapor suppression or blanketing does not change the nature of a hazardous material, it can greatly reduce the immediate hazard and danger associated with the presence of uncontrolled vapor. In addition, it buys additional time in which to undertake measures that will control and resolve the incident, under safer circumstances.

6-4.1.11   **Venting.** Venting is a process that is used to deal with liquids or liquefied compressed gases where a danger, such as an explosion or mechanical rupture of the container or vessel, is considered likely. The method of venting will depend on the nature of the hazardous material. In general, it involves the controlled release of the material to reduce and contain the pressure and diminish the probability of an explosion.

6-4.2*   **Chemical Methods.** Chemical methods of control involve the application of chemicals to treat spills of hazardous materials. Chemical methods may involve any one of several actions to reduce the involved area affected by the release of a hazardous material. In all cases, methods used should be acceptable to the incident commander. The selection of personal protective clothing should be based on the hazardous materials and/or conditions present and be appropriate for the hazards encountered. Refer to Table 6-4.2.

Figure 6.5   Chemical methods of hazardous materials mitigation include the use of neutralizing agents.

A-6-4.2   The procedures described in 6-4.2.1 through 6-4.2.10 should only be used by personnel trained in those procedures.

6-4.2.1*   **Adsorption.** Adsorption is the process in which a sorbate (hazardous liquid) interacts with a solid sorbent surface. See ASTM F726, *Method of Testing Sorbent Performance of Adsorbents*, for further information. The principal characteristics of this interaction are:

(a) The sorbent surface is rigid and no volume increase occurs as is the case with absorbents.

(b) The adsorption process is accompanied by heat of adsorption whereas absorption is not.

(c) Adsorption occurs only with activated surfaces, e.g., activated carbon, alumina, etc.

Table 6-4.2 Chemical Methods Mitigation

| Method | Chemical | | | | Biological | | | | Radiological | | | |
| | Gases | | Liq. | Sol. | Gases | | Liq. | Sol. | Gases | | Liq. | Sol. |
	LVP*	HVP**			LVP	HVP			LVP	HVP		
Adsorption	yes	yes	yes	no	yes³	yes	yes³	no	no	no	no	no
Burn	yes	yes	yes	yes	yes	yes	yes	yes	no	no	no	no
Dispersion/Emusification	no	no	yes	yes	no	no	yes³	no	no	no	no	no
Flare	yes	yes	yes	no	yes	yes	yes	no	no	no	no	no
Gelation	yes	no	yes	yes	yes³	no	yes³	yes³	no	no	no	no
Neutralization	yes¹	yes⁴	yes	yes²	no	no	no	no	no	no	no	no
Polymerization	yes	no	yes	yes	no	no	no	no	no	no	no	no
Solidification	no	no	yes	no	no	no	yes³	no	no	no	yes	no
Vapor Suppression	yes	yes	yes	yes	yes	yes	yes	yes	yes	yes	yes	yes
Vent/Burn	yes	yes	yes	no	yes	yes	yes	no	no	no	no	no

* Low Vapor Pressure
** High Vapor Pressure
1. Technique may be possible as a liquid or solid neutralizing agent and water can be applied.
2. When solid neutralizing agents are used, they must be used simultaneously with water.
3. Technique is permitted only if resulting material is hostile to the bacteria.
4. The use of this procedure requires special expertise and technique.

A-6-4.2.1 Spontaneous ignition can occur through the heat of adsorption of flammable materials, and caution should be exercised.

NOTE: Adsorbents saturated with volatile liquid chemicals can create a more severe vapor hazard than the spill alone because of the severely enlarged surface area for vapor release.

The term *sorbents* encompasses both absorbents and adsorbents. Adsorbents act in such a way that the internal structure of the material is not penetrated. They can be natural or synthetic materials and can be used on liquid spills on land and, to some degree, water. Adsorbents should be nonreactive to the spilled material. Porous clay and sand are examples of adsorbents.

6-4.2.2   Controlled Burning. For purposes of this practice, controlled combustion is considered a chemical method of control. However, it should only be used by qualified personnel trained specifically in this procedure.

In some emergency situations where extinguishing a fire will result in large, uncontained volumes of contaminated water, or threaten the safety of responders or the public, controlled burning is used as a technique. It is advised that consultation be made with the appropriate environmental authorities when this method is used.

There are some occasions when extinguishing a fire, which is generally accompanied by the run-off of large amounts of contaminated waste, is the improper approach. Such was the case with the Sandoz fire in Basel, Switzerland, which severely contaminated the Rhine River. At the Sherwin Williams fire in Dayton, Ohio, fire fighters decided not to use heavy fire streams because the building was essentially lost and because the contaminated run-off might pollute one of the city's major water supplies. This decision was praised by environmentalists and fire fighting experts alike.

Another aspect of the controlled burning approach is actually to incinerate the spilled hazardous material. Transportable incinerators are designed to promote combustion of spilled materials, especially oil.

6-4.2.3   Dispersion, Surface Active Agents, and Biological Additives. Certain chemical and biological agents can be used to disperse or break up the materials involved in liquid spills. The use of these agents results in a lack of containment and generally results in spreading the liquid over a much larger area. Dispersants are most often applied to spills of liquids on water. The dispersant breaks down a liquid spill into many fine droplets, thereby diluting the material to acceptable levels. Use of this method may require the prior approval of the appropriate environmental authority. See ASTM STP 659, *Chemical Dispersants for the Control of Oil Spills*, and ASTM STP 840, *Oil Spill Chemical Dispersants: Research, Experience, and Recommendations*, for further information.

Dispersants generally result in oil-in-water emulsions, since the chemicals used reduce the surface tension between oil and water. Because chemical dispersants should not be used in situations in which they might produce increased biological damage, it is essential that environmental authorities be consulted. Surface-active agents also result in increased emulsification and dispersion of a spill.

Surface cleaning equipment is available for soil surface cleaning, too. This equipment agitates the soil's surface with water to form a slurry. The contamination is then removed through a separation process that takes place in a specially designed sand separator.

Biological additives can be used to degrade by biochemical oxidation and biochemical accelerators certain hazardous materials spilled on land or in water.

6-4.2.4 Flaring. Flaring is a process that is used with high-vapor-pressure liquids or liquefied compressed gases for the safe disposal of the product. Flaring is the controlled burning of material in order to reduce or control pressure and/or dispose of a product.

6-4.2.5 Gelation. Gelation is the process of forming a gel. A gel is a colloidal system consisting of two phases, a solid and a liquid. The resulting gel is considered to be a hazardous material and must be disposed of properly.

Gelling agents used on hazardous chemicals produce a gel that is more easily cleaned up, either by mechanical or physical methods. Gels can be used on spilled liquids in water and, to a lesser degree, on land.

6-4.2.6 Neutralization. Neutralization is the process of applying acids or bases to a spill to form a neutral salt. The application of solids for neutralizing can often result in confinement of the spilled material. Special formulations are available that do not result in violent reactions or local heat generation during the neutralization process. In cases where special neutralizing formulations are not available, special considerations should be given to protecting persons applying the neutralizing agent, as heat is generated and violent reactions may occur. One of the advantages of neutralization is that a hazardous material may be rendered nonhazardous.

The pH scale is used to categorize compounds as acids or bases. A value of 7 on the scale is neutral, while descending values denote increasing acidity. Levels above 7 up to 14 denote bases, with the higher values indicating increasingly stronger bases. It is possible to neutralize acidic or basic spills by mixing the spilled material with a neutralizing agent.

6-4.2.7 Polymerization. A process in which a hazardous material is reacted in the presence of a catalyst, of heat or light, or with itself or another material to form a polymeric system.

6-4.2.8 Solidification. Solidification is the process whereby a hazardous liquid is treated chemically so that a solid material results. Adsorbents can be considered an example of a solidification process. There are other materials that can be used to convert hazardous liquids into nonhazardous solids. Examples are applications of special formulations designed to form a neutral salt in the case of spills of acids or caustics. The advantage of the solidification process is that a spill of small scale can be confined relatively quickly and treatment effected immediately.

Commercially available adsorbents can be used to solidify oily wastes that are water-insoluble. The spilled liquids are adsorbed to granules to form a solid, nonflowing mixture. The resulting product is safer than the spilled material and more easily transported to an appropriate disposal facility.

6-4.2.9 Vapor Suppression. The use of solid activated materials to treat hazardous materials so as to effect suppression of the vapor off-gasing from the materials. This process results in the formation of a solid that affords easier handling but that may result in a hazardous solid that must be disposed of properly.

6-4.2.10 Venting and Burning. This involves the use of shaped charges to vent the high vapor pressure at the top of the container and then with additional charges to release and burn the remaining liquid in the container in a controlled fashion.

In the Livingston, Louisiana, derailment of 1982, shaped explosive charges were used to "vent and burn" tank cars that were too badly damaged to attempt transfer of product. This was done some eight days into the incident. Vent and burn is a highly sophisticated technique that should be attempted only by adequately trained specialists under very controlled conditions.

Figure 6.6 *Venting and burning techniques have been used in at least one train derailment.*

References Cited in Commentary

1. Ludwig Benner, Jr., *Hazardous Materials Emergencies*, 2nd edition, 1978 (out of print).
2. Michael Hildebrand and Gregory Noll, "Gasoline Tank Truck Emergencies: Guidelines and Procedures" (Stillwater, OK: Oklahoma State University, Fire Protection Publications).
3. NFPA 70, *National Electrical Code*, 1993 edition.

7

Decontamination

7-1 Decontamination Plan.

At every incident involving hazardous materials there is a possibility that response personnel and their equipment will become contaminated. The contaminant poses a threat, not only to the persons contaminated, but to other personnel who may subsequently have contact with them or the equipment.

Personnel and equipment can be decontaminated by removing or neutralizing the contaminants that have accumulated on them. Decontamination requires an organized and well-ordered procedure, hence the need for a plan in order to carry it out successfully. The plan must take into account measures that will minimize contamination to begin with, and ultimate responsibility for implementing the plan falls to the incident commander.

The decontamination plan should address such factors as site layout, the methods to be used and the equipment needed, the number of personnel needed, the level of protective clothing and equipment that will have to be processed, disposal methods, run-off control, emergency medical requirements, and methods for collecting and disposing of contaminated clothing and equipment.

Figure 7.1 Dilution is one method of decontamination. It reduces the concentration of the contaminant to a level at which it is no longer harmful. Water is used, except when the potential exists for a chemical reaction.

7-1.1 Incident responders should have an established procedure to minimize contamination or contact, to limit migration of contaminants, and to properly dispose of contaminated materials. Decontamination procedures should begin upon arrival at the scene, should provide for an adequate number of decontamination personnel, and should continue until the incident commander determines that decontamination procedures are no longer required. Decontamination of victims may be required.

During the course of an incident, the decontamination plan will have to be revised to address changing conditions.

Good operating practices can reduce the extent of contamination to clothing and equipment. For example, sampling and monitoring equipment can be bagged for protection, leaving a limited number of openings as needed for operation. Disposable outer garments also help, and covering the source of the contamination where possible can greatly reduce the need for extensive decontamination procedures.

7-1.2 The following methods of decontamination are available for personnel and/or equipment:

(a) Absorption

(b) Adsorption

(c) Chemical degradation

(d) Dilution

(e) Disposal

(f) Isolation

(g) Neutralization

(h) Solidification.

The items listed above are the methods most often used to decontaminate personnel and/or equipment. Some methods are not as suitable for one as they are for the other. For example, some forms of decontamination may cause a chemical reaction that would not damage equipment but would injure a person. This may also vary, depending on the particular contaminant. The chemical and physical compatibility of any decontamination solution must be determined before the solution is used.

7-2 Personal Protective Equipment.

Before personal protective equipment is removed it should be decontaminated. During doffing of personal protective equipment, the clothing should be removed in a manner such that the outside surfaces do not touch or make contact with the wearer. A log of personal protective equipment used during the incident should be maintained. Personnel wearing disposable protective equipment should go through the decontamination process, and the disposable protective equipment should be disposed of in accordance with established procedures.

Some heavy contamination can be removed physically by wiping off, wetting down, and allowing to dry. Chemical means can also be used to deactivate a contaminant, as can a combination of physical and chemical means.

The log of personal protective equipment used during an incident should record the type of equipment used, the duration of use, the decontamination procedures used, and the types of chemicals to which the equipment was exposed. It should also record the name of the person using the equipment.

Disposable protective equipment should be placed in plastic bags or plastic trash cans pending final disposal.

Figure 7.2 For some contaminants, protective clothing can never be adequately decontaminated and must be disposed of; still, sufficient decontamination must take place to allow workers to remove the protective clothing safely.

7-3 Decontamination.

Decontamination consists of removing the contaminants by chemical or physical processes. The conservative action is always to assume contamination has occurred and to implement a thorough, technically sound decontamination procedure until it is determined or judged to be unnecessary.

There are established methods for testing the effectiveness of decontamination procedures, and they should be used. If the decontamination procedure is not working effectively, the process may have to be revised.

Test methods include visual examination in both natural and ultraviolet light, laboratory analysis of a wipe sample, analysis of the cleaning solution used, and laboratory analysis of a segment of protective clothing.

7-3.1   Procedures for all phases of decontamination must be developed and implemented to reduce the possibility of contamination to personnel and equipment. Reference guides for the development of decontamination procedures can be found in Appendix C-2.4(d) and C-2.4(f). Assuming protective equipment is grossly contaminated, use appropriate decontamination methods for the chemicals encountered.

Some decontamination procedures may actually present additional hazards. The decontamination solution may react with the chemical to which the clothing was exposed, it may itself permeate or degrade some protective clothing, or it may emit harmful vapors. Compatibility should be determined before use.

7-3.2   Outer clothing should be decontaminated prior to removal. The outer articles of clothing, after removal, should be placed in plastic bags for later additional decontamination, cleaning, and/or inspection. In some cases, they may have to be overpacked into containers for proper disposal. Water or other solutions used for washing or rinsing may have to be contained, collected, containerized, and analyzed prior to disposal.

Placing contaminated outer clothing on a plastic drop cloth is a common and useful procedure. It is also helpful to have lined containers on hand in which to pack the contaminated clothing and equipment. Metal or plastic drums are effective for storing washing or rinsing solutions.

7-3.3   Initial procedures should be upgraded or downgraded as additional information is obtained concerning the type of hazardous materials involved, the degree of hazard, and the probability of exposure of response personnel.

If the decontamination method being used is not effectively removing contaminants, different and additional measures must be taken to prevent further contamination. If necessary, specialists should be consulted.

The decontamination plan may have to be changed whenever the type of protective clothing and equipment changes, when site conditions change, or when new information is received.

7-3.4   Using solutions containing chemicals to alter or change contaminants to less hazardous materials should only be done after consultation with persons experienced and familiar with the hazards involved. The use of detergent-water washing solutions is more prevalent, but its effectiveness against certain contaminants may be low. It is less risky, however, than using chemical solutions.

If time is not an overriding factor and if technicians are not on site or immediately available, the detergent-water solution method of preliminary decontamination is the safest and most appropriate approach. See the commentary on 7-3.1 regarding hazards associated with the use of chemical solutions.

7-3.5 Many types of equipment are very difficult to decontaminate and may have to be discarded as hazardous wastes. Whenever possible, other pieces of small equipment should be disposable or made of nonporous material. Monitoring instruments and some types of sampling equipment can be placed in plastic bags (with only the detecting element exposed) to minimize potential contamination problems.

Apparatus can become contaminated in several ways. These include placing the apparatus downwind where vapors or smoke may contaminate it, placing the apparatus too close to the isolation area where it can be splashed or sprayed by hazardous materials, driving it through a spill, and placing contaminated equipment back on the apparatus. Tools should be inspected to determine how contaminated they are and to establish the best method of decontaminating them.

Figure 7.3 *In some cases, fire fighting equipment and apparatus may require decontamination as well.*

7-3.6 Large items of equipment, such as vehicles and trucks, should be subjected to decontamination by high pressure water washes, steam, or special solutions. Water or other solutions used for washing or rinsing may have to be contained, collected, containerized, and analyzed prior to disposal. Consultation with appropriate sources should be utilized to determine proper decontamination procedures.

To decontaminate large equipment, responders will have to have different supplies on hand. These might include adequately sized storage tanks or collection systems, pumps or drains, long-handled brushes, and wash booths or containers.

7-3.7 Personnel assigned to the decontamination team should wear an appropriate level of personal protective equipment and may require decontamination themselves.

The members of the decontamination team closest to the hot zone may require a higher level of protective clothing than those closest to the cold zone. The level of protection required varies with the decontamination equipment in use. Protective clothing should be selected by a qualified person.

8

Referenced Publications

The following documents or portions thereof are referenced within this recommended practice and should be considered part of the recommendations of this document. The edition indicated for each reference is the current edition as of the date of the NFPA issuance of this document.

8-1.1 NFPA Publications. National Fire Protection Association, 1 Batterymarch Park, P.O. Box 9101, Quincy, MA 02269-9101.

NFPA 70, *National Electrical Code*, 1990 edition

NFPA 472, *Standard for Professional Competence of Responders to Hazardous Materials Incidents*, 1992 edition

NFPA 1500, *Standard on Fire Department Occupational Safety and Health Program*, 1992 edition

NFPA 1981, *Standard on Open-Circuit Self-Contained Breathing Apparatus for Fire Fighters*, 1992 edition

NFPA 1982, *Standard on Personal Alert Safety Systems (PASS) for Fire Fighters*, 1988 edition

NFPA 1991, *Standard on Vapor-Protective Suits for Hazardous Chemical Emergencies*, 1990 edition

NFPA 1992, *Standard on Liquid Splash-Protective Suits for Chemical Emergencies*, 1990 edition

NFPA 1993, *Standard on Support Function Protective Garments for Hazardous Chemical Operations*, 1990 edition.

8-1.2 Other Publications.

8-1.2.1 ASTM Publications. American Society for Testing and Materials, 1916 Race Street, Philadelphia, PA 19103-1187.

ASTM F716-82, *Method of Testing Sorbent Performance of Absorbents*

ASTM F726-81, *Method of Testing Sorbent Performance of Adsorbents*

ASTM STP 659, *Chemical Dispersants for the Control of Oil Spills*

ASTM STP 840, *Oil Spill Chemical Dispersants: Research, Experience, and Recommendations*.

8-1.2.2 U.S. Government Publication. U.S. Government Printing Office, Superintendent of Documents, Washington, DC 20402.

Title 29 CFR Part 1910.120.

A

The material contained in Appendix A of this standard is included within the text of this handbook, and therefore is not repeated here.

Referenced Publications

B-1

The following documents or portions thereof are referenced within this recommended practice for informational purposes only and thus are not considered part of the recommendations of this document. The edition indicated for each reference is the current edition as of the date of the NFPA issuance of this document.

B-1.1 NFPA Publications. National Fire Protection Association, 1 Batterymarch Park, P.O. Box 9101, Quincy, MA 02269-9101.

NFPA 11, *Standard for Low Expansion Foam and Combined Agent Systems*, 1988 edition

NFPA 11A, *Standard for Medium- and High-Expansion Foam Systems*, 1988 edition

NFPA 16, *Standard on the Installation of Deluge Foam-Water Sprinkler and Foam-Water Spray Systems*, 1991 edition

NFPA 472, *Standard for Professional Competence of Responders to Hazardous Materials Incidents*, 1992 edition.

B-1.2 Other Publications.

B-1.2.1 **ASTM Publication.** American Society for Testing and Materials, 1916 Race Street, Philadelphia, PA 19103-1187.

ASTM F1129-88, *Using Aqueous Foams to Control the Vapor Hazard from Immiscible Volatile Liquids.*

B-1.2.2 **U.S. Government Publications.** U.S. Government Printing Office, Superintendent of Documents, Washington, DC 20402.

Title 29 CFR Parts 1910.120, 1910.134
Title 40 CFR Parts 261.33, 300, 302, and 355
Title 49 CFR Parts 170-179.

C

Suggested Reading List

This Appendix is not a part of the recommendations of this NFPA document, but is included for information purposes only.

C-1 Introduction.

This list provides the titles of references and organizations that may be of value to those responding to hazardous materials incidents. This list can be expanded based on personal preferences and requirements.

The references are categorized by subject. The title, author, publisher, and place of publication are given for each. The year of publication is not always given because many are revised annually. The user should attempt to obtain the most recent edition.

The last section lists sources of these references as well as other information that might be useful. Usually, these agencies or associations will provide a catalog on request. Where available, phone numbers are also listed.

C-2 References.

C-2.1 Industrial Hygiene (Air Sampling and Monitoring, Respiratory Protection, Toxicology).

(a) *Air Sampling Instruments for Evaluation of Atmospheric Contaminants*, American Conference of Governmental Industrial Hygienists, Cincinnati, OH.

(b) *Direct Reading Colorimetric Indicator Tubes Manual*, American Industrial Hygiene Association, Akron, OH.

(c) *Fundamentals of Industrial Hygiene*, National Safety Council, Chicago, IL.

(d) *Industrial Hygiene and Toxicology*, Frank A. Patty, John Wiley and Sons, Inc., New York, NY.

(e) *Manual of Recommended Practice for Combustible Gas Indicators and Portable, Direct Reading Hydrocarbon Detectors*, American Industrial Hygiene Association, Akron, OH.

(f) *NIOSH/OSHA Pocket Guide to Chemical Hazards*, DHHS No. 85-114, NIOSH, Department of Health and Human Services, Cincinnati, OH.

(g) *Occupational Health Guidelines for Chemical Hazards*, DHHS No. 81-123, NIOSH, Department of Health and Human Services, Cincinnati, OH.

(h) *Occupational Safety and Health Standards*, Title 29, *Code of Federal Regulations*, Part 1910.120, "Hazardous Waste Operations and Emergency Response Final Rule," U.S. Government Printing Office, Washington, DC.

(i) *TLVs Threshold Limit Values and Biological Exposure Indices (Threshold Limit Values for Chemical Substances and Physical Agents in the Workroom Environment)*, American Conference of Governmental Industrial Hygienists, Cincinnati, OH.

C-2.2 Chemical Data.

(a) *Chemical Hazard Response Information System (CHRIS)*, U.S. Coast Guard, Washington, DC, Commandant Instruction M.16565.12A.

(b) *CHRIS — A Condensed Guide to Chemical Hazards*, U.S. Coast Guard, Commandant Instruction M16565.11a.

(c) *The Condensed Chemical Dictionary*, G. Hawley, Van Nostrand Reinhold Co., New York, NY.

(d) *CRC Handbook of Chemistry and Physics*, CRC Press, Boca Raton, FL.

(e) *Dangerous Properties of Industrial Materials*, N. Irving Sax, Van Nostrand Reinhold Co., New York, NY.

(f) *Effects of Exposure to Toxic Gases*, Matheson.

(g) *Emergency Handling of Hazardous Materials in Surface Transportation*, Association of American Railroads, Washington, DC.

(h) *Farm Chemicals Handbook*, Farm Chemicals Magazine, Willoughby, OH.

(i) *Firefighter's Handbook of Hazardous Materials*, Baker, Charles J., Maltese Enterprises, Indianapolis, IN. .

(j) *Fire Protection Guide to Hazardous Materials*, National Fire Protection Association, Quincy, MA.

(k) *Hazardous Materials Handbook*, Meidl, J. H., Glencoe Press, Encino, CA.

(l) *The Merck Index*, Merck and Co., Inc., Rahway, NJ.

(m) *Emergency Action Guides*, Association of American Railroads, Washington, DC.

C-2.3   Safety and Personnel Protection.

(a) *A Guide to the Safe Handling of Hazardous Materials Accidents*, ASTM STP 825, American Society for Testing and Materials, Philadelphia, PA.

(b) *Fire Protection Handbook*, National Fire Protection Association, Quincy, MA.

(c) *Guidelines for Decontamination of Firefighters and Their Equipment Following Hazardous Materials Incidents*, Canadian Association of Fire Chiefs, Ottawa (May 1987).

(d) *Guidelines for the Selection of Chemical Protective Clothing*. Volume 1: Field Guide, A. D. Schwope, P. P. Costas, J. O. Jackson, D. J. Weitzman; Arthur D. Little, Inc., Cambridge, MA (March 1983).

(e) *Guidelines for the Selection of Chemical Protective Clothing*. Volume 2: Technical and Reference Manual, A. D. Schwope, P. P. Costas, J.O. Jackson, D.J. Weitzman, J. O. Stull; Arthur D. Little, Inc., Cambridge, MA, 3rd edition (February 1987).

(f) *Hazardous Materials*, Warren Isman and Gene Carlson, Glencoe Press, Encino, CA, 1981.

(g) *Hazardous Materials Emergencies Response and Control*, John R. Cashman, Technomic Publishing Company, Lancaster, PA (June 1983).

(h) *Hazardous Materials for the First Responder*, International Fire Service Training Association, Stillwater, OK (1988).

(i) *Hazardous Materials: Managing the Incident*, Gregory Noll, Michael Hildebrand, and James Yvorra, Fire Service Publications, Stillwater, OK (1988).

(j) *Handling Radiation Emergencies*, Purington and Patterson, National Fire Protection Association, Quincy, MA.

(k) *Hazardous Materials Injuries, A Handbook for Pre-Hospital Care*, Douglas R. Stutz, Robert C. Ricks, Michael F. Olsen, Bradford Communications Corp., Greenbelt, MD.

(l) *National Safety Council Safety Sheets*, National Safety Council, Chicago, IL.

(m) *Radiological Health — Preparedness and Response in Radiation Accidents*, U.S. Department of Health and Human Services, Washington, DC.

(n) *Standard First Aid and Personal Safety*, American Red Cross.

C-2.4 Planning Guides.

(a) *A Fire Department's Guide to Implementing Title III and the OSHA Hazardous Materials Standard* (August 1987), William H. Stringfield, International Society of Fire Service Instructors, Ashland, MA.

(b) *Federal Motor Carrier Safety Regulations Pocketbook*, U.S. Department of Transportation, J. J. Keller and Associates, Inc.

(c) *Hazardous Chemical Spill Cleanup*, Noyes Data Corporation, Ridge Park, New Jersey.

(d) *Occupational Safety and Health Guidance Manual for Hazardous Waste Site Activities*, NIOSH/OSHA/USCG/EPA, U.S. Department of Health and Human Services, NIOSH.

(e) *Hazardous Materials Emergency Planning Guide* (March 1987), National Response Team.

(f) *Standard Operating Safety Guides*, Environmental Response Branch, Office of Emergency and Remedial Response, U.S. Environmental Protection Agency.

C-3 Agencies and Associations.

Agency for Toxic Substances Disease Registry
Shamlee 28 S., Room 9
Centers for Disease Control
Atlanta, GA 30333
404/452-4100

American Conference of Governmental Industrial Hygienists
6500 Glenway Avenue — Building D-5
Cincinnati, OH 45211
513/661-7881

American Industrial Hygiene Association
475 Wolf Ledges Parkway
Akron, OH 44311-1087
216/762-7294

American National Standards Institute, Inc.
1430 Broadway
New York, NY 10018
212/354-3300

American Petroleum Institute (API)
1220 L Street N.W. 9th Floor
Washington, DC 20005
202/682-8000

Association of American Railroads
50 F Street N.W.
Washington, DC 20001
202/639-2100

Chemical Manufacturers' Association
2501 M Street N.W.
Washington, DC 20037
202/877-1100

CHEMTREC
Washington, DC
800/424-9300

The Chlorine Institute
2001 L Street N.W.
Washington, DC 20001
202/639-2100

Compressed Gas Association
1235 Jefferson Davis Highway
Arlington, VA 22202
703/979-0900

The Fertilizer Institute (TFI)
1015 18th Street N.W.
Washington, DC 20036
202/861-4900

International Society of Fire Service Instructors
30 Main Street
Ashland, MA 01721
617/881-5800

National Fire Protection Association
1 Batterymarch Park, P.O. Box 9101
Quincy, MA 02269-9101
617/770-3000

Spill Control Association of America
Suite 1575
100 Renaissance Center
Detroit, MI 48243-1075

U.S. Department of Transportation
Materials Transportation Bureau
DHM 51
RS PA
Washington, DC 20590
202/366-4555

U.S. EPA Office of Research & Development
Publications — CERI
Cincinnati, OH 45268
513/684-7562

U.S. EPA Office of Solid Waste (WH-562)
Superfund Hotline
401 M Street S.W.
Washington, DC 20460
800/424-9346

U.S. Mine Safety and Health Administration
Department of Labor
4015 Wilson Boulevard, Room 600
Arlington, VA 22203
703/235-1452

U.S. National Oceanic and Atmospheric Administration
Hazardous Materials Response Branch N/CMS 34
7600 Sand Point Way NE
Seattle, WA 98115

C-4 Computer Data Base Systems.

Hazardous Materials Information Exchange (HMIX)
Federal Emergency Management Agency
State and Local Programs Support Directory
Technological Hazards Division
500 C Street S.W.
Washington, DC 20472

Standard and Commentary for NFPA 472, *Standard for Professional Competence of Responders to Hazardous Materials Incidents*

1992 Edition

The text and illustrations that make up the commentary on the various sections of NFPA 472 are printed in black. The text of the standard itself is printed in blue.

Paragraphs that begin with the letter "A" are extracted from Appendix A of the standard. Appendix A material is not mandatory. It is designed to help users apply the provisions of the standard. In this handbook, material from Appendix A is integrated with the text, so that it follows the paragraph it explains. An asterisk (*) following a paragraph number indicates that explanatory material from Appendix A will follow.

1

Administration

It is important to note at the very beginning that this document does not apply only to a single segment of the emergency response community but to responders from the fire service, from law enforcement, from other public sector agencies, and from the private sector. As clearly indicated in the committee scope approved by NFPA's Standards Council, this standard is meant to apply to everyone who may be called upon to respond to, or take action at, a hazardous materials incident. In most cases, fire service personnel are the primary first responders to emergencies involving hazardous materials. Frequently, however, law enforcement agencies find themselves first on the scene at hazardous materials transportation accidents. And many other groups, including government and industrial responders, private fire brigades, and workers engaged in both mobile and fixed operations involving daily contact with hazardous materials, become involved in the mitigation of hazardous materials emergencies.

1-1 General.

1-1.1 Scope. This standard identifies the levels of competence required of responders to hazardous materials incidents. It specifically covers the competencies for first

responders at the awareness level, first responders at the operational level, hazardous materials technicians, incident commanders, and off-site specialist employees.

The key to achieving a given level of competence is training. A specified
number of hours of training will not provide equal results for every individual, since there are a number of variables involved, such as the ability of the
trainer and the trainee, and the quality, intensity, and frequency of the instruction. For this reason, the standard does not prescribe a certain number of
training hours. Instead, it deals with the objectives and abilities the responder
should attain.

This edition of the standard establishes competencies for five categories
of responders based on an analysis of the actions the responders must take
to control a hazardous materials incident. Chapter 2 addresses first responder
awareness; Chapter 3, the first responder operational; Chapter 4, the hazardous materials technician; Chapter 5, the incident commander; and Chapter 6, the off-site specialist employee.

Those familiar with the 1989 edition of NFPA 472 will notice that some
changes have been made. For example, the hazardous materials specialist no
longer exists. While conducting their task analysis of emergency response to
hazardous materials incidents, the committee found that there was very little difference between the technician level and the specialist level. As a result,
it was difficult to establish a clear distinction between the two. In this edition, most of the competencies that were previously contained in the specialist level are now found in the technician level. The knowledge and skills
responders need at the higher level have not been reduced, just presented in
a more logical fashion.

In addition, NFPA 472 now includes competencies for the off-site specialist employee. To establish competencies for personnel who may have to deal
with specific chemicals or containers, the committee has introduced Chapter 6, "Off-Site Specialist Employees."

Finally, NFPA 472 now contains competencies for the hazardous materials incident commander. The committee felt it was imperative to address
these competencies because the incident commander may have to manage
an incident at the operational level or at a higher level, and he or she must
understand the potential hazards involved, the capabilities of responding personnel, and the technical assistance available for dealing with complex
incidents.

Levels of competence, then, apply primarily to individuals rather than to response teams. A response team may be composed of members at several different levels. For example, a team may have personnel at the operations level and at the technician level, or it may be made up of personnel at off-site specialist employee A, B, or C levels.

It is important that we not let these designations create a barrier to developing an effective hazardous materials response. Rather, we must recognize that they provide a tool we can use to prepare personnel to handle such incidents adequately.

1-1.2 **Purpose.** The purpose of this standard is to specify minimum competencies for those who will respond to hazardous materials incidents. It is not the intent of this standard to restrict any jurisdiction from exceeding these minimum requirements.

One of the purposes of the competencies contained herein is to reduce the numbers of accidents, injuries, and illnesses during response to hazardous materials incidents and to help prevent exposure to hazardous materials to reduce the possibility of fatalities, illness, and disabilities affecting emergency response personnel.

Hazardous materials are a pervasive part of the world in which we live, and incidents involving such materials are inevitable. Most responders experience a complex hazardous materials incident only once in their careers. However, those dedicated to providing emergency services must be prepared to manage such an incident effectively and safely. It also imperative that those involved in manufacturing, using, and transporting hazardous materials are trained to safely undertake initial protective actions when an unplanned release occurs and to assist emergency responders who may be called upon to minimize and control the potential hazard.

The role of emergency responders in hazardous materials incidents is critical, and the better trained they are, the better they are able to curb injuries and damage. The purpose of this standard is to specify the minimum requirements those responders must fulfill in order to reduce the number of accidents, injuries, and illnesses, and the probability of fatalities.

Protecting the responders and ensuring their safety at the scene of an incident is one of the principal purposes of this document. Adhering to the provisions of this standard at each level of competency should provide a high degree of safety, despite the hazards encountered.

Figure 1.1 Hazardous materials incidents can happen at any time. Pre-planning and training allow the safe and effective handling of such situations.

Figure 1.2 There are more than 50,000 chemical compounds classified as hazardous. Approximately 500 new compounds enter the commercial market each year.

The committee has adopted a system that will be followed throughout this document. That system is represented by the diagram below.

DUTIES OF INITIAL RESPONSE PERSONNEL

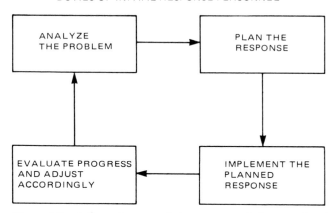

Figure 1.3   A flow diagram of the response duties associated with emergencies involving hazardous materials. (Source: Fire Protection Handbook, 17th edition.)

Each step in the figure above involves a series of tasks that must be considered and resolved. When supported by the response community, these steps are the framework for an appropriate, survival-oriented response to hazardous material incidents. Reasoned decisions based on this approach will keep any harm that may result from a hazardous materials incident to a minimum while reducing the risk to the responders.[1]

1-2 Definitions.

Approved.* Acceptable to the "authority having jurisdiction."

Acceptance by an authority having jurisdiction is usually based on tests or experience, rather than on an arbitrary decision.

A-1-2   Approved. The National Fire Protection Association does not approve, inspect or certify any installations, procedures, equipment, or materials nor does it approve or evaluate testing laboratories. In determining the acceptability of installations or procedures, equipment or materials, the authority having jurisdiction may base acceptance on compliance with NFPA or other appropriate standards. In the absence of such standards, said authority may require evidence of proper installation, procedure or use. The authority having jurisdiction may also refer to the listings or labeling practices of an organization concerned with product evaluations which is in a position to determine compliance with appropriate standards for the current production of listed items.

Authority Having Jurisdiction.* The "authority having jurisdiction" is the organization, office or individual responsible for "approving" equipment, an installation or a procedure.

In a document dealing with the very broad concept of hazardous materials response, many authorities will exercise jurisdiction. These authorities will include people at all levels of government, from federal to local, as well as manufacturers and shippers.

A-1-2 **Authority Having Jurisdiction.** The phrase "authority having jurisdiction" is used in NFPA documents in a broad manner since jurisdictions and "approval" agencies vary as do their responsibilities. Where public safety is primary, the "authority having jurisdiction" may be a federal, state, local or other regional department or individual such as a fire chief, fire marshal, chief of a fire prevention bureau, labor department, health department, building official, electrical inspector, or others having statutory authority. For insurance purposes, an insurance inspection department, rating bureau, or other insurance company representative may be the "authority having jurisdiction." In many circumstances the property owner or his designated agent assumes the role of the "authority having jurisdiction"; at government installations, the commanding officer or departmental official may be the "authority having jurisdiction."

CANUTEC. The Canadian Transport Emergency Center operated by Transport Canada. CANUTEC provides emergency response information and assistance on a 24-hr basis for responders to hazardous materials incidents.

CANUTEC is operated by the Transport Dangerous Goods Directorate of Transport Canada. It provides a national bilingual advisory service and is staffed by professional chemists experienced and trained in interpreting technical information and providing emergency response advice. In an emergency CANUTEC can be contacted by calling collect (613) 996-6666, 24 hours a day. For nonemergency information, CANUTEC can be contacted at (613) 992-4624.

Chemical. Regulated and nonregulated hazardous materials (solids, liquids, and gases; natural or manmade; including petroleum products) with the potential for creating harm to people, the environment, and property when released.

Chemical-Protective Clothing. Items made from chemical-resistive materials, such as clothing, hood, boots, and gloves, that are designed and configured to protect the wearer's torso, head, arms, legs, hands, and feet from hazardous materials. Chemical-protective clothing (garments) can be constructed as a single- or multipiece garment. The garment may completely enclose the wearer either by itself or in combination with the wearer's respiratory protection, attached or detachable hood, gloves, and boots.

Chemical-protective clothing (CPC) allows responders to work in or near hazardous atmospheres by isolating their bodies from the chemical or physical hazards. The type of CPC needed will vary depending on the type of hazardous materials and conditions present. At this time, no single type of CPC is suitable for all types of incidents.

CHEMTREC. The Chemical Transportation Emergency Center, a public service of the Chemical Manufacturers Association, in Washington, DC. CHEMTREC provides emergency response information and assistance on a 24-hr basis for responders to hazardous materials incidents.

You can call CHEMTREC toll-free throughout the U.S. and Canada by dialling (800) 424-9300. CHEMTREC can provide hazard information from their files, and they can contact the shipper for you. Responders may also teleconference with their technical specialists. For nonemergency information, call (800) 262-8200.

Cold Zone. The control zone of a hazardous materials incident that contains the command post and such other support functions as are deemed necessary to control the incident. This zone is also referred to as the clean zone or support zone in other documents.

It might appear that the cold zone has no outer boundary, but this is not the case. One can equate the outer boundary at a hazardous materials incident with the fire lines drawn at a major fire or some other emergency that are controlled by local law enforcement officers. The public does not have access to the cold zone under most circumstances.

Competence.   Possessing knowledge, skills, and judgment needed to perform indicated objectives satisfactorily.

Knowledge and skills can be measured, but one's judgment is not as easily evaluated—and one's judgment and decision-making skills can vary substantially with the circumstances. Nonetheless, training and experience can effectively improve the emergency decision-making process. Since the outcome of an incident is determined by the decisions that are made in handling it, proper and adequate training is of paramount importance.

Confined Space.   Refers to a space that by design has limited openings for entry and exit, that has unfavorable natural ventilation that could contain or produce dangerous concentrations of air contaminants, and that is not intended for continuous occupancy. Examples of confined spaces include, but are not limited to, storage tanks, compartments of ships, process vessels, pits, silos, vats, degreasers, reaction vessels, boilers, ventilation and exhaust ducts, sewers, tunnels, underground utility vaults, and pipelines.

The U.S. Occupational Safety and Health Administration (OSHA) is promulgating new regulations relating to worker safety in confined spaces. These regulations will introduce a system of working in such spaces that includes hazard recognition and risk assessment; testing, evaluation, and monitoring; and permits for entry, work, and rescue. It is anticipated that these regulations will become effective by the end of 1992.

Confinement.   Those procedures taken to keep a material in a defined or local area once released.

Confinement is one technique that can be used to control an incident. The material involved is assumed to be outside of its normal contained configuration.

Container.   Any vessel or receptacle that holds material, including storage vessels, pipelines, and packaging (see definition of packaging). Containers include:

(a) Nonbulk packaging, such as bags, bottles, boxes, carboys, cylinders, drums, jerricans, multicell packages, and wooden barrels;

(b) Bulk packaging, such as bulk bags, bulk boxes, cargo tanks, covered hopper cars, freight containers, gondolas, pneumatic hopper trailers, portable tanks and bins, protective overpacks for radioactive materials, tank cars, ton containers, and van trailers; and

(c) Fixed containers such as piping, reactors, storage bins, tanks, and storage vessels.

Although some codes define a container by placing size limitations on its capacity, this document does not. NFPA 472 says, in effect, that a container is anything designed or intended to hold a hazardous material.

Containment. The actions taken to keep a material in its container (e.g., stop a release of the material or reduce the amount being released).

Containment will often involve plugging or patching a container to stop a leak. Committing personnel to this type of operation must be carefully considered, and must take into account the level of training they have received. Many, if not most, containment activities can be considered to be "offensive" in nature and require training to the technician or the off-site specialist employee B level.

Contaminant. A hazardous material that physically remains on or in people, animals, the environment, or equipment, thereby creating a continuing risk of direct injury or a risk of exposure outside of the hot zone.

Inherent in the definition of a contaminant is the concept that the offending material is present where it does not belong and that it is somehow toxic or harmful.

Contamination. The process of transferring a hazardous material from its source to people, animals, the environment, or equipment, which may act as a carrier.

Contamination is one of the more important considerations for emergency responders from a health and life safety standpoint. The importance of determining whether personnel or equipment have been contaminated cannot be stressed too strongly. Because personnel are often unaware that contamination has occurred, it is imperative that procedures be established ahead of time to ensure proper monitoring and decontamination.

Contamination Reduction Corridor.   This area is usually located within the warm zone and is where decontamination procedures take place. This is also referred to as the decontamination area in other documents.

The purpose of the contamination reduction corridor is to keep contamination from spreading to the cold, or uncontaminated, zone. It is usually located in the warm zone because that is the transition area between the uncontaminated cold zone and the contaminated area of the hot zone.

Control.   The defensive or offensive procedures, techniques, and methods used in the mitigation of a hazardous materials incident, including containment, extinguishment, and confinement.

The word *control* can be used interchangeably with the word *mitigation*. Every measure taken to control a hazardous materials incident is part of the mitigation process. Limiting the degree of contamination by whatever means is available is also part of the control or mitigation process.

Control Zones.   The designation of areas at hazardous materials incidents based upon safety and the degree of hazard. Many terms are used to describe these control zones; however, for the purposes of this standard, these zones are defined as the hot, warm, and cold zones.

The committee's choice of basic terms such as *hot, warm*, and *cold* is based on the fact that the words are simple and easily understood and that they clearly suggest the nature of the situation one would expect to encounter in any area they designate.

Coordination.   The process used to get people, who may represent different agencies, to work together integrally and harmoniously in a common action or effort.

Determining who is in charge of a major hazardous materials incident can be one of the more difficult problems responders encounter. It can be solved by good local emergency planning, an essential element in the proper handling of any incident. Another, more remote concern involves the ultimate responsibility for the consequences associated with the incident. This particular problem may have to be solved by the courts.

Decontamination (Contamination Reduction).   The physical or chemical process of reducing and preventing the spread of contamination from persons and equipment used at a hazardous materials incident.

The broad concept of decontamination includes all actions taken at an incident that serve to reduce or prevent contamination. The narrower meaning suggests those steps taken to remove contaminants that may have accumulated on equipment or personnel at an incident. Good decontamination practices should incorporate and emphasize the notion of prevention as the first step in the process.

Degradation. (a) A chemical action involving the molecular breakdown of a protective clothing material or equipment due to contact with a chemical. (b) The molecular breakdown of the spilled or released material to render it less hazardous during control operations.

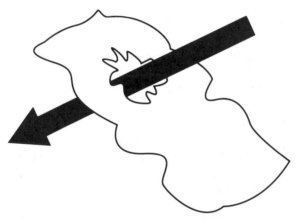

Figure 1.4 Degradation involves molecular breakdown of protective clothing material. (Illustration: Smizer Design.)

Protective clothing and equipment can be protected from chemical degradation by minimizing unnecessary contact with chemicals and by effectively decontaminating them after an incident. Protective clothing may also be subject to physical degradation, such as might occur when it is rubbed against a rough surface or even when it is used normally. The result of degradation, either chemical or physical, is an increased likelihood of permeation and penetration.

Demonstrate. To show by actual performance. This may be supplemented by simulation, explanation, illustration, or a combination of these.

Describe.   To explain verbally or in writing using standard terms recognized in the hazardous materials response community.

Emergency Decontamination.   The physical process of immediately reducing contamination of individuals in potentially life-threatening situations without the formal establishment of a contamination reduction corridor.

The need for emergency decontamination may arise when a sudden, unexpected release of a life-threatening material occurs or when the responders are not aware of the nature of the materials involved in an incident. In such instances, immediate measures, such as applying large amounts of water, must be taken. Once these initial steps have been taken, however, responders must proceed with the proper decontamination procedures.

Emergency Response Guidebook.   A reference book, written in plain language, to guide emergency responders in their initial actions at the incident scene. In the U.S., this book is published by the U.S. Department of Transportation. In Canada, this book is published by Transport Canada. Both books contain similar data in a similar format, with the exception of the table of isolation distances, which is found only in the U.S. version.

These books are the most widely used reference materials available to emergency responders in the U.S. and Canada. They provide basic guidance for initially identifying hazardous materials and the hazards involved in an incident, and they present recommendations for dealing with the incident.

Emergency Response Plan.   A plan that establishes guidelines for handling hazardous materials incidents as required by 29 CFR 1910.120.

Employers are required to develop emergency response plans. To avoid duplication, they may use the plan developed by the local authorities or the plan developed by the state, or both, as part of their own emergency response plan. In Supplement 6, there are two examples of fire department emergency response plans.

Endangered Area.   The actual or potential area of exposure from a hazardous material. This is sometimes referred to as the engulfed area.

The size of the endangered area is a key element in determining the complexity of a hazardous materials incident. Obviously, a hazardous material that poses a threat to 500 square feet will be much easier to manage than one that affects 1 square mile. The Table of Initial Isolation and Protective Action Distances found in the U.S. DOT *Emergency Response Guidebook* will provide guidance in determining the size of the endangered area.

Exposure. The process by which people, animals, the environment, and equipment are subjected to or come in contact with a hazardous material. The magnitude of exposure is dependent primarily upon the duration of exposure and the concentration of the hazardous material. Also used to describe a person, animal, the environment, or a piece of equipment.

Hazard/Hazardous. Capable of posing an unreasonable risk to health, safety, or the environment; capable of causing harm.

Hazard Sector. That function within an overall incident management system that deals with the mitigation of a hazardous materials incident. It is directed by a sector officer and principally deals with the technical aspects of the incident.

The hazard sector provides an organizational structure that allows the necessary supervision and control of the operations that are essential at a hazardous materials incident. These include the tactical objectives carried out in the hot zone, control of access to the site, and decontamination.

In some areas, the hazard sector is referred to as the *hazard division*.

Hazard Sector Officer. The person responsible for the management of the hazard sector.

The person holding this position is responsible for managing the three functional areas of a hazardous materials incident listed above. Rescue operations also come under the direction of this position, but activities outside the control zones do not.

This position is known as the *hazardous materials group supervisor* in some areas.

Hazardous Material.* A substance (solid, liquid, or gas) that when released is capable of creating harm to people, the environment, and property.

A-1-2 **Hazardous Material.** There are many definitions and descriptive names being used for the term hazardous material, each of which depends on the nature of the problem being addressed.

Unfortunately, there is no one list or definition that covers everything. The U.S. agencies involved, as well as state and local governments, have different purposes for regulating hazardous materials that, under certain circumstances, pose a risk to the public or the environment.

(a) *Hazardous Materials.* The U.S. Department of Transportation (DOT) uses the term *hazardous materials*, which covers eleven hazard classes, some of which have subcategories called divisions. DOT includes in its regulations hazardous substances and hazardous wastes as Class 9 (Miscellaneous Hazardous Materials), both of which are regulated by the U.S. Environmental Protection Agency (EPA), if their inherent properties would not otherwise be covered.

(b) *Hazardous Substances.* EPA uses the term *hazardous substance* for the chemicals that, if released into the environment above a certain amount, must be reported, and, depending on the threat to the environment, federal involvement in handling the incident can be authorized. A list of the hazardous substances is published in 40 CFR Part 302, Table 302.4. The U.S. Occupational Safety and Health Administration (OSHA) uses the term hazardous substance in 29 CFR Part 1910.120, which resulted from Title I of SARA and covers emergency response. OSHA uses the term differently than EPA. Hazardous substances, as used by OSHA, cover every chemical regulated by both DOT and EPA.

(c) *Extremely Hazardous Substances.* EPA uses the term *extremely hazardous substance* for the chemical that must be reported to the appropriate authorities if released above the threshold reporting quantity. Each substance has a threshold reporting quantity. The list of extremely hazardous substances is identified in Title III of the Superfund Amendments and Reauthorization Act (SARA) of 1986 (40 CFR Part 355).

(d) *Toxic Chemicals.* EPA uses the term *toxic chemical* for chemicals whose total emissions or releases must be reported annually by owners and operators of certain facilities that manufacture, process, or otherwise use a listed toxic chemical. The list of toxic chemicals is identified in Title III of SARA.

(e) *Hazardous Wastes.* EPA uses the term *hazardous wastes* for chemicals that are regulated under the Resource, Conservation, and Recovery Act (40 CFR Part 261.33). Hazardous wastes in transportation are regulated by DOT (49 CFR Parts 170–179).

(f) *Hazardous Chemicals.* OSHA uses the term *hazardous chemical* to denote any chemical that would be a risk to employees if exposed in the workplace. Hazardous chemicals cover a broader group of chemicals than the other chemical lists.

(g) *Dangerous Goods.* In Canadian transportation, hazardous materials are called *dangerous goods.*

Hazardous Materials Response Team. The hazardous materials response team is an organized group of trained response personnel operating under an emergency response plan and appropriate standard operating procedures, who are expected to perform work to handle and control actual or potential leaks or spills of hazardous materials requiring possible close approach to the material. The team members perform response to releases or potential releases of hazardous materials for the purpose of control or stabilization of the incident.

A hazardous materials response team may be made up of various arrangements of personnel. In some cases, they may all come from the same agency. In others, they may be drawn from various agencies and/or disciplines. However, all members of a hazardous materials response team must operate under one set of operating procedures, the team must have an organizational structure, and, most important, the team members must all train together on a regular basis. As stated earlier, the personnel on the team need not all be trained to the same level. It is important, though, that the duties and responsibilities of each member be commensurate with his or her level of training.

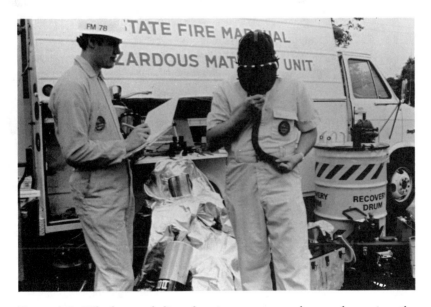

Figure 1.5 Whether a dedicated unit or a team that performs in other capacities, a hazardous materials response team should be properly equipped and fully trained.

High Temperature-Protective Clothing.   Protective clothing designed to protect the wearer for short-term high temperature exposures. This type of clothing is usually of limited use in dealing with chemical commodities.

Specialized high temperature clothing is designed to provide protection against brief radiant heat exposures to temperatures as high as 2000°F (1093°C). NFPA 1976, *Standard on Protective Clothing for Proximity Fire Fighting,*[2] provides performance criteria for proximity suits, which consist of a one- or two-piece overgarment with hood, gloves, and, often, boot covers of an aluminized or cotton fabric. There are also fire entry suits, specialized high temperature ensembles designed to protect the wearer against abnormally high temperatures for a brief exposure of two or three minutes.

Figure 1.6 Approach suits can be used in applications where temperatures reach as high as 2000°F. However, they are meant for brief exposures of three minutes or less. These suits will provide limited protection against both hot and cold liquids, mildly corrosive chemicals, and steam.

Hot Zone.   Area immediately surrounding a hazardous materials incident, which extends far enough to prevent adverse effects from hazardous materials releases to personnel outside the zone. This zone is also referred to as the exclusion zone or the restricted zone in other documents.

The hot zone is the area in which contamination does or could take place. It is also the area in which cleanup operations occur. The boundary between the hot and warm zones should be clearly indicated by some physical means, such as barrier tape, barricades, or some other type of obvious demarcation. Movement of personnel from one zone to another must be strictly regulated and supervised to keep contamination from being carried across zones. Regulating movement between zones will also limit the number of personnel exposed to the most hazardous conditions.

Identify. To select or indicate verbally or in writing using recognized standard terms. To establish the identity of; the fact of being the same as the one described.

Incident. An emergency involving the release or potential release of a hazardous material, with or without fire.

Incident Commander. The person responsible for all decisions relating to the management of the incident. The incident commander is in charge of the incident site. This is equivalent to the on-scene incident commander as defined by 29 CFR 1910.120.

The incident commander is responsible for the overall control of operations. Where multiple agencies are likely to be involved (which is generally the case when hazardous materials are involved), the roles and responsibilities of the various agencies should be clarified ahead of time. The actual command may change from one person to another as the incident develops and becomes more complex. Some incident command systems provide for a "unified command" when multiple agencies are involved. How the system is structured is less important than the fact that there is a system for command and control, that everyone involved is aware of his or her role, and that training exercises have been conducted to work out any problems.

Incident Management System. An organized system of roles, responsibilities, and standard operating procedures used to manage and direct emergency operations. Such systems are sometimes referred to as incident command systems (ICS).

An effective incident management system is one of the most important aspects of operating safely during an emergency. NFPA 1561, *Standard on Fire Department Incident Management Systems,*[3] provides an excellent guide for developing an effective incident management system and clearly addresses the components that must be included for personnel to operate at an incident in as organized and safe a manner as possible.

Figure 1.7 At *major incidents, the incident command post may be at a remote location.*

Individual Area of Specialization.   Refers to the qualifications or functions of a specific job(s) associated with chemicals and/or containers used within an organization.

This definition applies to those who have expertise in a specific product or container. This is typically someone from industry who, because of his or her work, has acquired an in-depth knowledge of such products or containers.

Liquid Splash-Protective Clothing.   The garment portion of a chemical-protective clothing ensemble that is designed and configured to protect the wearer against chemical liquid splashes but not against chemical vapors or gases. Liquid splash-protective clothing must meet the requirements of NFPA 1992. This type of protective clothing is a component of EPA Level B chemical protection.

This type of clothing consists of chemical-resistant coveralls or one- or two-piece splash suits. It is used in situations in which the products and hazards are known.

Listed.* Equipment or materials included in a list published by an organization acceptable to the "authority having jurisdiction" and concerned with product evaluation, that maintains periodic inspection of production of listed equipment or materials and whose listing states either that the equipment or material meets appropriate standards or has been tested and found suitable for use in a specified manner.

A-1-2 **Listed.** The means for identifying listed equipment may vary for each organization concerned with product evaluation, some of which do not recognize equipment as listed unless it is also labeled. The "authority having jurisdiction" should utilize the system employed by the listing organization to identify a listed product.

Local Emergency Response Plan. The plan promulgated by the authority having jurisdiction, e.g., as the local emergency planning committee for the community or a facility.

Whether it is designed for a community or a facility, the local emergency response plan must identify the hazards that are present and establish written procedures for handling incidents.

Match. To provide with a counterpart.

Material Safety Data Sheet (MSDS). A form, provided by manufacturers and compounders (blenders) of chemicals, containing information about chemical composition, physical and chemical properties, health and safety hazards, emergency response, and waste disposal of the material as required by 29 CFR 1910.1200.

A typical material safety data sheet is available from the U.S. Department of Labor Occupational Safety and Health Administration (OSHA).

Monitoring Equipment. Instruments and devices used to identify and quantify contaminants.

The use of monitoring equipment at an incident is important for several reasons. It can help determine the personal protective equipment that will be needed, it can help establish control zones, it can identify unknowns, and it can determine the level of contamination or decontamination.

Objective. A goal that is achieved through the attainment of a skill, knowledge, or both, and that can be observed or measured.

Off-site Specialist Employee A. Those persons who are specifically trained to handle incidents involving chemicals and/or containers for chemicals used in their organization's area of specialization. Consistent with the organization's emergency response plan and standard operating procedures, the off-site specialist employee A shall have the ability to analyze an incident involving chemicals within the organization's area of specialization, plan a response to that incident, implement the planned response within the capabilities of the resources available, and evaluate the progress of the planned response.

The individual at this level has the expertise and training to deal with a wide variety of products manufactured, used, or transported by his or her employer. An example of someone at this level is a member of an organized industrial response team that is trained to assist local authorities.

Off-site Specialist Employee B. Those persons who, in the course of their regular job duties, work with or are trained in the hazards of specific chemicals or containers within their individual area of specialization. Because of his or her education, training, or work experience, the off-site specialist employee B may be called upon to gather and record information, provide technical advice, and provide technical assistance (including work within the hot zone) at an incident involving chemicals consistent with his or her organization's emergency response plan and standard operating procedures and the local emergency response plan.

An off-site specialist employee B should know enough about, and be trained to deal with, a specific product or container to be able to provide the technical assistance necessary to conduct control operations inside the hot zone. For example, someone who has the equipment and training to control a leak in a chlorine container might be designated an off-site specialist employee B. This person may be part of an organized industrial response team or be an industrial specialist sent in to help local authorities.

Off-site Specialist Employee C. Those persons who may respond to incidents involving chemicals and/or containers within their organization's area of specialization. The off-site specialist employee C may be called upon to gather and record information, provide technical advice, and/or arrange for technical assistance consistent with his or her organizations's emergency response plan and standard operating procedures. The off-site specialist employee C is not expected to enter the hot/warm zone at an incident.

The off-site specialist employee C should be able to provide information on a specific chemical or container and have the organizational contacts needed to acquire additional technical assistance. This individual need not have the skills or training necessary to conduct control operations. He or she is generally found at the incident command post providing the incident commander or his or her designee with technical assistance.

Organization's Area of Specialization. Refers to any chemicals and containers used by the off-site specialist employee's employer.

This refers to the specific products or containers with which an individual has expertise. Even though a company may manufacture, use, or transport a wide variety of products, specific individuals may only have knowledge of certain of these products.

Packaging. Any container that holds a material (hazardous and nonhazardous). Packaging for hazardous materials includes nonbulk and bulk packaging.

Nonbulk Packaging. Any packaging having a capacity meeting one of the following criteria:

(a) Liquid — internal volume of 119 gal (450 L) or less;
(b) Solid — capacity of 882 lb (400 kg) or less; and
(c) Compressed gas — water capacity of 1,001 lb (454 kg) or less.

Figure 1.8   Railroad cars are an example of bulk packaging.

Bulk Packaging. Any packaging, including transport vehicles, having a capacity greater than described above under nonbulk packaging. Bulk packaging can be either placed on or in a transport vehicle or vessel, or constructed as an integral part of the transport vehicle.

(a) Liquid — internal volume of more than 119 gal (450 L);
(b) Solid — capacity of more than 882 lb (400 kg); and
(c) Compressed gas — water capacity of more than 1,001 lb (454 kg).

See Supplement 9 for more information about packaging.

Penetration. The movement of a material through a suit's closures, such as zippers, buttonholes, seams, flaps, or other design features of chemical-protective clothing, and through punctures, cuts, and tears.

A regular program of inspection can help uncover conditions that may eventually allow penetration.

Permeation. A chemical action involving the movement of chemical, on a molecular level, through intact material.

Different fabrics have different levels of resistance to chemical permeation, and all will absorb chemicals over a period of time. NFPA 1991,[4] 1992,[5] and 1993[6] provide guidelines for manufacturer permeation testing and certification of chemical-protective clothing (CPC). When purchasing CPC, it is important to be sure the garments meet the appropriate standard and have been certified as meeting that standard. In any event, the user should exercise extreme caution. Available data cannot cover every possible situation.

Personal Protective Equipment. The equipment provided to shield or isolate a person from the chemical, physical, and thermal hazards that may be encountered at a hazardous materials incident. Personal protective equipment includes both personal protective clothing and respiratory protection. Adequate personal protective equipment should protect the respiratory system, skin, eyes, face, hands, feet, head, body, and hearing.

The purpose of personal protective equipment is to protect the responder against every anticipated hazard. Although protective equipment has improved substantially in recent years, the nature of the hazards to which emergency personnel must respond has become much more complex. It is imperative

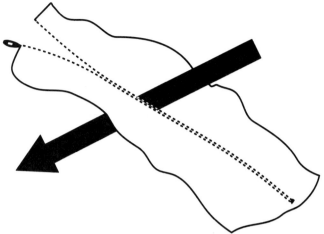

Figure 1.9 Penetration involves the movement of a material through a suit's closures, such as zippers, buttonholes, and seams. (Illustration: Smizer Design.)

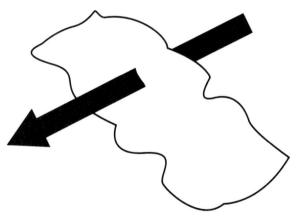

Figure 1.10 Permeation involves chemical movement, on a molecular level, through intact material. (Illustration: Smizer Design.)

for responders to be adequately trained to use, care for, and select appropriate personal protective equipment: Protective equipment will not prevent injuries or illness if improperly used.

Planned Response. The plan of action, with safety considerations, consistent with the local emergency response plan and an organization's standard operating procedures for a specific hazardous materials incident.

Appropriate preventive measures and control actions are identified based on the magnitude of the problem. This process considers available resources, then determines the direction the response effort must take to influence the events and to favorably change the outcome.

Protective Clothing.   Equipment designed to protect the wearer from heat and/or hazardous materials contacting the skin or eyes. Protective clothing is divided into four types:

(a) Structural fire fighting protective clothing;

Structural fire fighting protective clothing, as defined here, is the complete structural fire fighting ensemble, which includes appropriate footwear, trousers, a coat, gloves, a helmet, and respiratory protection. When properly worn, this ensemble provides protection from the hazards normally encountered in structural fire fighting.

A series of NFPA standards provides guidelines for fire fighters' protective clothing. They include:

- NFPA 1971, *Standard on Protective Clothing for Structural Fire Fighting*,[7]
- NFPA 1972, *Standard on Helmets for Structural Fire Fighting*,[8]
- NFPA 1973, *Standard on Gloves for Structural Fire Fighting*,[9]
- NFPA 1974, *Standard on Protective Footwear for Structural Fire Fighting*,[10] and
- NFPA 1981, *Standard on Open-Circuit Self-Contained Breathing Apparatus for Fire Fighters*.[11]

(b) Liquid splash-protective clothing;

NFPA 1992, *Standard on Liquid Splash-Protective Clothing for Hazardous Chemical Emergencies*, provides criteria for this type of protective clothing. This clothing protects against splash or spill hazards. It does not protect against hazardous vapors or gases.

(c) Vapor-protective clothing; and

NFPA 1991, *Standard on Vapor-Protective Suits for Hazardous Chemical Emergencies*, provides criteria for this type of protective clothing. Used with the proper respiratory protection, it provides the highest level of protection.

(d) High temperature-protective clothing.

NFPA 1976, *Standard on Protective Clothing for Proximity Fire Fighting,* provides criteria for the type of high temperature-protective clothing normally used in aircraft fire fighting and rescue, bulk flammable liquids fire fighting, flammable gas fire fighting, and similar situations in which high levels of radiant heat are released.

Qualified. Having satisfactorily completed the learning objectives.

Respiratory Protection. Equipment designed to protect the wearer from the inhalation of contaminants. Respiratory protection is divided into three types:

Since inhalation of toxics is one of the principal causes of serious injury to responders, respiratory protection is of the utmost importance.

(a) Positive pressure self-contained breathing apparatus;

This is the type of breathing apparatus fire fighters normally use. It does not restrict the wearer's mobility, and it provides the highest level of respiratory protection. The most commonly used units will provide 30 to 60 minutes of protection. However, units are available that provide longer periods of use. NFPA 1981, *Standard on Open-Circuit Self-Contained Breathing Apparatus for Fire Fighters,* provides performance criteria for SCBA.

(b) Positive pressure airline respirators; and

Positive pressure airline respirators give the user an unlimited supply of air and are lighter than SCBA. However, they restrict the travel distance of the user to a maximum of 300 feet and the path of travel must be kept clear of obstructions while the user is negotiating the incident scene. Nonetheless, they are an ideal method of supplying air to the emergency worker under certain conditions.

(c) Air purifying respirators.

Air purifying respirators (APR) use a filter or a sorbent to remove airborne contaminants. They should not be used by the initial responders to an incident involving hazardous materials. They should be used only when the hazard and the concentration of the hazardous material are known.

Response. That portion of incident management in which personnel are involved in controlling (defensively or offensively) a hazardous materials incident. The activities in the response portion of a hazardous materials incident include analyzing the incident, planning the response, implementing the planned response, and evaluating progress.

The actions that are taken, whether offensive or defensive, center on implementing the steps that will lead to containment, confinement, or extinguishment.

Safely.   To perform the assigned tasks without injury to self or others, to the environment, or to property.

It has been the committee's goal from the inception of this document to outline the minimum competencies responders need to increase their level of safety. Throughout the document, the committee stresses conducting operations safely.

Secondary Contamination.   The process by which a contaminant is carried out of the hot zone and contaminates people, animals, the environment, or equipment outside of the hot zone.

Controlling secondary contamination is a key factor in establishing control zones and contamination reduction areas. It is important not only to the health and safety of the on-scene responder, but also to the health and safety of the public at large: Contaminants can be carried far from the immediate incident site and have serious health consequences beyond the scene of the emergency. For example, ambulances, hospital emergency rooms, and fire stations may be contaminated. And contaminants carried home by responders may be hazardous to members of their families.

Shall.   Indicates a mandatory requirement.

When used in the context of any NFPA standard, the term *shall* refers to a requirement that must be met.

Should.   Indicates a recommendation or that which is advised but not required.

The term *should* is not used in the main body of NFPA standards, but it can be found in the appendices where the committee wants to make a recommendation but does not believe that it should be mandatory.

Stabilization.   The point in an incident at which the adverse behavior of the hazardous material is controlled.

Stabilization does not mean that the incident is over, but rather that the hazardous conditions should not escalate or intensify any further. It could well be that operations following stabilization are prolonged and that specialized types of equipment and expertise will be needed before the operations are completed.

State.   Where the noun "state" is used it shall also include by implication any outlying U.S. areas where this standard is in effect. Use of the noun "state" shall imply "provinces and territories" in Canada.

The committee does not intend to restrict the use of this document to the U.S. In countries where other terminology is used, references made to "state" regulations should be understood to mean whatever type of governmental authority is appropriate.

Structural Fire Fighting Protective Clothing.   This category of clothing, often called turnout or bunker gear, means the protective clothing normally worn by fire fighters during structural fire fighting operations. It includes a helmet, coat, pants, boots, gloves, and a hood to cover parts of the head not protected by the helmet and facepiece. Structural fire fighters' protective clothing provides limited protection from heat but may not provide adequate protection from the harmful gases, vapors, liquids, or dusts that are encountered during hazardous materials incidents.

With or without respiratory protection, this type of protective clothing provides the lowest level of protection, Level D, which is equivalent to that of an ordinary work uniform.

Termination.   That portion of incident management in which personnel are involved in documenting safety procedures, site operations, hazards faced, and lessons learned from the incident. Termination is divided into three phases: debriefing the incident, post-incident analysis, and critiquing the incident.

Debriefing the incident involves collecting all the pertinent information about the nature of the incident and about the emergency actions and operations that occurred while the incident was being controlled. The collected information is then evaluated during the post-incident analysis. Critiquing the incident is important because it teaches all responders valuable lessons they can use to enhance their performances during future operations at hazardous materials incidents.

UN/NA Identification Number.   UN/NA identification numbers are four-digit numbers assigned to a hazardous material. The number is used to identify and cross-reference the product.

This four-digit number is found in shipping papers and on all tank cars, cargo tanks, portable tanks, and bulk packages containing a hazardous material regulated by the U.S. DOT. The prefix "UN/NA" is found only in the

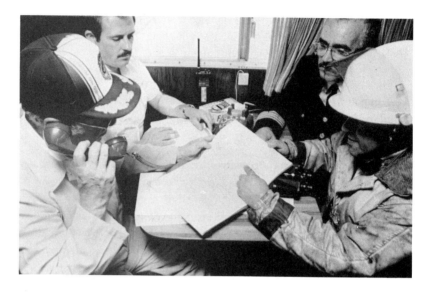

Figure 1.11 Debriefing, critiquing, and lessons learned are important elements of post-incident analysis.

shipping papers and indicates whether a product can be shipped internationally or only within North America.

Vapor-Protective Clothing.   The garment portion of a chemical-protective clothing ensemble that is designed and configured to protect the wearer against chemical vapors or gases. Vapor-protective clothing must meet the requirements of NFPA 1991. This type of protective clothing is a component of EPA Level A chemical protection.

This type of clothing must be worn when the greatest level of skin, respiratory, and eye protection is needed. It should be worn when the hazard will or could immediately endanger life and health.

Warm Zone.   The control zone at a hazardous materials incident site where personnel and equipment decontamination and hot zone support takes place. It includes control points for the access corridor, helping to reduce the spread of contamination. This zone is also referred to as the decontamination, contamination reduction, or limited access zone in other documents.

One function of the warm zone is to reduce the likelihood of contaminating the cold zone. To some extent, the warm zone serves as a buffer between the hot and cold zones. The intensity of contamination in the warm zone should decrease the closer one gets to the cold zone, not only because decontamination is taking place, but also because of the intervening space the warm zone affords between the other two zones.

An access corridor is a defined path between the hot and cold zones where personnel and equipment are decontaminated. Several access corridors may be needed at very large incidents. The access corridors must be tightly controlled and supervised to regulate movement between the zones. A site access control leader should be assigned to ensure that all people and equipment use the appropriate access routes and that records of entry are maintained for tracking purposes. Individuals entering the warm zone from the cold zone must wear the appropriate personal protective equipment.

References Cited in Commentary

1. Arthur E. Cote, ed., *Fire Protection Handbook*, 17th edition (Quincy, MA: National Fire Protection Association, 1991), p. 9-154.

2. NFPA 1976, *Standard on Protective Clothing for Proximity Fire Fighting*, 1992 edition.

3. NFPA 1561, *Standard on Fire Department Incident Management Systems*, 1990 edition.

4. NFPA 1991, *Standard on Vapor-Protective Suits for Hazardous Chemical Emergencies*, 1990 edition.

5. NFPA 1992, *Standard on Liquid Splash-Protective Suits for Hazardous Chemical Emergencies*, 1990 edition.

6. NFPA 1993, *Standard on Support Function Protective Garments for Hazardous Chemical Operations*, 1990 edition.

7. NFPA 1971, *Standard on Protective Clothing for Structural Fire Fighting*, 1991 edition.

8. NFPA 1972, *Standard on Helmets for Structural Fire Fighting*, 1992 edition.

9. NFPA 1973, *Standard on Gloves for Structural Fire Fighting*, 1988 edition.

10. NFPA 1974, *Standard on Protective Footwear for Structural Fire Fighting*, 1992 edition.

11. NFPA 1981, *Standard on Open-Circuit Self-Contained Breathing Apparatus for Fire Fighters*, 1992 edition.

2

Competencies for the First Responder at the Awareness Level

2-1 General.

2-1.1 **Introduction.** First responders at the awareness level shall be trained to meet all competencies of this chapter. In addition, first responders at the awareness level shall receive training to meet federal Occupational Safety and Health Administration (OSHA), local occupational health and safety regulatory, or U.S. Environmental Protection Agency (EPA) requirements, whichever are appropriate for their jurisdiction.

2-1.2 **Definition.** First responders at the awareness level are those persons who, in the course of their normal duties, may be the first on the scene of an emergency involving hazardous materials. First responders at the awareness level are expected to recognize hazardous materials presence, protect themselves, call for trained personnel, and secure the area.

Responders at this level are those who are likely to witness or discover a hazardous materials release and who would be expected, as part of their responsibilities, to begin emergency response procedures. Awareness level responders include truck drivers, train crews, law enforcement officers, and others whose duties require them to work in facilities where hazardous materials are transported, stored, or used. Responders at the awareness level are not expected to take any action that would require a great deal of training and experience. Rather, their actions will be basic and limited.

2-1.3 **Goal.** The goal of the competencies at the awareness level shall be to provide first responders with the knowledge and skills to perform the following tasks safely. Therefore, when first on the scene of an emergency involving hazardous materials, the first responder at the awareness level shall be able to:

(a) Analyze the incident to determine both the hazardous materials present and the basic hazard and response information for each hazardous material by completing the following tasks:

It is important that responders at all levels analyze the problem at hand and determine the specific hazards present and the potential consequences of those hazards. A systematic approach will help the responder make accurate decisions about the potential outcomes.

At each level, the same analysis system is used, although the skills of the responder and his or her ability to weigh more factors and make more complex decisions are expected to increase.

1. Detect the presence of hazardous materials;

The first step in analyzing the problem is to determine whether hazardous materials are present. Placards, the type of containers involved, and the presence of fires or explosions are a few typical indicators. The standard clearly indicates that training in recognizing and identifying a hazardous material is a given and that the employer, whether in the public or private sector, is responsible for seeing that his or her employees receive such training.

2. Survey a hazardous materials incident, from a safe location, to identify the name, UN/NA identification number, or type placard applied for any hazardous materials involved; and

All actions must be taken with appropriate safety precautions. The responder at this level will not have any specialized protective clothing or equipment and must therefore exercise extreme caution when confirming the presence of hazardous materials.

3. Collect hazard information from the current edition of the *Emergency Response Guidebook*.

Figure 2.1 *First responders need to know labels and placards associated with hazardous materials.*

The DOT *Emergency Response Guidebook*, published in the U.S., and the *Dangerous Goods Initial Response Guide*, published in Canada, provide the responder with an excellent tool for beginning the initial assessment of the hazards present.

(b) Implement actions consistent with the local emergency response plan, the organization's standard operating procedures, and the current edition of the *Emergency Response Guidebook* by completing the following tasks:

1. Initiate protective actions consistent with the local emergency response plan, the organization's standard operating procedures, and the current edition of the *Emergency Response Guidebook;* and

The responder must be trained in the community's response procedures and know how to initiate them. Responders should also be familiar with their own organization's response plan and with their role in it. In many cases, these actions will consist of notifying the local emergency responders, such as the fire department, and securing the immediate area to prevent exposure to others.

2. Initiate the notification process specified in the local emergency response plan and the organization's standard operating procedures.

It is important for the responder to know how to begin the emergency notification process. This applies to incidents at fixed sites, as well as to those that occur off-site.

2-2 Competencies — Analyzing the Incident.

2-2.1 **Detecting the Presence of Hazardous Materials.** The first responder at the awareness level shall, given various facility and/or transportation situations, with and without hazardous materials present, identify those situations where hazardous materials are present. The first responder at the awareness level shall be able to:

2-2.1.1* Identify the definition of hazardous materials (or dangerous goods, in Canada).

The responder must be able to identify what the hazardous materials are and what hazards they present. The standard also implies that responders must take care to avoid making the situation worse than it is.

A-2-2.1.1 See A-1-2, Hazardous Materials.

2-2.1.2* Identify the DOT hazard classes and divisions of hazardous materials and identify common examples of materials in each hazard class or division.

Because proper identification of hazardous materials is extremely important, the actions of the awareness level responder can be critical to a successful emergency response. By being able to identify the hazard classes, the responder will have a better understanding of potential problems that may arise.

A-2-2.1.2 **Definitions of Department of Transportation Hazard Classes and Divisions.** Department of Transportation, Research and Special Programs Administration, "Performance-Oriented Packaging Standards; Changes to Classification, Hazard Communication, Packaging and Handling Requirements Based on UN Standards and Agency Initiative; Final Rule," *Federal Register*, Vol. 55, No. 246, December 21, 1990, pages 52, 402–52, 729.

Class 1 (Explosives)

Explosive means any substance or article, including a device, that is designed to function by explosion (i.e., an extremely rapid release of gas and heat) or that, by chemical reaction within itself, is able to function in a similar manner even if not designed to function by explosion. Explosives in Class 1 are divided into six divisions. Each division will have a letter designation.

Division 1.1 consists of explosives that have a mass explosion hazard. A mass explosion is one that affects almost the entire load instantaneously.

Examples of Division 1.1 explosives include black powder, dynamite, and TNT.

Division 1.2 consists of explosives that have a projection hazard but not a mass explosion hazard.

Examples of Division 1.2 explosives include aerial flares, detonating cord, and power device cartridges.

Division 1.3 consists of explosives that have a fire hazard and either a minor blast hazard or a minor projection hazard, or both, but not a mass explosion hazard.

Examples of Division 1.3 explosives include liquid-fueled rocket motors and propellant explosives.

Division 1.4 consists of explosive devices that present a minor explosion hazard. No device in the division may contain more than 25 g (0.9 oz) of a detonating material. The explosive effects are largely confined to the package and no projection of fragments of appreciable size or range are expected. An external fire must not cause virtually instantaneous explosion of almost the entire contents of the package.

Examples of Division 1.4 explosives include line-throwing rockets, practice ammunition, and signal cartridges.

Division 1.5 consists of very insensitive explosives. This division is comprised of substances that have a mass explosion hazard but are so insensitive that there is very little probability of initiation or of transition from burning to detonation under normal conditions of transport.

Examples of Division 1.5 explosives include prilled ammonium nitrate fertilizer-fuel oil mixtures (blasting agents).

Division 1.6 consists of extremely insensitive articles that do not have a mass explosive hazard. This division is comprised of articles that contain only extremely insensitive detonating substances and that demonstrate a negligible probability of accidental initiation or propagation.

Class 2

Division 2.1 (Flammable Gas) means any material that is a gas at 20°C (68°F) or less and 101.3 kPa (14.7 psi) of pressure, a material that has a boiling point of 20°C (68°F) or less at 101.3 kPa (14.7 psi) and that:

(a) Is ignitable at 101.3 kPa (14.7 psi) when in a mixture of 13 percent or less by volume with air; or

(b) Has a flammable range at 101.3 kPa (14.7 psi) with air of at least 12 percent regardless of the lower limit.

Examples of Division 2.1 gases include inhibited butadienes, methyl chloride, and propane.

Division 2.2 (Nonflammable, Nonpoisonous Compressed Gas, Including Compressed Gas, Liquefied Gas, Pressurized Cryogenic Gas, and Compressed Gas in Solution) A nonflammable, nonpoisonous compressed gas means any material (or mixture) that exerts in the packaging an absolute pressure of 280 kPa (41 psia) at 20°C (68°F).

A cryogenic liquid means a refrigerated liquefied gas having a boiling point colder than –90°C (–130°F) at 101.3 kPa (14.7 psi) absolute.

Examples of Division 2.2 gases include anhydrous ammonia, cryogenic argon, carbon dioxide, and compressed nitrogen.

Division 2.3 (Poisonous Gas) means a material that is a gas at 20°C (68°F) or less and a pressure of 101.3 kPa (14.7 psi or 1 atm), a material that has a boiling point of 20°C (68°F) or less at 101.3 kPa (14.7 psi), and that:

(a) Is known to be so toxic to humans as to pose a hazard to health during transportation; or

(b) In the absence of adequate data on human toxicity, is presumed to be toxic to humans because, when tested on laboratory animals, it has an LC_{50} value of not more than 5,000 ppm.

Examples of Division 2.3 gases include anhydrous hydrogen fluoride, arsine, chlorine, and methyl bromide.

Hazard zones are associated with Division 2.3 materials:

Hazard zone A — LC_{50} less than or equal to 200 ppm.

Hazard zone B — LC_{50} greater than 200 ppm and less than or equal to 1,000 ppm.

Hazard zone C — LC_{50} greater than 1,000 ppm and less than or equal to 3,000 ppm.

Hazard zone D — LC_{50} greater than 3,000 ppm and less than or equal to 5,000 ppm.

Class 3 (Flammable Liquid)

Flammable liquid means any liquid having a flash point of not more than 60.5°C (141°F).

Examples of Class 3 liquids include acetone, amyl acetate, gasoline, methyl alcohol, and toluene.

Hazard zones are associated with Class 3 materials:

Hazard zone A — LC_{50} less than or equal to 200 ppm.

Hazard zone B — LC_{50} greater than 200 ppm and less than or equal to 1,000 ppm.

Combustible Liquid

Combustible liquid means any liquid that does not meet the definition of any other hazard class and has a flash point above 60°C (140°F) and below 93°C (200°F). Flammable liquids with a flash point above 38°C (100°F) may be reclassified as a combustible liquid.

Examples of combustible liquids include mineral oil, peanut oil, and No. 6 fuel oil.

Class 4

Division 4.1 (Flammable Solid) means any of the following three types of materials:

(a) Wetted explosives — explosives wetted with sufficient water, alcohol, or plasticizers to suppress explosive properties.

(b) Self-reactive materials — materials that are liable to undergo, at normal or elevated temperatures, a strongly exothermic decomposition caused by excessively high transport temperatures or by contamination.

(c) Readily combustible solids — solids that may cause a fire through friction and any metal powders that can be ignited.

Examples of Division 4.1 materials include magnesium (pellets, turnings, or ribbons) and nitrocellulose.

Division 4.2 (Spontaneously Combustible Material) means any of the following materials:

(a) Pyrophoric material — a liquid or solid that, even in small quantities and without an external ignition source, can ignite within 5 minutes after coming in contact with air.

(b) Self-heating material — a material that, when in contact with air and without an energy supply, is liable to self-heat.

Examples of Division 4.2 materials include aluminum alkyls, charcoal briquettes, magnesium alkyls, and phosphorus.

Division 4.3 (Dangerous When Wet Material) means a material that, by contact with water, is liable to become spontaneously flammable or to give off flammable or toxic gas at a rate greater than 1 L per kg of the material, per hour.

Examples of Division 4.3 materials include calcium carbide, magnesium powder, potassium metal alloys, and sodium hydride.

Class 5

Division 5.1 (Oxidizer) means a material that may, generally by yielding oxygen, cause or enhance the combustion of other materials.

Examples of Division 5.1 materials include ammonium nitrate, bromine trifluoride, and calcium hypochlorite.

Division 5.2 (Organic Peroxide) means any organic compound containing oxygen (O) in the bivalent -O-O- structure that may be considered a derivative of hydrogen peroxide, where one or more of the hydrogen atoms have been replaced by organic radicals.

Division 5.2 (Organic Peroxide) materials are assigned to one of seven types:

Type A — organic peroxide that can detonate or deflagrate rapidly as packaged for transport. Transportation of type A organic peroxides is forbidden.

Type B — organic peroxide that neither detonates nor deflagrates rapidly, but that can undergo a thermal explosion.

Type C — organic peroxide that neither detonates nor deflagrates rapidly and cannot undergo a thermal explosion.

Type D — organic peroxide that detonates only partially or deflagrates slowly, with medium to no effect when heated under confinement.

Type E — organic peroxide that neither detonates nor deflagrates and shows low, or no, effect when heated under confinement.

Type F — organic peroxide that will not detonate, does not deflagrate, shows only a low, or no, effect if heated when confined, and has low or no explosive power.

Type G — organic peroxide that will not detonate, does not deflagrate, shows no effect if heated when confined, and has no explosive power, is thermally stable, and is desensitized.

Examples of Division 5.2 materials include dibenzoyl peroxide, methyl ethyl ketone peroxide, and peroxyacetic acid.

Class 6

Division 6.1 (Poisonous Material) means a material, other than a gas, that is either known to be so toxic to humans as to afford a hazard to health during transportation, or in the absence of adequate data on human toxicity, is presumed to be toxic to humans, including irritating materials that cause irritation.

Examples of Division 6.1 materials include aniline, arsenic compounds, carbon tetrachloride, hydrocyanic acid, and tear gas.

Division 6.2 (Infectious Substance) means a viable microorganism, or its toxin, that causes or may cause disease in humans or animals. Infectious substance and etiologic agent are synonymous.

Examples of Division 6.2 materials include anthrax, botulism, rabies, and tetanus.

Class 7

Radioactive material means any material having a specific activity greater than 0.002 microcuries per gram (μCi/g).

Examples of Class 7 materials include cobalt, uranium hexafluoride, and "yellow cake."

Class 8

Corrosive material means a liquid or solid that causes visible destruction or irreversible alterations in human skin tissue at the site of contact, or a liquid that has a severe corrosion rate on steel or aluminum.

Examples of Class 8 materials include nitric acid, phosphorus trichloride, sodium hydroxide, and sulfuric acid.

Class 9

Miscellaneous hazardous material means a material that presents a hazard during transport, but that is not included in another hazard class, including:

(1) Any material that has an anesthetic, noxious, or other similar property that could cause extreme annoyance or discomfort to a flight crew member so as to prevent the correct performance of assigned duties; and
(2) Any material that is not included in any other hazard class, but is subject to the DOT requirements (a hazardous substance or a hazardous waste).

Examples of Class 9 materials include adipic acid, hazardous substances (e.g., PCBs), and molten sulfur.

ORM-D Material

An ORM-D material is a material that presents a limited hazard during transportation due to its form, quantity, and packaging.

Examples of ORM-D materials include consumer commodities and small arms ammunition.

Forbidden

Forbidden means prohibited from being offered or accepted for transportation. Prohibition does not apply if these materials are diluted, stabilized, or incorporated in devices.

2-2.1.3* Identify the primary hazards associated with each of the DOT hazard classes and divisions of hazardous materials by hazard class or division.

By being able to accurately identify the type of hazardous materials present and the primary hazards they involve, the awareness level responder can begin to take the correct protective actions early. He or she can give this information to the emergency responders, who will then understand the type of incident they are responding to and be able to request any specialized equipment or additional resources they might need. For example, one would initiate considerably different actions if one were dealing with a Class 1, Division 1.1 material (mass detonating explosives) than one would if one were dealing with a Class 2, Division 2.2 (nonflammable gases) material.

In 1991 HM-181, which modified the existing U.S. DOT Hazard Classification system, was adopted by the U.S. Department of Transportation. (*See Table 2.1 for a comparison of the pre- and post-1991 requirements.*) This new system establishes new placarding and labeling requirements. By 1994 only the placards and labels outlined in HM-181 will be allowed. Until then, however, either the pre-1991 labels and placards or the ones adopted by HM-181 may be used.

A-2-2.1.3 See A-2-2.1.2.

2-2.1.4 Identify the difference between hazardous materials incidents and other emergencies.

The adverse consequences of exposure to a hazardous material can be far-reaching and severe. Hazardous materials emergencies thus stand apart from

Table 2.1 Comparison of U.N./U.S. DOT Hazard Classes

U.N. Classes and Divisions After January 1991	U.S. Classes Pre-January 1991	Examples	General Hazard Properties (Not All-Inclusive)
Class 1—EXPLOSIVES			Explosive; exposure to heat, shock, or contamination could result in thermal and mechanical hazards.
Division 1.1–Mass denotating	Class A Explosives	Dynamite, TNT, Black Powder	
Division 1.2–Mass detonating w/fragments	Class A Explosives/ Class B Explosives		
Division 1.3–Fire hazard w/minor blast or projectile hazard	Class B Explosives	Propellant Explosives, Rocket Motors, Special Fireworks	
Division 1.4–Substances that present no significant hazard	Class C Explosives	Common Fireworks, Small Arms Ammunition	
Division 1.5–Very insensitive explosives	Blasting Agents	Ammonium Nitrate-Fuel Oil Mixtures	
Division 1.6–Extremely insensitive explosives			
Class 2—GASES			Under pressure; container may rupture violently (fire and nonfire); may be a flammable, poisonous, a corrosive, an asphyxiant and/or an oxidizer; may cause frost-bite.
Division 2.1–Flammable Gases	Flammable Gases	Propane, Butadiene (inhibited), Acetylene	
Division 2.2–Nonflammable Gases	Nonflammable Gases	Carbon Dioxide, Methyl Chloride, Anhydrous Ammonia	
Division 2.3–Poisonous Gases	Poison A	Arsine, Phosgene, Hydrogen Fluoride Chlorine	
Class 3—FLAMMABLE LIQUIDS			Flammable; container may rupture violently from heat/fire; may be corrosive, toxic, and/or thermally unstable.
*Division 3.1–flash point <0°F	Flammable Liquids	Gasoline	
*Division 3.2–flash point ≥0°F to <73°F	Flammable Liquids	Acetone, Methyl Alcohol, Toluene	
*Division 3.3–flash point ≥73° to <141°F	Flammable/Combustible Liquids	Amyl Acetate Fuel Oils	
Combustible Liquids	Combustible Liquids		

HAZARDOUS MATERIALS RESPONSE HANDBOOK

Table 2.1, Continued

U.N. Classes and Divisions After January 1991	U.S. Classes Pre-January 1991	Examples	General Hazard Properties (Not All-Inclusive)
Class 4—FLAMMABLE SOLIDS OR SUBSTANCES			Flammable, some spontaneously; may be water reactive, toxic, and/or corrosive; may be extremely difficult to extinguish.
Division 4.1–Flammable Solids	Flammable Solids	Pyroxylin Plastics, Magnesium Phosphorus	
Division 4.2–Spontaneously Combustible/Pyrophoric Liquids	Flammable Solids and Liquids		
Division 4.3–Dangerous When Wet	Flammable Solids and Liquids	Metalic Sodium, Potassium, Calcium Carbide	
Class 5—OXIDIZING SUBSTANCES			Supplies oxygen to support combustion; sensitive to heat, shock, friction, and/or contamination.
Division 5.1–Oxidizing Substances	Oxidizers	Ammonium Nitrite Fertilizer	
Division 5.2–Organic Peroxides	Organic Peroxides	Benzoyl Peroxide, Peracetic Acid, Acetal Peroxide Solution	
Class 6—POISONOUS AND INFECTIOUS SUBSTANCES			Toxic by inhalation, ingestion, and skin and eye absorption; may be flammable.
Division 6.1–Poisons	Poisons B, Irritants, ORM-A	Aniline, Arsenic Tear Gas, Dry Ice, Carbon Tetrachloride	
Division 6.2–Infectious Substances	Etiologic Agents	Anthrax, Botulism, Rabies, Tetanus	
Class 7—RADIOACTIVE SUBSTANCES			May cause burns and biologic effects; energy and matter.
Radioactive Material	Radioactive Materials	Plutonium, Cobalt, Uranium Hexafluoride	

Table 2.1, Continued

U.N. Classes and Divisions After January 1991	U.S. Classes Pre-January 1991	Examples	General Hazard Properties (Not All-Inclusive)
Class 8—CORROSIVES			
Corrosive Material	Corrosive Materials	Hydrochloric Acid, Sulfuric Acid, Sodium Hydroxide, Nitric Acid	Disintegration of contacted tissues; may be fuming, water reactive.
	ORM-B	Unslaked Lime, Metallic Mercury	
Class 9—MISCELLANEOUS HAZARDOUS MATERIALS			
	ORM-C	Bleaching Powder, Molten Sulfur	
	ORM-E	Hazardous Substances (e.g. PCBs)	
	ORM-D	Consumer commodities	
ORM-D			

*International only.
(Source: *Fire Protection Handbook*, 17th edition.)

HAZARDOUS MATERIALS RESPONSE HANDBOOK

other types of emergencies because they present such a large potential for doing great harm and because responders must be specifically trained and equipped to deal with them properly.

Figure 2.2   What may seem like a simple truck fire could become a hazardous materials incident.

2-2.1.5 Identify typical occupancies and locations in the community where hazardous materials are manufactured, transported, stored, used, or disposed of.

Individuals involved with hazardous materials should understand where they can be found so that they can recognize the potential harm and take the appropriate emergency actions. Local planning for response should be based, in part, on the knowledge that particular hazardous materials may be present at certain fixed sites.

2-2.1.6 Identify typical container shapes that may indicate hazardous materials.

The configuration of some containers is so unusual that it signals the presence of some hazardous materials. Containers that provide clues include those used for radioactive materials, pressurized products, cryogenics, and corrosives.

2-2.1.7 Identify facility and transportation markings and colors that indicate hazardous materials, including:

(a) UN/NA identification numbers;

Several numbers are used on placards. The first, which denotes the class or division number, may be displayed at the bottom of the placard (for example, 1 or 1.3). In addition, the four-digit product identification number, such as 1203, may be used. In some cases, both numbers are used on the same placard.

The colors of the placards also indicate the hazard class. For example, yellow is used for Class 5, Oxidizing Substances, and black-and-white is used to denote Class 8, Corrosives.

(b) NFPA 704 markings;

See commentary to 2-2.1.7.1.

(c) Military hazardous materials markings;

The military uses different types of markings for shipments on military facilities. The military marking system establishes four hazard classes:

- Class 1, Division 1—Materials that present a mass detonation hazard.
- Class 1, Division 2—Materials that present an explosive with fragmentation hazard.
- Class 1, Division 3—Materials with a mass fire hazard.
- Class 1, Division 4—Materials that present a moderate fire hazard.

In addition, there are three special warnings that are indicated separately. These include:

- Chemical Hazard
  Highly Toxic
  Harassing Agents
  White Phosphorus Munitions
- Apply No Water
- Wear Protective Breathing Apparatus

(d) Special hazard communication markings;

These may be found in some facilities and identify hazardous materials.

Figure 2.3(a)   Military class markings. (Illustration: Smizer Design.)

Figure 2.3(b)   Military hazard symbols. (Illustration: Smizer Design.)

(e) Pipeline marker; and

Pipeline markers are usually metal signs placed adjacent to a hazardous materials pipeline. They contain information about the location and ownership of the line.

(f) Container markings.

Often, markings on a container will provide some indication as to the type of product it holds. These markings include product names, such as "chlorine."

2-2.1.7.1 Given an NFPA 704 marking, identify the significance of the colors, numbers, and special symbols.

The system is based on the "704 diamond" (*see Figure 2.5*), which visually presents information on three categories of hazard: health, flammability, and self-reactivity, as well as the degree of severity of each hazard. It also indicates two special hazards: reactivity with water and oxidizing ability. The NFPA 704 diamond symbol is intended to provide immediacy at some sacrifice of adequacy, and there is a tendency to read more into it than it says.

Figure 2.4 *Example of a pipeline marker.*

The five degrees of hazard, in descending order, have these general meanings to fire fighters:

4—Fire is too dangerous to approach with standard fire fighting equipment and procedures. Withdraw and obtain expert advice on how to handle.
3—Fire can be fought using methods intended for extremely hazardous situations, such as remote-control monitors or personal protective equipment that prevents all bodily contact.

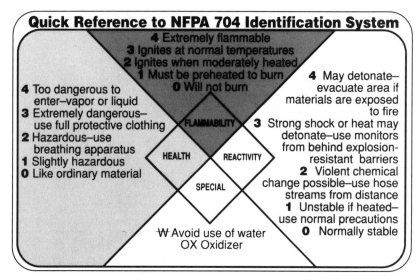

Figure 2.5 The NFPA 704 diamond. (Source: Fire Protection Handbook, 17th edition.)

2—Fire can be fought with standard procedures, but hazards are present that can be handled safely only with certain special equipment or procedures.

1—Nuisance hazards are present that require some care, but standard fire fighting procedures can be used.

0—No special hazards are present; therefore, no special measures are needed.

NFPA 704 describes the hazard categories and the security levels that the various numbers indicate for the three hazards. The following, adapted from Appendix B of the standard, summarizes the hazard information and recommends protective actions.[2]

The numbers from 0 through 4 are placed in the three upper squares of the diamond to show the degree of hazard present for each of the three hazard categories. The 0 indicates the lowest degree of hazard, and 4, the highest. The fourth square, at the bottom, is used for special information. Two symbols for this bottom space are recognized by NFPA 704 (*see Figure 2.6*). They are:

(a) A letter W with a bar through it (W̶) to indicate that a material may have a hazardous reaction with water. This does not mean "do not use water," since some forms of water—fog or fine spray—may be used in many cases. What it does say is: "Water may cause a hazard, so use it very cautiously until you have proper information."

(b) The letters OX to indicate an oxidizer.

Although not recognized by NFPA 704, some users will insert the letters *ALK* for alkaline materials and *ACID* for acidic materials.

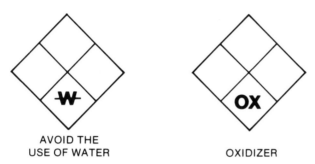

AVOID THE
USE OF WATER   OXIDIZER

Figure 2.6 Special information is presented in the bottom square of the NFPA 704 diamond. The square is color-coded white. (Source: Fire Protection Handbook, 17th edition.)

Health Hazards

In general, the health hazard in fire fighting is that of a single exposure, the duration of which may vary from a few seconds up to an hour. The physical exertion demanded in fire fighting or other emergencies may be expected to intensify the effects of any exposure. In assigning degrees of danger, local conditions must be considered. The following explanation is based on use of the protective equipment normally worn by fire fighters (*see Figure 2.7*).

4—Materials that are too dangerous to health for fire fighters to be exposed. A few whiffs of the vapor could cause death, or the vapor or liquid could be fatal on penetrating the fire fighter's normal protective clothing. The normal full protective clothing and breathing apparatus available to the average fire department will not provide adequate protection against inhalation or skin contact with these materials.

3—Materials that are extremely hazardous to health, but fire areas may be entered with extreme care. Full protective clothing, self-contained breathing apparatus (SCBA), rubber gloves, boots, and bands around legs, arms, and waist should be provided. No skin surface should be exposed.

2—Materials that are hazardous to health, but fire areas may be entered freely with self-contained breathing apparatus.

1—Materials that are only slightly hazardous to health.

0—Materials that on exposure under fire conditions would offer no health hazard beyond that of ordinary combustible material.

Flammability Hazards

Susceptibility to burning is the basis for assigning degrees within this category (*see Figure 2.8*). The method of attacking the fire is influenced by this susceptibility factor.

4—Very flammable gases or very volatile flammable liquids. If possible, shut on/off flow and keep cooling water streams on exposed tanks or containers. Withdrawal may be necessary.

3—Materials that can be ignited under almost all normal temperature conditions. Water may be ineffective because of the low flash point of the materials.

2—Materials that must be moderately heated before ignition will occur. Water spray may be used to extinguish the fire because the material can be cooled below its flash point.

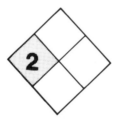

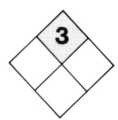

Figure 2.7   Figure 2.8

Health hazards appear in the left-hand square of the NFPA diamond, color-coded blue (Figure 2.7). Flammability hazards appear in the top square of the NFPA diamond, color-coded red (Figure 2.8).

1—Materials that must be preheated before ignition can occur. Water may cause frothing if it gets below the surface of the liquid and turns to steam. If this is the case, water fog gently applied to the surface will cause a frothing that will extinguish the fire.

0—Materials that will not burn.

Reactivity (Stability) Hazards

The assignment of relative degrees of hazard in the reactivity category is based on the susceptibility of materials to release energy either by themselves or in combination with other materials (*see Figure 2.9*). Fire exposure was one of the factors considered along with conditions of shock and pressure.

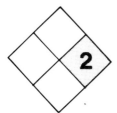

Figure 2.9 Reactivity (stability) hazards are presented in the right-hand square of the NFPA diamond. This square is color-coded yellow.

4—Materials that are readily capable of detonation or explosive decomposition at normal temperatures and pressures. If they are involved in a massive fire, vacate the area.

3—Materials that when heated and under confinement are capable of detonation or explosive decomposition and that may react violently with water. Fire fighting should be conducted from behind explosion-resistant locations.

2—Materials that will undergo a violent chemical change at elevated temperature and pressures but do not detonate. Use portable monitors, hose holders, or straight hose streams from a distance to cool the tanks and the material in them. Use caution.

1—Materials that are normally stable but may become unstable in combination with other materials or at elevated temperatures and pressures. Use normal precautions as in approaching any fire.

0—Materials that are normally stable and therefore do not present any reactivity hazard to fire fighters.

Special Information

When W appears at the bottom in the fourth space (*see Figure 2.6*):

4—W is not used with reactivity hazard 4.

3—In addition to hazards above, these materials can react explosively with water. Explosion protection is essential if water is to be used.

2—In addition to hazards above, these materials may react with water or form potentially explosive mixtures with water.

1—In addition to hazards above, these materials may react vigorously but not violently with water.

0—W is not used with reactivity hazard 0.

Methods of Presentation

Considerable leeway is allowed in the presentation of the numbers. The only basic requirement is that numbers be spaced as though they were in the diamond outline. Several methods that have been used are shown in Figure 2.10. Chapter 6 of NFPA 704 presents recommended layout and sizes for the symbol, a distance-legibility table, and several examples using the symbol.

Assigning Degrees of Hazard

Numbers (degrees of hazard) for use in the diamond are assigned on the basis of the worst hazard expected in the area, whether it be from hazards of the original material or of its combustion or breakdown products. The effects of local conditions must be considered. For instance, a drum of carbon tetrachloride sitting in a well-ventilated storage shed presents a different hazard than a drum sitting in an unventilated basement.

Advantages of the NFPA 704 System

The NFPA 704 system can warn against hazards under fire conditions of materials that other information systems class as nonhazardous. For example, edible tallow produces toxic and irritating combustion products. It would be given a "2" degree of health hazard, indicating the need for air-supplied respiratory equipment.

NFPA 704 also can warn against overall fire hazards in an area. On the door of a laboratory or storage room, it can warn of the worst hazards likely in a fire situation. Such information is useful both in preplanning and in actual fires.

NFPA 704 can be used without a supplementary manual. Because of its simplicity, the general meanings of the numbers are easily understood and the whole symbol is read and interpreted quickly on the spot, in poor light, and at a distance.

Disadvantages of the NFPA 704 System

The NFPA 704 system supplies only minimum information on the hazards themselves. Since the system informs on protective measures, the same

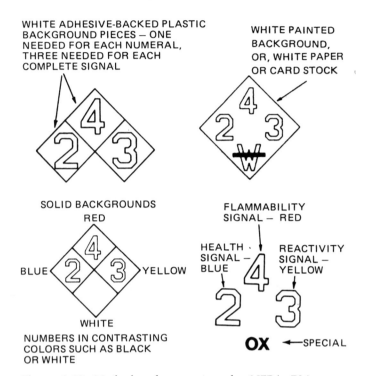

WHITE ADHESIVE-BACKED PLASTIC BACKGROUND PIECES – ONE NEEDED FOR EACH NUMERAL, THREE NEEDED FOR EACH COMPLETE SIGNAL

WHITE PAINTED BACKGROUND, OR, WHITE PAPER OR CARD STOCK

SOLID BACKGROUNDS
RED

BLUE   YELLOW

WHITE

NUMBERS IN CONTRASTING COLORS SUCH AS BLACK OR WHITE

FLAMMABILITY SIGNAL – RED

HEALTH SIGNAL – BLUE

REACTIVITY SIGNAL – YELLOW

OX ←—SPECIAL

Figure 2.10 Methods of presenting the NFPA 704 system hazard information. (Source: Fire Protection Handbook, 17th edition.)

number may be used for various types of hazards so that, for instance, a health hazard "3" means "no contact" without saying whether the hazard is corrosiveness to the skin or toxicity by absorption through the skin. Thus, the symbol is most useful only to trained or informed persons.

2-2.1.8 Identify U.S. and Canadian placards and labels that indicate hazardous materials.

See Figure 2.11 and color insert.

2-2.1.9 Identify the basic information on material safety data sheets (MSDS) and shipping papers that indicates hazardous materials.

An MSDS contains information on the following major categories: the manufacturer's name and location; the name and family of the chemical; the

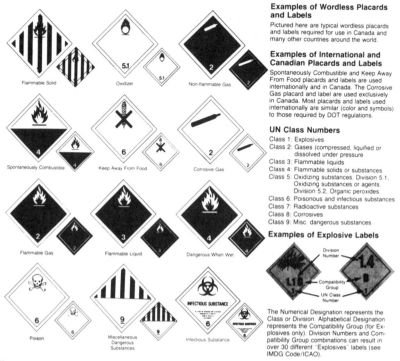

Examples of Wordless Placards and Labels
Pictured here are typical wordless placards and labels required for use in Canada and many other countries around the world.

Examples of International and Canadian Placards and Labels
Spontaneously Combustible and Keep Away From Food placards and labels are used internationally and in Canada. The Corrosive Gas placard and label are used exclusively in Canada. Most placards and labels used internationally are similar (color and symbols) to those required by DOT regulations.

UN Class Numbers
Class 1: Explosives
Class 2: Gases (compressed, liquified or dissolved under pressure
Class 3: Flammable liquids
Class 4: Flammable solids or substances
Class 5: Oxidizing substances. Division 5.1, Oxidizing substances or agents. Division 5.2, Organic peroxides.
Class 6: Poisonous and infectious substances
Class 7: Radioactive substances
Class 8: Corrosives
Class 9: Misc. dangerous substances

Examples of Explosive Labels

The Numerical Designation represents the Class or Division. Alphabetical Designation represents the Compatibility Group (for Explosives only). Division Numbers and Compatibility Group combinations can result in over 30 different "Explosives" labels (see IMDG Code/ICAO).

Figure 2.11 Examples of Canadian and international placards and labels. (Source: Fire Protection Handbook, 17th edition.)

hazardous ingredients; physical data; fire and explosion hazard data; health hazard data; spill or leak procedures; special protection information; and special precautions that must be taken when dealing with the material.

2-2.1.9.1 Identify where to find material safety data sheets (MSDS).

Employers are required to maintain material safety data sheets for all hazardous chemicals used at their facilities. This is part of the Hazard Communication Standard 29 CFR 1910.1200.[3]

2-2.1.9.2 Identify entries on a material safety data sheet that indicate the presence of hazardous materials.

The major categories on an MSDS that indicate the presence of hazardous materials include hazardous ingredients, fire and explosion hazard data, health hazard data, reactivity data, spill or leak procedures, special protection information, and special precautions.

2-2.1.9.3 Identify the entries on shipping papers that indicate the presence of hazardous materials.

Shipping papers may contain several pieces of information that indicate the presence of a hazardous material. These include the material's proper shipping name, its hazard class or division, and its product identification number.

Figure 2.12 Information identifying the materials should also be available from the shipping papers with the vehicle.

2-2.1.9.4 Match the name of the shipping papers found in transportation (air, highway, rail, and water) with the mode of transportation.

2-2.1.9.5 Identify the person responsible for having the shipping papers in each mode of transportation.

See Table 2.2.

Table 2.2   A List of the Shipping Papers by Mode of Transportation,
Title of Shipping Paper, Location, and Responsible Person

Mode of Transportation	Title of Shipping Paper	Location of Shipping Papers	Responsible Person
Highway	Bill of Lading or Freight Bill	Cab or vehicle	Driver
Rail	Waybill and/or Consist	With member of train crew (conductor or engineer)	Conductor
Water	Dangerous Cargo Manifest	Wheelhouse or pipe-like container or barge	Captain or Master
Air	Air Bill with Shippers Certification for Restricted Articles	Cockpit (may also be found attached to the outside of packages)	Pilot

(Source: *Fire Protection Handbook*, 17th edition.)

2-2.1.9.6 Identify where the shipping papers are found in each mode of transportation.

See Table 2.2.

2-2.1.9.7 Identify where the papers may be found in an emergency in each mode of transportation.

See Table 2.2.

2-2.1.10* Identify examples of clues (other than occupancy/location, container shape, markings/color, placards/labels, and shipping papers) that use the senses of sight, sound, and odor to indicate hazardous materials.

A-2-2.1.10 These clues would include odors, gas leaks, fire or vapor cloud, visible corrosive actions or chemical reactions, pooled liquids, hissing of pressure releases, condensation lines on pressure tanks, injured victims, or casualties.

2-2.1.11 Describe the limitations of using the senses in determining the presence or absence of hazardous materials.

One problem in using your senses to evaluate the presence of hazardous materials is the fact that, if you are close enough to use them, you may already have endangered yourself.

2-2.2 **Surveying the Hazardous Materials Incident from a Safe Location.** The first responder at the awareness level shall, given examples of facility and transportation situations involving hazardous materials, identify the hazardous material(s) in each situation by name, UN/NA identification number, and/or type placard applied. The first responder at the awareness level shall be able to:

2-2.2.1 Identify difficulties encountered in determining the specific names of hazardous materials in both facilities and transportation.

Even if the awareness level responder knows placarding and labeling systems and is familiar with other methods of identifying the presence of hazardous materials, he or she may still have difficulty determining which materials are involved in a specific incident. In some cases, the labels or placards will be missing; the placards or labels will list only the class or division, not the specific product identifier; shipments may contain mixed loads of hazardous materials and require only the "dangerous" placard; the shipper may err in the labeling and placarding; or the shipping papers may be inaccessible.

2-2.2.2 Identify sources for obtaining the names of, UN/NA identification numbers for, or types of placard associated with hazardous materials in transportation.

One of the best ways to identify products or types of placards is to use the U.S. DOT *Emergency Response Guidebook* or the Canadian *Dangerous Goods Guide to Initial Emergency Response*. The shipping papers will also contain both the four-digit identification number and the proper shipping name of the material.

2-2.2.3 Identify sources for obtaining the names of hazardous materials in a facility.

At fixed facilities, the names of hazardous materials are found on the MSDS and in the emergency planning documents, and there may be signs or other markings on storage containers.

2-2.3* Collecting Hazard Information. The first responder at the awareness level shall, given the identity of various hazardous materials (name, UN/NA identification number, or type placard), identify the fire, explosion, and health hazard information for each material using the current edition of the *Emergency Response Guidebook*. The first responder at the awareness level shall be able to:

A-2-2.3 It is the intent of this standard that the first responder at the awareness level be taught the noted competency to a specific task level. This task level would be to have knowledge of the contents of the current edition of the *Emergency Response Guidebook* or other reference material provided. Awareness level responders should be familiar with the information provided in those documents so that they can use it to assist with accurate notification of an incident and take protective actions.

If other sources of response information [including the material safety data sheet (MSDS)] are provided to the hazardous materials responder at the awareness level in lieu of the current edition of the *Emergency Response Guidebook*, the responder shall identify hazard information similar to that found in the current edition of the *Emergency Response Guidebook*.

The information responders at the awareness level can use to identify products is generally limited. The most likely sources of this information are the U.S. DOT *Emergency Response Guidebook* and the Canadian *Dangerous Goods Guide*. It is therefore important that the awareness level responder be familiar with these books and be able to look up information in them quickly.

If the responders use other types of documents, such as the MSDS, to identify products and locate information about handling a spill, leak, fire, or explosion, they must be able to find the information quickly. Speed is important when dealing with an emergency.

2-2.3.1* Identify the ways hazardous materials are harmful to people, the environment, and property at hazardous materials incidents.

A-2-2.3.1 These would include thermal, mechanical, poisonous, corrosive, asphyxiation, radiation, and etiologic. There may also be psychological harm.

Although psychological harm might be considered an ever-present phenomenon, hazardous materials do not always present thermal, mechanical, poisonous, corrosive, asphyxiation, radiation, and etiologic threats at every hazardous materials incident. The responder may be exposed to only one of these threats at any given emergency or to a combination of several. Because these various life-threatening hazards may be present at any hazardous materials incident, however, responders must be adequately trained to minimize the potential danger these hazards present to themselves, to the public, and to the environment.

2-2.3.2* Identify the general routes of entry for human exposure to hazardous materials.

A-2-2.3.2 These are: contact, absorption, inhalation, and ingestion. Absorption includes entry through the eyes and through punctures.

Contact: A corrosive material will damage skin or body tissue on contact. Acids and alkalis can cause severe burns. If the skin is broken or the responder has an open wound, another entry route exists.

Absorption: In this process, one substance penetrates the inner structure of another. Hydrogen cyanide, for example, can be absorbed through the skin with fatal results.

Inhalation: Many substances cause severe damage if inhaled. Chlorine is an example of such a substance.

Ingestion: Toxic substances can be present in drinking water and in food.

Obviously, proper personal protective equipment is important, and the responder must fully understand both the potential hazards and the appropriate safeguards.

2-2.3.3 Given the current edition of the *Emergency Response Guidebook*, identify the three methods for determining the appropriate guide page for a specific hazardous material.

To find this information in the U.S. DOT *Emergency Response Guidebook*, the responder must:

(a) Identify the material by finding the four-digit identification number on a placard or orange panel, in a shipping paper, or on the package, then locate the number in the yellow-bordered pages and determine the appropriate guide page, or
(b) Locate the name of the material in the shipping papers or on the placard or package, then locate the material in the alphabetical listing of products on the blue-bordered pages and determine the appropriate guide page, or

(c) If a name or identification number is not available but the responder can see the placard, he or she should locate a matching placard in the table of placards and consult the two-digit guide number found next to the look-alike placard. If the material is an explosive, the responder should consult one of the four guides listed inside the front cover of the U.S. DOT *Emergency Response Guidebook*. [*See Figures 2.13(a), (b), and (c)*].

2-2.3.4 Given the current edition of the *Emergency Response Guidebook*, identify the two general types of hazards found on each guide page.

Each guide page of the U.S. DOT *Emergency Response Guidebook* contains information on the fire and explosion hazard and on the health hazard.

2-3 Competencies — Planning the Response.

(No competencies currently required at this level.)

The committee felt that the responder at this level would not be involved in planning for an emergency response but would apply standard operating procedures established by the organization or found in the local emergency response plan. At this level, the responder's responsibilities are to identify a hazardous material, notify the authorities, and isolate the material.

2-4 Competencies — Implementing the Planned Response.

2-4.1* **Initiating Protective Actions.** First responders at the awareness level shall, given examples of facility and transportation hazardous materials incidents, identify the actions to be taken to protect themselves and others and to control access to the scene using the local emergency response plan, the organization's standard operating procedures, or the current edition of the *Emergency Response Guidebook.* The first responder at the awareness level shall be able to:

The competencies in this section are designed to ensure that the awareness level responder can implement the appropriate protective actions based on the knowledge he or she has acquired while analyzing the incident.

ID No.	Guide No.	Name of Material	ID No.	Guide No.	Name of Material
1246	27	METHYLPROPENYL KETONE, inhibited	1266	26	PERFUMERY PRODUCTS, with flammable solvent
1247	27	METHYL METHACRYLATE, monomer, inhibited	1267	27	PETROLEUM CRUDE OIL
			1268	27	NAPHTHA DISTILLATE
1248	26	METHYL PROPIONATE	1268	27	PETROLEUM DISTILLATE, n.o.s.
1249	26	METHYL PROPYL KETONE			
1250	29	METHYL TRICHLOROSILANE	1268	27	ROAD OIL
1251	28	METHYL VINYL KETONE	1270	27	OIL, petroleum, n.o.s.
1255	27	NAPHTHA, PETROLEUM	1270	27	PETROLEUM OIL
1256	27	NAPHTHA, SOLVENT	1271	26	PETROLEUM ETHER
1257	27	CASINGHEAD GASOLINE	1271	26	PETROLEUM SPIRIT
1257	27	NATURAL GASOLINE	1272	26	PINE OIL
1259	28	NICKEL CARBONYL	1274	26	PROPANOL
1261	26	NITROMETHANE	1274	26	PROPYL ALCOHOL
1262	27	ISOOCTANE	1275	26	PROPIONALDEHYDE
1262	27	OCTANE	1276	26	PROPYL ACETATE
1263	26	COMPOUND, PAINT, etc., removing, reducing, or thinning liquid	1277	68	MONOPROPYLAMINE
			1277	68	PROPYLAMINE
			1278	26	PROPYL CHLORIDE
1263	26	ENAMEL	1279	27	DICHLOROPROPANE
1263	26	LACQUER	1279	27	PROPYLENE DICHLORIDE
1263	26	LACQUER BASE, liquid	1280	26	PROPYLENE OXIDE, inhibited
1263	26	PAINT, etc., flammable liquid	1281	26	PROPYL FORMATE
1263	26	PAINT RELATED MATERIAL, flammable liquid	1282	26	PYRIDINE
			1286	26	RESIN OIL
1263	26	POLISH, liquid	1286	26	ROSIN OIL
1263	26	SHELLAC	1287	26	RUBBER SOLUTION
1263	26	STAIN	1288	27	SHALE OIL
1263	26	THINNER	1289	26	SODIUM METHYLATE, solutions in alcohol
1263	26	VARNISH			
1263	26	WOOD FILLER, liquid	1292	29	ETHYL SILICATE
1264	26	PARALDEHYDE	1292	29	TETRAETHYL SILICATE
1265	27	AMYL HYDRIDE	1293	26	TINCTURE, medicinal
1265	27	ISOPENTANE	1294	27	TOLUENE
1265	27	PENTANE	1295	30	TRICHLOROSILANE
			1296	68	TRIETHYLAMINE

SEE "HOW TO USE THIS GUIDEBOOK" ON THE FIRST PAGE, IF YOU HAVE NOT YET BECOME FAMILIAR WITH THE DETAILS OF USING THESE INDEXES TO THE GUIDES.

Figure 2.13(a)   Sample page from the ID Number Index. The four-digit ID number 1294 indicates that the cargo is toluene, and that the correct guide number is 27. (Source: Fire Protection Handbook, 17th edition.)

HAZARDOUS MATERIALS RESPONSE HANDBOOK

Name of Material	Guide No.	ID No.	Name of Material	Guide No.	ID No.
TETRAHYDROPHTHALIC ANHYDRIDE	60	2698	THIOUREA	53	2877
TETRAHYDROPYRIDINE	26	2410	THIRAM	55	2771
TETRAHYDROTHIOPHENE	26	2412	THORIUM METAL, pyrophoric	65	2975
TETRALIN HYDROPEROXIDE, technical pure	48	2136	THORIUM METAL, pyrophoric	65	9170
			THORIUM NITRATE, solid	64	2976
TETRAMETHYL AMMONIUM HYDROXIDE	60	1835	THORIUM NITRATE, solid	64	9171
			TIN CHLORIDE, fuming	39	1827
1,1,3,3-TETRAMETHYLBUTYL HYDROPEROXIDE, technical pure	48	2160	TINCTURE, medicinal	26	1293
			TIN TETRACHLORIDE	39	1827
1,1,3,3-TETRAMETHYLBUTYL-PEROXY-2-ETHYL HEXA-NOATE, technical pure	52	2161	TITANIUM, metal, powder, dry	37	2546
TETRAMETHYL LEAD	56	1649	TITANIUM, metal, powder, wet with not less than 20% water	32	1352
TETRAMETHYLMETHYLENE-DIAMINE	58	9069	TITANIUM HYDRIDE	32	1871
TETRAMETHYL SILANE	29	2749	TITANIUM SPONGE, granules or powder	32	2878
TETRAPROPYL-ortho-TITANATE	27	2413	TITANIUM SULFATE SOLUTION	60	1760
TETRANITROMETHANE	47	1510	**TITANIUM TETRACHLORIDE** *	39	1838
TEXTILE TREATING COMPOUND	60	1760	TITANIUM TRICHLORIDE, pyrophoric	37	2441
TEXTILE WASTE, wet, n.o.s.	32	1857	TITANIUM TRICHLORIDE MIXTURE	60	2869
THALLIUM CHLORATE	42	2573	TITANIUM TRICHLORIDE MIXTURE, pyrophoric	37	2441
THALLIUM COMPOUND, n.o.s.	53	1707	TOE PUFFS, nitrocellulose base	32	1353
THALLIUM NITRATE	42	2727	TOLUENE	27	1294
THALLIUM SALT, n.o.s.	53	1707	TOLUENE DI-ISOCYANATE (T.D.I.)	57	2078
THALLIUM SULFATE, solid	53	1707			
THIAPENTANAL	55	2785	TOLUENE SULFONIC ACID, liquid	60	2584
THINNER	26	1263	TOLUENE SULFONIC ACID, liquid	60	2586
THIOACETIC ACID	26	2436	TOLUENE SULFONIC ACID, solid	60	2583
THIOGLYCOL	53	2966			
THIOGLYCOLIC ACID	60	1940	TOLUENE SULFONIC ACID, solid	60	2585
THIOLACTIC ACID	59	2936	TOLUIDINES (o-, m-, and p-)	55	1708
THIONYL CHLORIDE	39	1836			
THIOPHENE	27	2414			
THIOPHOSGENE	55	2474			
THIOPHOSPHORYL CHLORIDE	60	1837			

* Look for information next to this **NAME** in the TABLE OF EVACUATION DISTANCES in the back of this book. Use this in addition to the Guide Page if there is NO FIRE.

Figure 2.13(b)   Sample page from the Materials Name Index. Once again the Guide number for toluene is 27. (Source: Fire Protection Handbook, 17th edition.)

Guide 27

POTENTIAL HAZARDS

FIRE OR EXPLOSION

Flammable/combustible material; may be ignited by heat, sparks or flames.
Vapors may travel to a source of ignition and flash back.
Container may explode in heat of fire.
Vapor explosion hazard indoors, outdoors or in sewers.
Runoff to sewer may create fire or explosion hazard.

HEALTH HAZARDS

May be poisonous if inhaled or absorbed through skin.
Vapors may cause dizziness or suffocation.
Contact may irritate or burn skin and eyes.
Fire may produce irritating or poisonous gases.
Runoff from fire control or dilution water may cause pollution.

EMERGENCY ACTION

Keep unnecessary people away; isolate hazard area and deny entry.
Stay upwind; keep out of low areas.
Wear self-contained (positive pressure if available) breathing apparatus and full protective clothing.
Isolate for 1/2 mile in all directions if tank car or truck is involved in fire.
FOR EMERGENCY ASSISTANCE CALL CHEMTREC **(800) 424-9300.**
If water pollution occurs, notify appropriate authorities.

FIRE

Small Fires: Dry chemical, CO_2, water spray or foam.
Large Fires: Water spray, fog or foam.
Move container from fire area if you can do it without risk.
Cool containers that are exposed to flames with water from the side until well after fire is out.
For massive fire in cargo area, use unmanned hose holder or monitor nozzles; if this is impossible, withdraw from area and let fire burn.
Withdraw immediately in case of rising sound from venting safety device or any discoloration of tank due to fire.

SPILL OR LEAK

Shut off ignition sources; no flares, smoking or flames in hazard area.
Stop leak if you can do it without risk.
Use water spray to reduce vapors.
Small Spills: Take up with sand or other noncombustible absorbent material and place into containers for later disposal.
Large Spills: Dike far ahead of spill for later disposal.

FIRST AID

Move victim to fresh air; call emergency medical care.
If not breathing, give artificial respiration.
If breathing is difficult, give oxygen.
In case of contact with material, immediately flush eyes with running water for at least 15 minutes. Wash skin with soap and water.
Remove and isolate contaminated clothing and shoes at the site.

Figure 2.13(c)   Guide number 27 refers to a set of detailed instructions. (Source: Fire Protection Handbook, 17th edition.)

A-2-4.1 Those jurisdictions that have not developed an emergency response plan may refer to the document NRT-1, *Hazardous Materials Emergency Planning Guide*, developed by the National Response Team.

The National Response Team, composed of fourteen federal agencies having major responsibilities in environmental, transportation, emergency management, worker safety, and public health areas, is the national body responsible for coordinating federal planning, preparedness, and response actions related to oil discharges and hazardous substance releases. Under the Superfund Amendments and Reauthorization Act of 1986, the NRT is responsible for publishing guidance documents for the preparation and implementation of hazardous substance emergency plans.

NRT-1 was first published and distributed in March 1987. It discusses the elements required to develop an effective hazardous materials emergency response plan and addresses the planning process.

2-4.1.1 Identify the location of both the local emergency response plan and the organization's standard operating procedures.

2-4.1.2 Given a copy of the current edition of the DOT *Emergency Response Guidebook*, describe the difference between the protective action distances in the orange-bordered guide pages and the green-bordered pages in the document.

The orange-bordered, or emergency action guide, pages in the "Emergency Action Section" of the ERG provide isolation distances for selected materials that are involved in a fire. These distances may extend up to 1 mile in all directions. The protective action distances listed in the green-bordered pages (that is, in the Table of Initial Isolation and Protective Action Distances) are to be used only if the material is *not involved* in a fire. The distances found in this table are the recommended downwind protective action distances for materials whose poisonous vapors present inhalation hazards.

For example, ethylene oxide has an isolation distance, as found in the orange-bordered pages, of 1 mile when involved in a fire. When it is not involved in a fire, this same material has a recommended isolation distance of 0.8 miles for a small spill from a container not larger than a 55-gallon drum and 2 miles for a large spill.

Another point to remember is that the distances recommended in the Table of Initial Isolation and Protective Action Distances are to be used only during the first 30 minutes of an incident.

This competency points out how important it is that responders understand how to use the U.S. DOT *Emergency Response Guidebook* or similar guidebooks.

Table 2.3   Sample of the Table of Initial Isolation and Protective Distances from the Department of Transportation's 1990 Emergency Response Guidebook

ID No.	Name of Material	Use This Table When the Material Is Not on Fire	Small Spills (Leak or spill from a small package or small leak from a large package)		Large Spills (Leak or spill from a large package or spill from many small packages)	
			First ISOLATE in all directions (Feet)	Then, PROTECT those persons in the DOWNWIND direction (Miles)	First, ISOLATE in all directions (Feet)	Then, PROTECT those persons in the DOWNWIND direction (Miles)
1005	Ammonia		150	0.2	300	1
1005	Ammonia, Anhydrous, liquefied		150	0.2	300	1
1005	Ammonia Solution with more than 50% ammonia		150	0.2	300	1
1005	Anhydrous Ammonia		150	0.2	300	1
1008	Boron Trifluoride		1500	5	1500	5
1016	Carbon Monoxide		150	0.4	150	0.8
1017	Chlorine		900	3	1500	5
1023	Coal Gas		300	1	1500	5
1026	Cyanogen		300	1	300	1
1026	Cyanogen, liquefied		300	1	300	1
1026	Cyanogen Gas		300	1	300	1
1040	Ethylene Oxide		150	0.8	600	2
1041	Carbon Dioxide–Ethylene Oxide Mixture, with more than 6% Ethylene Oxide		150	0.8	1200	4

(Source: *Fire Protection Handbook*, 17th edition.)

2-4.1.3 Given the local emergency response plan or the organization's standard operating procedures, identify the role of the first responder at the awareness level during a hazardous materials incident.

Emergency response plans should establish the methods and procedures that facility owners and operators, as well as local emergency and medical response personnel, are to follow. Responders at all levels must understand their role and its importance. If a response is to be handled effectively, the first responder on the scene must accurately assess the situation and initiate the appropriate response measures.

2-4.1.4 Given the local emergency response plan or the organization's standard operating procedures, identify the basic precautions to be taken to protect himself/herself and others in a hazardous materials incident.

At the awareness level, responders are generally expected to take protective actions to isolate the hazard and to evacuate threatened persons from the immediate area. If evacuation is not possible, the responders are to provide in-place protection until additional resources become available. In some cases, in-place protection may be all that is required. The U.S. DOT ERG recommends that persons protected in place be warned to stay far away from windows with a direct line of sight of the scene because such windows may explode during a fire or explosion and shower them with glass or metal fragments.

2-4.1.4.1 Identify the precautions necessary when providing emergency medical care to victims of hazardous materials incidents.

The awareness level responder need not worry unduly about this matter. Suffice it to say that hazards exist for both the victim and the responder. The victim may well be contaminated, and decontamination measures must be considered. In addition, many awareness level responders will not be wearing respiratory protection or any other personal protective clothing that would protect them from the more severe hazards. The number of times both victim and responder may be exposed is limited only by the specific circumstances of the incident. Responders must understand these problems so that they do not become victims themselves while attempting to rescue someone else.

2-4.1.4.2 Identify typical ignition sources found at the scenes of hazardous materials incidents.

Sources of ignition may include open flames; smoking materials; cutting and welding operations; heated surfaces; frictional heat; radiant heat; static, electrical, and mechanical sparks; and spontaneous ignition, such as occurs during heat-producing chemical reactions or is produced by pyrophoric materials. Lightning should not be overlooked as a source of ignition, either.

The responder can control or eliminate some, but not all, ignition sources. If flammable vapors are present, he or she should take measures to reduce or eliminate them or to disperse them into the atmosphere.

2-4.1.5* Given the identity of various hazardous materials (name, UN/NA identification number, or type placard), identify the following response information using the current edition of the *Emergency Response Guidebook*:

(a) Emergency action (fire, spill, or leak and first aid);
(b) Personal protective equipment necessary; and
(c) Initial isolation and protective action distances.

A-2-4.1.5 If other sources of response information [including the material safety data sheet (MSDS)] are provided to the hazardous materials responder at the awareness level in lieu of the current edition of the *Emergency Response Guidebook*, the responder shall identify response information similar to that found in the current edition of the DOT *Emergency Response Guidebook*.

Again, it is important that responders demonstrate their proficiency in using the various emergency response guides they may be required to consult. The U.S. DOT ERG and the Canadian *Dangerous Goods Guide* provide general information about the potential hazards of a number of products and outline the emergency procedures to be used in handling incidents involving these products. Responders should be given several exercises that will allow them to demonstrate their skill at locating and interpreting the appropriate information.

2-4.1.5.1 Given the current edition of the *Emergency Response Guidebook* and the name of a hazardous material, identify the recommended personal protective equipment for the particular incident from the following list of protective equipment:

(a) Street clothing and work uniforms;
(b) Structural fire fighters' protective clothing;

(c) Positive pressure self-contained breathing apparatus; and

(d) Chemical-protective clothing and equipment.

This competency allows awareness level responders to demonstrate their ability to locate information in emergency response guides about the type of protective clothing they should use when dealing with various products. The U.S. DOT ERG, for example, tells responders what type of protective clothing to wear for various incidents and advises them whether they need respiratory protection.

2-4.1.5.2 Given the current edition of the *Emergency Response Guidebook*, identify the definitions for each of the following protective actions:

Protective actions are those steps taken to preserve the health and safety of emergency responders and the public during an incident involving the release of hazardous materials.

(a) Isolate hazard area and deny entry;

Everybody not directly involved in the emergency response operations should be kept away from the affected area, and unprotected emergency responders should not be allowed within the isolation area.

(b) Evacuate; and

Evacuation is the movement of everyone from a threatened area to a safer place. To perform an evacuation, there must be enough time to warn people, to get them ready to go, and to leave the area. If there is enough time for evacuation, it is likely to be the best protective action.

Begin evacuating people who are nearby and those outdoors in direct view of the scene. Send evacuees upwind by a specific route to a definite place far enough away from the contaminated area that they will not have to be moved again if the wind shifts. As you acquire additional help, expand the area to be evacuated downwind and crosswind at least to the extent recommended in the Table of Isolation and Protective Action Distances. Even after people move to the distances recommended, they are not completely safe from harm, and they should not be permitted to congregate at such distances.

(c) In-place protection.

In-place protection is used when an evacuation cannot be performed or when evacuating the public would put them at greater risk than directing them to stay where they are. When using in-place protection, the responder directs people to go quickly inside a building and remain there until the danger has passed. The people inside the building should be told to close all doors and windows and to shut off all ventilating, heating, and cooling systems.

In-place protection may not be the best option if there are explosive vapors present, if it will take a long time to clear the area of gas, or if the building cannot be tightly closed. Vehicles are not as effective as buildings for in-place protection, but they can offer some protection for a short period if the windows are closed and the ventilating system is shut off.

2-4.1.5.3 Given the current edition of the DOT *Emergency Response Guidebook,* identify the shapes of recommended initial isolation and protective action zones.

The U.S. DOT ERG provides initial isolation zones and protective action distances for hazardous materials whose vapors may produce poisonous effects. The shapes of those areas are shown in the figures below.

These distances are provided in the Table of Initial Isolation and Protective Actions on the green-bordered pages (*see Table 2.3*).

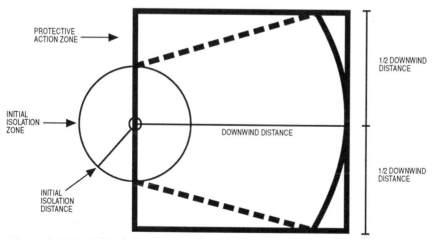

Figure 2.14 Initial isolation zone and protective action zone. (Illustration: Smizer Design.)

2-4.1.5.4 Given the current edition of the DOT *Emergency Response Guidebook*, describe the difference between small and large spills as found in the table of isolation distances.

The U.S. DOT ERG defines a small spill as one that involves a single, small package, up to the size of a 55-gallon drum, or a small cylinder. A small spill may also be a small leak from a large package. A large spill involves a big spill from an opening in a large package or spills from many small packages. A large spill would come from a 1-ton cylinder, a tank truck, or a railcar.

In determining the isolation and protective action distances, the DOT assumed that the maximum pool size for a small spill that formed a liquid pool was 48 feet in diameter. A large spill pool was assumed to be a maximum of 60 feet in diameter. The distances were calculated following a 30-minute period from the start of the release. This is the reason the DOT cautions that the distances are valid only for the 30-minute period following a spill.

2-4.1.5.5 Given the current edition of the DOT *Emergency Response Guidebook*, identify the circumstances under which the following distances are used at a hazardous materials incident:

(a) Table of initial isolation and protective action distances; and

These distances are used only for products whose vapors present an inhalation hazard, where the release does not involve a fire, and *where no more than 30 minutes have elapsed* between the spill and the response.

(b) Isolation distances in the numbered guides.

The isolation distances in the guide pages are to be used when a hazardous material or its container is exposed to fire or when the product's vapors have potentially poisonous effects.

2-4.1.6 Identify the techniques used to isolate the hazard area and deny entry to unauthorized persons at hazardous materials incidents.

At this level, responders do not have many resources available to them. However, there are several steps they may be able to take. They could use a vehicle to block a road or driveway, or they might place a rope or some other type of barricade across the entrance to the area to block access. They may also notify law enforcement officials to begin diverting traffic from the scene. In a fixed facility, responders may close a door or gate, use the public address system to announce the problem to the facility's occupants, or notify security.

2-4.2 Initiating the Notification Process. The first responder at the awareness level shall, given either a facility or transportation scenario of hazardous materials incidents, identify the appropriate notifications to be made and how to make them, consistent with the local emergency response plan or the organization's standard operating procedures. The first responder at the awareness level shall be able to:

2-4.2.1 Identify the initial notification procedures for hazardous materials incidents in the local emergency response plan or the organization's standard operating procedures.

Awareness level responders must be familiar with the notification process they must follow to begin an effective response to a hazardous materials incident. This may only involve notifying the local fire or police department. In some fixed facilities, internal notification procedures may be used to initiate the response of on-site specialists, the plant fire brigade, or security personnel. Whatever the procedures, the responder must rapidly set the proper notification process in motion.

References Cited in Commentary

1. NFPA 704, *Standard System for the Identification of the Fire Hazards of Materials*, 1990 edition.
2. Ibid.
3. Title 29 CFR Part 1910.1200 (Washington, DC: U.S. Government Printing Office).

3

Competencies for the First Responder at the Operational Level

3-1 General.

3-1.1 **Introduction.** First responders at the operational level shall be trained to meet all requirements at the awareness and operational levels. In addition, first responders at the operational level shall receive training to meet federal Occupational Safety and Health Administration (OSHA), local occupational health and safety regulatory, or U.S. Environmental Protection Agency (EPA) requirements, whichever are appropriate for their jurisdiction.

It is important to note here that the competencies for the operational level responder build on those for the awareness level responder. Therefore, the responder at this level is assumed to have successfully demonstrated competency at the awareness level.

3-1.2 **Definition.** First responders at the operational level are those persons who respond to releases or potential releases of hazardous materials as part of the initial response to the incident for the purpose of protecting nearby persons, the environment, or property from the effects of the release. They shall be trained to respond in a defensive fashion to control the release from a safe distance and keep it from spreading.

The wording clearly indicates that the responder at this level belongs to some emergency response service or organization, be it public or private. These organizations include such groups as the fire service, emergency medical services, law enforcement, special chemical response teams, industrial fire brigades, and regional response teams. First responders at the operational level are assumed to have a certain level of experience and skill in dealing with emergencies. However, they are not in the category of hazardous materials response teams, since they are not expected to use specialized chemical-protective clothing or specialized control equipment.

Operational level responders are expected to take defensive control measures to protect people, the environment, or property from the effects of an unplanned hazardous materials release. Their function is to contain the release from a safe distance, while protecting exposures. They will not ordinarily come in direct contact with the hazardous material in the normal course of their work.

3-1.3 Goal. The goal of the competencies at the operational level shall be to provide first responders with the knowledge and skills to perform the following tasks safely. Therefore, in addition to being competent at the awareness level, the first responder at the operational level shall be able to:

Safety must be considered in every action taken at an incident. By demonstrating the knowledge and skills that follow, a responder should be able to reduce the risk inherent in managing a hazardous materials incident. Risks will always be present, however, and it is important that the responder maintain a positive attitude toward safety and follow the basic procedures that he or she has been taught.

(a) Analyze a hazardous materials incident to determine the magnitude of the problem in terms of outcomes by completing the following tasks:

As Charles Wright notes in Section 9, Chapter 14 of the 17th edition of the *Fire Protection Handbook*:

"The analysis process begins when a responder receives notification of a problem and continues throughout the incident, typically at the scene threatened by any hazardous materials involved."[1]

Outcomes are the direct and indirect results or consequences associated with an emergency. Direct outcomes are considered in terms of people, property, and/or the environment.

Figure 3.1   Hazardous materials incidents can vary widely.

1. Survey the hazardous materials incident to determine the containers and materials involved, whether hazardous materials have been released, and the surrounding conditions;

According to Charles Wright, the incident survey should be done from a safe distance so that the responder is not exposed to any released materials.

The responder should complete the following steps:

1. Identify the containers involved. Containers will fit into one of three categories: nonbulk, bulk, or facility containment systems (*see Table 3.1*). Identifying a container will allow the responder to track it throughout the incident and will give him or her information about the quantities of material involved, thus indicating the possible magnitude of the problem.
2. Identify the name, four-digit product identification number, or placard on each container to help determine the particular hazardous materials involved.
3. Note any leaking containers, smells, vapor clouds, fires, or explosions, and determine whether the safety relief valves have operated. These factors will help a responder establish whether there has been a hazardous materials release at the site.

Figure 3.2   Risk assessment should be started before entering the scene.

Table 3.1   A Listing of the Various Types of Nonbulk, Bulk, and Facility Containment Systems

Transportation		Facility Containment Systems
Nonbulk	Bulk	
Bags	Bulk Bags	Buildings
Bottles	Bulk Boxes	Machinery
Boxes	Cargo Tanks	Open Piles (outdoors and indoors)
Carboys	Covered Hopper Cars	Piping
Cylinders	Freight Containers	Reactors (chemical and nuclear)
Drums	Gondolas	Storage Bins, Cabinets, or Shelves
Jerricans	Pneumatic Hopper Trailers	
Multicell	Portable Tanks and Bins	
Tanks and Storage Vessels Packages	Protective Overpacks or Radioactive Materials	
Wooden Barrels	Tank Cars	
	Ton Containers	
	Van Trailers	

(Source: *Fire Protection Handbook*, 17th edition.)

4. Assess the conditions at the site of the incident. This includes the topography, land use, accessibility, weather conditions, bodies of water, public exposure potential, and nature and extent of injuries. The responder may use a hazardous materials survey to help collect this information (*see Figure 3.3*).[2]

2. Collect hazard and response information from material safety data sheets (MSDS), CHEMTREC/CANUTEC, and shipper/manufacturer contacts;

It is very important for the responder to know what resources are available for providing technical assistance during hazardous materials emergencies and to know how to use those resources.

3. Predict the likely behavior of a material and its container; and

The responder must be able to assess the potential behavior of a hazardous material and its container. Is it likely to explode? Is it nonflammable? Is it corrosive? Will the container rupture violently?

4. Estimate the potential harm at a hazardous materials incident.

As the responder collects the information necessary to predict the behavior of the material and its container, he or she can begin to estimate the potential harm it presents. Generally, initial assessments should be very conservative and should look at the worst possible event that could occur. As more information becomes available, these predictions can be modified.

(b) Plan an initial response within the capabilities and competencies of available personnel, personal protective equipment, and control equipment by completing the following tasks:

A plan of action must be developed using the information gathered and the estimates of potential harm. This plan will establish the responders' objectives for controlling the incident. Control operations include containment, extinguishment, and confinement. Keep in mind that the responder at the operations level is expected to carry out defensive control actions and that any plan of action must be formulated based on the resources that can be brought to bear on the mitigation process.

It is conceivable that responders at this level could completely control an incident if it were relatively minor. At more severe incidents, they would merely perform to the extent their training and equipment allowed and would summon appropriate help.

Hazardous Material Incident Survey Form

Location _____

Date ____-____-____ Time: ____ : ____ ____ Weather Conditions _____

Terrain _____

Closest Populated Buildings _____

Closest Bodies of Water _____

Closest Other Buildings _____

No. Deaths _____ No. Injuries _____ Remedial Actions Taken _____

Containment System			Material Identification			Release		
Container Type	Container ID No.	QTY	DOT ID No.	Placard	Name of Material	Yes	Form	Location
						☑	☐ Solid ☐ Liquid ☐ Gas	
						☑	☐ Solid ☐ Liquid ☐ Gas	
						☑	☐ Solid ☐ Liquid ☐ Gas	
						☑	☐ Solid ☐ Liquid ☐ Gas	
						☑	☐ Solid ☐ Liquid ☐ Gas	
						☑	☐ Solid ☐ Liquid ☐ Gas	

Sketch of scene   ↑
                                                                              ⊕
                                                                              N

Figure 3.3 This worksheet is used for recording information collected during the task of surveying a hazardous materials incident. (Source: Fire Protection Handbook, 17th edition.)

1. Describe the response objectives for hazardous materials incidents;

The response objectives for a hazardous materials incident focus on controlling events as they occur or keeping even more severe events from occurring.

2. Describe the defensive options available for a given response objective;

Basically, defensive options are those actions the responders can take safely without coming in direct contact with the hazardous material involved in the incident. These options may include constructing a dike to contain a material, placing hose streams to protect exposures, even placing a pail under a leaking valve, if it can be done with the protective clothing and equipment available to the responder at this level.

3. Determine whether the personal protective equipment provided is appropriate for implementing each defensive option; and

Individuals involved in emergency response must understand their limitations, particularly where personal protective clothing and equipment are concerned. Responders must also understand that clothing and equipment requirements vary depending on the material involved in an incident. Fire fighters responding to structural fires generally use the same type of protective clothing each time. In hazardous materials incidents, however, what is appropriate for one material may be totally unacceptable for another, so the defensive tactics responders use may be very different.

4. Identify the emergency decontamination procedures.

Emergency decontamination may be necessary if a life-threatening exposure has occurred and the individual needs immediate medical treatment. It may also be necessary if life-threatening contamination has occurred and a decontamination process has not yet been set up.

(c) Implement the planned response to favorably change the outcomes consistent with the local emergency response plan and the organization's standard operating procedures by completing the following tasks:

Once the responder has analyzed the incident and planned the initial response, he or she must implement that response. Although some might expect that this is a lengthy process, it frequently takes only a few minutes. For example, a responder may be able to analyze, plan, and implement the response to a small spill of home heating fuel in a nonthreatening location in minutes. If the incident involved a gasoline tank truck overturned on a congested highway interchange, however, the process would take a little longer. No matter how severe the incident, the responder should go through the same planning process to ensure that nothing has been overlooked.

1. Establish and enforce scene control procedures including control zones, emergency decontamination, and communications;

Scene control is critical in keeping both responders and the public safe, so it should be established immediately. Responders can do this by establishing control zones and an exclusion perimeter to keep the public away from the emergency responder's working areas.

2. Initiate the incident management system (IMS) for hazardous materials incidents;

When arriving on the scene, operational level responders must begin implementing the incident management system developed in the local emergency response plan. An incident management plan identifies the roles and responsibilities that will help personnel control the incident safely and effectively.

3. Don, work in, and doff personal protective equipment provided by the authority having jurisdiction; and

Response personnel must be able to use correctly the protective clothing and equipment that is provided. An effective training program will give them the opportunity to demonstrate this ability during various training exercises.

4. Perform defensive control functions identified in the plan of action.

The responder at the operational level must be able to select and use the equipment necessary and to implement the defensive control functions chosen.

(d) Evaluate the progress of the actions taken to ensure that the response objectives are being met safely, effectively, and efficiently by completing the following tasks:

Part of an effective response is the on-going evaluation of the actions that have been undertaken. Responders must not base their actions solely on their initial assessment of conditions at the site because these conditions can, and do, change—the wind may shift, it may begin to rain, or resources may become unavailable.

1. Evaluate the status of the defensive actions taken in accomplishing the response objectives; and

Are the defensive actions being taken having the desired results? Is the incident stabilizing, or is it intensifying? Responders may find that the actions they chose initially are no longer correct because they no longer suit the circumstances. The weather may have changed, for example, or the arrival of additional personnel may have been delayed.

2. Communicate the status of the planned response.

The incident commander must be kept informed of the effectiveness of the defensive actions being taken. An effective response cannot be carried out without frequent status reports. Everyone must be aware of this and provide the necessary information through the appropriate channels.

3-2 Competencies — Analyzing the Incident.

3-2.1* **Surveying the Hazardous Materials Incident.** The first responder at the operational level shall, given examples of both facility and transportation situations involving hazardous materials, survey the hazardous materials incident to identify the containers and materials involved, whether hazardous materials have been released, and the surrounding conditions. The first responder at the operational level shall be able to:

A-3-2.1 The survey of the incident includes an inventory of the type of containers involved, identification markings on containers, quantity in or capacity of containers, materials involved, release information, and surrounding conditions. The accuracy of the data must be verified.

As Charles Wright notes in the *Fire Protection Handbook*:

"After detecting the presence of hazardous materials in an emergency, and while initiating command and control activities, the next task is to survey the hazardous materials incident. Completion of this task provides an inventory of the containment systems and materials involved, materials released, and surrounding conditions. This incident survey should be conducted from a safe distance, without exposure to the released materials."[3]

3-2.1.1* Given examples of various hazardous materials containers, identify the general shapes of containers for liquids, gases, and solids.

A-3-2.1.1 Examples should include all containers including nonbulk packaging, bulk packaging, vessels, and facility containers such as piping, open piles, reactors, and storage bins. Refer to the Chemical Manufacturers Association/Association of American Railroads Hazardous Materials Technical Bulletin *Packaging for Transporting Hazardous and Non-hazardous Materials*, issued June 1989.

Recognizing the shapes of various containers and knowing what each normally holds helps the responder verify the presence of hazardous materials and could help him or her identify the particular materials involved in an incident. See Supplement 9.

Figure 3.4(a) The responder should be able to identify various containers and have a general idea of the products they contain.

3-2.1.1.1 Given examples of the following tank cars, identify each tank car by type:

Roger Fitch and Charles Wright note in Chapter 35, Section 8, "Rail Transportation Systems," of the *Fire Protection Handbook* that "Tank cars are classed according to their construction, features, and fittings. The specification of the tank determines the product it may transport."[4] (*See Figure 3.5.*)

Figure 3.4(b)   Liquefied gas containers.

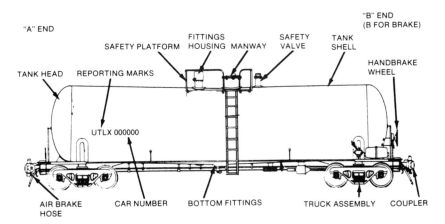

Figure 3.5 Diagram of tank car tank with some basic features identified. (Source: Union Tank Car Company.)

(a) Nonpressure tank cars with and without expansion domes;

Fitch and Wright note that nonpressure tank cars, also known as general-service or low-pressure tank cars, carry a wide variety of hazardous and nonhazardous materials at low pressures. They may transport such materials as flammable and combustible liquids, flammable solids, oxidizers, organic peroxides, poison B materials, corrosive materials, molten solids, and certain flammable and nonflammable gases. Their capacities range from 4,000 to 45,000 gallons (15 to 17m³), and they are cylindrical with rounded heads (*see Figure 3.6*).[5]

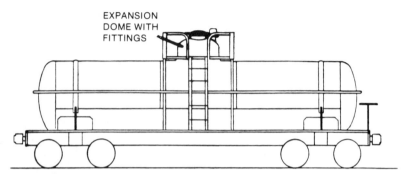

Figure 3.6 Diagram of a typical nonpresssure tank car with an expansion dome. (Source: Union Pacific Railroad.)

(b) Pressure tank cars; and

According to Fitch and Wright, pressure tank cars typically transport hazardous materials, including flammable, nonflammable, or poison gases, at higher pressures. Among the products pressure tank cars may transport are ethylene oxide, pyrophoric liquids, sodium metal, motor fuel anti-knock compounds, bromine, anhydrous hydrofluoric acid, and acrolein. Tank pressures range from 100 to 600 psi (689 to 4137 kPa), and tank capacities range from 4,000 to 45,000 gal (15 to 170m³). Tanks are cylindrical, noncompartmented, and made of steel or aluminum with rounded heads. They may be insulated or thermally protected, and the top two-thirds of the cars, which are not insulated and have no thermal protection, will be painted white (*see Figure 3.7*).[6]

(c) Cryogenic liquid tank cars.

Fitch and Wright note that cryogenic tank cars carry low-pressure (usually 25 psig or lower) liquids refrigerated to –155°F (–93.5°C) and below. Typically, these liquids include argon, ethylene, hydrogen, nitrogen, and oxygen. A cryogenic tank car is actually a tank-within-a-tank, the inner tank made of stainless steel or nickel. The space between the inner and outer tanks is filled with insulation and is under a vacuum (*see Figure 3.8*).[7]

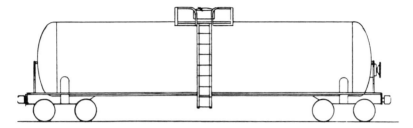

Figure 3.7 Diagram of a typical pressure tank car. (Source: Union Pacific Railroad.)

Figure 3.8 Diagram of a typical cryogenic liquid tank car. (Source: Union Pacific Railroad.)

3-2.1.1.2 Given examples of the following intermodal tank containers, identify each intermodal tank container by type:

According to Fitch and Wright, intermodal tank containers, often referred to as tank containers, are being used more and more frequently in North America to transport a wide range of commodities, including an increasing number of hazardous materials. Among the factors that account for this increased use are their improved safety, their portability, and the lower transportation costs they offer. Tank containers also offer the benefits of a multi-modal transport system. Because they consist of a single metal tank mounted inside a sturdy metal supporting structure, they can be used interchangeably on several modes of transport, such as railroad cars, tank trucks, and ships (*see Figure 3.9*).[8]

(a) Nonpressure intermodal tank containers; and

According to Fitch and Wright, nonpressure tank containers, sometimes referred to as intermodal portable tanks or IM portable tanks, are usually used to transport liquid or solid materials at pressures of up to 100 psi (6,890 kPa) (*see Figure 3.10*).[9]

Figure 3.9 Diagrams of tank containers in various modes of transportation. (Source: Union Pacific Railroad.)

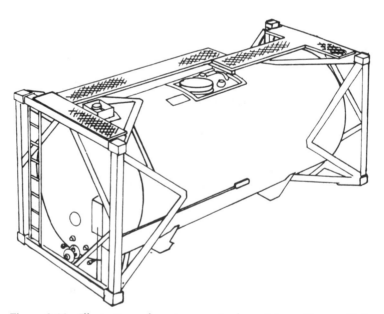

Figure 3.10 Illustration of nonpressure tank container. (Source: Union Pacific Railroad.)

(b) Pressure intermodal tank containers.

Fitch and Wright note that pressure tank containers are designed to accommodate internal pressures of 100 to 500 psig (6,890 to 34,450 kPa) and are generally used to transport gases liquefied under pressure, such as LP-Gas and anhydrous ammonia. They may also carry such liquids as motor fuel anti-knock compound and aluminum alkyls (*see Figure 3.11*).[10]

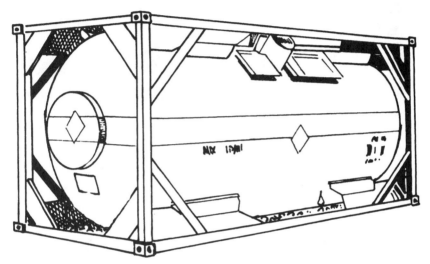

Figure 3.11 Diagram of a typical pressure tank container. (Source: Union Pacific Railroad.)

3-2.1.1.3 Given examples of the following cargo tanks, identify each cargo tank by type:

Cargo tanks are large containers used for hauling liquids in bulk. Sometimes referred to as tank motor vehicles or tank trucks, they are the most common vehicles used to transport combustible, flammable, and corrosive materials, as well as flammable and nonflammable compressed gases. See Figures 3.12(a)–3.12(d) and Supplement 9.

(a) MC-306/DOT 406 cargo tanks;

(b) MC-307/DOT-407 cargo tanks;

(c) MC-312/DOT-412 cargo tanks;

(d) MC-331 cargo tanks;

(e) MC-338 cargo tanks; and

(f) Dry bulk cargo tanks.

Figure 3.12(a) An MC-306 cargo tank. (Courtesy of Fruehauf Trailer Corporation.)

Figure 3.12(b) An MC-307 cargo tank. (Courtesy of Fruehauf Trailer Corporation.)

Figure 3.12(c) An MC-312 cargo tank. (Courtesy of Fruehauf Trailer Corporation.)

Figure 3.12(d) An MC-331 pressure tank. (Courtesy of Harry Abraham.)

3-2.1.1.4 Given examples of the following facility tanks, identify each fixed facility tank by type:

In Chapter 25, Section 2, "Storage of Flammable and Combustible Liquids," of the *Fire Protection Handbook*, Orville M. Sly, Jr., notes that tanks may be installed above ground, below ground, or, under certain conditions, inside buildings. It is important that the responder be able to identify the difference between pressure and nonpressure tanks.[11]

(a) Nonpressure facility tanks; and

Nonpressure tanks, also called atmospheric tanks, are designed for pressures of 0 to 0.5 psig (0 to 4 kPa) (*see Figure 3.13*).

ORDINARY CONE ROOF TANK

FLOATING ROOF TANK

Roof deck rests upon liquid and moves upward and downward with level changes.

LIFTER ROOF TANK

Liquid sealed roof moves upward and downward with vapor volume changes.

VAPORDOME ROOF TANK

Flexible diaphragm in hemispherical roof moves in accordance with vapor volume changes.

Figure 3.13   Common types of atmospheric storage tanks. (Source: Fire Protection Handbook, 17th edition.)

(b) Pressure facility tanks.

Pressure tanks are divided into low-pressure storage tanks, for pressures of 0.5 to 15 psig (4 to 103 kPa), and pressure vessels, for pressures above 15 psig (103 kPa) (*see Figure 3.14*).

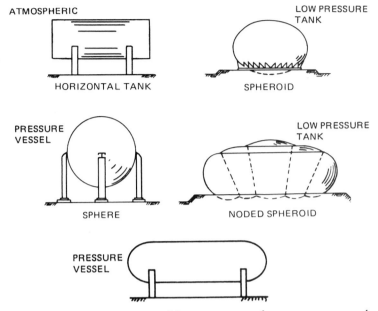

Figure 3.14 *Common types of low-pressure tanks or pressure vessels. (Source: Fire Protection Handbook, 17th edition.)*

3-2.1.2 Given examples of facility and transportation containers, identify the markings that differentiate one container from another.

Containers at fixed facilities may be marked with the NFPA 704 system, and transportation vehicles may be marked with DOT placards or identification numbers. Particular tanks or storage areas at fixed facilities may also be identified by labeling and pre-emergency planning documents, as may the products they contain.

3-2.1.2.1 Given examples of the following transport vehicles and their corresponding shipping papers, identify the vehicle or tank identification marking in all applicable locations:

(a) Rail transport vehicles, including tank cars;
(b) Intermodal equipment including tank containers; and
(c) Highway transport vehicles, including cargo tanks.

There will be an identification marking on each transport vehicle that will be included on the shipping papers. This allows the responders to ensure that the shipping papers and vehicles match. The identification number also provides a way to contact the shipper for information about a specific vehicle.

3-2.1.2.2 Given examples of facility containers, identify the markings indicating container size, product contained, and/or site identification numbers.

Containers at fixed facilities are often stencilled with a product name or some type of identification number that refers to a site plan or an emergency operations plan that identifies the product and the quantity stored.

3-2.1.3 Given examples of facility and transportation situations involving hazardous materials, identify the name(s) of the hazardous material(s) in each situation.

In order to develop the appropriate action options, the operational responder should be able to gather whatever information is necessary to identify the hazardous materials at an incident by name.

3-2.1.3.1 Identify the following information on a pipeline marker:

(a) Product;
(b) Owner; and
(c) Emergency telephone number.

Pipelines that carry hazardous materials must be identified, and the above information provided (*see Figure 3.15*).

3-2.1.3.2 Given a pesticide label, identify each of the following pieces of information; then match the piece of information to its significance in surveying the hazardous materials incident:

According to William J. Keffer, writing in Chapter 16, Section 3, "Pesticides," of the *Fire Protection Handbook*, pesticides are classified according to their primary or specific control purposes or to reflect the manner in which they are used. Among the pesticides classified by control purposes are insecticides, fungicides, herbicides, nematocides, and rodenticides. Among those classified by the manner in which they are used are fumigants.[12]

(a) Name of pesticide;

The label contains the manufacturer's name for the pesticide.

Figure 3.15 Typical markings on a pipeline marker.

(b) Signal word;

Pesticide labels must have a "signal word" that indicates the relative hazard of the product. Table 3.2 lists the current warnings the EPA requires, based on the hazard of the active ingredient.

(c) Pest control product (PCP) number (in Canada);

In Canada, labels carry a pest control number that can be used to acquire additional information regarding a specific product. In the U.S., labels carry an EPA registration number.

Table 3.2   EPA Toxicity Categories

Category 1:   POISON/DANGER

All pesticide products meeting the following criteria:
  Oral LD_{50} up to and including 50 mg/kg,
  Inhalation LD_{50} up to and including 0.2 mg/liter,
  Dermal LD_{50} up to and including 200 mg/kg,
  Eye effects—corneal opacity not reversible within 7 days, and
  Skin effects corrosive.
Must bear on the front panel the signal word "Danger." In addition, if the product was assigned to Category 1 on the basis of its oral, inhalation, or dermal toxicity (as distinct from skin and eye local effects), the word "Poison" must appear in red on a background of distinctly contrasting color and the skull and crossbones must appear in immediate proximity to the word "Poison."

Category 2:   WARNING

All pesticides meeting the following criteria:
  Oral LD_{50} from 50 through 500 mg/kg,
  Inhalation LD_{50} from 0.2 through 2.0 mg/liter,
  Dermal LD_{50} from 200 through 2,000 mg/kg,
  Eye effects—corneal opacity reversible within 7 days (irritation persisting for
    7 days) and
  Skin effects—severe irritation at 72 hours
Must bear on the front panel the signal word "Warning."

Categories 3 and 4:   CAUTION

All pesticide products meeting the following criteria:
  Oral LD_{50} greater than 500 mg/kg,
  Inhalation LD_{50} greater than 2.0 mg/liter,
  Dermal LD_{50} greater than 2,000 mg/kg,
  Eye effects—no corneal opacity, and
  Skin effects—moderate irritation at 72 hours
Must bear on the front panel the signal word "Caution."

(Source: *Fire Protection Handbook*, 17th edition.)

(d) Precautionary statement;

Labels also carry a caution statement indicating the care that must be taken when using the product. Such statements include "Keep Out of Reach of Children," "Restricted Use Pesticide," or "Hazard to Humans and Domestic Animals."

(e) Hazard statement; and

Typically, a hazard statement indicates that the product poses an environmental hazard and advises against contaminating water supplies.

(f) Active ingredient.

Each active ingredient in the pesticide is identified and the percentage indicated. Inert ingredients are also shown, but only by percentage.

3-2.1.4* Identify and list the surrounding conditions that should be noted when surveying hazardous materials incidents.

A-3-2.1.4 The list of surrounding conditions should include: topography; land use; accessibility; weather conditions; bodies of water; public exposure potential; overhead and underground wires and pipelines; storm and sewer drains; possible ignition sources; adjacent land use such as rail lines, highways, and airports; and nature and extent of injuries. Also, include building information such as floor drains, ventilation ducts, air returns, etc., when appropriate.

Surrounding conditions are important to the responder in that they influence the options available to him or her. It is important for the responder to think very broadly when considering the surrounding conditions.

3-2.1.5 Give examples of ways to verify information obtained from the survey of a hazardous materials incident.

Responders should collect information about an incident continuously so that they can validate the information collected earlier. This information may be verified by, among other things, contacting CHEMTREC/CANUTEC to verify the hazard information found in emergency response guides, contacting the shipper to verify the products listed on shipping papers, and using additional references to confirm the emergency handling procedures.

3-2.2 **Collecting Hazard and Response Information.** The first responder at the operational level shall, given known hazardous materials, collect hazard and response information using material safety data sheets (MSDS), CHEMTREC/CANUTEC, and contacts with the shipper/manufacturer. The first responder at the operational level shall be able to:

In the *Fire Protection Handbook*, Charles Wright notes that:
Once a hazardous material is identified, information about the material's hazards, behavior characteristics, and response should be collected. This information, which may be obtained simultaneously with determining the extent

of container/packaging damage, is used to predict the behavior of the material. The information to be collected is divided into six groups: material identification information, physical properties, chemical properties, physical hazards, health hazards, and response information. The task of obtaining, recording, and interpreting hazardous materials information can be lengthy and rigorous. Various forms are being used to record hazard and response information (*see Figure 3.3*).[13]

The information the responder at the operational level collects allows him or her to determine the defensive options that can be performed safely, given the personnel and equipment available. At the technician level, this same process continues, and the responder uses the information collected to determine whether it is feasible to conduct offensive operations.

3-2.2.1 Match the definitions associated with the DOT hazard classes and divisions of hazardous materials, including refrigerated liquefied gases and cryogenic liquids, with the class or division.

The responder should be able to match the hazard class or division of a hazardous material with the appropriate definition of that material. For example, the responder should be able to match the definition "consists of explosives that have a mass explosion hazard" with a Class 1, Division 1.1 material.

3-2.2.2 Identify two ways to obtain a material safety data sheet (MSDS) in an emergency.

MSDSs are available at fixed facilities and can often be found in transporting vehicles, as well. A responder may obtain MSDSs from CHEMTREC or from the shipper, who may fax them to dispatch offices or to portable fax machines available to field personnel.

3-2.2.3 Using a material safety data sheet (MSDS) for a specified material, identify the following hazard and response information:

In the U.S., MSDSs are required by the Occupational Safety and Health Administration (OSHA). Although OSHA provides a standard form, manufacturers may use similar forms of their own design that OSHA has approved.

Certain basic information specified by OSHA must appear on all forms. Responders should be familiar with this information and know how to locate it on an MSDS, which is a valuable source of product information. The responder should understand that, although the location of certain information on the MSDS may vary, the basic information outlined below can be found somewhere on the sheet. For example, information relating to storage found on one form in a section called "Precautions for Safe Handling and Use" may be found on another in a section called "Special Protection Information."

(a) Physical and chemical characteristics;

The information in this section provides the responder with information about such physical characteristics of the hazardous material as its vapor density, boiling point, specific gravity, water solubility, pH, and physical appearance. For example, a substance may be described as "white to pale yellow sticks, granules or powder, no odor."

(b) Physical hazards of the material;

This section includes information about a material's fire and explosion hazards, including its flash point, autoignition temperature, and flammability limits, as well as information about the extinguishing agents that may be used on the material. This section may also provide information about hazards associated with fire control operations. For example, an entry may read ". . . a water stream directed at molten material can scatter the material, increasing the flammability of any combustible material it contacts." This section may also recommend appropriate personal protective clothing and respiratory protection.

A separate reactivity section generally provides information about the material's stability and notes what an unstable material will react with. For instance, an entry may read ". . . it is a strong oxidizing agent that will increase the flammability of all combustible materials it contacts."

(c) Health hazards of the material;
(d) Signs and symptoms of exposure;
(e) Routes of entry;
(f) Permissible exposure limits;

The above information is generally contained in the health hazard section, which provides the responder with important data on the health hazards a material presents, including the threshold limit value (TLV), the routes of exposure, and the material's effects. Also provided is information about emergency first aid measures.

(g) Responsible party contact;

Information about the material's manufacturer and possibly the names of its distributors is also provided, as are telephone numbers for emergency contacts.

(h) Precautions for safe handling (including hygiene practices, protective measures, procedures for cleanup of spills or leaks);

The MSDS also tells the responder what steps should be taken in the event of a spill or leak and how to dispose of such spilled or leaked material. For example, this section might instruct the responder to remove ignition sources or to cover the material with soda ash. And to dispose of the material, the responder might be advised to ". . . mix slurry with wet sand and dispose of suitably in an approved landfill."

(i) Applicable control measures including personal protective equipment; and

This information will help the responder choose the appropriate respiratory protection, eye protection, protective gloves, and so on for working with the hazardous material. It may also indicate how the material should be stored. Should it be kept a in well-ventilated area, for example, or in a cool, dry area? And with what incompatible materials should it not be stored?

(j) Emergency and first aid procedures.

Information about emergency and first aid procedures is often found with other health-related data. It will detail the actions that should be taken immediately if an individual is exposed to a hazardous material and recommend when to seek additional medical attention.

3-2.2.4 Identify the following:

(a) The type of assistance provided by CHEMTREC/CANUTEC;
(b) How to contact CHEMTREC/CANUTEC; and
(c) The information to be furnished to CHEMTREC/CANUTEC.

"CHEMTREC" stands for Chemical Transportation Emergency Center and is a public service of the Chemical Manufacturers Association, located in Washington, DC. CHEMTREC provides the on-scene commander with immediate advice by telephone and contacts the involved shipper for detailed assistance and response follow-up. It can also notify the National Response

Center (NRC) of significant incidents and bridge a caller to the NRC to report a spill. CHEMTREC operates 24 hours a day and can be contacted throughout the U.S. and Canada by calling 1-800-424-9300.

CHEMTREC can usually provide hazard information warnings and guidance when given a material's four-digit identification number, the name of the product, and the nature of the problem. If the product is unknown or more detailed information and assistance is needed, the caller should attempt to provide as much of the following information as possible:

The caller's name and a call-back number
The guide number being used
The name of the shipper or manufacturer
The rail car or truck number
The carrier's name
The consignee
Local conditions.

At an incident, the caller should try to keep a phone line open to CHEMTREC so that they can provide guidance and assistance. CHEMTREC can also provide a teleconferencing bridge that allows them to connect technical experts to the caller's line as necessary.

CANUTEC is the Canadian Transport Emergency Center located in Ottawa, Canada, and operated by the Transport Dangerous Goods Directorate of Transport Canada. It provides technical assistance to emergency responders much the same as CHEMTREC. Personnel will provide technical information regarding the physical, chemical, toxicological, and other properties of the products involved in an incident; recommend remedial actions for fires, spills, or leaks; provide advice on protective clothing and emergency first aid; and contact the shipper, manufacturer, or others who are deemed necessary. CANUTEC can be reached by calling 1-613-996-6666 collect 24 hours a day.

3-2.2.5 Identify two methods of contacting the manufacturer or shipper to obtain hazard and response information.

The responder may contact the manufacturer or shipper through CHEMTREC/CANUTEC, as described above, or by using information provided on the shipping papers or on the MSDS.

3-2.3* **Predicting the Behavior of a Material and Its Container.** The first responder at the operational level shall, given examples of facility and transportation hazardous materials incidents involving a single hazardous material, predict the likely behavior of the material and its container in each incident. The first responder at the operational level shall be able to:

A-3-2.3 Predicting the likely behavior of a hazardous material and its container requires the following skills: the ability to identify the types of stress involved and the ability to predict the type of breach, release, dispersion pattern, length of contact, and the health and physical hazards associated with the material and its container. Reference can be made to Benner, Ludwig, Jr., *A Textbook for Use in the Study of Hazardous Materials Emergencies,* 2nd edition, Lufred Industries, Inc., Oakton, VA, 1978; the National Fire Academy's training program, *Hazardous Materials Incident Analysis,* 1984; Noll, Gregory G., et al., *Hazardous Materials: Managing the Incident,* Fire Protection Publications, Stillwater, OK, 1988; or Wright, Charles J., "Managing the Hazardous Materials Incident," *Fire Protection Handbook,* 17th edition, National Fire Protection Association, Quincy, MA, 1991.

Ludwig Benner, Jr., describes the process of predicting hazardous materials behavior as the ". . . visualization of an events sequences in a 'mental movie' framework. . . The responder needs to think in terms of events and then relate them to the prediction of the emergency events." The responder ". . . needs to focus on what the hazardous material is going to do. . . in order to influence the sequence of events."[14]

3-2.3.1 Given situations involving known hazardous materials, interpret the hazard and response information obtained from the current edition of the *Emergency Response Guidebook,* material safety data sheets (MSDS), CHEMTREC/CANUTEC, and shipper/manufacturer contacts.

Not only must the responder know where to find response information, but he or she must also be able to interpret that information in order to decide what actions are appropriate. The responder must also recognize that different emergency response guides may present conflicting information or emphasize one area more than another. It is necessary to gather, interpret, and choose the information that is most appropriate for a given situation.

3-2.3.1.1 Match the following chemical and physical properties with their significance and impact on the behavior of the container and/or its contents:

(a) Corrosivity (pH);

Corrosivity is a measure of a substance's tendency to deteriorate in the presence of another substance or in a particular environment. U.S. federal regulations define a corrosive material as a liquid or solid that causes visible destruction or irreversible alterations in human skin tissue at the site of contact or that causes steel to corrode at a severely accelerated rate. The degree of corrosiveness is measured by pH, which ranges from 1 to 14. A pH of 7 is neutral, while a pH below 7 is acidic and a pH above 7 represents a base. The responder must understand the type of corrosive with which he or she is dealing.

(b) Flammable (explosive) range;

A material's flammable or explosive range is the difference between its upper and lower flammable limits. The lower flammable limit (LFL) is the minimum concentration of vapor to air below which a flame will not propagate in the presence of an ignition source. The upper flammable limit (UFL) is the maximum vapor-to-air concentration above which a flame will not propagate. If a vapor-to-air mixture is below the LFL, it is described as being "too lean" to burn; if it is above the UFL, it is "too rich" to burn. When the vapor-to-air ratio is somewhere between the LFL and the UFL, fires and explosions can occur, and the mixture is said to be in the flammable range. The flammable range for gasoline is 1.4 percent to 7.6 percent and the flammable range for carbon monoxide is 12.5 percent to 74 percent. It is important that the responder be aware of the range, as well as the LFL.

If the responder suspects or knows that flammable vapors are present, he or she must determine the concentration of vapor in air. Combustible gas-detection instruments are used for this purpose.

(c) Flash point;

The flash point of a liquid is the minimum temperature at which it gives off vapor in sufficient concentration to form an ignitable mixture with air. A liquid's flash point is the primary property or characteristic used to determine its relative degree of flammability. Since it is the vapors of flammable liquids that burn, vapor generation is a primary factor in determining the liquid's fire hazard.

(d) Form (solid, liquid, gas);

Hazardous materials are either solids, liquids, or gases, and the responder should understand the difference form makes on the hazards a material presents. For example, gases present significantly different hazards than solids.

(e) Ignition (autoignition) temperature;

"Ignition temperature" and "autoignition temperature" are interchangeable terms. The ignition temperature of a substance, whether solid, liquid, or gaseous, is the minimum temperature required to cause self-sustained combustion in the absence of any source of ignition. The responder should look upon assigned ignition temperatures as approximations.

Ignition temperatures can be quite high, especially in relation to a liquid's flash point. For example, the flash point of gasoline is –45°F (–43°C), while its ignition temperature is well over 500°F (156°C).

(f) Reactivity;

Reactivity is the susceptibility of a material to release energy either by itself or in combination with other materials.

(g) Specific gravity;

Specific gravity is the ratio of the weight of a volume of liquid or solid to the weight of an equal volume of water, with the gravity of water being 1.0. A substance with a specific gravity of less than 1.0 will float on water, and one with a specific gravity greater than 1.0 will sink.

(h) Toxic products of combustion;

All products of combustion should be considered toxic, but those produced by hazardous materials may be more toxic than those produced by nonhazardous materials. The danger that the products of combustion of hazardous materials are contaminated to a greater degree with higher levels of toxins than one would find in a regular structure fire increases the need for respiratory protection and for evacuation downwind of the fire. Fire fighters must be aware that this is as true for outside fires as it is for structural fires.

(i) Vapor density; and

Vapor density measures the weight of a given vapor as compared with an equal volume of air, with air having a value of 1.0. A vapor density greater than 1.0 indicates it is heavier than air; a value less than 1.0 indicates it is lighter.

Vapor density can be important to the responder, since it will determine the behavior of free vapor at the scene of a liquid spill or gas release.

(j) Water solubility.

Water solubility, or the degree to which a substance is soluble in water, can be useful in determining effective extinguishing agents and methods. Response personnel should consider the property of solubility along with that of specific gravity.

3-2.3.1.2 Identify the differences among the following terms:

(a) Exposure and hazard;

A person may be exposed to large quantities of a hazardous material in concentrations that do not present a hazard or to small amounts of a hazardous material that present a very high hazard.

(b) Exposure and contamination; and

A person exposed to a hazardous material may not necessarily be contaminated by it.

(c) Contamination and secondary contamination.

Responders working in the hot zone may become contaminated during control operations. If they then carry that contamination outside the hot zone on their equipment, clothing, skin, or hair in sufficient quantities and are not adequately decontaminated, they may contaminate others. Substances that pose a high risk of secondary contamination include asbestos, mercaptans, pesticides, and PCBs. Materials that present little risk of secondary contamination include carbon monoxide, weak acids, and gasoline.

3-2.3.2* Identify three types of stress that could cause a container system to release its contents.

A-3-2.3.2 The three types of stress that could cause a container to release its contents are: thermal, mechanical, and chemical.

Stress is defined by Ludwig Benner, Jr., as "the applied force or system of forces that tend to strain or deform a body."[15] The "body" in question could refer to either a container or its contents. Responders are more likely to encounter this type of stress on the job than they are to encounter the two other types of stress to which containers are subject, irradiation and etiologic agents, or active microorganisms that move around.

3-2.3.3*   Identify five ways in which containers can breach.

A-3-2.3.3   The five ways in which containers can breach are: disintegration, run-away cracking, closures opening up, punctures, and splits or tears.

The performance objectives contained in 3-2.3.3 through 3-2.3.5 should be taught in a manner and language understandable to the audience. The intent is to convey the simple concepts that containers of hazardous materials under stress will open up and allow the contents to escape. This refers to both pressurized and nonpressurized containers. This content release will vary in type and speed. A pattern will be formed by the escaping product that will possibly expose people, the environment, or property, creating physical or health hazards. This overall concept is often referred to as a general behavior model and is used to estimate the behavior of the container and its contents under emergency conditions.

When a hazardous materials container loses its integrity, the incident will often escalate. The timing of such a release cannot always be predicted, and it will vary with the duration, intensity, and type of stress to which the container is subjected.

The responder must understand that a container may be stressed or its contents may be stressed and that there is a difference between the two phenomena. A container may degrade under stress. The material inside a container can either degrade the container or breach a container that has not been degraded.[16]

According to Charles Wright, factors that affect the intensity of the breach include ". . . type and duration of the stress being applied, behavior of the container under the stresses being applied, behavior of the contents, location of the stresses, force of opening of the containment system, size of breach, and speed of breach event."[17]

3-2.3.4*   Identify four ways in which containers can release their contents.

A-3-2.3.4   The four ways in which containment systems can release their contents are: detonation, violent rupture, rapid relief, and spill or leak.

According to Ludwig Benner, Jr.:

The types of release include: detonation, disintegration of the container, and/or detonation of the contents; violent massive failure behavior, the runaway cracking of the container, and rapid-acceleration polymerization or oxidizing hazardous materials reactions that burst the container abruptly; rapid relief behavior, including pressure ruptures or safety valve operation; and spill or leak behavior, including gradual flow through openings, tears or splits, and punctures.[18]

3-2.3.5*   Identify at least four dispersion patterns that can be created upon release of a hazardous material.

A-3-2.3.5   The seven dispersion patterns that can be created upon release of hazardous materials are: hemisphere, cloud, plume, cone, stream, pool, and irregular.

Knowing how hazardous materials behave when they are released is important in determining the area potentially endangered. The way the material is released influences the dispersion pattern.

3-2.3.6*   Identify the three general time frames for predicting the length of time that exposures may be in contact with hazardous materials in an endangered area.

A-3-2.3.6   The three general time frames for predicting the length of time that an exposure may be in contact with hazardous materials in an endangered area are: short-term (minutes and hours), medium-term (days, weeks, and months), and long-term (years and generations).

Factors that influence the length of time an exposure may last in an endangered area include the quantity of the material released, the method of dispersion, and the speed at which it is released. For example, did the container leak or did it detonate? The presence of secondary reactions will also influence the length of an exposure.

3-2.3.7*   Identify the health and physical hazards that could cause harm.

A-3-2.3.7   The health and physical hazards that could cause harm in a hazardous materials incident are: thermal, mechanical, poisonous, corrosive, asphyxiation, radiation, and etiologic.

Harm is the injury or damage caused by being exposed to the hazards of the released contents.

As Charles Wright notes in the *Fire Protection Handbook*, "Considerations include: hazards associated with the released contents, concentrations of released materials, duration of contact, and how often the exposures will come in contact with the released materials."[19]

3-2.3.7.1*   Identify the health hazards associated with the following terms:

(a) Asphyxiant;
(b) Irritant/corrosive;

(c) Sensitizer/allergen;

(d) Convulsant; and

(e)* Chronic health hazard.

A-3-2.3.7.1 Health Hazard Definitions.

(a) *Carcinogen:* A chemical is considered to be a carcinogen if:

1. It has been evaluated by the International Agency for Research on Cancer (IARC), and found to be a carcinogen or potential carcinogen; or

2. It is listed as a carcinogen or potential carcinogen in the *Annual Report on Carcinogens* published by the National Toxicology Program (NTP) (latest edition); or,

3. It is regulated by federal OSHA as a carcinogen (may be regulated additionally by states).

(b) *Corrosive:* A chemical that causes visible destruction of, or irreversible alterations in, living tissue by chemical action at the site of contact.

(c) *Highly toxic:* A chemical falling within any of the following categories:

1. A chemical that has a median lethal dose (LD_{50}) of 50 mg or less per kg of body weight when administered orally to albino rats weighing between 200 and 300 g each.

2. A chemical that has a median lethal dose (LD_{50}) of 200 mg or less per kg of body weight when administered by continuous contact for 24 hr (or less if death occurs within 24 hr) with the bare skin of albino rabbits weighing between 2 and 3 kg each.

3. A chemical that has a median lethal concentration (LD_{50}) in air of 200 parts per million by volume or less of gas or vapor, or 2 mg per L or less of mist, fume, or dust, when administered by continuous inhalation for 1 hr (or less if death occurs within 1 hr) to albino rats weighing between 200 and 300 g each.

(d) *Irritant:* A chemical that is not corrosive but that causes a reversible inflammatory effect on living tissue by chemical action at the site of contact.

(e) *Sensitizer:* A chemical that causes a substantial proportion of exposed people or animals to develop an allergic reaction in normal tissue after repeated exposure to the chemicals.

(f) *Toxic:* A chemical falling within any of the following categories:

1. A chemical that has a median lethal dose (LD_{50}) or more than 50 mg per kg but not more than 500 mg per kg of body weight when administered orally to albino rats weighting between 200 and 300 g each.

2. A chemical that has a median lethal dose (LD_{50}) of more than 200 mg per kg but not more than 1,000 mg per kg of body weight when administered by continuous contact for 24 hr (or less if death occurs within 24 hr) with the bare skin of albino rabbits weighing between 2 and 3 kg each.

3. A chemical that has a median lethal concentration (LD_{50}) in air of more than 200 parts per million but not more than 3,000 parts per million by volume of gas or vapor, or more than 2 mg per L but not more than 200 mg per L of mist, fume, or dust, when administered by continuous inhalation for 1 hr (or less if death occurs within 1 hr) to albino rats weighing between 200 and 300 g each.

(g) *Target organ effects:* The following is a target organ categorization of effects that may occur, including examples of signs and symptoms and chemicals that have been found to cause such effects. These examples are presented to illustrate the range and diversity of effects and hazards that may be encountered but are not intended to be all-inclusive.

1. *Hepatotoxins.* Chemicals that produce liver damage.
   Signs and Symptoms: Jaundice; liver enlargement.
   Chemicals: Carbontetrachloride; nitorsamines.
2. *Nephroxtoxins.* Chemicals that produce kidney damage.
   Signs and Symptoms: Edema; protein urea.
   Chemicals: Halogenated hydrocarbons; uranium.
3. *Neruotoxins.* Chemicals that produce their primary toxic effects on the nervous system.

   a. *Central Nervous System Hazards:* Chemicals that cause depression or stimulation of consciousness or otherwise injure the brain.

   b. *Peripheral Nervous System:* Chemicals that damage the nerves that transmit messages to and from the brain and the rest of the body.
   Signs and Symptoms: Numbness, tingling, decreased sensation; change in reflexes; decreased motor strength.

   Examples: Arsenic, lead, toluene, styrene.

4. Agents that decrease hemoglobin in the blood of function; deprive the hematopolatic body tissues of oxygen system.
   Signs and Symptoms: Cyanosis; loss of consciousness.
   Chemicals: carbon monoxide; benzene.
5. Agents that irritate the lung or damage the pulmonary tissue.
   Signs and Symptoms: Cough; tightness in chest; shortness of breath.
   Chemicals: Silica; asbestos; HCL.

6. *Reproductive toxins.* Chemicals that affect the reproductive capabilities including chromosomal damage (mutations) and effects on fetuses (teratogenesis).
   Signs and Symptoms: Birth defects; sterility.
   Chemicals: Lead; DBCP.
7. *Cutaneous hazards.* Chemicals that affect the dermal layer of the body.
   Signs and Symptoms: Defatting of the skin; rashes; irritation.
   Chemicals: Ketones; chlorinated compounds.
8. *Eye hazards.* Chemicals that affect the eye or visual capacity.
   Signs and Symptoms: Conjunctivitis; corneal damage.
   Chemicals: Organic solvents; acids.

A-3-2.3.7.1(e)   Chronic health hazards would include carcinogen, mutagen, and teratogen.

Responders need to be able to recognize the effect(s) that hazardous materials can have on human beings. Knowing the class of health hazard a particular chemical falls into will help the responder evaluate the risk and readily identify possible reactions to the exposure. The product may have more than one type of reaction on an individual, and reaction to a particular product may vary with the individual. Depending on the type of health hazard the product possesses, the reaction may be immediate, delayed, or chronic.

3-2.4*   **Estimating the Potential Harm.** The first responder at the operational level shall estimate the potential harm within the endangered area at a hazardous materials incident. The first responder at the operational level shall be able to:

A-3-2.4   The process for estimating the potential outcomes within an endangered area at a hazardous materials incident includes: determining the dimensions of the endangered area, estimating the number of exposures within the endangered area, measuring or predicting concentrations of materials within the endangered area, estimating the physical, health, and safety hazards within the endangered area, identifying the areas of potential harm within the endangered area, and estimating the potential outcomes within the endangered area.

The operational level responder must be able to estimate the area of potential harm. Responders at this level are not expected to engage in intricate calculations or even to use sophisticated computer modeling, but they should be able to make general estimates based on the hazard information collected so that they can begin evacuations and safely implement defensive operations.

As an incident becomes more complex, technician level responders would probably become involved, and they would be able to make more definitive predictions about the endangered area.

3-2.4.1* Identify a resource for determining the size of an endangered area of a hazardous materials incident.

A-3-2.4.1 One resource for determining the size of an endangered area of a hazardous materials incident is the current edition of the *Emergency Response Guide*.

The U.S. DOT *Emergency Response Guidebook* recommends isolation and evacuation distances for some hazardous materials. CHEMTREC or CANUTEC may provide additional information.

3-2.4.2 Given the dimensions of the endangered area and the surrounding conditions at a hazardous materials incident, estimate the number and type of exposures within that endangered area.

Exposures include people, the environment, and property. Once the responder has estimated the size and location of the endangered area, he or she can determine what is inside the perimeter. Factors influencing his or her decision include the time of day, the type of occupancies involved, and the type of area involved. Is it rush hour, for example? Is the occupancy a petrochemical plant? Is the area congested or is it rural?

3-2.4.3 Identify resources available for determining the concentrations of a released hazardous material within an endangered area.

Various types of monitoring equipment and dispersion modeling programs are available for determining the concentrations of a released hazardous material. In most cases, however, responders at the operational level would not have this type of equipment. Nonetheless, they should understand what can be used and know how to acquire it and the technical assistance needed to operate it. For example, they should be aware of any company in their area that handles hazardous materials and know whether the company has personnel and equipment available to monitor hazardous conditions.

Responders should also know whether they can ask a regional hazardous materials response team to assist them. State or county environmental agencies or health departments can often provide monitoring equipment and personnel. However, responders should plan ahead and make these arrangements before an incident occurs, not at 3:00 a.m. on a rainy night when an over-turned cargo tanker begins leaking hazardous materials all over the highway.

3-2.4.4* Identify the factors for determining the extent of physical, health, and safety hazards within the endangered area of a hazardous materials incident given the concentrations of the released material.

A-3-2.4.4 The factors for determining the extent of physical, health, and safety hazards within an endangered area at a hazardous materials incident are: surrounding conditions, an indication of the behavior of the hazardous material and its container, and the degree of hazard.

Responders must determine the extent of physical harm that can be expected in an endangered area and compare the gains they may receive from intervening—or not intervening—before they can determine the actions they should or should not take. Factors that influence this decision are the quantity and concentration of the hazardous materials released, the number of exposures in the endangered area, and the manner in which the exposures will be subjected to the hazardous material. Is the material a liquid or gas, for example? How far away from the source are the exposures? How fast is the material being released?

3-3 Competencies — Planning the Response.

3-3.1 Describing Response Objectives for Hazardous Materials Incidents. The first responder at the operational level shall, given simulated facility and transportation hazardous materials problems, describe the first responder's response objectives for each problem. The first responder at the operational level shall be able to:

Up to this point, we have addressed the competencies that allow the responder to gather the information necessary to make appropriate decisions when determining response objectives. This section outlines the competencies necessary for using this information to develop defensive options that can be safely implemented to influence the outcomes of the incident. Remember, actions are limited by the resources available. Responders should not take any action until they have a clear idea of what they are trying to accomplish.

In the *Fire Protection Handbook*, Charles Wright notes that, "The planning process begins as part of the pre-emergency response planning efforts prior to the incident and continues on the scene of the incident, again from a safe location. Federal, state, and local agencies; industry; and carrier personnel may be called upon to help.

The process of response planning is based on the following tasks:

1. Determine the response objectives,
2. Determine the available response options that could favorably change the outcomes,
3. Identify the personal protective equipment for the response objectives,
4. Identify an appropriate decontamination process for each response option,
5. Select response options within the response community's capabilities that will most favorably change the outcomes, and
6. Develop a plan of action including safety considerations."[20]

3-3.1.1 Identify the steps for determining the number of exposures that could be saved by the first responder with the resources provided by the authority having jurisdiction and operating in a defensive fashion, given an analysis of a hazardous materials problem and the exposures already lost.

To determine the number of exposures that might be saved, the responder must first determine the number of exposures there are and the number that have already been lost, then estimate the effectiveness of the chosen action options.

3-3.1.2 Describe the steps for determining defensive response objectives given an analysis of a hazardous materials incident.

Charles Wright notes that, "Response objectives, based on the stage of the incident, are the strategic goals for stopping the event now occurring or keeping future events from occurring. Decisions should focus on changing the actions of the stressors, the containment system, and the hazardous material."[21]

3-3.2 Identifying Defensive Options. The first responder at the operational level shall, given simulated facility and transportation hazardous materials problems, identify the defensive options for each response objective. The first responder at the operational level shall be able to:

After the responder has determined what to protect, he or she must determine how to protect it. Operational level responders must remember that their actions are expected to be defensive in nature.

3-3.2.1 Identify the defensive options to accomplish a given response objective.

The available defensive options fall into two categories: containment and confinement. This is true for offensive operations, too, but the responder at this level is expected to take actions without actually stopping the release.

Some broad general control methods that might be available at this level include absorption; dilution with water; dams, dikes, and diversions; and vapor dispersion.

3-3.2.2 Identify the purpose for, and the procedures, equipment, and safety precautions used with, each of the following control techniques:

(a) Absorption;

Absorption can be described as the process by which one substance, such as a liquid, combines with another substance, such as a solid, by entering the interior of that substance.

Two things should be kept in mind when using sorbents. First, the sorbent must be compatible with the material being absorbed. And second, the sorbent material must be handled as a hazardous waste and disposed of correctly once it has been contaminated.

A number of sorbents are available commercially in the form of booms, pads, pillows, and so on. Other materials, such as sand, soil, sawdust, and vermiculite, may also be used as sorbents. The responder should know where large quantities of these materials are available in the community and establish a method for obtaining them 24 hours a day in the event of an emergency.

(b) Dike, dam, diversion, retention;

Although substantial hazardous liquids spills can pose insurmountable problems, many lesser spills can be controlled by diking, damming, diversion, or retention. These techniques are the most commonly used methods of controlling releases because they can be improvised with a little ingenuity. Earthen dikes or dams can be erected quickly under favorable conditions, and sandbags can be used in the damming effort. Trenches can be used to collect spilled liquids. And commercial booms can be used to control spills on waterways. For more information, see NFPA 471 in this handbook.

(c) Dilution;

Dilution refers to the application of water to water-miscible hazardous materials to reduce to safe levels the hazard they represent. Responders must take care when using water to achieve the desired effect. They must also remember that dilution will increase the total volume of liquid with which they will have to deal.

(d) Vapor dispersion; and

Water spray can be used to disperse or move vapors away from certain materials. The gas concentration of some materials, such as liquefied petroleum gas, can be reduced below the lower flammable limit.

(e) Vapor suppression.

Vapor suppression refers to the reduction or elimination of vapors emanating from a spilled or released material. While vapor suppression does not change the nature of the hazardous material, it is an important mitigation procedure because it can greatly reduce the immediate hazard associated with the presence of an uncontrolled vapor. For more information, see NFPA 471 in this handbook.

3-3.3 Determining Appropriateness of Personal Protective Equipment. The first responder at the operational level shall, given the name of the hazardous material involved and the anticipated type of exposure, determine whether available personal protective equipment is appropriate for implementing a defensive option. The first responder at the operational level shall be able to:

Responders at this level are not expected to use specialized chemical-protective clothing. Rather, they are expected to use the type of protective clothing they normally wear in their working environment. For example, fire fighters would wear structural fire fighting protective clothing, while the employee of an industrial facility might use liquid splash-protective clothing.

Responders must understand the differences among the various types of protective clothing and the levels of protection they afford. The level of protective clothing available to operational level responders is a significant factor when considering what type of defensive operations they can safely undertake and when they should obtain additional expertise and specialized equipment.

3-3.3.1* Identify the appropriate respiratory protection required for a given defensive option.

A-3-3.3.1 The minimum requirement for respiratory protection at hazardous materials incidents (emergency operations until concentrations have been determined) is positive pressure self-contained breathing apparatus. Therefore, the minimum for the first responder at the operational level is positive pressure self-contained breathing apparatus.

OSHA 1910.120, Subpart L, 4iiiD, requires that "Employees engaged in emergency response and exposed to hazardous substances shall wear positive pressure self-contained breathing apparatus while engaged in emergency response until such time that the individual in charge of the ICS determines through the use of air monitoring that a decreased level of respiratory protection will not result in hazardous exposure to employees."[22]

This does not mean that the responder at the operational level cannot use other types of respiratory protection. Once the material has been identified and the appropriate type of protection has been determined, it may be used even if the response is still in the "emergency" phase. However, the incident commander should make this decision only after he or she has consulted the incident safety officer.

3-3.3.1.1 Identify the three types of respiratory protection and the advantages and limitations presented by the use of each at hazardous materials incidents.

The three types of respiratory protection are the air purifying respirator (APR), the supplied-air respirator (SAR), and the self-contained breathing apparatus (SCBA). APRs are lightweight, and they restrict users' movement and travel less than the two other types of respiratory protection. However, APRs cannot be used in oxygen-deficient atmospheres. They must be worn only in atmospheres in which the hazard has been identified and the concentrations are within allowable limits.

SARs are supplied with air by an external source, usually a compressor or compressed air cylinders located away from the actual work site. SARs can be used in oxygen-deficient atmospheres, they can operate longer than SCBA, and they are lighter to wear than SCBA. However, the SARs' hoses limit the distance the wearer can travel and may become tangled or twisted. In addition, the wearer must enter and leave the site at the same point.

Positive pressure SCBA provides the highest level of respiratory protection available. It can be used in oxygen-deficient atmospheres and does not restrict the distance the wearer may travel or path he or she may take. However, the positive pressure SCBA's limited air supply—2 hours at the most—will limit the length of time it can be used, thus controlling the distance the wearer may travel. In addition, the weight of the SCBA may cause the wearer to exert additional physical effort, which may shorten his or her period of effectiveness (*see Table 3.3*).

Table 3.3 Respirator, Protection Equipment

Type	Advantages	Limitations
1. Positive pressure self-contained breathing apparatus	• Provides highest level of protecton against airborne contaminants and oxygen deficiency.	• Bulky, heavy. • Finite air supply limits work duration. • May impair movement in confined spaces.
2. Positive pressure supplied-air respirators	• Longer work periods than with SCBA. • Less bulky and heavy than SCBA. • Protects against airborne contaminants.	• Impairs mobility. • MSHA/NIOSH limits hose to 300 feet. • As length of hose is increased, minimum approved airflow may not be delivered at the face-piece. • Air line is vulnerable to damage, chemical contamination, and degradation. Decontamination of hoses may be difficult. • Worker must retrace steps to leave work area. • Requires supervision/monitoring of the air supply line. • Not approved for use in atmospheres immediately dangerous to life or health (IDLH) or in oxygen-deficient atmospheres unless equipped with an emergency egress unit such as an escape-only SCBA that can provide immediate emergency respiratory protection in case of air line failure.
3. Air purifying respirator [including powered air purifying respirators (PAPRs)]	• Enhance mobility. • Lighter in weight than an SCBA.	• Cannot be used in IDLH or oxygen-deficient atmosphere (less than 19.5 percent oxygen at sea level). • Limited duration of protection. May be hard to gauge safe operating time in field conditions. • Only protects against specific chemicals and up to specific concentrations. • Use requires monitoring of contaminants and oxygen levels. • Can only be used: (1) against gas and vapor contaminants with adequate warning properties, or (2) for specific gases or vapors provided that the service is known and a safety factor is applied or if the unit has an end-of-service-life indicator (ESLI).

3-3.3.1.2 Identify the required physical capabilities and limitations of personnel working in positive pressure self-contained breathing apparatus.

There are certain limitations on the use of SCBA. Positive pressure SCBA places a strain on the wearer's cardiovascular system because inhaling and exhaling requires increased effort. And some individuals are claustrophobic and simply cannot wear SCBA. In addition, OSHA regulations require that an individual be medically certified to wear respiratory protection safely.

NFPA 1500, *Standard on Fire Department Occupational Safety and Health Program*, requires that all responders using SCBA be medically certified annually by a physician.[23] The certifying physician should consult ANSI Z88.6, *Standard for Respiratory Protection-Respirator Use—Physical Qualifications for Personnel*, to determine which medical review is appropriate.[24] NFPA 1500 also requires that SCBA users be trained, tested, and certified regularly in the safe and proper use of the equipment (*see Table 3.4*).

3-3.3.2 Identify the appropriate personal protective equipment required for a given defensive option.

Given the name of the material involved in an incident and the type of exposure, the responder should be able to determine what type of personal protective equipment is required when implementing defensive options.

3-3.3.2.1 Identify skin contact hazards encountered at hazardous materials incidents.

There are a number of skin contact hazards, including burns, rashes, and absorption, so protecting exposed skin is critical. Responders must be made aware that their skin may be exposed to a hazardous material even when they are wearing protective clothing if the clothing in contact with their skin becomes saturated with the material.

3-3.3.2.2 Identify the purpose, advantages, and limitations of the following levels of protective clothing at hazardous materials incidents:

(a) Structural fire fighting clothing;
(b) High temperature-protective clothing; and
(c) Chemical-protective clothing:

    1. Liquid splash-protective clothing; and
    2. Vapor-protective clothing.

Table 3.4 Relative Advantages and Disadvantages of SCBA

Type	Description	Advantages	Disadvantages	Comments
Entry-and-Escape SCBA Open-Circuit SCBA	Supplies clean air to wearer from cylinder. Wearer exhales air directly to atmosphere.	• Operated in positive pressure mode, provides highest respiratory protection now available. Warning alarm signals when 20 to 25 percent of air supply remains.	• Shorter operating time (30, 60 minutes), heavier weight [up to 35 lb (13.6 kg)] than closed-circuit SCBA.	• Operating time may vary depending on size of air tank and work rate of individual.
Closed-Circuit SCBA (Rebreather)	Recycles exhaled gases (CO_2, O_2, nitrogen) by removing CO_2 with alkaline scrubber and replenishing consumed oxygen with oxygen from liquid or gaseous source.	• Longer operating time (up to 4 hours), lighter weight [21 to 30 lbs (9.5 to 13.6 kg)] than open-circuit apparatus. • Warning alarm signals when 20 to 25 percent of oxygen supply remains. • Oxygen supply is depleted before CO_2 sorbent scrubber supply, protecting wearer from CO_2 breakthrough.	• At very cold temperatures, scrubber efficiency may be reduced, CO_2 breakthrough may occur. • Units retain heat exchanged in exhalation, generate heat in CO_2 scrubbing operations, adding to danger of heat stress. • Auxiliary cooling devices may be required. • When worn outside encapsulating suit, breathing bag may be permeated by chemicals, contaminating breathing apparatus and respirable air. • Decontamination of breathing bag may be difficult.	• Positive pressure units offer more protection than negative-pressure units, which are not recommended on hazardous waste sites. While these devices may be certified as closed-circuit SCBAs, NIOSH cannot certify closed-circuit SCBAs as positive pressure devices due to limitations in certification procedures defined in 30 CFR Part 11.
Escape-Only SCBA	Supplies clean air from an air cylinder or from oxygen-generating chemical. Approved for escape only.	• Lightweight [10 lb (4.5 kg) or less], low bulk, easy to carry. • Available in pressure-demand, continuous flow modes.	• Cannot be used for entry.	• Provides only 5 to 15 minutes of respiratory protection, depending on model, wearer breathing rate.

It is important to remember that no single type of protective clothing will protect the wearer against all possible hazards. This is true even of chemical-protective clothing. See definitions for specific type of protective clothing in Section 1-2 of NFPA 472 and Chapter 5 of NFPA 471.

3-3.4* Identifying Emergency Decontamination Procedures. The first responder at the operational level shall identify emergency decontamination procedures. The first responder at the operational level shall be able to:

A-3-3.4 Refer to the following publication: *Hazardous Materials Response Handbook*, National Fire Protection Association, Quincy, MA.

Decontamination, or contamination reduction, is critical to health and safety at a hazardous materials incident. It protects responders from hazardous substances that may contaminate and eventually permeate their personal protective equipment and the other equipment they use at the incident. Decontamination also minimizes the transmission of harmful substances from one control zone to another and helps protect the environment. Contamination reduction procedures must be specific for the type of hazard encountered. See Supplement 8, "Guidelines for Decontamination of Fire Fighters and Their Equipment."

3-3.4.1 Identify ways that personnel, personal protective equipment, apparatus, and tools and equipment become contaminated.

Contamination occurs when responders come in contact with hazardous substances at an incident. If the responders are adequately protected, only the garments they are wearing or the equipment they are using in the hot zone will be contaminated. All contaminants must be removed from the responders' personal protective clothing before the clothing itself is removed if the clothing is to maintain the level of protection it is designed to afford.

3-3.4.2 Describe how the potential for secondary contamination determines the need for emergency decontamination procedures.

Personnel can carry contaminants from the hot zone on their protective clothing or equipment. Decontamination procedures must be implemented to minimize the threat of secondary contamination.

3-3.4.3 Identify the purpose of emergency decontamination procedures at hazardous materials incidents.

Emergency, or gross, decontamination is designed to remove immediately those contaminants that pose an immediate threat to life.

3-3.4.4 Identify the advantages and limitations of emergency decontamination procedures.

Emergency decontamination can be implemented without establishing a formal contamination reduction corridor. However, it only provides gross decontamination, so the victim may still be exposed to contaminants and may pose a threat of secondary contamination.

3-4 Competencies — Implementing the Planned Response.

3-4.1   Establishing and Enforcing Scene Control Procedures. The first responder at the operational level shall, given scenarios for facility and/or transportation hazardous materials incidents, identify how to establish and enforce scene control including control zones, emergency decontamination, and communications. The first responder at the operational level shall be able to:

3-4.1.1 Identify the procedures for establishing scene control through control zones.

The site control process must be put into place very quickly at each hazardous materials incident to maintain control of the scene. The size of the control zone is based on the degree of hazard present (*see Section 4-3 of NFPA 471*).

3-4.1.1.1 Identify the criteria for determining the locations of the control zones at hazardous materials incidents.

Initially, the responder will probably determine where to locate the control zones using recommendations from emergency response guides, such as the U.S. DOT's initial isolation and evacuation distances, or the advice of such organizations as CHEMTREC or CANUTEC. In addition, the responder will rely on his or her own observations of the incident and on his or her assessment of related information.

As the incident progresses, the responder may adjust the zones based on sampling and monitoring results, on evaluations of the extent of contamination and the path it might take in case of a leak, and on the space needed to support control operations (*see NFPA 471, Section 4-4*).

3-4.1.2 Identify the basic techniques for the following protective actions at hazardous materials incidents:

(a) Evacuation; and

Evacuation is defined by the U.S. DOT *Emergency Response Guidebook* as the process of moving "all the people from the area threatened to a safer place." For an evacuation to be successful, Charles Wright notes, there must be enough time to warn the people in the affected area and enough time for them to get ready and leave that area. Evacuees should not be allowed to congregate on the perimeter of the control zones. Rather, they should be sent upwind of the threatened area by a specific route to a specific place far enough away from the incident that they will not have to be moved again if the conditions change.[25]

(b) In-place protection.

The U.S. DOT *Emergency Response Guidebook* defines in-place protection as a "means to direct people to quickly go inside a building and remain inside until the danger passes." Certain protective actions must be taken inside that building, such as closing windows and doors, shutting down HVAC equipment (which draws in outside air), staying away from windows, etc.

3-4.1.3 Identify the considerations associated with locating emergency decontamination areas.

Such considerations include the ability to control runoff and to minimize exposure to others.

3-4.1.4* Demonstrate the ability to perform emergency decontamination.

A-3-4.1.4 See A-3-3.4.

Also see NFPA 471 for more information relating to decontamination.

3-4.1.5* Identify the items to be considered in a safety briefing prior to allowing personnel to work on a hazardous materials incident.

A-3-4.1.5 Refer to the following publication: NIOSH/OSHA/USCG/EPA *Occupational Safety and Health Guidance Manual for Hazardous Waste Site Activities*, October 1985.

The following items should be presented during a safety briefing:

- Preliminary evaluation
- Hazard identification
- Description of the site
- Task(s) to be performed
- Length of time for task(s)
- Required personal protective clothing
- Monitoring requirements
- Notification of identified risks.

These are some of the key elements, but there may be others that are important to include depending on the type of incident.

3-4.2*   Initiating the Incident Management System (IMS). The first responder at the operational level shall, given simulated facility and/or transportation hazardous materials incidents, initiate the incident management system (IMS) specified in the local emergency response plan and the organization's standard operating procedures. The first responder at the operational level shall be able to:

A-3-4.2   See A-2-4.1.

One of the key elements of conducting a safe and effective control operation at a hazardous materials incident is the implementation of an effective incident management system. For an incident management system to be effective, it must be part of a pre-emergency plan and be adopted as a standard operating procedure for emergency responders in a given area. NFPA 1561, *Standard on Incident Management Systems*,[26] provides guidance on what an effective incident management system should provide in order to operate as effectively as possible. The National Inter-Agency Incident Management System (NIIMS) lists the following components of its incident command system:

> Common terminology
> Modular organizations
> Integrated communications
> Unified command structure
> Consolidated action plans
> Manageable span of control.

3-4.2.1   Identify the role of the first responder at the operational level during hazardous materials incidents as specified in the local emergency response plan and the organization's standard operating procedures.

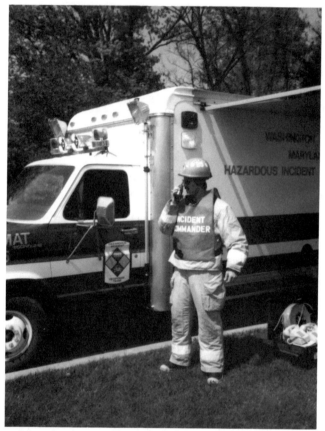

Figure 3.16 The first-arriving senior officer at the scene of the emergency should assume the position of incident commander and institute the incident command system. (Source: Fire Protection Handbook, 17th edition.)

The local emergency response plan should outline the role of emergency responders at the operational level. Primarily, that role includes responding to an emergency, assessing the nature of the incident, implementing initial defensive actions, notifying other involved parties, and asking for additional assistance when needed.

3-4.2.2 Identify the levels of hazardous materials incidents as defined in the local emergency response plan.

See NFPA 471, Chapter 3, "Response Levels."

3-4.2.3 Identify the purpose, need, benefits, and elements of an incident management system (IMS) at hazardous materials incidents.

See commentary on 3-4.2.

3-4.2.4 Identify the considerations for determining the location of the command post for a hazardous materials incident.

The initial command post may well be the first-arriving unit at the incident. As more equipment and personnel arrive, however, the incident commander may use a designated command post, which may be a specially designed vehicle. The IC may also use radios or telephones to disseminate information. At more complex incidents, an emergency operations center may actually be set up.

Each incident should have only one command post. That post should be clearly marked, and access to it should be controlled. The responder should consider locating the command post in an area from which it is unlikely to be moved.

3-4.2.5 Identify the procedures for requesting additional resources at a hazardous materials incident.

It is extremely important that responders at every level know what type of resources are available and understand how to request them. It is not uncommon for a responder to reply, "Gee, I didn't know it was available," when asked why he or she did not ask for a particular type of assistance.

3-4.2.6* Identify the responsibilities of the safety officer.

A-3-4.2.6 The hazardous materials safety officer should meet all the competencies for the responder at the level of operations being performed.

A hazardous materials safety officer is an individual who directs the safety of operations within the hot and warm zones. A hazardous materials safety officer must be designated specifically at all hazardous material incidents (CFR 1910.120). The hazardous materials safety officer has the following responsibilities:

(a) Obtains a briefing from the incident commander or incident safety officer and the hazard sector officer;

(b) Participates in the preparation of and monitors the implementation of the incident safety plan (including medical monitoring of entry team personnel before and after entry);

(c) Advises the incident commander/sector officer of deviations from the incident safety plan and of any dangerous situations; and

(d) Alters, suspends, or terminates any activity that is judged to be unsafe.

The position of safety officer at a hazardous materials incident is critical, as can be seen by the position's responsibilities as outlined in A-3-4.2.6. The incident commander may be responsible for implementing tasks at an incident, but the safety officer has the authority to see that they are accomplished safely. At most incidents, there is only one safety officer. At more complex incidents, however, there may be additional assistants.

Training and experience are vital. The person serving as safety officer should be qualified to operate at the level of the incident. If the incident is being controlled at the operational level, for example, the safety officer should be trained to that level. If offensive control operations are being conducted, the safety officer should be trained to the level of technician.

3-4.3 **Using Personal Protective Equipment.** The first responder at the operational level shall demonstrate the ability to don, work in, and doff the personal protective equipment provided by the authority having jurisdiction. The first responder at the operational level shall be able to:

Not only should responders know all about the protective clothing and equipment they are given, but they must also be able to use it.

3-4.3.1 Identify the importance of the buddy system in implementing the planned defensive options.

One purpose of the buddy system is to keep track of everyone working at an incident and to make sure they are all safe. No one involved in operations at an incident works alone: everyone is either part of a team or has a partner.

3-4.3.2 Identify the importance of the back-up personnel in implementing the planned defensive options.

During emergency operations, responders often work in conditions that may deteriorate rapidly and unexpectedly, even with the best of planning. Thus, it is vitally important that back-up personnel are available, that they are equipped with the same level of personal protective equipment as the personnel they are backing up, and that they can be deployed immediately in the event of an emergency.

3-4.3.3 Identify the safety precautions to be observed when approaching and working at hazardous materials incidents.

An incident should be approached from upwind and uphill whenever possible, and the approach should be calculated and deliberate. Binoculars can help responders identify the material involved, and monitoring equipment can help them assess the hazard and determine what protective equipment they should use.

Responders working at an incident should be aware of what is happening around them. Is there an increase in the rate of venting? Does the fire seem bigger than it did? Does the problem seem to be lessening? Is there a change in atmospheric conditions? Is it getting windier? These are a few examples of the questions responders should ask themselves while working at a hazardous materials incident.

3-4.3.4 Identify the symptoms of heat and cold stress.

The two most serious heat problems are heat exhaustion and heat stroke. Heat exhaustion is characterized by fatigue, headache, nausea, dizziness, pallor, and profuse sweating. It usually occurs when a person is dehydrated from not having adequate liquid intake. Heat stroke is characterized by hot, dry skin because of the inability to sweat. It may occur quickly, and the victim will exhibit confusion and impaired judgment.

Cold stress may occur if responders are exposed to prolonged cold temperatures. Hypothermia is one of the more severe manifestations of cold stress. Victims will exhibit shivering, apathy, listlessness, drowsiness, slow pulse, a low respiratory rate, and possible freezing of the extremities. If the victim is exposed to water, either by sweating or if clothing becomes wet, this will increase the possibility of cold stress.

3-4.3.5 Identify the physical capabilities required for and the limitations of personnel working in the personal protective equipment as provided by the authority having jurisdiction.

The physical and medical requirements for personnel wearing respiratory protection were discussed in 3-3.3.1.2. In addition to those requirements, responders must have the physical stamina to wear the protective clothing and equipment provided and to work under strenuous conditions. That is why it is so important to conduct realistic training exercises that require the responder to demonstrate his or her true abilities.

3-4.3.6 Match the function of the operational components of the positive pressure self-contained breathing apparatus provided the hazardous materials responder to the name of the component.

This allows responders to demonstrate their knowledge of the positive pressure self-contained breathing apparatus, which consists of a full facepiece, an exhalation valve, a breathing tube, a regulator, an air-supply hose, a cylinder, a cylinder valve, a bypass valve, a main line valve, and a harness assembly.

3-4.3.7 Identify the procedures for cleaning, sanitizing, and inspecting respiratory protective equipment.

Manufacturers and suppliers of the various types of respiratory equipment furnish guidelines and instructions for their use, maintenance, and cleaning. These instructions should be followed.

3-4.3.8 Identify the procedures for donning, working in, and doffing positive pressure self-contained breathing apparatus.

3-4.3.9 Demonstrate donning, working in, and doffing positive pressure self-contained breathing apparatus.

3-4.4 Performing Defensive Control Actions. The first responder at the operational level shall, given a plan of action for a hazardous materials incident within his or her capabilities, demonstrate the ability to perform the defensive control actions set out in the plan. The first responder at the operational level shall be able to:

3-4.4.1 Using the type of fire fighting foam or vapor suppressing agent and foam equipment furnished by the authority having jurisdiction, demonstrate the proper application of the fire fighting foam(s) or vapor suppressing agent(s) on a spill or fire involving hazardous materials.

3-4.4.1.1 Identify the characteristics and applicability of the following foams:

(a) Protein;

Protein foam concentrates produce dense, viscous foams of high stability, high heat resistance, and good resistance to burnback, but they are less resistant to breakdown by fuel saturation than are AFFF and fluoroprotein foams. They are nontoxic and biodegradable after dilution.

(b) Fluoroprotein;

Fluoroprotein foams are very effective in fire fighting situations where the foams become coated with fuel because of their "fuel shedding" properties. They are used in subsurface injection systems such as in storage tank fires and other in-depth crude petroleum or hydrocarbon fuel fires. They possess superior vapor securing and burnback resistance characteristics. In addition, they demonstrate better compatibility with dry chemical agents than do regular protein-type foams. They are biodegradable and nontoxic after dilution.

(c) Special purpose:

1. Polar solvent alcohol-resistant concentrates;

These types of foam concentrates produce foam that is resistant to breakdown when used on water-soluble or water-miscible materials or on polar solvents, such as alcohols, lacquer or enamel thinners, and acetone. The most common exhibit AFF characteristics on hydrocarbons and produce a floating gel-like mass for foam buildup on water-miscible fuels.

2. Hazardous materials concentrates.

Vapor-mitigating foams have been developed that are relatively stable on many hazardous liquids. One type is an alcohol-resistant fire fighting foam that has been additionally stabilized to lengthen its effectiveness as a foam and its stability on toxic, flammable, or corrosive liquids. Another type is medium-expansion foams that are specially formulated to be stable on either acidic or alkaline hazards. In general, these foams should be applied to spills of toxic liquids only by trained personnel.

(d) Aqueous film-forming foam (AFFF); and

AFFFs are composed of synthetically produced materials that form air-foams similar to protein-based foams. In addition, they are capable of forming water solution films on the surface of flammable liquids. AFFFs have low viscosity and have fast spreading and leveling characteristics. They also produce a continuous aqueous layer of solution under the foam. This film is self-healing following mechanical disruption and continues to spread because there is a reservoir of nearby foam. Film effectiveness may be reduced on hot surfaces and aromatic hydrocarbons. AFFFs are nontoxic and biodegradable.

(e) High expansion.

High-expansion foams are agents for control of Class A and Class B fires and are particularly suited for total flooding of confined spaces. They are suitable for transporting wet foam masses to inaccessible places. Their use outdoors may be limited by the effects of weather.

3-4.4.2 Given the appropriate tools and equipment, describe how to perform the following defensive control activities:

(a) Absorption;
(b) Dike, dam, diversion, and retention;
(c) Dilution;
(d) Vapor dispersion; and
(e) Vapor suppression.

The best way to prepare personnel to conduct the above operations is to subject them to a training exercise that allows them to perform these operations while wearing the appropriate personal protective clothing and equipment. See 3-3.3.2.2 for additional information.

3-4.4.3 Identify the location and use of the mechanical, hydraulic, and air emergency remote shutoff devices as found on MC-306/DOT-406 and MC-331 cargo tanks.

Emergency shutoff valves on MC-306/DOT-406 tanks are usually mechanical and are located at both the front and rear of the cargo tanks.

On the MC-331 there are two remote methods of closure. Both closures are required to operate by mechanical and thermal means. One is located at the front of the cargo tank and the other is located at the rear of the vehicle.

3-4.4.4 Describe the objectives and dangers of search and rescue missions at hazardous materials incidents.

This broadly stated competency requires only a general response, since search and rescue operations are so varied in nature and in complexity. If circumstances mandate such an operation, responders are obliged to take whatever actions are necessary to locate and rescue or account for the victims.

In performing such an operation, responders will expand their size-up of the incident. The hazards associated with search and rescue operations are obvious: rescuers expose themselves to increased hazards and risks.

Figure 3.17 A manual shutoff at the rear of an MC-306 cargo tank.

3-5 Competencies — Evaluating Progress.

3-5.1   **Evaluating the Status of Defensive Actions.** The first responder at the operational level shall, given simulated facility and/or transportation hazardous materials incidents, evaluate the status of the defensive actions taken in accomplishing the response objectives. The first responder at the operational level shall be able to:

All responders should understand why their efforts must be evaluated. If they are not making progress, the plan must be reevaluated to determine why progress is not being made.

3-5.1.1   Identify the considerations for evaluating whether defensive options are effective in accomplishing the objectives.

To decide whether the actions being taken at an incident are effective and the objectives are being achieved, the responder must determine whether the incident is stabilizing or increasing in intensity.

3-5.1.2 Describe the circumstances under which it would be prudent to pull back from a hazardous materials incident.

There is no reason to remain in the immediate vicinity of an incident when nothing can be done to mitigate it and the situation may be about to deteriorate. If flames are impinging on an LP-Gas vessel, for example, and there is no way to provide the necessary volume of water to cool it, it would be prudent to withdraw to a safe distance. Many other examples can be used to illustrate this point.

3-5.2 **Communicating the Status of the Planned Response.** The first responder at the operational level shall communicate the status of the planned response to the incident commander and other response personnel. The first responder at the operational level shall be able to:

3-5.2.1 Identify the methods for communicating the status of the planned response to the incident commander through the normal chain of command.

The incident management system should establish procedures for communicating through the chain of command. The responder should know this and know how to do it.

3-5.2.2 Identify the methods for immediate notification of the incident commander and other response personnel about critical emergency conditions at the incident.

A procedure should be established to allow responders to notify the incident commander immediately when conditions become critical and personnel are threatened. This may take the form of a preestablished emergency radio message or tone that signifies danger, for example, or it might be repeated blasts on an air horn. The message should not be delayed while responders try to locate a specific person in the chain of command.

References Cited in Commentary

1. Arthur E. Cote, ed., *Fire Protection Handbook*, 17th edition (Quincy, MA: National Fire Protection Association, 1991), p. 9-158.
2. Ibid., p. 9-165.
3. Ibid., p. 9-160.
4. Ibid., p. 8-269.
5. Ibid., p. 8-270.

6. Ibid., p. 8-271.

7. Ibid., p. 8-271.

8. Ibid., p. 8-273.

9. Ibid., p. 8-275.

10. Ibid., p. 8-275.

11. Ibid., p. 2-231.

12. Ibid., p. 3-171.

13. Ibid., p. 9-167.

14. Benner, Ludwig, Jr., *A Textbook for Use in the Study of Hazardous Materials Emergencies*, 2nd edition, Lufred Industries, Inc., Oakton, VA, 1978.

15. Ibid.

16. Ibid.

17. Cote, *Fire Protection Handbook*, p. 9-172.

18. Benner, *Hazardous Materials Emergencies*.

19. Cote, *Fire Protection Handbook*, p. 9-172.

20. Ibid., p. 9-175.

21. Ibid., p. 9-175.

22. OSHA 1910.120, Subpart L, 4iiiD (Washington, DC: U.S. Government Printing Office).

23. NFPA 1500, *Standard on Fire Department Occupational Safety and Health Program*, 1992 edition.

24. ANSI Z88.6, *Standard for Respiratory Protection-Respirator Use— Physical Qualifications for Personnel*, 1984 edition.

25. Cote, *Fire Protection Handbook*, p. 9-180.

26. NFPA 1561, *Standard on Fire Department Incident Management System*, 1990 edition.

4

Competencies for the Hazardous Materials Technician

4-1 General.

4-1.1 **Introduction.** Hazardous materials technicians shall be trained to meet all requirements at the first responder awareness and operational levels and at the technician level. In addition, hazardous materials technicians shall meet the training requirements and be provided medical surveillance in accordance with federal Occupational Safety and Health Administration (OSHA), local occupational health and safety regulatory, or U.S. Environmental Protection Agency (EPA) requirements, whichever are appropriate for their jurisdiction.

The technician level competencies build upon the skills the responder acquires at the awareness and operational levels. Training programs at this level must ensure that trainees have acquired the requisite knowledge and skills.

4-1.2 **Definition.** Hazardous materials technicians are those persons who respond to releases or potential releases of hazardous materials for the purpose of controlling the release. Hazardous materials technicians are expected to use specialized chemical-protective clothing and specialized control equipment.

The role of the responder at the technician level is significantly different from the roles of the responder at the two previous levels. At those levels, individuals were likely to be those first on the scene or those whose normal duties include initial response to emergencies that involve hazardous materials. Their actions are limited, and their primary duty is to secure the immediate area and minimize the potential harm of an unplanned release of hazardous materials.

The technician, on the other hand, is asked to assume a more aggressive role in controlling an incident. The competencies in this chapter are intended to provide the technician with the skills he or she will need to approach the point of release to plug, patch, or otherwise stop a hazardous substance spill. Because of the additional dangers the responder faces at this level, federal regulations require medical surveillance.

4-1.3 Goal. The goal of training at the technician level shall be to provide the hazardous materials technician with the knowledge and skills to perform the following tasks safely. Therefore, in addition to being competent at both the first responder awareness and operational levels, the hazardous materials technician shall be able to:

(a) Analyze a hazardous materials incident to determine the magnitude of the problem in terms of outcomes by completing the following tasks:

1. Survey the hazardous materials incident to identify special containers involved, to identify or classify unknown materials, and to verify the presence and concentrations of hazardous materials through the use of monitoring equipment;
2. Collect and interpret hazard and response information from printed resources, technical resources, computer data bases, and monitoring equipment;
3. Determine the extent of damage to containers;
4. Predict the likely behavior of materials when released; and
5. Estimate the size of an endangered area using computer modeling, monitoring equipment, or specialists in this area.

The competencies at this level clearly separate the operational level responder from the technician level responder. The HMT is expected to be able to operate monitoring equipment in order to identify the presence and concentrations of hazardous materials. He or she also must be capable of identifying unknowns and must have a more in-depth knowledge of containers. In addition, the HMT should be able to access and interpret a wide variety of resources for hazard and response information.

At the operational level, the responder is expected to be able to make a general estimate of the potential harm a release presents. At the technician level, the responder is expected to be able to predict more definitively how the materials will behave when released.

Finally, the HMT should know either what modeling systems are available or how to get help from someone who can conduct modeling simulations of the dispersion patterns for various hazardous materials.

(b) Plan a response within the capabilities of available personnel, personal protective equipment, and control equipment by completing the following tasks:

1. Identify the response objectives for hazardous materials incidents;

It is important to note the difference between this competency and the corresponding competency at the operational level. At the operational level, the responder is asked to "describe" response objectives; at this level, he or she is asked to "identify" response objectives. Identifying a response objective requires a higher level of competency than describing a response objective because the responder must first acquire the information necessary to evaluate the objective and then determine the appropriate course of action.

The response objectives available at this level are defensive, offensive, or non-intervention. The responder must identify clear objectives, i.e., determine the harmful results one wants to prevent or reduce.

2. Identify the potential action options available by response objective;

At this level, the responder is again asked to identify the options available to him or her to achieve the response objectives that have been identified. This is done by considering the choices, given the resources and time available and the risk at hand. Some of these options include separating the hazard from potential exposures, establishing barriers or dikes, diluting the material, or transferring the product to an undamaged container.

3. Select the personal protective equipment required for a given action option;

Responders at this level must know more about personal protective equipment than responders at the previous levels because they are expected to be involved in offensive operations. This may mean that they have to work in specialized chemical-protective clothing, such as a totally encapsulating suit.

4. Select the appropriate decontamination procedures; and

It is imperative that responders at this level be able to select the appropriate decontamination procedures because they may come in contact with

Figure 4.1   *Hazardous materials technicians control a leak from a drum.*

hazardous materials while conducting such offensive operations as plugging or patching.

5. Develop a plan of action, including safety considerations, consistent with the local emergency response plan and the organization's standard operating procedures, and within the capability of the available personnel, personal protective equipment, and control equipment.

A plan of action is developed after the response objectives have been identified. The plan of action outlines these objectives and identifies the personnel and equipment necessary to accomplish them.

(c) Implement the planned response to favorably change the outcomes consistent with the organization's standard operating procedures and/or a site safety plan by completing the following tasks:

1. Perform the duties of an assigned position within the local incident management system (IMS);

At this level, the responder should be able to perform assigned roles in the incident management system. Generally, the hazardous materials technician is expected to fill a role related to his or her special expertise. Supplement 7 outlines an incident management system designed for hazardous materials response.

2. Don, work in, and doff appropriate personal protective clothing including, but not limited to, both liquid splash and vapor protective clothing with appropriate respiratory protection; and

3. Perform the control functions identified in the plan of action.

Responders at the technician level must be able to demonstrate their ability to select and use the appropriate equipment and to implement the defensive control functions chosen.

4-2 Competencies — Analyzing the Incident.

4-2.1   Surveying the Hazardous Materials Incident. The hazardous materials technician shall identify special containers involved and, given the appropriate equipment, identify or classify unknown materials, verify the identity of hazardous materials, and determine the concentration of hazardous materials. The hazardous materials technician shall be able to:

At this level, the responder should be able to identify the characteristics of special containers that might indicate the presence of hazardous materials. The responder is also expected to be able to conduct an analysis that will identify and classify unknown materials. Special containers include high-pressure containers, which can be identified by their rounded ends; cryogenic cargo tanks or cylinders; and casks for radioactive materials.

4-2.1.1   Given examples of various specialized containers, identify each container by name and match the hazard class of the materials typically found inside the container.

4-2.1.1.1   Given examples of the following tank cars, identify each tank car by type:

(a) Cryogenic liquid tank cars;

Cryogenic tank cars carry low-pressure liquids—usually 25 psig (172 kPa) or lower—that are refrigerated to −155°F (−93.5°C) and below. The cryogenic tank car is a tank-within-a-tank. The space between the inner and outer tanks is filled with insulation and normally maintained under a vacuum.

Cryogenic cars are distinguished by the absence of top fittings: the fittings are enclosed in cabinets either at ground level on both sides or at the end of the car. Among the materials that may be shipped in cryogenic tank cars are argon, ethylene, hydrogen, nitrogen, and oxygen (*see Figure 3.8*).

Another type of cryogenic tank car is the box tank. This type of tank is built inside a 40-foot (12-m) box car, and its fittings are located inside the doors on both sides (*see Supplement 9, Section IV for illustration*).

(b) High-pressure tube cars; and

In the *Fire Protection Handbook*, Roger Fitch and Charles Wright note: "The high pressure tube car is a 40-foot (12 m) box-type open-frame car. Inside the frame is a visible cluster of 30 stainless steel, uninsulated cylinders, arranged horizontally and permanently attached to the car. Tank test pressures of these tube cars range up to 5000 psi (345,000 kPa). These tube cars transport helium, hydrogen, or oxygen in the gaseous form.

Each end of the tube car is enclosed in a steel case. Loading and unloading fittings and safety devices are located in a walk-in cabinet at one end of the car. Safety relief devices, either safety relief valves or safety vents, are found on each cylinder. For cars transporting flammable gases, safety relief devices are equipped with ignition devices to burn off any released material."[1]

All cylinders contain the same material (*see Figure 4.2*).

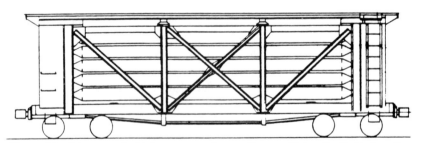

Figure 4.2   Diagram of a high-pressure tube car. (Source: Union Pacific Railroad.)

(c) Pneumatically unloaded hopper car.

According to Fitch and Wright, "Pneumatically unloaded covered hopper cars are built to tank car specifications. It is a covered hopper car that is unloaded through pressure differential, or pneumatics, by the application of air pressure. Even though the pressure is used only during loading and unloading, tank test pressures for the car range from 20 to 80 psi (138 to 551 kPa). Dry caustic soda is one of the commodities transported in this type of car."[2]

Polyvinyl chloride pellets are also transported in this type of car (*see illustration in Supplement 9, Section IV*).

4-2.1.1.2 Given examples of the following intermodal tank containers, identify each intermodal tank container by type:

(a) IM-101 portable tanks;

These tanks are built to withstand maximum allowable working pressures (MAWP) of 25.4 to 100 psig (172 to 6890 kPa). They may be used to transport both nonhazardous and hazardous materials, including toxics, corrosives, and flammables with a flash point below 32°F (0°C). Internationally, an IM-101 portable tank is called an IMO Type 1 tank container. Also see 3-2.1.1.2(a).

(b) IM-102 portable tanks; and

Fitch and Wright note in the *Fire Protection Handbook*:
"These tanks are designed to handle lower MAWP, that is from 14.5 to 25.4 psig (103 to 172 kPa). They transport materials such as liquor, alcohols, some corrosives, pesticides, insecticides, resins, industrial solvents, and flammables with flash points between 32 and 140°F (0 and 60°C). More often, they transport various nonregulated materials, such as food commodities. Internationally, an IM-102 tank is called an IMO Type 2 tank container."[3]
Also see 3-2.1.1.2(a).

(c) Specialized intermodal tank containers:

1. Cryogenic intermodal tank containers;

Cryogenic tank containers carry refrigerated liquid gases, such as liquefied argon, oxygen, helium, ethylene, and nitrogen. They consist of a tank-within-a-tank, with insulation between the inner and outer tanks. The space between the tanks is normally maintained under a vacuum (*see illustration in Supplement 9, Section III*).

2. Tube modules.

While they are not actually portable tanks, tube modules are used to transport bulk gases. This rigid bulk packaging consists of several horizontal seamless steel cylinders, from 9 to 48 inches in diameter, that are permanently mounted inside an open frame with a boxlike compartment at one end enclosing the loading and unloading fittings and safety devices. Service pressures range from 3,000 to 5,000 psi (20,685 to 345,000 kPa). Among the materials shipped in tube modules are nonliquefied gases such as helium, nitrogen, and oxygen (*see Figure 4.3*).

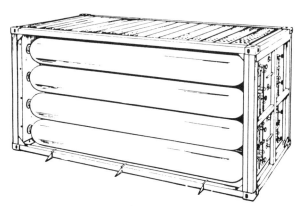

Figure 4.3 *Diagram of a tube module. (Fire Protection Handbook, 17th edition.)*

4-2.1.2 Given examples of both facility and transportation containers, identify the approximate quantity in or capacity of each container.

Facility containers are often marked with their capacities. Transportation containers are required to be marked with their capacities.

4-2.1.2.1 Given examples of the following transport vehicles, identify the capacity (by weight and/or volume) of each transport vehicle using the markings on the vehicle:

(a) Tank cars;

Tank cars must be marked by the U.S. Department of Transportation. Figure 4.4 shows an example of tank car markings.

(b) Tank containers; and

The U.S. DOT also requires that tank containers have standardized markings (*see Figure 4.5*).

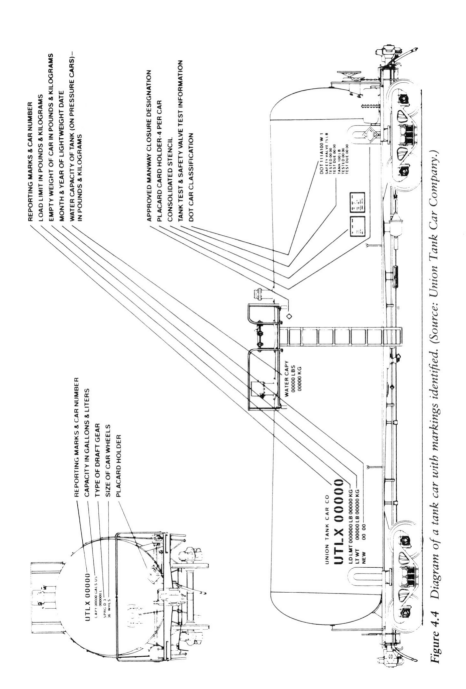

REPORTING MARKS & CAR NUMBER
LOAD LIMIT IN POUNDS & KILOGRAMS
EMPTY WEIGHT OF CAR IN POUNDS & KILOGRAMS
MONTH & YEAR OF LIGHTWEIGHT DATE
WATER CAPACITY OF TANK (ON PRESSURE CARS)—
IN POUNDS & KILOGRAMS

APPROVED MANWAY CLOSURE DESIGNATION
PLACARD CARD HOLDER-4 PER CAR
CONSOLIDATED STENCIL
TANK TEST & SAFETY VALVE TEST INFORMATION
DOT CAR CLASSIFICATION

REPORTING MARKS & CAR NUMBER
CAPACITY IN GALLONS & LITERS
TYPE OF DRAFT GEAR
SIZE OF CAR WHEELS
PLACARD HOLDER

WATER CAPY
00000 LBS
00000 KG

DOT 111A100 W 1
SAFETY VALVE 75 LB
TEST LD 00 00
TANK 100 LB
TEST LD 00 00
TEST DUE 00 00

UNION TANK CAR CO
UTLX 00000
LD LMT 000000 LB 00000 KG
LT WT   00000 LB 00000 KG
NEW   00 00

UTLX 00000
CAPY 00000 GALS U S
00000 L
SPRG D 3
36 WHLS

Figure 4.4   Diagram of a tank car with markings identified. (Source: Union Tank Car Company.)

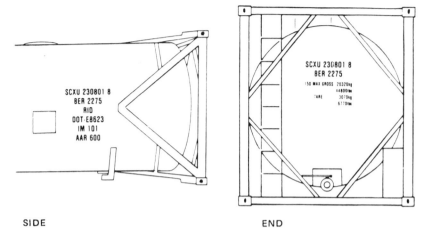

SIDE   END

Figure 4.5 Typical markings on tank containers. (Source: Fire Protection Handbook, 17th edition.)

(c) Cargo tanks.

A data plate and a specification plate must be affixed to cargo tanks. These plates provide the volume and/or weight of the container.

4-2.1.3   Given at least three unknown materials, one of which is a solid, one a liquid, and one a gas, identify or classify by hazard each unknown material.

The responder at this level is expected to be able to identify unknown materials in the event that a container's shipping papers, placards, MSDS, or other identifying items have been destroyed or are unavailable.

4-2.1.3.1   Identify steps in an analysis process for identifying unknown materials.

Unknown materials must be identified cautiously. Monitoring equipment can help identify the hazard class and, in some cases, even specific chemicals. Kits are also available to provide assistance in identification.

If the responder must take samples in order to conduct an analysis or approach the site to monitor it, he or she must use the proper level of protective clothing. In most cases, Level A should be used to ensure adequate protection. Responders must approach the site from upwind, wearing the appropriate level of protective clothing and working in at least pairs, with adequate backup personnel.

Kenneth York and Gerry Grey, in their text *Hazardous Materials/Waste Handling for the Emergency Responder*, recommend that ". . . unless you are certain, beyond any doubt, as to what your hazard is, you should approach the incident employing the information listed below. Assuming that you are not certain what all the hazards are, you should measure, in the order shown below, for:

- Radioactivity
- Combustibility
- Oxygen availability (deficiency)
- pH (if liquid)
- Hydrogen sulfide (if in areas of, or adjacent to, petroleum refining activities)
- Carbon monoxide (in a fire or post-fire incident)
- Organic vapor."[4]

4-2.1.3.2 Identify the type(s) of monitoring equipment used to determine the following hazards:

Hazardous materials response teams should develop protocols for monitoring the air during operations at hazardous materials incidents. A wide variety of monitoring equipment is available, and response teams must determine what is appropriate for the type of hazards they are likely to encounter. The response teams must also become proficient in calibrating, operating, and maintaining this equipment. Each instrument has its own operating characteristics and limitations, and responders must be familiar with them.

(a) Corrosivity (pH);

Corrosivity is measured by determining the pH of a material. This can be done using pH paper or pH meters. pH paper changes color, indicating the pH level. By comparing the color of the pH paper with the color chart provided, one can determine the pH of the material in question. pH meters have a probe that is inserted into the material; the pH is indicated on the meter's display. pH meters provide a more accurate reading than pH paper.

(b) Flammability;

Combustible gas indicators (CGI) can be used to determine the presence of flammable vapors of hydrocarbon products. Certain instruments are specifically designed to monitor methane vapors only; they measure the flammable vapors as a percentage of the lower explosive limit. Flash point testers are also available for field use. These allow the responder to determine fairly accurately the flammability of an unknown material and the class of flammable or combustible liquid with which they are dealing.

(c) Oxidizing potential;

(d) Oxygen deficiency;

Equipment used to monitor oxygen concentrations generally measures over a range of 0 to 25 percent oxygen in air. Some models contain an alarm that sounds if the oxygen level drops below 19.5, which is the minimum adequate percentage of oxygen availability established by the U.S. Occupational Safety and Health Administration. Below this level, the responder must have an air-supplied respirator.

(e) Radioactivity; and

Radiation detectors are available to monitor alpha particles, beta particles, gamma rays, and neutron particles. Generally, such detectors measure two or more types of radiation. For example, the Geiger Counter, probably the most common type of radiation detector, can detect both gamma and beta radiation. However, it cannot effectively measure the amounts of beta radiation. Ion chambers can also measure radiation, but they are not effective on low-level radiation, either *(see Figure 4.6)*.

(f) Toxic exposures.

Several types of instruments can be used to measure toxic exposures. These include photoionization detectors, flame ionization detectors, infrared spectrophotometers, and detector tubes. Some of these instruments are designed to measure specific chemicals, such as hydrogen sulfide, and some may measure more than one. Detector tubes, which allow responders to evaluate potential hazards quickly, operate by drawing an air sample through a tube. This causes the material inside the tube to change color, indicating the concentration of the material in the air.

4-2.1.3.3* Identify the limiting factors associated with the selection and use of the following monitoring equipment:

(a) Carbon monoxide meter;

The carbon monoxide meter is limited in that it will measure only carbon monoxide. It may not indicate whether an area is oxygen-deficient, and it will not indicate the percent of the lower explosive limit to warn responders that they have entered a flammable atmosphere. *(See Figure 4.7.)*

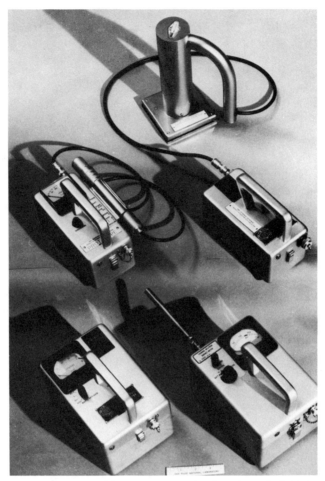

Figure 4.6 Portable instruments developed at Oak Ridge National Laboratory for measuring radioactive emissions. The meters are for fast neutrons (bottom row left), thermal neutrons (bottom row right), beta-gamma (top row left), and alpha particles (top row right). An alpha scintillation detector is at the top of the picture.

(b) Colorimetric tubes;

Since colorimetric tubes are designed to read a specific material, responders need to know the material they are sampling. The tubes also have a specific shelf life and must be monitored to ensure that they are still usable. In

Figure 4.7 This type of carbon monoxide meter was developed for use by an individual. It contains an alarm that sounds if safe limits are exceeded. (Courtesy Mine Safety Appliance Co.)

addition, they are affected by temperature and humidity. Colorimetric tubes may not be very accurate, giving responders only a general idea of the presence of the material for which they are testing. And it may be difficult to interpret the color change to determine the concentration of the material present. (*See Figure 4.8.*)

(c) Combustible gas meter;

The combustible gas indicator (CGI) (*see Figure 4.9*) only measures whether the atmosphere contains a flammable material that is within a percentage of the lower explosive limit (LEL). It cannot identify other hazards, such as toxicity. In an atmosphere that contains more than 100 percent of

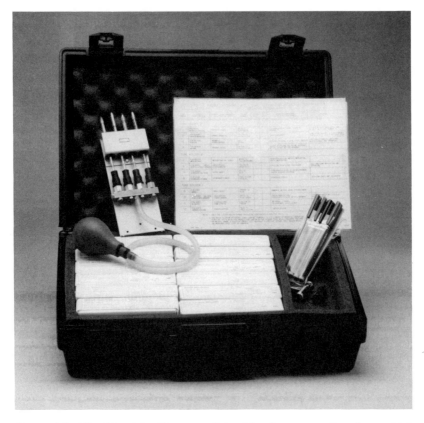

Figure 4.8 The Haz Mat Detector Tube Kit allows sampling for multiple gases and vapors simultaneously. (Courtesy Mine Safety Appliance Co.)

the LEL, the needle on an analog CGI will go to the maximum, then return quickly to zero. If the operator is not watching closely, he or she may miss this and come away with a false impression of the hazard the material presents.

Since the CGI only measures the LEL, the U.S. Environmental Protection Agency has established the following guidelines for working in areas that contain flammable vapors:

- <10 percent of LEL—continue working with caution
- 10 to 25 percent of LEL—continue working with continuous monitoring
- >25 percent of LEL—explosion hazard, withdraw from the area.

Figure 4.9 A combustible gas indicator that uses an aspirator bulb to draw air samples from either immediate areas or remote locations. (Courtesy Mine Safety Appliance Co.)

(d) Oxygen meter;

Oxygen meters do not indicate what type of material may be replacing the oxygen, nor do they indicate the toxicity of such materials. The sensing unit on an oxygen meter may be adversely affected by high concentrations of carbon monoxide or carbon dioxide. The sensing units also deteriorate over time and must be changed. Oxygen meters may be affected by changes in temperature, humidity, and atmospheric pressure, as well.

(e) Passive dosimeter;

Passive dosimeters, or passive samplers, are limited in several ways. For example, the responder must know what material he or she is going to sample, and other chemicals present in the area may interfere with the reading.

Figure 4.10 This monitoring device records oxygen levels and combustible gas levels simultaneously. (Courtesy Mine Safety Appliance Co.)

(f) pH papers, pH meters, and strips; and

pH paper and strips may be difficult to read if the material sampled has been contaminated with oil, mud, or some other material. In addition, the personnel interpreting the paper or strips must have accurate color perception. pH meters can also be affected by oils or other contaminants.

To measure the corrosivity of a solid, the responder may place it in water and measure the pH of the water. Since measuring pH usually requires coming near or touching the material being sampled, responders must exercise extreme caution until they have determined the degree of hazard.

(g) Radiation detection instruments.

It is important that the proper radiation detection equipment be available to determine the type of radiation hazard present. The more common instruments measure only gamma rays and will detect, but not measure, other types of radiation, such as beta radiation.

A-4-2.1.3.3 These factors include but are not limited to: operation, calibration, response time, detection range, relative response, sensitivity, selectivity, inherent safety, environmental conditions, and nature of hazard. Also refer to the following publication: NIOSH/OSHA/USCG/EPA *Occupational Safety and Health Guidance Manual for Hazardous Waste Site Activities*, October 1985.

The key to successfully selecting and using sampling and monitoring instruments is intensive training. Personnel must thoroughly understand the operating characteristics and limitations of all the instruments available to them.

4-2.1.3.4 Given examples of various hazardous materials and the following monitoring equipment, in addition to other monitoring and detection equipment provided by the authority having jurisdiction, select the appropriate monitoring equipment to identify and quantify the materials:

(a) Carbon monoxide meter;
(b) Colorimetric tubes;
(c) Combustible gas meter;
(d) Oxygen meter;
(e) pH papers, pH meters, and strips; and
(f) Radiation detection instruments.

The purpose of this competency is to have responders at this level demonstrate their ability to select the proper instrument and demonstrate its use.

4-2.1.3.5 Demonstrate the field maintenance and testing procedures for the monitoring equipment provided by the authority having jurisdiction.

Responders who operate monitoring equipment must know how to field-calibrate it in order to get accurate readings and must be able to perform minor field maintenance, such as replacing the sensing device in a combustible gas indicator.

4-2.1.4 Given a label for a radioactive material, identify vertical bars, contents, activity, and transport index, then match the label item to its significance in surveying a hazardous materials incident.

The radioactive label provides responders with information about the level of radiation hazard a material presents. Radioactive White I, with a single vertical bar, is the lowest level hazard. Radioactive Yellow II, with two vertical bars, is the next highest level, and Radioactive Yellow III, with three vertical bars, is the highest level of hazard. These levels are based on the radiation activity of the material. The radiation level must be written on the label. (*see Figure 4.11*).

Figure 4.11   Radioactive materials include cobalt, which is used in medicine, and plutonium and uranium, which are transported by the nuclear power industry.

4-2.2   Collecting and Interpreting Hazard and Response Information. The hazardous materials technician shall, given access to printed resources, technical resources, computer data bases, and monitoring equipment, collect and interpret hazard and response information not available from the current edition of the *Emergency Response Guidebook* or a material safety data sheet (MSDS). The hazardous materials technician shall be able to:

At the operational level, the responder must be able to collect hazard and response information from sources that, in general, are readily available, such as the U.S. DOT *Emergency Response Guidebook* or MSDSs. At this level, however, the responder must be able to collect and interpret information from a broader range of sources.

Responders at this level must understand the importance of using multiple resources for gathering hazard and response information. They should collect information from a variety of resources, compare that information, then make their decisions based on this comparison. Responders will generally give more weight to the more conservative information and base their actions on that information. It is important for them to remember that there will be different information, and some of it may conflict.

4-2.2.1 Identify the types of hazard and response information available from each of the following resources and explain the advantages and disadvantages of each resource:

(a) Reference manuals;

There are a number of reference manuals available, such as *The Condensed Chemical Dictionary,*[5] the *Manual of Hazardous Chemical Reactions* (NFPA 491M),[6] and *Dangerous Properties of Industrial Materials*[7] by Irving Sax, to name just a few. Each manual tends to emphasize information a little differently. For example, some will present quite a bit of information about medical hazards but very little about handling emergency incidents. The responder must understand this and recognize the value of gathering information from more than one source and comparing that information for a broader picture of the potential hazards.

(b) Hazardous materials data bases;

As with reference manuals, a number of data bases are available. CAMEO 3.0 (Computer-Aided Management of Emergency Operations) is probably the most widely used. It is available in both DOS and Macintosh applications. Other public-domain data bases are also available to the responder, including OHM/TADS (Oil and Hazardous Materials Technical Assistance Database), RTECS (Registry of Toxic Effects of Chemical Substances), and CHRIS (Chemical Hazard Response Information System). Each data base presents information a little differently and concentrates on different areas. Responders at this level should understand the differences among them and use those that are the most appropriate for handling a specific incident.

(c) Technical information centers (for example CHEMTREC/CANUTEC, NRC);

Technical information centers such as CHEMTREC can provide the responder with valuable information during hazardous materials incidents, and it is important that responders know what assistance is available from these sources.

Using its data base, CHEMTREC can provide initial response information on more than one million product-specific MSDSs. CHEMTREC can put the responder at the scene in contact with the shipper and can help the responder identify the materials involved in an incident using waybill numbers and other sources. In addition to contacting shippers, CHEMTREC can help the responder contact manufacturers and other technical specialists. If needed, CHEMTREC also can activate its emergency response mutual aid network, composed of more than 250 emergency response teams from chemical companies and private contractors that can respond to the scene of an incident and help those on site.

(d) Technical information specialists; and

Technical specialists can be a valuable asset to the responder, who may want to keep a directory of individuals who are able to provide technical assistance. However, the responder must remember that no individual is likely to have all the answers. For example, there are chemists who specialize in formulating perfumes and chemists who specialize in formulating explosives. Depending on the type of incident the responder is involved in, the assistance of one would be more appropriate than the assistance of the other.

Developing a network of people with technical knowledge is one of the most important things a responder can do. Because incidents differ, responders cannot rely on finding books or data bases that address all the possibilities.

(e) Monitoring equipment.

Monitoring equipment again provides a source of information regarding the hazards that are present. The responder should not rely on a single means of monitoring at any incident because the equipment could be affected by unknown materials and give a false reading. Responders should have a sound understanding of the instruments they are using and recognize the value of not relying on a single source for determining the level of hazard present.

4-2.2.2 Describe the following chemical and physical properties and their significance in a hazardous materials release:

(a) Boiling point;

Boiling point is the temperature at which the transition from liquid to gaseous occurs. At this temperature, the vapor pressure of a liquid equals the surrounding atmospheric pressure so that the liquid rapidly becomes a vapor. Flammable materials with low boiling points generally present greater problems than those with high boiling points. For example, the boiling point of acetone is 133°F (56°C), and the boiling points for jet fuels range from 400 to 550°F (204 to 288°C).

(b) Concentration;

When dealing with corrosives, the amount of acid or base is compared to the amount of water present. Concentrated acids are not the same as strong acids. It is possible to have a high concentration of a weak acid and a low concentration of a strong acid.

(c) Corrosivity (pH);

Corrosivity is measured by pH, which indicates the concentration of hydrogen ions in the material being tested. 40 CFR 261.22 defines a corrosive material as one with a pH of 2 or less, or of 12.5 or more.[8] Those with a pH of 2 or less are acidic, and those with a pH at or over 12.5 are bases.

(d) Expansion ratio;

The expansion ratio is the amount of gas produced by a given volume of liquid at a given temperature. For instance, liquid propane has an expansion ratio of liquid to gas of 270 to 1, while liquefied natural gas has an expansion ratio of 635 to 1. Obviously, the greater the expansion ratio, the more gas is produced, and the larger the endangered area becomes.

(e) Flammable (explosive) range;

There are several important things to remember about the flammable range of a material. One is the lower limit, and another is the width of the range. For example, gasoline has a lower flammable limit (LFL) of 1.4 percent and an upper flammable limit (UFL) of 7.6 percent. Carbon monoxide, on the other hand, has an LFL of 12.5 percent and a UFL of 74 percent.

(f) Flash point;

It is important to take flash point into account when determining the level of hazard of flammable liquids. Aviation-grade gasoline (100–130) has a flash point of –50°F (–45.5°C), and kerosene has a flash point of 100°F (37.8°C). A kerosene spill that occurs when the temperature is 25°F (–3.8°C) will present less danger than a spill that occurs when the temperature is 110°F (44°C). Gasoline would present a serious hazard even at –25°F (–3.8°C).

(g) Form (solid, liquid, gas);

The form a hazardous material takes plays a significant role in determining not only the measures that will be used to control a spill but often the hazards that it presents. A solid is much easier to contain than a liquid, and liquids that have to be diked or dammed while in a gaseous state generally cannot be contained.

(h) Ignition (autoignition) temperature;

The ignition temperature is the minimum temperature to which a material must be raised before it will ignite; it is also the temperature the ignition source must be. Carbon disulfide has an ignition temperature of 194°F (90°C), while ammonia has an ignition temperature of 1,204°F (651°C). Products with lower ignition temperatures are in greater danger of igniting than those with higher ignition temperatures.

(i) Melting point;

The melting point is the temperature at which a solid becomes a liquid. Materials with low melting points present problems because they easily become liquid and spread more readily. The reverse is also true, however: If the temperature of a liquid can be lowered, the responder may be able to convert it to a solid.

(j) Reactivity;

Reactivity describes a substance's propensity to release energy or undergo change. Some materials are self-reactive or can polymerize. Others undergo violent reactions if they come in contact with other materials. Substances that are air-reactive will ignite or release energy when exposed to air.

Organic peroxides are examples of highly reactive materials. Other examples are corrosives, radioactive materials, oxidizing materials, pyrophoric substances, explosives, and water-reactive materials.

(k) Specific gravity;

Specific gravity is the weight of a solid or liquid compared to an equal volume of water. If a material has a specific gravity greater than 1.0 and it does not dissolve in water, it will sink. If its specific gravity is less than 1.0, it will float on water. This becomes important when conducting some types of damming or booming operations and when dealing with flammable liquids, which generally have specific gravities less than 1.0 and will be spread around when water is applied.

(l) Temperature of product;

The temperature of a product will influence the measures taken to control an incident that involves that product. A product's temperature may also present hazards. An incident involving molten sulfur, for example, raises a different set of concerns than one involving a cryogenic material such as liquefied natural gas.

(m) Toxic products of combustion;

All products of combustion, from cigarette smoke to the smoke from a fire involving pesticides, have some toxic effects. Some materials generate more highly toxic products of combustion than others, and appropriate levels of protective clothing and equipment must be used to counter them.

(n) Vapor density;

Vapor density is the relative density of a vapor compared to air. The vapor density of air is 1.0. If a material has a vapor density higher than 1.0, it is heavier than air and will settle. Toluene, for example, has a vapor density of 3.14, and it will settle and pool in low-lying areas. If a vapor's density is less than 1.0, it is lighter than air and will rise and tend to dissipate.

(o) Vapor pressure; and

Vapor pressure is the pressure exerted on the inside of a closed container by the vapor in the space above the liquid in the container. Products with high vapor pressures have a greater potential to breach their containers when heated, since the pressure increases as the temperature rises. Products with high vapor pressures are more volatile.

Vapor pressure is measured in mm of mercury. The vapor pressure of water is 21 mm mercury, and that of chlorine is 4,800 mm mercury.

(p) Water solubility.

The ability of a substance to form a solution with water can be important when determining control methods. For example, gasoline is insoluble, while anhydrous ammonia is soluble.

4-2.2.3 Match the following chemical and physical terms with their significance and impact on the behavior of the container and/or its contents:

The following examples illustrate the significance and possible impacts of the terms listed. They are not intended to be all-inclusive, and the user should think of additional examples.

(a) Acid, caustic;

Acids or caustics may cause the pressure within a container to rise, particularly if they become contaminated.

(b) Air reactivity;

Materials that are potentially air-reactive can ignite if they are exposed to air, and the potential for container failure due to overpressurization exists.

(c) Catalyst;

Catalysts are used to control the rate of a chemical reaction by either speeding it up or slowing it down. If used improperly, catalysts can speed up a reaction and cause failure of a container that cannot withstand either the pressure or the heat build-up.

(d) Chemical interactions;

The chemical interaction of materials in a container may result in a build-up of heat that, in turn, causes an increase in pressure. The combined materials may be more corrosive than the material the container was originally designed to withstand, and the container may fail.

(e) Compound, mixture;

Compounds have a tendency to break down into their component parts, sometimes in an explosive manner. If the compound nitroglycerine has been contaminated, for example, it can decompose explosively when heated or shocked.

(f) Critical temperatures and pressure;

Critical temperature and pressure relate to the process of liquefying gases. The critical temperature is the minimum temperature at which a gas can be liquefied no matter how much pressure is applied. The critical pressure is the pressure that must be applied to bring a gas to its liquid state.

A gas cannot be liquefied above its critical temperature. The lower the critical temperature, the less pressure is required to bring a gas to its liquid state. If a liquefied gas container exceeds its critical temperature, the liquid will convert instantaneously to gas, which may cause the container to fail violently.

(g) Halogenated hydrocarbon;

A hydrocarbon with halon atoms attached is a halogenated hydrocarbon. Halogenated hydrocarbons are used to produce such things as flammable liquids, combustible liquids, and liquids used as extinguishing agents. They are often more toxic than naturally occurring organic chemicals, and they all decompose into smaller, more harmful elements when exposed to high temperatures for long periods of time.

(h) Inhibitor;

Inhibitors are added to products to control their chemical reaction with other products. For example, an inhibitor is added to monomers, such as ethylene, when they are being shipped to keep them from polymerizing. If the inhibitor is not added or escapes during an incident, the material will begin to polymerize, which creates a very dangerous situation. The final result may be a violent rupture of the container.

(i) Instability;

Materials that decompose spontaneously, polymerize, or otherwise self-react are generally considered unstable. They do not need to mix with other chemicals to react. Organic peroxides, for example, exhibit this characteristic. At low temperatures, they can be fairly stable, but they begin to decompose rapidly when exposed to higher temperatures, and once this reaction has begun, it cannot be stopped. The term *instability* is often used interchangeably with the term *reactivity*.

(j) Organic and inorganic;

Organic materials are derived from materials that are living or were once living, such as plants or decayed products, and they contain chains of two or more carbon atoms. An example of an organic material is methane (CH_4). Inorganic materials lack carbon chains, but they may contain a carbon atom. An example of an inorganic material that contains a carbon atom is carbon dioxide (CO_2).

Knowing whether a material is organic or inorganic can be helpful in choosing the proper instrumentation. Some organic materials are reactive with oxidizers. In addition, inorganic acids are generally stronger than organic acids at the same concentration. Organic acids are generally flammable, however; as a rule, they are also the toxic and explosive acids.

(k) Oxidation ability;

Combining anything with oxygen is called *oxidation*. The rusting of steel is slow oxidation, while the burning of wood is rapid oxidation. The ability of a substance to oxidize is a measure of its propensity to yield oxygen. Oxygen is easily released, especially when heated, and it will accelerate the burning of combustible materials. The more readily a material gives up its oxygen molecule, the greater the hazard it presents. This is the case with oxidizing agents.

(l) pH;

The pH of a substance is a numerical measure of its relative acidity or alkalinity. pH is an accurate determinant of a solution's hydrogen ion concentration.

The level of pH determines the type of container in which a material may be stored or transported. With a low pH of 2 or less, or a high pH of 12.5 or more, special containers are needed. Containers that do not meet the requirements will fail and release the product.

(m) Polymerization;

Polymerization is a chemical reaction in which small molecules combine to form larger molecules. A hazardous polymerization takes place at a rate that releases large amounts of energy, which can cause a fire or explosion or burst a container. Materials that polymerize usually contain inhibitors that delay the reaction. [*See 4-2.2.3(i), Inhibitors.*]

(n) Radioactivity;

The level and type of radioactivity determines the type of packaging a radioactive material needs. The greater the radioactivity, the greater the packaging requirements.

HAZARDOUS MATERIALS RESPONSE HANDBOOK

(o) Salt, nonsalt;

A chemical reaction that combines metal elements with nonmetal elements produces a compound called a *salt*. The type of bonding that occurs is called *ionic bonding*.

(p) Saturated, unsaturated, and aromatic hydrocarbons;

Saturated hydrocarbons (*alkanes*) are those in which the carbon atoms are linked by only single covalent bonds. In saturated hydrocarbons, all the carbon atoms are saturated with hydrogen. Examples include methane (CH_4) and ethane (C_2H_6). Unsaturated hydrocarbons (*alkenes* and *alkynes*) have at least one multiple bond between two carbon atoms somewhere in the molecule. Examples are ethylene (C_2H_4) and acetylene (C_2H_2). Generally speaking, unsaturated hydrocarbons are more active chemically than saturated hydrocarbons. As a result, they are considered more hazardous.

Aromatic hydrocarbons contain the benzene "ring," which is formed by six carbon atoms and contains double bonds. Examples are benzene (C_6H_6) and toluene (C_7H_8).

(q) Solution, slurry;

A solution is a mixture in which all of the ingredients are completely dissolved. It is a homogeneous mixture of the molecules, atoms, or ions of two or more different substances.

A slurry is a pourable mixture of a solid and a liquid.

(r) Strength;

Strength is a term used to describe the concentration of a solution. In corrosives, it refers to the degree of ionization of the acid or base in water. Hydrochloric acid is a strong acid, for example, and acetic acid is a weak acid.

(s) Sublimation;

In sublimation, a substance passes directly from the solid state to the vapor state without passing through the liquid state. Solids such as naphthalene, used in mothballs, are an example. An increase in temperature increases the rate of sublimation. During an incident, a responder should assess the toxicity and flammability of the vapors of any spilled material that sublimes.

The opposite of sublimation is deposition.

(t) Viscosity;

Viscosity, a measure of the thickness of a liquid, determines how easily it flows. Liquids with high viscosity, such as heavy oils, must be heated to increase their fluidity. Liquids that are more viscous tend to flow more slowly, while those that are less viscous will spread more easily. During an incident, liquids that are less viscous are likely to flow away from a leaking container, expanding the endangered area.

(u) Volatility;

Volatility describes the ease with which a liquid or solid can pass into the vapor state. The higher a material's volatility, the greater its rate of evaporation. Vapor pressure is a measure of a liquid's propensity to evaporate. Thus, the higher a liquid's vapor pressure, the more volatile it is. During an incident, a volatile material will disperse in air and expand the endangered area.

(v) Water miscible, immiscible; and

The term *miscible* refers to the tendency or ability of two or more liquids to form a uniform blend, or to dissolve in each other. Liquids may be totally miscible, partially miscible, or not at all miscible.

(w) Water reactivity.

Water reactivity describes the sensitivity of a material to water without the addition of heat or confinement. The more sensitive materials release heat or flammable or toxic gases. Some materials even react explosively when they are exposed to water. Examples of water-reactive substances are sulfuric acid and sodium and aluminum chloride.

4-2.2.4 Given various hazardous materials and appropriate reference materials, identify the signs and symptoms of exposure to each material and the target organ effects of exposure to that material.

The responder at this level should be able to use various references to determine what effect various chemicals have on target organs. OSHA has categorized the possible health effects on target organs in a table that includes examples of signs and symptoms and notes the chemicals that may have caused the effects.

4-2.3 **Describing the Condition of the Container Involved in the Incident.** The hazardous materials technician shall, given simulated facility and transportation container damage, describe the damage found using one of the following terms:

(a) Undamaged, no product release;
(b) Damaged, no product release;
(c) Damaged, product release; and
(d) Undamaged, product release.

These terms allow the responder to use standard terminology when assessing containers to determine how badly they have been damaged.

4-2.3.1 Given examples of the following containers, identify the basic design and construction features of each bulk packaging and storage vessel:

(a) Fixed tanks, storage tanks;

According to Orville Sly, Jr., writing in the *Fire Protection Handbook:* "The thickness of the metal used in tank construction is based not only on the strength required to hold the weight of the liquid, but also on an added allowance for corrosion. When intended for storing corrosive liquids, the specifications for the thickness of the tank shell are increased to provide additional metal and allow for the expected service life of the tank. In some cases, special tank linings are used to reduce corrosion. . .

All aboveground storage tanks should be built of steel or concrete, unless the character of the liquid necessitates the use of other materials. Both steel and concrete tanks resist heat from exposure fires. . .

The nominal thickness of shell plates, including shell extensions for floating roof tanks, must be no less than those values given in Table 4.1.

Concrete tanks require special engineering, and unlined tanks should be used only for the storage of liquids with a specific gravity of 40° API or heavier.'"[9]

For additional information see NFPA 30, *Flammable and Combustible Liquids Code*[10], which provides specifications for fixed containers designed to store flammable and combustible liquids.

Table 4.1   Shell Plate Thicknesses

Nominal Tank Diameter in Feet*	Nominal Thickness in Inches†
Smaller than 50	$3/16$
50 to, but not including, 120	$1/4$
120 to 200, inclusive	$5/16$
Over 200	$3/8$

* 1 ft = 0.304 m
† 1 in. = 25.4 mm
(Source: *Fire Protection Handbook*, 17th edition.)

(b) Tank containers (intermodal portable tanks);

In the *Fire Protection Handbook*, Charles Wright and Roger Fitch note that:
"Tank containers consist of a single, metal tank mounted inside a sturdy, metal supporting frame. This unique frame structure, built to international standards, makes tank containers multi-modal (intermodal). . .

The tank container tank is generally built as a cylinder enclosed at the ends by heads. Other tank shapes . . . and configurations . . . exist but they are rare . . .

Ninety percent or more of the tanks are built of stainless steel, and the rest are constructed of mild steel . . . Aluminum and magnesium alloy tanks are available, but they cannot be used in water transport mode.

Minimum head and shell thickness are measured in terms of 'equivalent thickness in mild steel' after forming . . . For regulated materials, the minimum thickness is $3/8$ in. (9.5 mm). For stainless steel tanks, the minimum thickness for nonregulated materials is slightly less than $1/8$ in. (3.2 mm). For regulated commodities, the minimum thickness is just under $3/16$ in. (4.8 mm).

Most tanks are built according to the pressure-vessel standards of the American Society of Mechanical Engineers (ASME), and the welds are x-rayed."[11]

See the 17th edition of NFPA's *Fire Protection Handbook* for more information on tank containers.

(c) Piping;

Pipelines are constructed according to standards established by the American Society of Mechanical Engineers (ASME).

(d) Tank cars; and

Roger Fitch and Charles Wright also note in the *Fire Protection Handbook* that:

"Tank cars generally consist of a single tank mounted on the car structure, although there are exceptions, such as a multi-unit tank car tank commonly known as the 'ton container,' the high pressure tube car, and tank cars with multiple compartments.

Carbon steel is used in over 90 percent of the tank car tanks in use, with aluminum making up most of the remainder... The plate thickness of materials used to construct tank car tanks is specified by regulations. (*See Table 4.2*)."[12]

For additional information about tank car construction, see the 17th edition of NFPA's *Fire Protection Handbook*.

(e) Cargo tanks (tank trucks and trailers).

Cargo tank specifications are found in 49 CFR, Part 178.[13] The U.S. Department of Transportation has established five classifications for cargo tanks: MC-306 (DOT- 406), MC-307 (DOT-407), MC-312 (DOT-412), MC-331, and MC-338. The classifications in parentheses are new classifications that will become effective October 1, 1993. There will be some changes to the specifications, but both classifications of cargo tanks will have the same general appearance.

4-2.3.1.1 Given DOT specification markings for nonbulk or bulk packaging (including tank cars, tank containers, and cargo tanks) and the appropriate reference guide, identify the design and construction of the packaging and identify examples of the likely materials found in the packaging.

See Supplement 9, "Packaging for Transporting Hazardous and Non-Hazardous Materials."

4-2.3.2 Given examples of the following containers, identify the closures found on each container by name and match the purpose of each closure to the name of the closure:

(a) Cylinders;

See Supplement 9.

Table 4.2   Minimum Tank and Jacket Plate Thickness

Minimum Plate Thickness After Forming	Common Use of Plate Thickness
Steel	
11 gage (approximately $\frac{1}{8}$ inch) also aluminum	Jacket of insulated tank cars; or jacket for thermally protected cars.
$\frac{7}{16}$ in.	Tank for nonpressure tank cars; outer tank for nonpressure tank within a tank; or shell portion of outer tank for cryogenic liquid tank cars.
$\frac{1}{2}$ in.	Head puncture resistance (head shield); or head portion of outer tank for cryogenic liquid tank cars.
$\frac{9}{16}$ in.	Tank for steel pressure tank cars with tank test pressures of 200 psi and below.
$\frac{11}{16}$ in.	Tank for steel pressure tank cars with tank test pressure of 300 psi and greater.
$\frac{3}{4}$ in.	Tank for steel pressure tank cars in chlorine service.
Aluminum	
$\frac{1}{2}$ in.	Tank for nonpressure aluminum tank cars.
$\frac{5}{8}$ in.	Tank for aluminum pressure tank cars.

Notes:
1. If high tensile strength steels are used, the plate thickness for pressure tank car tanks may be reduced, but in no case should that thickness be less than $\frac{1}{2}$ in.
2. The plate thickness for nonpressure steel tank cars with expansion domes is a function of where the plate is used in the tank and the diameter of the tank. Thickness ranges from $\frac{1}{4}$ to $\frac{1}{2}$ in. For tank cars built after 1969, a minimum plate thickness is $\frac{7}{16}$ in., except for tank cars with a diameter of 112 to 122 in. where the thickness is $\frac{1}{2}$ in.
3. For SI units: 1 in. = 2.54 cm; 1 psi = 6.9 kPa.
(Source: *Fire Protection Handbook*, 17th edition.)

(b) Drums;

See Supplement 9.

(c) Fixed tanks, storage tanks;

The flow of materials through the piping of fixed storage tanks is controlled by valves. These tanks also have manholes that provide access to the tank's interior. These are often fastened by a "dogging" type of closure.

(d) Tank containers, intermodal portable tanks;

See Supplement 9.

(e) Piping;

Valves are often attached to piping so that the line can be opened or closed. Often, these valves are remotely controlled. Sensors that indicate either an increase or a decrease in flow may be used to alert technicians to closing valves.

(f) Tank cars; and

See Supplement 9.

(g) Cargo tanks (tank trucks and trailers).

See Supplement 9.

4-2.3.3 Identify how a liquid pipeline may carry different products.

Many different products can be transported through a single pipeline. This is done by means of a "pig" that isolates a product that may have just been transported from a different product that is to follow.

4-2.3.4 Given an example of a ruptured pipeline, identify the following:

(a) Ownership of the line;
(b) Type of product in the line;
(c) Procedures for checking for gas migration; and
(d) Procedure for shutting down the line or controlling the leak.

The ownership of a line may be indicated with a pipeline marker. The responder can contact the company identified on the marker for information on the type of product in the line and for help in shutting down the line. Responders should also consider acquiring pipeline maps for their area so they can plan for pipeline incidents.

4-2.3.5 Given an example of a domestic gas line break and the readings from a combustible gas indicator, determine the area of evacuation.

The intent here is to have the responder demonstrate that he or she can interpret the readings from a combustible gas indicator and establish an appropriate evacuation area.

4-2.3.6 Identify the method for determining the pressure in bulk packaging or facility containers using both a pressure gauge and the temperature of the contents.

4-2.3.7 Identify the method for determining the amount of lading in bulk packaging or facility containers.

The amount of lading can be determined from shipping papers, container specification markings, facility documents, and the like.

4-2.3.8* Identify the types of damage that a container could incur.

A-4-2.3.8 Some of the types of damage that containers could incur include:

The types of damage listed below would be most common in tank cars and tank trucks.

(a) *Cracks.* A crack is a narrow split or break in the container metal that may penetrate through the metal of the container.

A cracked pressurized container should be considered to be critically damaged since it is generally difficult to determine the depth of a crack and thus difficult to establish the safest course of action to take. Cracks in the base metal, no matter how small, warrant offloading the container. If the crack is associated with a dent, the tank car should not be moved until it has been offloaded.

(b) *Scores.* A score is a reduction in the thickness of the container shell. It is an indentation in the container made by a relatively blunt object. A score is characterized by the relocation of the container or weld metal in such a way that the metal is pushed aside along the track of contact with the blunt object.

Scores that are not accompanied by a dent are not critical. Nor is a score critical if it crosses a welded seam and does not cut into the heat-affected area of the weld. If a score does cut into the heat-affected area of a welded seam, however, the situation is critical and the product should be offloaded.

(c) *Gouges.* A gouge is a reduction in the thickness of the container. It is an indentation in the shell made by a sharp, chisel-like object. A gouge is characterized by the cutting and complete removal of the container or weld metal along the track of contact.

Gouges are similar to scores, and the same considerations apply.

(d) *Dents.* A dent is a deformation of the container metal. It is caused by impact with a relatively blunt object. With a sharp radius, there is the possibility of cracking.

Large dents are not serious unless gouges are also present. Longitudinal dents along the long axis of a tank are the most serious. Dents with a radius of less than 4 in. (10 cm) that occur in tank cars built before 1966 are critical, and tanks that are so damaged should be offloaded without being moved. The situation is also critical if the tank car was built in 1966 or later and the radius of the dent is less than 2 in. (5 cm); in such cases, the tank car should be offloaded without being moved.

4-2.3.8.1 Given examples of tank car damage, identify the type of damage in each example by name.

4-2.3.8.2 Identify the basic design and construction features of the following nonbulk packages used to store or transport hazardous materials:

(a) Carboys;
(b) Cylinders; and
(c) Drums.

See Supplement 9.

4-2.4 **Predicting Behavior of Containers and Contents Where Multiple Materials Are Involved.** The hazardous materials technician shall, given examples of both facility and transportation incidents involving multiple hazardous materials, predict the likely behavior of the contents in each case. The hazardous materials technician shall be able to:

This competency requires the hazardous materials technician to be able to predict the behavior of containers and contents when multiple hazardous materials are involved. This is a clear distinction between the operational level responder and the hazardous materials technician.

4-2.4.1 Identify at least three resources available that indicate the effects of mixing various chemicals.

NFPA's *Hazardous Chemical Reactions*; Bretherick's *Handbook of Reactive Chemical Hazards*; and Irving Sax' *Hazardous Chemicals Desk Reference* are examples of references that provide information about the mixture of various chemicals.[14, 15, 16] This is by no means a complete listing of this type of reference.

4-2.4.2 Describe the heat transfer processes that occur as a result of a cryogenic liquid spill.

Because cryogenic liquids are kept at temperatures below –150°F (–100°C), cryogenic liquid spills will vaporize rapidly when exposed to the higher ambient temperatures of the atmosphere outside the tank. Expansion ratios for common cryogenics range from 560 to 1,445 to 1.

4-2.4.3 Identify the impact of the following fire and safety features on the behavior of the products during an incident at a bulk storage facility:

(a) Tank spacing;

Adequate tank spacing will minimize the hazard to uninvolved tanks. See NFPA 30, *Flammable and Combustible Liquids Code*, for information about tank spacing requirements.[17]

(b) Product spillage and control (impoundment and diking);

Dikes and other impoundment features are designed to contain spilled product and minimize the exposure to adjoining tanks. See NFPA 30, *Flammable and Combustible Liquids Code*, for information about diking requirements.[18]

(c) Tank venting and flaring systems;

Tank vent and flaring systems allow a tank to release pressure, thus minimizing the threat of a rupture and an increase in the size of the endangered area. See NFPA 30, *Flammable and Combustible Liquids Code*, for information about venting requirements.[19]

(d) Transfer operations;

Transferring product from one tank to another minimizes the danger to surrounding containers.

(e) Monitoring and detection systems; and

Monitoring and detection systems permit early notification of potential problems and allow responders to initiate control actions while an incident is still relatively small, thereby limiting the threat to other containers.

(f) Fire protection systems.

Fire protection systems allow responders to apply fire extinguishing agents sooner and to control an incident in its initial stages, thus reducing the threat to adjoining containers.

4-2.5 **Estimating the Size of an Endangered Area.** The hazardous materials technician shall, given various facility and transportation hazardous materials incidents, estimate the size, shape, and concentrations associated with the materials involved in the incident using computer modeling, monitoring equipment, or specialists in this field. The hazardous materials technician shall be able to:

4-2.5.1 Identify local resources for dispersion pattern prediction and modeling including computers, monitoring equipment, or specialists in the field.

The responder should be able to identify resources that will help him or her predict dispersion patterns. These resources include the weather service; industrial facilities; colleges or universities; county, state, or federal agencies, such as health departments; environmental protection agencies; and the U.S. Coast Guard, among others. Responders must be able to predict dispersion patterns to determine which areas are likely to become endangered by a spill.

4-2.5.2 Identify the steps for determining the extent of physical, health, and safety hazards within the endangered area of a hazardous materials incident given the concentrations of the released material.

Once the responder has determined the concentrations of the materials that have been released, he or she must determine the acceptable exposure limits for those materials. This can be done using resources that list acceptable exposure values. A number of texts and government publications provide this type of information.

4-2.5.2.1 Match the following toxological terms and exposure values with their significance in predicting the extent of health hazards in a hazardous materials incident:

The reader will notice that the various exposure values in the section that follows are similar. This is due to the fact that these values are established by different organizations, including the American Council of Governmental and Industrial Hygienists (ACGIH), the U.S. Occupational Safety and Health Administration (OSHA), and the National Institute of Occupational Safety and Health (NIOSH). Responders should be familiar with all these terms because different references may use different values, and responders must understand the differences and similarities so that they can make the appropriate comparisons.

(a) Immediately dangerous to life and health value (IDLH);

This is the maximum level to which a healthy worker can be exposed for 30 minutes and escape without suffering irreversible health effects or impairment. If at all possible, exposure to this level should be avoided. If that is not possible, responders should wear Level A or Level B protection with positive pressure self-contained breathing apparatus or a positive pressure supplied-air respirator with an auxiliary escape system. This limit is established by OSHA and NIOSH.

(b) Lethal concentrations (LC_{50});

The LC is the median lethal concentration of a hazardous material. It is defined as the concentration of a material in air that, on the basis of laboratory tests (inhalation route), is expected to kill 50 percent of a group of test animals when administered in a specific time period.

(c) Lethal dose (LD_{50});

The LD of a substance is a single dose that will cause the death of 50 percent of a group of test animals exposed to it by any route other than inhalation.

(d) Permissible exposure limit (PEL);

This is a term OSHA uses in its health standards covering exposures to hazardous chemicals. It is similar to the TLV/TWA established by the ACGIH.

PEL, which generally relates to legally enforceable TLV limits, is the maximum concentration, averaged over 8 hours, to which 95 percent of healthy adults can be repeatedly exposed for 8 hours per day, 40 hours per week.

(e) Threshold limit value ceiling (TLV-C);

This is the maximum concentration to which a healthy adult can be exposed without risk of injury. It is comparable to the IDLH, and exposures to higher concentrations should not occur.

(f) Threshold limit value short-term exposure limit (TLV-STEL);

This is the maximum average concentration, averaged over a 15-minute period, to which healthy adults can be safely exposed for up to 15 minutes continuously. Exposure should not occur more than four times a day with at least 1 hour between exposures.

(g) Threshold limit value time-weighted average (TLV-TWA);

This is the maximum concentration, averaged over 8 hours, to which a healthy adult can be repeatedly exposed for 8 hours per day, 40 hours per week.

(h) Parts per million (ppm), parts per billion (ppb); and

The values used to establish the exposure limits above are quantified in parts per million or parts per billion. A good reference to remember is that 1 percent equals 10,000 ppm, 1 percent equals 1,000 ppb. So if you obtain a reading from a sampling instrument of 0.5 percent, that is equivalent to 5,000 ppm, or 500 ppb. If you then determine the TLV is 7,500 ppm, you can relate the reading from the instrument to determine the degree of hazard.

(i) Emergency response planning guide value (ERPG).

4-2.5.2.2 Match the following terms associated with radioactive materials with their significance in predicting the extent of health hazards in a hazardous materials incident:

(a) Alpha radiation;

Alpha radiation involves the alpha particle, a positively charged particle emitted by some radioactive materials. It is less penetrating than beta and gamma radiation and is not considered dangerous unless ingested. If ingested, alpha radiation will attack internal organs.

(b) Beta radiation;

Beta radiation involves the beta particle, which is much smaller but more penetrating than the alpha particle. Beta particles can damage skin tissue, and they can damage internal organs if they enter the body. Full protective clothing, including positive pressure self-contained breathing apparatus, will protect against most beta radiation.

(c) Gamma radiation;

Gamma radiation is especially harmful since it has great penetrating power. Gamma rays are a form of ionizing radiation with high energy that travels at the speed of light. It can cause skin burns and can severely injure internal organs. Protective clothing is inadequate in preventing gamma radiation from harming the body.

(d) Half-life; and

Half-life is a measure of the rate of decay of a radioactive material. It indicates the time needed for one half of a given amount of radioactive material to change to another nuclear form or element.

(e) Time, distance, and shielding.

Time, distance, and shielding are methods of protecting oneself from harmful exposures to radiation. The shorter the time of exposure, the lower the dosage. The farther the distance from the exposure, the lower the amount of radiation; this is calculated as inverse to the square of the distance. Shielding refers to blocking radiation by using varying thicknesses of different materials.

4-2.5.3 Identify the method for estimating the outcomes within an endangered area of a hazardous materials incident.

An estimate is a series of predictions that attempts to provide an overall picture of potential outcomes. Responders must assess the information gathered during analysis and predict the outcome based on that assessment.

According to the National Fire Academy's (NFA) *Initial Response to Hazardous Materials Incidents, Course II: Concept and Implementation*, it is necessary to break an incident into three components: the product, the container, and the environment. Each of these can then be broken into three subgroups: damage, hazard, and vulnerability and risk. In addition, incidents may have three elements that may occur separately or in conjunction with one another: a spill, a leak, or a fire. The estimate identifies the relationships between the three components of an incident and the three elements of an incident.

The NFA course states, "An estimate is made by analyzing the physical, cognitive, and technical information that has been gathered. Then, by breaking the incident into the components dealing with product, container, environment and their respective sub-groups, a conclusion can be drawn[,. . . a] conclusion with some measure of quantifiable accuracy that suggests what the full impact(s) of the relationships will be." [20]

It is important to understand that this analysis continues throughout an incident. As new information is gathered, old estimates should be verified or new estimates made. Predictions, which should be based on worst-case scenarios, will allow the responder to develop an overall estimate of the incident's potential outcomes.

Incident commanders should keep in mind that the safety of both emergency personnel and the public is their primary objective. There may very well be times when the most prudent action is no action and the establishment of an appropriate evacuation area is the best possible course to take.

4-3 Competencies — Planning the Response.

4-3.1 Identifying Response Objectives. The hazardous materials technician shall, given simulated facility and transportation problems, describe the response objectives for each problem. The hazardous materials technician shall be able to:

4-3.1.1 Describe the steps for determining response objectives (defensive, offensive, nonintervention) given an analysis of a hazardous materials incident.

In 3-3.1, the operational level responder is required to determine the appropriate response objectives for responders at that level. The options available are either defensive or nonintervention. The response objectives for the hazardous materials technician (HMT) may also include offensive operations.

In the *Fire Protection Handbook*, Charles Wright notes that:
"The first task is to determine the response objectives (strategy) based on the estimated outcomes. The response objectives, based on the stage of incident, are the strategic goals for stopping the event now occurring or keeping future events from occurring. Two basic principles apply to these decisions.

1. You cannot influence events that have already happened or change the outcomes of those events; and
2. The earlier that the event sequence can be interrupted, the more acceptable the loss.

Response objectives can include modifying the stress being applied to the container, changing the size of the breach, changing the quantity being released, changing the size of the endangered area, reducing exposures, and reducing the level of harm. These objectives can be met either defensively, offensively, or through non-intervention. Obviously, the potential loss will be reduced if actions can be safely taken that will terminate the incident in a shorter period time (*see Figure 4.12*)."[21]

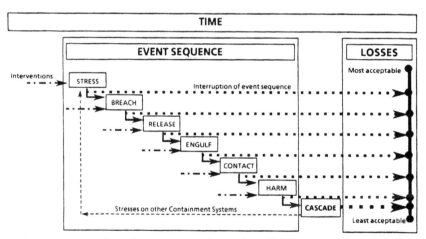

Figure 4.12   A chart depicting the relationship of the events sequence and the losses. (Source: Fire Protection Handbook, 17th edition.)

4-3.2   Identifying Potential Action Options. The hazardous materials technician shall, given simulated facility and transportation hazardous materials incidents, identify the possible action options (defensive, offensive, and nonintervention) by response objective for each problem. The hazardous materials technician shall be able to:

4-3.2.1 Identify the possible action options to accomplish a given response objective.

Because they can conduct offensive operations, HMTs can take actions that are not available to the responder at the operational level. Figure 4.13 shows options available to the HMT to accomplish response objectives.

Response Objective Analysis Form				Containment System ID.	
				Material	

Event Sequence

Stress	Breach	Release	Engulf	Contact	Harm

Response Objectives

Change Applied Stresses	Change Breach Size	Change Quantity Released	Change Size of Danger Zone	Change Exposures Contacted	Change Severity of Harm

Sample Response Options

Move stressor Move stressed system Shield stressed system	Chill contents Limit Stress levels Activate venting devices Mechanical Repair	Change container position Minimize pressure differential Cap off breach Remove contents	Barriers Dikes and Dams Adsorbents Absorbents Diluents Reactants Overpack	Provide Sheltering Begin evacuation Personal Protective Equipment	Rinse off contaminant Increase distance from source Provide shielding Provide Prompt medical attention

(Adapted from Ludwig Benner's *Hazardous Materials Emergencies*, 1978.)

Figure 4.13 Response Objective Analysis Worksheet. This worksheet is used for identifying response options in a hazardous materials incident by response objective. (Source: Fire Protection Handbook, 17th edition.)

The responder should note that many of the options available can be used either offensively or defensively. For example, the size of the endangered area may be changed defensively by placing barriers around the endangered area to contain a spill or offensively by entering the hot zone and plugging the leak. It should also be noted that the HMT may choose defensive operations as the most prudent method of dealing with an incident.

4-3.2.2 Identify the purpose and the procedures, equipment, and safety precautions for each of the following control techniques:

(a) Adsorption;

See 6-4.2.1 of NFPA 471.

(b) Neutralization;

See 6-4.2.6 of NFPA 471.

(c) Overpacking; and

See 6-4.1.5 of NFPA 471.

(d) Patch and plug.

See 6-4.1.6 of NFPA 471.

4-3.3 **Selecting Personal Protective Equipment.** The hazardous materials technician shall, given situations with known and unknown hazardous materials, determine the appropriate personal protective equipment for the action options specified in the plan of action in each situation. The hazardous materials technician shall be able to:

4-3.3.1 Identify the four levels of chemical protection (EPA/NIOSH) and match both the equipment required for each level and the conditions under which each level is used.

See Section 5-5 and Table 5.1 in NFPA 471.

4-3.3.2 Identify the factors to be considered in selecting the proper respiratory protection for a specified action option.

The type of respiratory protection required depends on many factors. A key factor is the level of protective clothing necessary to protect the HMT conducting a specific action. Level A protection limits the user to positive pressure self-contained breathing apparatus or positive pressure supplied-air respirators. The responder should always use the highest level of protection until the levels of concentration have been determined.

See Section 1-2, "Respiratory Protection," for additional information.

4-3.3.2.1 Describe the advantages, limitations, and proper use of the following types of respiratory protection at hazardous materials incidents:

(a) Air purifying respirator; and

Air purifying respirators (APR), which use a filter or sorbent to remove airborne contaminants, are designed for use only in atmospheres that contain enough oxygen to sustain life. They should be used only when the hazard and concentration of the hazardous material is known, and the level must be within the limitations of the filter being used. The life of the filtering cartridges depends on the concentrations present, the type of filter material being used, and the breathing volume of the user.

APRs should not be used by initial responders to an emergency involving hazardous materials.

(b) Supplied air respirator (air line respirator).

Positive pressure air line respirators, which are lighter than SCBA, allow the user an unlimited supply of air. However, the air line supply unit restricts the user's travel distance to a maximum of 300 ft (92 m), and it must be kept clear of obstructions while the user is negotiating the incident scene. Under certain conditions, though, it is an ideal method of supplying air to the emergency worker.

It should be noted that, in some instances (when working in IDLH atmospheres, for example), the person wearing an air line respirator must have an emergency egress system available.

4-3.3.2.2 Identify the process for selecting the proper respiratory protection at hazardous materials incidents.

To determine the appropriate respiratory protection to wear during a specific incident, the responder must establish the name of the chemicals involved in the incident, the concentration of those chemicals, and the hazards they present. The responder must then determine the types of exposure that he or she would face based on the action he or she chooses to take.

The minimum level of protection OSHA requires for emergency response is positive pressure self-contained breathing apparatus. This should be worn at least until the material has been identified and its concentrations determined.

4-3.3.2.3 Identify the operational components of the air purifying respirators and supplied air respirators by name and match the function to the component.

The intent of this competency is to allow the HMT to demonstrate his or her knowledge of SARs, APRs, and their components.

4-3.3.3 Identify the factors to be considered in selecting the proper chemical-protective clothing for a specified action option.

4-3.3.3.1 Match the following terms with their definitions and explain their impact and significance on the selection of chemical-protective clothing:

(a) Degradation;

Degradation of chemical-protective clothing (CPC) can be either chemical or physical. The result of degradation is an increased likelihood that a hazardous material will permeate and penetrate the garments, thus endangering the health of the responder.

Chemical degradation can be minimized by avoiding unnecessary contact with chemicals and by undergoing effective decontamination procedures. It is important that the garments a responder wears are chosen based on their compatibility with the chemicals involved in an incident and that they have breakthrough times consistent with their expected use.

Protective clothing can also degrade physically, such as might occur when the garment rubs against a rough surface. CPC wearers should recognize the physical limitations of their garments and make every effort to avoid circumstances that may cause the material to be damaged physically.

NFPA 1991, 1992, and 1993[22] contain criteria for abrasion or tear testing and for manufacturers' certification of CPC. When purchasing CPC, the responder should ascertain whether the garments can be certified to the appropriate NFPA CPC standard.

See Supplement 10 for additional information.

(b) Penetration; and

Penetration is the movement of a material through a suit's closures; these include zippers, buttonholes, seams, flaps, and other design features of CPC. Hazardous materials may also penetrate CPC through cracks or tears in the suit's fabric.

Protection against penetration is vital. A regular and routine program of inspection can help uncover conditions that could lead to penetration. CPC must also be properly stored and regularly maintained and tested in order to ensure that it can still provide the proper level of protection. NFPA 1991, 1992, and 1993 provide criteria for testing CPC for penetration resistance and for manufacturer's certification of CPC.[23]

(c) Permeation.

Different fabrics have different resistance levels to chemical permeation, and all will absorb chemicals over a period of time. NFPA 1991, 1992, and 1993 provide guidelines for manufacturer permeation testing and certification.

NFPA 1991, *Standard on Vapor-Protective Suits for Hazardous Chemical Emergencies,* requires the manufacturer to provide documentation on a garment's permeation resistance for *three* hours against at least the following chemicals:[24]

- Acetone*
- Acetonitrile
- Anhydrous ammonia
- Carbon disulfide
- Chlorine
- Dichloromethane
- Diethyl amine*
- Dimethyl formamide
- Ethyl acetate
- Hexane*
- Methanol
- Nitrobenzene
- Sodium hydroxide*
- Sulfuric acid*
- Tetrachloroethylene*
- Tetrahydrofuran*
- Toluene.*

NFPA 1992, *Standard on Liquid Splash-Protective Suits for Hazardous Chemical Emergencies,* and NFPA 1993, *Standard on Support Function Protective Garments for Hazardous Chemical Operations,* require that the manufacturer provide documentation on a garment's permeation resistance for one hour, to at least the chemicals noted with an asterisk in the list above.[25]

Before buying CPC, the HMT should make sure that it meets, and has been certified as meeting, the appropriate standard. When choosing CPC for use at an incident, the HMT must be sure that the garment is compatible with the type of material to which it is going to be exposed. In any event, the wearer is advised to use extreme caution. Available data does not cover every situation the responder may encounter.

See Supplement 10 for more information.

4-3.3.3.2 Identify at least three indications of material degradation of chemical-protective clothing.

Indications of material degradation are stiffness or excess pliability, tears, cuts or abrasions, and damage to zippers or other closures. This list is not all-inclusive, and users should check manufacturers' recommendations for the inspection of CPC.

4-3.3.3.3* Identify the three types of vapor-protective and splash-protective clothing and describe the advantages and disadvantages of each type.

A-4-3.3.3.3 Refer to the Chemical Manufacturers Association and Association of American Railroads Hazardous Materials Technical Bulletin *Recommended Terms for Personal Protective Equipment*, issued October 1985.

4-3.3.3.4 Identify the relative advantages and disadvantages of: heat exchange units, air-cooled jackets, water-cooled jackets, and ice vests used for the cooling of personnel in chemical-protective clothing.

Wearing CPC can cause wearers to suffer increased heat stress. Thus, it is important for responders to be monitored closely while working in CPC.

Some CPC garments have temperature control features. Some have air-cooling systems that require an air line and large quantities of breathable air. Others incorporate water-cooling systems that require an ice supply or refrigeration units and a pump. This adds additional weight and bulk to the suit. Users should conduct a thorough evaluation of such units to make sure they are appropriate for the intended use.

There are also vests that can hold coolant packs. This requires a supply of the frozen coolant packs or an ice source at the scene. The vests add additional weight to the CPC.

4-3.3.3.5 Identify the process for selecting the proper protective clothing at hazardous materials incidents.

The HMT should determine the appropriate level of protection based on the criteria established by the U.S. Environmental Protection Agency (*see Section 5-5 and Table 5.1 of NFPA 471*). The chemical-protective clothing chosen must be compatible with the chemicals to which it will be exposed. The HMT must determine whether the breakthrough times of the chosen garment will allow him or her enough time to enter the contaminated area safely, do the necessary work, leave the area, and undergo decontamination. In some cases, layering materials may provide increased protection. It is important to remember that no single garment will protect a responder against all chemicals. The HMT should know the manufacturer's recommendations regarding the use of its CPC.

4-3.3.3.6 Given examples of various hazardous materials, determine the appropriate protective clothing construction materials for a given action option using chemical compatibility charts.

This competency allows the HMT to demonstrate his or her ability to interpret chemical compatibility charts to determine the appropriate type of CPC to use for a chosen action option.

4-3.3.3.7 Identify the physical and psychological stresses that can affect users of specialized protective clothing.

Specialized protective clothing, particularly fully encapsulating garments, increases the stress a responder may feel when responding to a hazardous materials incident. Persons wearing CPC usually experience a loss of dexterity and mobility; the higher the level of protection, the greater this loss will be. Their visibility is also restricted, and their communications are affected. In addition, wearing CPC increases the likelihood of heat stress and heat exhaustion. Reductions in dexterity, mobility, visibility, and communication, in turn, create additional physical and mental stresses.

It is important that the HMT be aware of these additional stresses and receive adequate rest and rehabilitation by way of compensation. It has been found that drinking fluids such as water before one dons CPC will reduce some of the effects of excess heat.

4-3.4 Developing Appropriate Decontamination Procedures. The hazardous materials technician shall, given a simulated hazardous materials incident, select an appropriate decontamination procedure and determine the equipment required to implement that procedure. The hazardous materials technician shall be able to:

4-3.4.1 Identify the advantages and limitations and describe an example where each of the following decontamination methods would be used:

There are basically two ways to decontaminate something: physically and chemically. Physical methods manually separate the chemical from the material being decontaminated by scrubbing or washing the material, or both. Physical decontamination is often easier than chemical decontamination, but it may not completely remove all the contaminants.

Chemical methods involve changing one chemical into another or into a form that will facilitate its removal. Unfortunately, the chemical process involved could introduce other hazards, and the HMT should be aware of this. Care must be taken to collect all the contamination that has been removed by either method and to dispose of it properly.

(a) Absorption;

Figure 4.14 Decontamination equipment. (Courtesy of Mike Callan.)

Absorption is the process by which materials hold liquids. Many types of commercial absorbents are available. Sand or soil can also be used for this purpose, although they are more suited for decontaminating equipment or the area surrounding a spill than they are for decontaminating personnel. Absorbents are often readily available, but they must be disposed of properly, and they may retain the properties of the material that they absorb.

(b) Adsorption;

Adsorption is a chemical method of decontamination involving the interaction of a hazardous liquid and a solid sorbent surface. Examples of adsorbents are activated charcoal, silica or aluminum gel, fuller's earth, and other clays. Adsorption produces heat and can cause spontaneous combustion. Adsorbents must also be disposed of properly.

(c) Chemical and physical degradation;

Chemical degradation is the natural breakdown of the contaminants as they age. Physical degradation is the reduction of contamination over a period of time because of physical wear. An example of chemical degradation is the evaporation of a flammable liquid spill. The decontamination of an oil spill on a beach because of manual (pressure washer) or natural (wave action) actions is an example of physical degradation. Either of the two methods has limitations depending on such factors as the location of the spill and the toxicity of the material. In some cases, however, they are the most practical methods.

(d) Dilution;

Dilution, which simply reduces the concentration of a contaminant, is best used on materials that are soluble or miscible in water, such as chlorine and ammonia. An advantage of dilution is that solutes, especially water, are generally available in large quantities. A disadvantage is that the run-off must be collected and disposed of.

(e) Disposal;

Disposal is the direct removal of a contaminant from a carrier. An example is the removal of a contaminated object from a piece of equipment. This type of decontamination may not entirely remove all contamination.

(f) Neutralization;

Neutralizers alter a contaminant chemically so that the resulting chemical is harmless. For example, the addition of soda ash to an acidic solution can increase the pH, making it a chemically harmless substance. Many neutralizing chemicals present hazards of their own, however, and should only be used by HMTs who are fully aware of the consequences. One advantage of neutralizers is that by rendering the remaining material harmless, they reduce the problem of disposal.

(g) Solidification;

Commercial products are available that cause certain liquids to solidify. One advantage of solidification is that it allows responders to confine a small spill relatively quickly. As with other decontaminants, however, the resulting solid must be disposed of properly when the incident is over.

(h) Evaporation;

In some cases, responders may allow a hazardous material simply to evaporate, particularly if the vapors do not present a hazard. This may be the case with a small spill of gasoline, for example, as long as it does not present a vapor problem. Evaporation is an easy operation and requires minimal personnel. It is not as effective on porous surfaces as it is on nonporous surfaces, however, and it could take quite a while, depending on the quantity of the chemical involved.

(i) Washing; and

A very effective decontamination process for many materials involves washing the contaminated person, building, or equipment. Materials that are not soluble in water, such as oil-based contaminants, can be washed with detergent solutions. Washing equipment, protective clothing, and personnel is one of the easiest ways to decontaminate them. In many cases, it is necessary to collect and properly dispose of the run-off.

(j) Vacuuming.

Vacuuming allows for the collection of materials, either liquid or solid, into containers. It is important that the equipment being used is appropriate for the material being vacuumed. If it is corrosive or flammable, for example, specialized equipment is needed.

4-3.4.2 Identify the sources of technical information for selecting appropriate decontamination procedures and identify how to contact those sources in an emergency.

Among the sources of technical information about decontamination are CHEMTREC/CANUTEC, MSDSs, product manufacturers, the National Response Center, and local or regional poison control centers.

4-3.5 Developing a Plan of Action. The hazardous materials technician shall, given simulated hazardous materials incidents in facility and transportation settings, develop a plan of action, including safety considerations. The plan shall be consistent with the local emergency response plan and the organization's standard operating procedures and be within the capability of available personnel, personal protective equipment, and control equipment for that incident. The hazardous materials technician shall be able to:

According to Charles Wright, writing in the *Fire Protection Handbook*: "After selecting the response option for a hazardous materials incident, a plan of action including safety and health considerations should be developed. This plan of action describes the response objectives and options and the personnel and equipment required to accomplish the objectives. The plan also provides a permanent record of the decisions made at the incident. An organization's standard operating procedures provide the basis of this plan of action.

Components for a typical plan of action would include the following:

1. Site description;
2. Entry objectives;
3. On-scene organization;
4. On-scene control;
5. Hazard evaluation;
6. Personal protective equipment;
7. On-scene work assignments;
8. Communications procedures;
9. Decontamination procedures;
10. On-scene safety and health considerations including designation of the safety officer, emergency medical care procedures, environmental monitoring, emergency procedures, and personnel monitoring."[26]

Obviously, the complexity of an incident will determine the detail identified in the plan of action. Each of the above items must be considered, however, in order to ensure that nothing is being overlooked.

Figure 4.15 *Developing a plan of action is an important step in hazardous materials incident mitigation.*

Figure 4.16 *Examples of plugging and patching materials (see next page). (Courtesy of Mike Callan.)*

4-3.5.1 Describe the purpose of, procedures for, equipment required, and safety precautions used with the following techniques for hazardous materials control:

(a) Adsorption;
(b) Neutralization;
(c) Overpacking; and
(d) Patch and plug.

See NFPA 471, Section 6-4.

4-3.5.1.1 Given MC-306/DOT-406, MC-307/DOT-407, MC-312/DOT-412, MC-331, and MC-338 cargo tanks, identify the common methods for product transfer from each type of cargo tank.

Transfer operations are complex and can be hazardous. The hazardous materials technician should be very cautious when considering this type of operation. Usually transfer operations are conducted by a cleanup company or by personnel from the shipper or manufacturer.

As Hildebrand and Noll point out in their text, *Handling Gasoline Tank Truck Emergencies: Guidelines and Procedures*, the most common methods of product transfer for MC-306/DOT-406 cargo tanks are vacuum pumps, vehicles with power take-off (PTO) pumps, or air-driven portable pumps. This true for most nonpressure cargo tanks.[27]

Transfer operations involving pressure tanks present additional hazards and should be undertaken only by personnel with the necessary training and skills.

4-3.5.2 Develop a site safety plan for a hazardous materials incident.

4-3.5.2.1 Describe the components of a site safety plan for a hazardous materials incident.

According to 29 CFR 1910.120, Appendix C, comprehensive site safety and control plans should address the following: analysis of hazards on the site and a risk analysis of those hazards, site map or sketch, site work (control) zones, use of buddy system, site communications, command post, standard operating procedures and safe work practices, medical assistance and triage area, and other relevant topics. This plan should be a part of the employer's emergency response plan or an extension of it to the specific site.

See also NFPA 471, Chapter 4.

4-3.5.2.2 Given a simulated hazardous materials incident, demonstrate the ability to develop a site safety plan.

The HMT should be able to develop an appropriate site safety plan that includes the components listed above.

4-3.5.2.3 Given a plan of action for a simulated hazardous materials incident, identify the points that should be made in a safety briefing prior to working on the scene.

See Chapter 4 of NFPA 471.

4-4 Competencies — Implementing the Planned Response.

4-4.1 Performing Incident Management Duties. The hazardous materials technician shall, given a role within the local incident management system for hazardous materials incidents, demonstrate how to perform the functions and responsibilities of that role. The hazardous materials technician shall be able to:

This section is intended to ensure that the HMT understands his or her role as it is outlined in the standard operating procedures of his or her organization.

4-4.1.1 Identify the role, specified in the local emergency response plan and the organization's standard operating procedures, of the hazardous materials technician during an incident involving hazardous materials.

The emergency response plan is the link between the community's response plans and the operational personnel who are expected to implement those plans. Emergency response plans incorporate standard operating procedures (SOPs), which may identify the type of response appropriate to a particular type of incident, as well as site-specific procedures. The emergency response plan should also address alerting procedures, response and coordination procedures, personnel, the command structure, communications, and training.

4-4.1.2 Given the local emergency response plan or organization's standard operating procedures, identify the duties and responsibilities of the following hazard sector functions within the incident management system, including:

(a) Safety;

The safety officer is designated by the incident commander. He or she must be knowledgeable in the operations being conducted at the emergency response site, with the specific responsibility of identifying and evaluating hazards and providing direction with respect to the safety of operations for the emergency. See A-3-4.2.6 for safety responsibilities and Supplement 7 for additional information.

(b) Entry/reconnaissance;

Those assigned to reconnaissance should gather information about the layout of the incident scene and any other factors that may have an influence on the incident. They may develop a checklist that includes information about such things as access and egress routes, the weather, utilities, drainage, topography, water supplies, exposures, and the number, type, and condition of containers.

Entry may be supervised by an entry officer/leader, who supervises entry operations and makes recommendations concerning control activities in the hot zone.

(c) Information/research;

Personnel assigned to information/research are responsible for developing, documenting, and coordinating the information gathered during an incident. This information will be used in the hazard and risk assessment and should include, but not be limited to, evacuation considerations and protective clothing and equipment selection.

(d) Resources;

The resources officer keeps track of available personnel and equipment, which should be kept at a staging area until assigned. The resources officer must work very closely and coordinate with the operations officer. (This is also referred to as "logistics" in some incident management systems.)

(e) Decontamination; and

The decontamination officer is responsible for the operations of the decontamination unit. He or she reports to the hazard sector/division officer and supervises decontamination operations.

(f) Operations.

Operations manages the activities that apply directly to the primary mission, and the operations officer is responsible for allocating and assigning resources to control the incident. The incident commander may perform this function during small incidents.

4-4.1.3   Given the local emergency response plan or organization's standard operating procedures, identify the duties and responsibilities of the hazard sector officer and describe how to coordinate all activities of that sector.

The hazard sector officer, who may also be referred to as the hazardous materials group supervisor, is responsible for implementing the plan of action for hazardous materials control operations. The HMT must understand the functions of this position and be able to establish the management system needed to direct operations based on the size and complexity of the incident.

4-4.1.4   Given a simulated hazardous materials incident, demonstrate set-up of the contamination reduction corridor as specified in the planned response.

See Chapter 7 of NFPA 471.

Figure 4.17   A decontamination corridor set up for training purposes.

4-4.1.5 Given a simulated hazardous materials incident, demonstrate how to perform the decontamination process specified in the planned response.

See Chapter 7 of NFPA 471.

4-4.2   Using Protective Clothing and Respiratory Protection. The hazardous materials technician shall demonstrate the ability to don, work in, and doff both liquid splash- and vapor-protective clothing and any other specialized personal protective equipment provided by the authority having jurisdiction, with the appropriate respiratory protection. The hazardous materials technician shall be able to:

4-4.2.1 Identify the safety and emergency procedures for personnel wearing vapor-protective clothing.

Because personnel wearing vapor-protective clothing may experience a loss of mobility, dexterity, vision, and communications capability, it is important for them to be closely monitored. Back-up personnel wearing the same level of protective clothing must be available, and hand signals should be established to aid in communications. Personnel must also be monitored for the effects of heat, and a proper rehabilitation program should be in place to replenish fluids and allow for rest and recovery.

Figure 4.18 It is important that all crews operating in a hostile environment have both a primary and a secondary form of communication. Do not overlook the need for effective hand signals when operating in contaminated atmospheres.

4-4.2.2* Identify the procedures for donning, working in, and doffing the following types of respiratory protection.

(a) Air purifying respirator; and

(b) Air line respirator and required escape unit.

A-4-4.2.2 Competency for positive pressure self-contained breathing apparatus was met as part of Chapter 3.

4-4.2.3 Demonstrate donning, working in, and doffing chemical-protective clothing in addition to any other specialized protective equipment provided by the authority having jurisdiction.

HMTs should practice donning and doffing chemical-protective clothing to become proficient. One of the more effective ways to evaluate an HMT's ability to don and doff the protective clothing provided is to conduct training exercises that require him or her to put on the personal protective equipment and to conduct simulated control activities, followed by simulated decontamination.

Because some types of chemical-protective clothing are very costly, it may be advisable to use garments that are no longer adequate for emergency response. Haz mat training suits that are less expensive are also available; these allow for cost-effective training in the use of totally encapsulating suits.

4-4.2.4 Demonstrate the ability to record the use, repair, and testing of chemical-protective clothing according to manufacturer's specifications and recommendations.

It is very important to keep accurate records on the use, repair, and testing of chemical-protective clothing. Proper records will help identify possible problems or potential failures.

4-4.2.5 Describe the maintenance, testing, inspection, and storage procedures for personal protective equipment provided by the authority having jurisdiction according to the manufacturer's specifications and recommendations.

Chemical-protective clothing must be maintained properly according to the manufacturer's specifications. It is also important that garments be completely decontaminated before they are stored and that they be stored properly after use. Compatible components should be stored together so that the wrong item will not be used by mistake. Chemical-protective clothing should be inspected visually before it is worn to ensure that it is still in usable condition.

4-4.3 **Performing Control Functions Identified in Plan of Action.** The hazardous materials technician shall, given various simulated hazardous materials incidents involving nonbulk and bulk packaging and facility containers, select the tools, equipment, and materials for the control of hazardous materials incidents and identify the precautions for controlling releases from those packagings/containers. The hazardous materials technician shall be able to:

4-4.3.1* Given a nonbulk and a bulk pressure vessel/container, select the appropriate material or equipment and demonstrate a method(s) to contain the following leaks:

The purpose of this competency is to allow the HMT to demonstrate ability to choose the appropriate equipment and methods to control the above situations. The equipment necessary to conduct the following operations may vary depending on the type of material that is leaking. For example, nonsparking tools should be used when working with flammables. And in some cases, special tools may be needed for certain valves. Such situations should be identified during pre-incident planning.

(a) Valve gland;

Leaks from valve glands may be controlled by tightening the packing nut. It may also be possible to cap the outlet.

(b) Valve seat;

The valve should be tightened first to ensure that it is closed all the way. It may also be necessary to open and reclose the valve to clear debris that is preventing the valve from seating properly. In addition, it may be possible to cap the outlet.

(c) Valve inlet threads;

If there is a leak around the inlet threads of a valve, it may be possible to tighten the valve assembly.

(d) Valve blowout;

If a valve has blown out, it may be possible to use a wooden plug to stop the flow temporarily until the line can be shut down or the container drained for replacement.

Figures 4.19(a) and (b)   Some companies offer training programs that allow hands-on exercises, which may be provided locally with mobile training props.

(e) Fusible plug threads;

(f) Fusible metal of plug;

(g) Valve stem assembly blowout; and

If the valve assembly has blown out, it may be possible to drive a wooden plug into the hole to gain temporary control depending on the pressure of the product in the container.

(h) Side wall of cylinder.

A patch placed over a leak in the side wall of a container and secured may control the leak.

A-4-4.3.1   Contact the Chlorine Institute for assistance in obtaining training on the use of the various chlorine kits.

4-4.3.2*   Given the fittings on a pressure container, demonstrate the ability to:

(a) Close open valves;

(b) Tighten loose plugs; and

(c) Replace missing plugs.

This competency is intended to allow the HMT to demonstrate the ability to choose the appropriate equipment and methods to control the above situations. For the competency to be most effective, the HMT should demonstrate these skills while wearing chemical-protective clothing and equipment.

A-4-4.3.2   See A-4-4.3.1.

4-4.3.3   Given a 55-gal drum, demonstrate the ability to contain the following leaks using appropriate tools and materials:

(a) Bung leak;

Bung leaks can often be stopped by tightening the bung with a bung wrench.

(b) Chime leak;

(c) Nail puncture; and

Punctures can often be stopped with a wooden plug. A sheet metal screw and a gasket may also be used.

(d) Forklift puncture.

Because a puncture made by a forklift may be large and irregularly shaped, it may be plugged with a number of wooden plugs of different sizes or shapes. While an appropriate plugging or patching device is being assembled, it may be possible to stop the drum from leaking by turning it on its side, with the opening at the top.

4-4.3.4 Given a 55-gal drum and an overpack drum, demonstrate the ability to place the 55-gal drum into the overpack drum using the following methods:

When conducting overpacking operations manually, responders must take care to lift and move the drum properly so as to avoid back strain and injuries to hands or feet.

(a) Slide-in;

This involves laying a leaking drum on its side and sliding it into an overpack drum.

(b) Rolling slide-in; and

This method again involves laying the drum on its side. In this instance, however, rollers are put underneath the drum, and the drum is rolled into the overpack drum. Items that may be used as rollers include lengths of pipe and other rounded materials.

(c) Slip-over.

This method involves placing the overpack drum over the top of the leaking drum and manually rotating it upright. A device called a drum upender is available commercially to assist with this type of operation.

4-4.3.5 Identify the maintenance and inspection procedures for the tools and equipment provided for the control of hazardous materials releases according to the manufacturer's specifications and recommendations.

4-4.3.6 Identify three considerations for assessing a leak or spill inside a confined space without entering the area.

Confined space operations are extremely dangerous, and the HMT must exercise the utmost caution during such operations. One of the most critical considerations is whether the confined space is oxygen-deficient. Another is whether it contains a flammable or toxic atmosphere. And if the material in the container is loose or granular, the HMT must be aware that it could give way and engulf him or her.

4-4.3.7 Identify the safety considerations for product transfer operations, including bonding, grounding, elimination of ignition sources, and shock hazards.

Whenever flammables are transferred, responders must ensure that appropriate safety precautions have been taken. Obviously, all ignition sources must be eliminated. In addition, proper grounding and bonding connections must be made. Bonding is the process of connecting two or more objects by means of a conductor. Grounding, a specific form of bonding, is the process of connecting one or more conductive objects to the ground.

4-4.3.8 Given an MC-306/DOT-406 cargo tank and a dome cover clamp, demonstrate the ability to install the clamp on the dome properly.

Because this type of cargo tank is used so frequently, the HMT should become familiar with it and know how to secure the dome cover clamp. Arrangements can sometimes be made with a local distributor to use a cargo tank for training purposes.

4-4.3.9 Identify the methods and precautions used when controlling a fire involving an MC-306/DOT-406 aluminum shell cargo tank.

When exposed to fire, an MC-306/ DOT-406 tank will melt, preventing the build-up of excessive pressure. Melting will occur only above the liquid level of the product.

According to Hildebrand and Noll in *Handling Gasoline Tank Truck Emergencies: Guidelines and Procedures*, "A fully involved MC306/DOT406 cargo tank truck fire will require a substantial amount of foam for final control and extinguishment. Attempting to apply 95 gpm (360 liters/minute) of foam from a single 5-gallon (18.9 liters) foam concentrate container onto a 9,000-gallon (34,065 liters) gasoline spill fire will ruin your day.

"There may be situations where a controlled burn of the fire may be [appropriate] . . . gasoline burns at the rate of approximately one foot per hour. Although this may extend the duration of the incident, it will usually minimize both groundwater and surface water contamination. . . "[28]

Figure 4.20(a) *A dome cover on an MC-306/DOT-406 cargo tank.*

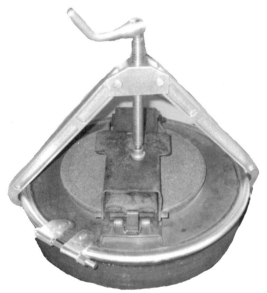

Figure 4.20(b) *A dome cover clamp applied to an MC-306/DOT-406 cargo tank.*

HAZARDOUS MATERIALS RESPONSE HANDBOOK

4-4.3.10 Describe methods for containing the following leaks in MC-306/DOT-406, MC-307/DOT-407, and MC-312/DOT-412 cargo tanks:

(a) Puncture;
(b) Irregular-shaped hole;
(c) Split or tear; and
(d) Dome cover leak.

There are a variety of methods to control leaks in cargo tanks. These may include the use of wooden plugs, patches, specially designed dome clamps, and others. The important thing is for the responder to be aware of the type of product he or she is dealing with and that the patching materials are compatible. It is also important that the responder have an opportunity to train with the above vehicles in order to have an understanding of the characteristics of each type of cargo tank.

4-4.3.11 Describe product removal and transfer considerations for overturned MC-306/DOT-406, MC-307/DOT-407, MC-312/DOT-412, MC-331, and MC-338 cargo tanks, including:

(a) Inherent risks associated with such operations;
(b) Procedures and safety precautions; and
(c) Equipment required.

Each of the above cargo tanks will present different problems if they are overturned and product must be transferred. A cargo tank hauling gasoline will require different procedures and equipment than a cargo tank of liquefied petroleum gas. Again, it is important for the responder to be familiar with the various types of cargo tanks common in his or her response area and to work with shippers to gain an understanding of the considerations for conducting product transfer operations. In many cases, if not most, product transfer operations will require the equipment and expertise of the shipper or a company involved in hazardous materials cleanup. The HMT needs to know where such assistance can be obtained and must have an understanding of the procedures and safety considerations involved. See also 4-3.5.1.1.

4-5* Competencies — Evaluating Progress.

The hazardous materials technician shall be capable of evaluating the effectiveness of any control functions identified in the plan of action.

A-4-5 The Committee feels that the evaluation competencies required at the operational level (Section 3-5) meet the requirements of the technician level.

See Section 3-5.

References Cited in Commentary

1. Arthur E. Cote, ed., *Fire Protection Handbook,* 17th edition (Quincy, MA: National Fire Protection Association, 1991), p. 8-272.
2. Ibid., p. 8-272.
3. Ibid., p. 8-272.
4. Kenneth J. York and Gerald L. Grey, *Hazardous Materials/Waste Handling for the Emergency Responder* (New York: Fire Engineering, 1989), p. 281.
5. G. Hawley, *The Condensed Chemical Dictionary* (New York: Van Nostrand Reinhold Co., 1987).
6. NFPA 491M, *Manual of Hazardous Chemical Reactions,* 1991 edition.
7. N. Irving Sax, *Dangerous Properties of Industrial Materials* (New York: Van Nostrand Reinhold Co.).
8. Title 40 CFR, Part 261 (Washington, DC: U.S. Government Printing Office).
9. Cote, *Fire Protection Handbook,* p. 2-232.
10. NFPA 30, *Flammable and Combustible Liquids Code,* 1990 edition.
11. Ibid., p. 8-274.
12. Ibid., p. 8-267.
13. Title 49 CFR, Part 178 (Washington, DC: U.S. Government Printing Office).
14. NFPA 491M.
15. L. Bretherick, *Handbook of Reactive Chemical Hazards* (Boston: Butterworths, 1990).
16. N. Irving Sax and Richard J. Lewis, Sr., *Hazardous Chemicals Desk Reference* (New York: Van Nostrand Reinhold, 1987).
17. NFPA 30.
18. Ibid.
19. Ibid.
20. National Fire Academy, *Initial Response to Hazardous Materials Incidents, Course II: Concept and Implementation* (Emmitsburg, MD).
21. Cote, *Fire Protection Handbook,* p. 9-176.

22. NFPA 1991, *Standard on Vapor-Protective Suits for Hazardous Chemical Emergencies,* 1990 edition; NFPA 1992, *Standard on Liquid Splash-Protective Suits for Hazardous Chemical Emergencies,* 1990 edition; NFPA 1993, *Standard on Support Function Protective Garments for Hazardous Chemical Operations,* 1990 edition.
23. Ibid.
24. Ibid.
25. Ibid.
26. Cote, *Fire Protection Handbook,* p. 9-179.
27. Michael S. Hildebrand and Gregory C. Noll, *Handling Gasoline Tank Truck Emergencies: Guidelines and Procedures* (Stillwater, OK: Fire Protection Publications, 1991).
28. Ibid.

5

Competencies for the Incident Commander

5-1 General.

5-1.1 Introduction. The incident commander shall be trained to meet all requirements indicated for the first responder at the awareness (Chapter 2) and operational (Chapter 3) levels and the requirements of this chapter. In addition, the incident commander shall receive any additional training to meet federal Occupational Safety and Health Administration (OSHA), local occupational health and safety regulatory, or U.S. Environmental Protection Agency (EPA) requirements, whichever are appropriate for his or her jurisdiction.

This chapter is new to the 1992 edition of NFPA 472. The committee believed that it was important to outline the competencies required of the incident commander (IC), who may also be referred to as the on-scene incident commander.

The IC assumes control of an incident that goes beyond the capabilities of the awareness level responder. OSHA 1910.120 requires that the IC have training at least equal to that of the operational level responder and have additional training relating to hazardous materials incident management. Although operations at many incidents are conducted only by operational level personnel operating defensively, the IC must have additional skills, even when he or she is managing operational personnel. No matter what the level of the incident or the personnel operating at the incident, the IC must be trained and competent in the incident management system, emergency operations plans, the hazards and risks of operating at the scene, state and federal resources and plans, and the importance of decontamination.

The committee recognized that in complex incidents involving the response of hazardous materials technicians or off-site specialist employees (*see Chapter 6*), the IC would use their expertise in formulating response objectives, action options, and the plan of action. Although the IC may be trained to the HMT level, this may not always be the case.

5-1.2* **Definition.** The incident commander is that person who is responsible for directing and coordinating all aspects of a hazardous materials incident.

A-5-1.2 The following are the typical duties and responsibilities of the incident commander at a hazardous materials incident. These duties and responsibilities may be performed directly or delegated to other response personnel as necessary.

(a) Analysis Activities

- Classify and identify unknown materials
- Verify known materials
- Monitor changes in climatic conditions
- Identify contaminated people and equipment
- Establish environmental monitoring
- Interpret the data collected from environmental monitoring

(b) Planning Activities

- Develop a plan of action for the incident
- Develop a plan of action for activities in the control zones
- Develop an incident safety plan
- Seek technical advice
- Evaluate and recommend public protective actions
- Coordinate handling, storage, and transfer of contaminants
- Determine personal protective equipment compatibility
- Organize and supervise assigned personnel to control site access
- Provide required emergency medical services

(c) Implementation Activities

- Conduct safety briefings
- Implement the plan of action for the incident
- Implement the incident safety plan
- Oversee placement of control zones
- Supervise entry operations
- Direct rescue operations
- Maintain communications and coordination during the incident
- Provide medical monitoring of entry personnel before and after entry
- Protect personnel from physical, environmental, and safety hazards/exposures
- Provide information for public and private agencies
- Enforce recognized safe operational practices
- Ensure that injured or exposed individuals are decontaminated prior to departure from the hazard site
- Separate and keep track of potentially contaminated persons
- Track persons passing through the contamination reduction corridor
- Ensure that decontamination activities are conducted
- Coordinate transfer of decontaminated patients

(d) Evaluation Activities

- Evaluate progress of the actions taken and modify as necessary
- Recognize deviations from the incident safety plan and any dangerous situations
- Alter, suspend, or terminate any activity that may be judged unsafe
- Keep required records for litigation and documentation
- Ensure that medical-related exposure records are maintained

It is extremely important from an incident management standpoint that a designated person be in charge and that he or she be clearly identified. The IC works from the strategic level and develops the overall response objectives; he or she should not become involved in tactical operations. Broadly speaking, the IC is responsible for the safety of response personnel and the public, for controlling the incident, and for minimizing harm to the environment and property.

Even though the IC is responsible for directing and coordinating the response, some management functions may have to be delegated to others. In some cases, the IC will establish a hazard sector or a hazardous materials group to manage activities around the control zones. These are managed by a hazard sector officer or hazardous materials group supervisor, who reports to the IC or to Operations, respectively.

A number of different agencies may be responsible for controlling and cleaning up the more complex hazardous materials incidents. These agencies may be local, county, state, or federal agencies; some incidents involve more than one agency from each jurisdiction. It is absolutely imperative that these agencies be identified during pre-incident planning and that the plans establish protocols spelling out which agency is to be the lead agency during an incident. Some incident management systems establish what is called a "unified command" for multiagency or multijurisdictional incidents. When a unified command has been established, representatives from each agency may be identified as the lead person for that agency. Collectively, these lead persons will manage the incident. Generally, one if them is designated as the spokesperson for the unified command.

Figure 5.1 An incident command system enables the incident commander to manage and direct operations.

5-1.3 Goal. The goal of this chapter shall be to provide the incident commander with the knowledge and skills to perform the following tasks safely. Therefore, in addition to being competent at the awareness and operational levels, the incident commander shall be able to:

(a) Analyze a hazardous materials incident to determine the magnitude of the problem in terms of outcomes by completing the following tasks:

1. Collect and interpret hazard and response information from printed resources, technical resources, computer data bases, and monitoring equipment, and;
2. Estimate the potential outcomes within the endangered area at a hazardous materials incident.

(b) Plan a response within the capabilities and competencies of available personnel, personal protective equipment, and control equipment by completing the following tasks:

1. Identify the response objectives for hazardous materials incidents;
2. Identify the potential action options (defensive, offensive, and nonintervention) available by response objective;
3. Approve the level of personal protective equipment required for a given action option; and
4. Develop a plan of action consistent with the local emergency response plan and the organization's standard operating procedures and within the capability of available personnel, personal protective equipment, and control equipment.

(c) Implement a response to favorably change the outcomes consistent with the local emergency response plan and the organization's standard operating procedures by completing the following tasks:

1. Implement the incident management system including the specified procedures for notification and utilization of nonlocal resources, e.g., private, state, and federal government personnel;
2. Direct resources (private, governmental, and others) with expected task assignments and on-scene activities, and provide management overview, technical review, and logistical support to private and governmental sector personnel; and
3. Provide a focal point for information transfer to media and local elected officials.

(d) Evaluate the progress of the planned response to ensure that the response objectives are being met safely, effectively, and efficiently and adjust the plan of action accordingly by completing the following tasks:

1. Evaluate the progress of the plan of action;
2. Report and document the hazardous materials incident, and;
3. Conduct a multiagency critique.

5-2 Competencies — Analyzing the Problem.

5-2.1 Collecting and Interpreting Hazard and Response Information. The incident commander shall, given access to printed resources, technical resources, computer data bases, and monitoring equipment, collect and interpret hazard and response information not available from the current edition of the *Emergency Response Guidebook* or a material safety data sheet (MSDS). The incident commander shall be able to:

5-2.1.1 Identify the types of hazard and response information available from each of the following resources and explain the advantages and disadvantages of each resource:

(a) Reference manuals;
(b) Hazardous materials data bases;
(c) Technical information centers;
(d) Technical information specialists; and
(e) Monitoring equipment.

The IC is expected to understand the types of resources that are available and the types of information each can provide. The IC is not expected to have an in-depth knowledge of computer data bases, for example, but he or she should know the type of information that can be accessed from them. Likewise, the IC is not expected to be able to operate the various types of monitoring equipment, but he or she should understand its function and its limitations.

5-2.2 Estimating Potential Outcomes. The incident commander shall, given simulated facility or transportation incidents involving hazardous materials, the surrounding conditions, and the predicted behavior of the container and its contents, estimate the potential outcomes within the endangered area. The incident commander shall be able to:

5-2.2.1 Given the dimensions and the surrounding conditions of an endangered area of a hazardous materials incident, identify the steps for estimating the number of exposures within the endangered area.

See 4-2.5.3.

5-2.2.2 Match the following toxological terms and exposure values with their significance in predicting the extent of health hazards in a hazardous materials incident:

(a) Immediately dangerous to life and health value (IDLH);
(b) Lethal concentrations (LC_{50});
(c) Lethal dose (LD_{50});
(d) Permissible exposure limit (PEL);
(e) Threshold limit value ceiling (TLV-C);
(f) Threshold limit value short-term exposure limit (TLV-STEL);
(g) Threshold limit value time-weighted average (TLV-TWA);
(h) Parts per million (ppm), parts per billion (ppb); and
(i) Emergency response planning guide value (ERPG).

See 4-2.5.2.1. The IC must have a level of knowledge beyond that required at the operational level in order to understand clearly the potential hazards of the products that may be involved in a hazardous materials incident.

5-2.2.3 Match the following terms associated with radioactive materials with their significance in predicting the extent of health hazards in a hazardous materials incident.

(a) Alpha radiation;
(b) Beta radiation;
(c) Gamma radiation;
(d) Half-life; and
(e) Time, distance, and shielding.

See 4-2.5.2.2.

5-2.2.4 Identify the method for predicting the areas of potential harm within the endangered area of a hazardous materials incident.

Predicting the areas of potential harm in the endangered area involves determining the potential concentrations of the hazardous material that has been released. This includes the toxicity of the concentrations, the length of time that persons in the endangered area would be exposed, and the intensity of that exposure.

5-3 Competencies — Planning the Response.

5-3.1 **Identifying Response Objectives.** The incident commander shall, given simulated facility and transportation hazardous materials incidents, describe the response objectives for each problem. The incident commander shall be able to:

5-3.1.1 Describe the steps for determining response objectives (defensive, offensive, and nonintervention) given an analysis of a hazardous materials incident.

See 4-3.1.1.

5-3.2 **Identifying the Potential Action Options.** The incident commander shall, given simulated facility and transportation hazardous materials incidents, identify the possible action options (defensive, offensive, and nonintervention) by response objective for each problem. The incident commander shall be able to:

5-3.2.1 Identify the possible action options to accomplish a given response objective.

See 4-3.2.1.

5-3.2.2 Identify the purpose of each of the following techniques for hazardous materials control:

(a) Adsorption;

See 6-4.2.1 of NFPA 471 and 4-3.4.1 of NFPA 472.

(b) Neutralization;

See 6-4.2.6 of NFPA 471 and 4-3.4.1 of NFPA 472.

(c) Overpacking; and

See 6-4.1.5 of NFPA 471 and 4-3.5.1 of NFPA 472.

(d) Patch and plug.

See 6-4.1.6 of NFPA 471 and 4-3.5.1 of NFPA 472.

5-3.3 **Approving the Level of Personal Protective Equipment.** The incident commander shall, given situations with known and unknown hazardous materials, approve the appropriate personal protective equipment for the action options specified in the plan of action in each situation. The incident commander shall be able to:

It is important to note here the difference between this competency for the IC and the similar competency for the HMT. At this level, the IC is expected to "approve" the personal protective equipment chosen. The HMT is expected to be more knowledgeable and to be able to "select" the appropriate personal protective clothing.

5-3.3.1 Identify the four levels of chemical protection (EPA/NIOSH) and match the equipment required for each level with the conditions under which each level is used.

See Chapter 5 of NFPA 471.

5-3.3.2 Match the following terms with their impact and significance on the selection of chemical-protective clothing:

(a) Degradation;
(b) Penetration; and
(c) Permeation.

See 4-3.3.3.1.

5-3.3.3 Identify the safety considerations for personnel wearing vapor-protective, liquid splash-protective, and high temperature-protective clothing.

See 4-4.2.1.

5-3.3.4 Identify the physiological and psychological stresses that can affect users of specialized protective clothing.

See 4-3.3.3.7.

5-3.4 **Developing a Plan of Action.** The incident commander shall, given simulated facility and transportation hazardous materials incidents, develop a plan of action, consistent with the local emergency response plan and the organization's standard operating procedures, that is within the capability of the available personnel, personal protective equipment, and control equipment. The incident commander shall be able to:

See 4-3.5.

5-3.4.1 Identify the order of steps for developing a plan of action consistent with the local emergency response plan and the organization's standard operating procedures and within the capability of available personnel, personal protective equipment, and control equipment.

See 4-3.5.

5-3.4.2 Identify the factors to be evaluated in selecting public protective actions including evacuation and in-place protection.

See 3-4.1.2.

5-3.4.3 Given the local emergency response plan or the organization's standard operating procedures, identify which agency will:

(a) Receive the initial notification;
(b) Provide secondary notification and activation of response agencies;
(c) Make on-going assessments of the situation;
(d) Command on-scene personnel (incident management system);
(e) Coordinate support and mutual aid;
(f) Provide law enforcement and on-scene security (crowd control);
(g) Provide traffic control and rerouting;
(h) Provide resources for public safety protective action (evacuation or in-place protection);
(i) Provide fire suppression services when appropriate;
(j) Provide on-scene medical assistance (ambulance) and medical treatment (hospital);
(k) Provide public notification (warning);
(l) Provide public information (news media statements);
(m) Provide on-scene communications support;
(n) Provide on-scene decontamination when appropriate;
(o) Provide operational-level hazard control services;
(p) Provide technician-level hazard mitigation services; and
(q) Provide environmental remedial action ("cleanup") services.

The Emergency Planning and Community Right-to-Know Act of 1986 requires local emergency planning committees (LEPCs) to develop local plans for emergency response to hazardous materials incidents. Persons who may be expected to serve as incident commanders must understand the local emergency response plan, which should address the actions listed above.

Emergency response agencies should develop their own standard operating procedures based on their roles in the local emergency plan. These procedures should address the actions listed and should indicate who is responsible for these actions, both inside and outside the agency, and how they will be accomplished.

5-3.4.4 Identify the process for determining the effectiveness of an action option on the potential outcomes.

"Before a response option is selected, the effect of that response option or combination of response options on the sequence of events and ultimately the outcomes, should be reviewed (*see Table 5.1*). Prioritize the response options based on their effect on the outcomes."[1]

5-3.4.5 Identify the procedures for presenting a safety briefing prior to allowing personnel to work on a hazardous materials incident.

See Chapter 4 of NFPA 471 and 4-3.5.2 – 4-3.5.2.3 of NFPA 472.

5-4 Competencies — Implementing the Planned Response.

5-4.1 Implementing the Incident Management System. The incident commander shall, given a copy of the local emergency response plan, identify the requirements of the plan including the required procedures for notification and utilization of nonlocal resources (private, state, and federal government personnel). The incident commander shall be able to:

5-4.1.1 Identify the process and procedures for obtaining cleanup and restoration services in the local emergency response plan or organization's standard operating procedures.

The IC should know how to obtain cleanup and restoration services. In some cases, this simply involves contacting the local communications center and asking that those services be dispatched to the scene. In others, the IC may have to contact a different agency to activate those services.

Table 5.1   Worksheet for Estimating the Effect of Each Response
            Option

Exposures/ Harm	Amount that Could Be Saved	Amount Saved with Option 1	Amount Saved with Option 2	Amount Saved with Option 3	Amount Saved with Option 4	Amount Saved with Option 5
People Deaths	#	#	#	#	#	#
People Injuries	#	#	#	#	#	#
Property Damage	$	$	$	$	$	$
Environmental Damage	$	$	$	$	$	$

(Source: *Fire Protection Handbook*, 17th edition.)

5-4.1.2   Identify the steps for implementing the local and related emergency response plans as required under SARA Title III Section 303 of the federal regulations or other local emergency response planning legislation.

Normally, emergency response plans are set in motion when someone notifies an emergency operations center that an incident has occurred. The emergency response plan then identifies the type of resources that are to be dispatched and determines whether there are any additional entities that should be notified. The IC should understand this process and be aware of which officials and/or agencies must be notified.

5-4.1.3   Given the local emergency response planning documents, identify the elements of each of the documents.

Title III of SARA requires that an emergency plan contain certain elements. These issues must be addressed in the planning process.

The emergency plan must address the following items if they are not covered elsewhere:

1. Pre-emergency planning and coordination with outside parties
2. Personnel roles, lines of authority, training, and communications

3. Emergency recognition and prevention
4. Safe distances and places of refuge
5. Site security and control
6. Evacuation routes and procedures
7. Decontamination
8. Emergency medical treatment and first aid
9. Emergency alerting and response procedures
10. Critique response and follow-up
11. Personal protective equipment (PPE) and emergency equipment.

If the SOPs adequately cover these elements, as is sometimes the case, they need not be included in the emergency plan.

For a detailed discussion of these elements, see Supplement 4, Chapter 5, "Hazardous Materials Planning Elements."

5-4.1.4 Identify the elements of the incident management system necessary to coordinate response activities at hazardous materials incidents.

There are several different models for hazardous materials incident command systems, some of which are presented in other sections. For example, Supplement 7 provides an example of the system used in California's Firescope system. However, each system generally has the same components, even though they may use some different titles. For example, hazard sector officers and hazardous materials group supervisors generally perform the same functions. Basically, both are responsible for implementing the incident action plan that deals with operations intended to control the hazardous materials portion of an incident.

Within the incident management system, there are generally five primary functional areas: the incident commander, operations, planning, logistics, and, in some cases, finance. Usually, there will also be a public information officer and a safety officer. For hazardous materials incidents, additional functions are identified, including the hazard sector officer, the decontamination officer, and the entry officer (*see 4-4.1.2*).

It is extremely important that the IC be familiar with the incident management system established in the local emergency response plan and that he or she be able to implement it.

5-4.1.5 Identify the primary local, state, regional, and federal government agencies and identify the scope of their regulatory authority (including the regulations) pertaining to the production, transportation, storage, and use of hazardous materials and the disposal of hazardous wastes.

The IC must know which agencies may become involved in a hazardous materials incident and know what their regulatory authority is. The local emergency response plan should identify these agencies and delineate their roles, authority, and functions.

5-4.1.6 Identify the governmental agencies and private sector resources offering assistance during a hazardous materials incident, and identify their role and the type of assistance or resources available.

Some government agencies provide response or technical assistance. For example, the U.S. Coast Guard staffs three strategically located National Strike Teams that are trained and equipped to respond to major oil spills and chemical releases. The U.S. Environmental Protection Agency also has a response component called the Environmental Response Team (ERT). The ERT is a group of scientists and engineers who are trained in multimedia sampling and analysis, hazard assessment, cleanup techniques, and other technical support. In addition, many state and local governments can provide technical assistance.

Figure 5.2   The U.S. Coast Guard can provide assistance at incidents.

HAZARDOUS MATERIALS RESPONSE HANDBOOK

The private sector also has many resources available to emergency responders, from providing technical advice and assisting with on-scene monitoring to providing specialized equipment. In most areas, there are private-sector companies that provide cleanup and disposal services. Some of these companies may even stage equipment in an area and fly personnel to the scene when a hazardous materials incident occurs. CHEMTREC/CANUTEC can put responders in touch with private-sector resources if they are unfamiliar with any in their area.

Figure 5.3 Private industry provides specialized response personnel to assist in controlling hazardous materials incidents.

The IC must know how to get specialized assistance when it is needed and must be familiar with the type of assistance available. These resources, both public and private, should be identified in the local emergency response plan. In situations involving services that must be contracted for, arrangements should be addressed during the pre-incident planning phase so as not to complicate matters during an actual emergency.

5-4.2 **Directing Resources (Private and Governmental).** Given a simulated hazardous materials incident and the necessary resources to implement the planned response, the incident commander shall demonstrate the ability to direct the resources in a safe and efficient manner consistent with the capabilities of those resources. The incident commander shall be able to:

5-4.2.1 Given a hazardous materials incident, terminate the emergency phase of the incident.

The termination of the emergency phase of an incident involves documenting incident activities. This information will help responders comply with local, state, and federal regulations pertaining to hazardous materials incidents.

5-4.2.1.1 Identify the steps required in terminating the emergency phase of a hazardous materials incident.

The steps involved in terminating the emergency phase of an incident are incident debriefing, critiquing the incident, and after-action activities.

The incident debriefing involves gathering information from response personnel. This information should then be used to develop a chronological report of emergency activities.

The critique is a review of the incident intended to identify and document lessons learned. It should allow the participants to review response activities and determine what worked well and what didn't. The critique should be a positive process that allows responders to modify response procedures if problems are identified.

After-action activities include analyzing the information gathered during the debriefing and the critique, documenting that analysis, and following up as necessary to ensure that any recommendations made to improve emergency operations are implemented.

5-4.2.1.2 Identify the procedures for conducting incident debriefings at a hazardous materials incident.

The incident debriefing is not a critique; it is the gathering of information intended to provide an overall summary of the activities of each sector/division during an incident. The objectives of a debriefing are to identify who responded, what they did, when they did it, and how effective their operations were. The debriefing should also document any injuries suffered, note the type of treatment given, and indicate whether any medical follow-up is needed. Equipment that has been damaged should be reported and any unsafe conditions noted.

It is important that responders are told what materials they were exposed to, warned of any symptoms those materials might produce, and told what decontamination procedures they should undergo. These procedures may include showering and washing or disposing of clothing, for example.

5-4.2.1.3 Identify the steps in transferring authority as prescribed in the local emergency response plan or the organization's standard operating procedures.

Transferring authority at an incident generally means transferring command, or the role of IC, from one person to another. This process should be identified in the incident management system's SOPs. Authority may be transferred from one officer to another officer of higher rank or, in the case of some agencies, from one person to another person with a higher authority and more responsibility.

Authority may also be transferred when the emergency phase has ended and a nonemergency phase begins, as when cleanup and remediation activities must continue at the site. In such cases, it is likely that some agency, be it local, state, or federal, will be chosen to manage this phase. Whatever the case, procedures should be developed to identify who may be responsible for overseeing these operations.

5-4.3 Providing a Focal Point for Information Transfer to Media and Elected Officials. The incident commander shall, given a simulated hazardous materials incident, identify appropriate information to provide to the media and local, state, and federal officials. The incident commander shall be able to:

5-4.3.1 Identify the local policy for providing information to the media.

The emergency response plan and/or an agency's SOPs should establish a procedure for providing information to the media, which the media can then make available to the public. If the public is not to panic, it must receive accurate information. The media can also help responders alert the public to possible evacuations or any other protective actions that may be necessary.

5-4.3.2 Identify the responsibilities of the public information officer at a hazardous materials incident.

The public information officer (PIO) should function as a part of the IC's staff. One person should be chosen to serve as spokesperson at an incident. This person should have training and experience in public information and media relations.

The PIO should establish a press area in a safe location and regularly provide the media with accurate information about the incident. There may be areas to which the press is not allowed access because of the hazards present. If the media is allowed to move about, however, the PIO should provide escorts for them or identify safe areas into which they may go unescorted.

5-5 Competencies — Evaluating Progress.

5-5.1 **Evaluating Progress of the Plan of Action.** The incident commander shall, given simulated facility and transportation hazardous materials incidents, evaluate the progress of the plan of action to determine whether the efforts are accomplishing the response objectives. The incident commander shall be able to:

5-5.1.1 Identify the procedures for evaluating whether the action options are effective in accomplishing the objectives.

To determine whether the actions being taken at an incident are effective and the objectives are being met, responders must determine whether the incident is stabilizing or increasing in intensity. Feedback will allow responders to modify either their strategic goals or the action options being implemented. This feedback should include information on the effectiveness of personnel, on personal protective clothing and equipment, on control zones, on decontamination procedures, and on the action options being implemented.

5-5.1.2 Identify the steps for comparing actual behavior of the material and the container to that predicted in the analysis process.

When comparing actual behavior to predicted behavior, the IC should determine whether events at an incident are happening as predicted, are occurring out of sequence, or are different than expected. The IC should also determine whether events that were predicted to occur are not happening at all. This process should continue until the incident has been terminated so that there are no "surprises" during the cleanup or overhaul phase.

5-5.1.3 Given a simulated hazardous materials incident, determine the effectiveness of:

(a) Personnel being used;
(b) Personal protective equipment;
(c) Established control zones; and
(d) Decontamination process.

This competency is intended to allow the IC to show that he or she can analyze an incident to determine how effective the chosen action options were in achieving their strategic goals. The competencies in 5-5.1.1 and 5-5.1.2 provide the basis for this analysis.

5-5.2 **Reporting and Documenting the Hazardous Materials Incident.** The incident commander shall, given a simulated hazardous materials incident, demonstrate the ability to report and document the incident consistent with the local, state, and federal requirements. The incident commander shall be able to:

5-5.2.1 Identify the reporting requirements of federal, state, and local agencies.

The IC must be aware of the reporting requirements necessary to deal with a hazardous materials incident. In many cases, the responsibility for reporting to specific agencies is outlined in an agency's or an organization's standard operating procedures. However, the IC must ensure that the proper agencies have been notified and that the proper reports have been made.

5-5.2.2 Identify the importance of documentation for a hazardous materials incident including training records, exposure records, incident reports, and critique reports.

Questions about an incident may not arise until someone files a claim some time after the incident is over, and if information documenting the incident is not available, it could have serious ramifications. Thus, documenting information about personnel training and exposure and keeping incident and critique reports on file is critical to ensuring that answers are available to any questions that might arise about the handling of an incident. The IC may be asked to explain the operations and the use of personnel during the incident, and asked why certain decisions were made or not made.

5-5.2.3 Identify the steps in keeping an activity log and exposure records for hazardous materials incidents.

The IC should assign someone to maintain a record of incident events. This will be helpful in completing the incident analysis and conducting the critique.

Personnel exposure records should also be maintained, as they are required by federal and, in many cases, state law. The IC should assign someone to gather the necessary information about the type of exposure to which personnel were subjected, the exposure level, the length of the exposure, the type of personal protective clothing and equipment personnel were using, and the type of decontamination personnel underwent. Any on-scene medical assistance that personnel received should also be documented.

5-5.2.4 Identify the requirements for compiling hazardous materials incident reports found in the local emergency response plan and the organization's standard operating procedures.

The IC must understand that he or she is responsible for completing the required incident reports.

5-5.2.5 Identify the requirements for filing documents and maintaining records found in the local emergency response plan and the organization's standard operating procedures.

Again, it is important for the proper records and reports to be completed and properly filed and stored for future reference.

5-5.3   Conducting a Multiagency Critique. The incident commander shall, given the details of a simulated hazardous materials incident, conduct a critique of the incident. The incident commander shall be able to:

5-5.3.1 Identify the procedure for conducting a critique of a hazardous materials incident.

The procedure for conducting a multiagency critique is similar to that outlined in 5-4.2.1.1, although the process is a little more complicated. It is advisable to have an initial meeting to identify who should be involved in the critique in order to ensure that no one is overlooked and that each agency is properly represented.

References Cited in Commentary

1. Arthur E. Cote, ed., *Fire Protection Handbook*, 17th edition (Quincy, MA: National Fire Protection Association, 1991), p. 9-175.

6

Competencies for Off-site Specialist Employees

6-1 General.

6-1.1 Introduction. Off-site specialist employees are those who, in the course of their regular job duties, work with or are trained in the hazards of specific materials and/or containers. In response to incidents involving chemicals, they may be called upon to provide technical advice or assistance to the incident commander relative to their area of specialization. Off-site specialist employees shall receive training or demonstrate competency in their area of specialization annually. In addition, off-site specialist employees shall receive training to meet any applicable federal Department of Transportation, Occupational Safety and Health Administration, Environmental Protection Agency, or local occupational health and safety regulatory agency requirements applicable to their area of specialization.

The committee determined that there is a need for recognized competencies for off-site specialist employees. Off-site specialist employees generally, although not always, work for private-sector companies that send them to the scene of hazardous materials incidents to provide incident commanders with technical advice or assistance. Among the companies these individuals may represent are chemical manufacturers, transportation companies, users of products, and container manufacturers.

The committee also recognized that such individuals are not all equally skilled, and designated three different levels to describe their varying qualifications. These are Level A, Level B, and Level C. Off-site specialist employees at Level C have only a general knowledge of the products or containers made or used by their company, but they know how to obtain additional technical assistance. Off-site specialist employees at Level B know more about the specific products and containers their companies manufacture or use and are trained to deal with them in an offensive mode. Level A is the highest specialist employee level. These individuals have a broad knowledge of their companies' product lines and containers and have been trained to assist the incident commander in offensive control measures.

All competencies for off-site specialist employees at every level apply to the individuals' areas of specialization and include only those chemicals and containers to which the individual is expected to respond. The off-site specialist employee must perform all activities in a manner consistent with the organization's emergency response (E/R) plan and standard operating procedures (SOPs), and with the available resources.

Regardless of their areas of expertise, all off-site specialist employees must at least be familiar with:

(a) The Department of Transportation's *Emergency Response Guidebook* (U.S. DOT ERG) or the Canadian *Dangerous Goods Initial Emergency Response Guide*;

(b) His or her organization's E/R plan and SOPs, including the person or persons to contact for additional assistance;

(c) The format and contents of his or her organization's material safety data sheets (MSDSs);

(d) His or her role in the E/R plan, which specifically defines the individual actions permitted;

(e) His or her role in the implementation of the incident management system (IMS); and

(f) The process required to contact CHEMTREC or CANUTEC.

6-1.2 Scope. This chapter will address competencies for the following off-site specialist employees:

(a) Off-site specialist employee C;

(b) Off-site specialist employee B; and

(c) Off-site specialist employee A.

6-2 Off-site Specialist Employee C.

6-2.1 General.

6-2.1.1 Introduction. The off-site specialist employee C shall be trained to meet the competencies at the first responder awareness level (Chapter 2) relative to his or her organization's area of specialization and the additional competencies in Section 6-2 of this chapter. In addition, the off-site specialist employee C shall receive training to meet any applicable federal Department of Transportation (DOT), Occupational Safety and Health Administration (OSHA), Environmental Protection Agency (EPA), or local occupational health and safety regulatory agency requirements.

6-2.1.2 Definition. Level C off-site specialist employees are those persons who may respond to incidents involving chemicals and/or containers within their organization's area of specialization. The off-site specialist employee C may be called upon to gather and record information, provide technical advice, and/or arrange for technical assistance consistent with his or her organization's emergency response plan and standard operating procedures. The off-site specialist employee C is not expected to enter the hot/warm zone at an incident.

6-2.1.3 Goal. The goal of these competencies is to ensure that the off-site specialist employee C has the knowledge and skills to safely perform the duties and responsibilities assigned in his or her organization's emergency response plan and standard operating procedures. Therefore, in addition to being trained at the first responder awareness level relative to his or her organization's area of specialization, the off-site specialist employee C shall also be able to:

(a) Assist the incident commander in analyzing the magnitude of an incident involving chemicals and/or containers for chemicals used in his or her organization's area of specialization by completing the following tasks:

1. Provide information on the hazards and harmful effects of specific chemicals used in his or her organization's area of specialization.
2. Provide information on the characteristics of specific containers for chemicals used in his or her organization's area of specialization.

(b) Assist the incident commander in planning a response to an incident involving chemicals and/or containers for chemicals used in his or her organization's area of specialization by completing the following task:

1. Provide information on the potential response options for chemicals and/or containers for chemicals used in his or her organization's area of specialization.

6-2.2 Competencies — Analyzing the Problem.

6-2.2.1 Providing Information on Hazards of Specific Chemicals. Given a specific chemical used in his or her organization's area of specialization and an appropriate material safety data sheet (MSDS) or other appropriate resource, the off-site specialist employee C shall advise the incident commander of the chemical's hazards and harmful effects. The off-site specialist employee C shall be able to:

Off-site specialist employees at Level C are expected to provide information about the chemical involved in an incident. The items listed in 6-2.2.1.1 represent the minimum requirements needed to determine proficiency in this area. Additional resources include the U.S. DOT'S ERG and the Association of American Railroads' *Emergency Handling of Hazardous Materials and Surface Transportation*, among others.[1]

6-2.2.1.1 Given a specific chemical within his or her organization's area of specialization and an appropriate material safety data sheet (MSDS), identify the following hazard information from the MSDS:

(a) Physical and chemical characteristics;
(b) Physical hazards of the chemical (including fire and explosion hazards);
(c) Health hazards of the chemical;
(d) Signs and symptoms of exposure;
(e) Routes of entry;
(f) Permissible exposure limits;
(g) Reactivity hazards; and
(h) Environmental concerns.

Off-site specialist employees at Level C should be able to point out the required information on the MSDS. This requirement can be fulfilled by HAZCOM training (OSHA 29 CFR 1910.1200).[2]

6-2.2.1.2 Identify how to contact CHEMTREC/CANUTEC.

This requirement can be met by determining whether the responder can find the correct telephone number, which is listed in the U.S. DOT's ERG and in Transport Canada's *Dangerous Goods Initial Emergency Response*

HAZARDOUS MATERIALS RESPONSE HANDBOOK

Guide. The company to which the product belongs may include the number in MSDSs, on labels, on containers, and in shipping papers. The number must be included in the E/R plan or SOPs of the specialist employee's organization.

6-2.2.1.3 Identify the resources available from CHEMTREC/CANUTEC.

CHEMTREC/CANUTEC can provide information on products, fax MSDSs, and contact manufacturers, mutual aid responders, and contractors for hire. See 3-2.2.4 for additional information.

6-2.2.1.4 Given his or her organization's emergency response plan and standard operating procedures, identify additional resources of hazard information including a method of contact.

Off-site specialist employees at Level C must be able to contact knowledgeable personnel, as prescribed in the organization's E/R plan or SOPs. He or she should also know where in his or her organization's plans to locate the appropriate information.

6-2.2.2 Providing Information on Characteristics of Specific Containers. Given a specific container for a chemical used in his or her organization's area of specialization, the off-site specialist employee C shall advise the incident commander of the characteristics of the containers. The off-site specialist employee C shall be able to:

6-2.2.2.1 Given examples of various containers for chemicals used in his or her organization's area of specialization, identify each container by name.

Off-site specialist employees at Level C must be able to identify the containers, such as tank cars, cargo tanks, drums, and cylinders, to which they will be expected to respond in the event of an emergency. Specialist employees may already know about these containers, or they may learn about them from such resources as other company specialists, the CMA's *Packaging for Transporting Hazardous and Non-hazardous Materials,*[3] and their own companies' packaging guides. See Supplement 9 for additional information on packaging.

6-2.2.2.2 Given examples of facility and transportation containers for chemicals in his or her organization's area of specialization, identify the markings that differentiate one container from another.

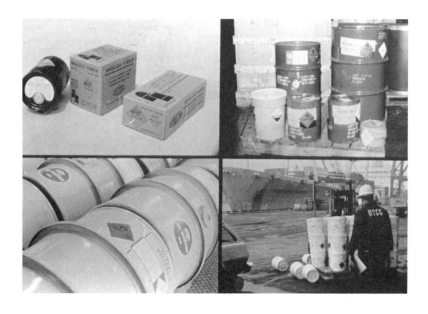

Figures 6.1(a) and (b)   *Hazardous materials are transported by truck, railroad car, ship, and airplane in a variety of containers, including wooden boxes; metal, plastic, or fiber board drums; and plastic pails [Figure 6.1(a)] and glass carboys in plywood boxes or polystyrene cases, shipping tubes, and multiwall paper bags [Figure 6.1(b)].*

Off-site specialist employees at Level C should be able to identify various container markings including tank car reporting marks and numbers, cargo tank numbers, portable tank numbers, and fixed facility tank markings. They may already know about these markings, or they may learn about them from sources in their organizations.

6-2.2.2.3 Given his or her organization's emergency response plan and standard operating procedures, identify the resources available that can provide information about the characteristics of the container.

Among the resources that can provide the responder with information about the characteristics of the container are knowledgeable persons in the responder's organization, container manufacturers, the Association of American Railroads (AAR), CHEMTREC, and various carriers.

6-2.3 Competencies — Planning the Response.

6-2.3.1 Providing Information on Potential Response Options for Specific Chemicals. Given a specific chemical used in his or her organization's area of specialization and an appropriate material safety data sheet (MSDS) or other appropriate resource, the off-site employee C shall advise the incident commander of the response information for that chemical. The off-site specialist employee C shall be able to:

By virtue of their job duties, their knowledge of specific chemicals, or their access to appropriate resources within their organizations, off-site specialist employees at Level C must be able to provide the incident commander with response information. This information may include the physical and chemical properties of the hazardous material involved in an incident, health and environmental data, and containment and reactivity data.

6-2.3.1.1 Given a specific chemical used in his or her organization's area of specialization and an appropriate material safety data sheet (MSDS), obtain the following response information:

(a) Precautions for safe handling, including hygiene practices, protective measures, and procedures for cleanup of spills/leaks;
(b) Applicable control measures, including personal protective equipment; and
(c) Emergency and first aid procedures.

This requirement should be fulfilled by HAZCOM training.

6-2.3.1.2 Given his or her organization's emergency response plan and standard operating procedures, identify additional resources for obtaining response information for a chemical used in his or her organization's area of specialization.

Additional resources the off-site specialist employee might use include the U.S. DOT's ERG, the AAR's EAG, knowledgeable persons in his or her organization, CHEMTREC/CANUTEC, and the National Fire Protection Association (NFPA).

6-3 Off-site Specialist Employee B.

6-3.1 General.

6-3.1.1 Introduction. The off-site specialist employee B shall be trained to meet the competencies at the off-site specialist employee C level and the additional competencies in Section 6-3 of this chapter. In addition, the off-site specialist employee B shall receive training to meet any applicable federal Department of Transportation (DOT), Occupational Safety and Health Administration (OSHA), Environmental Protection Agency (EPA), or local occupational health and safety regulatory agency requirements.

6-3.1.2 Definition. Level B off-site specialist employees are those persons who, in the course of their regular job duties, work with or are trained in the hazards of specific chemicals or containers within their individual area of specialization. Because of his or her education, training, or work experience, the off-site specialist employee B may be called upon to respond to incidents involving chemicals. The off-site specialist employee B may be used to gather and record information, provide technical advice, and provide technical assistance (including work within the hot zone) at the incident consistent with his or her organization's emergency response plan and standard operating procedures and the local emergency response plan.

Off-site specialist employees at Level B are expected by their organization to respond *only* within their areas of expertise. They may have expertise in a number of fields, including:

- Product specialties
- Industrial hygiene/toxicological specialties

- Environmental waste/remediation specialties
- Monitoring specialties
- Container specialties
- Material handling, such as loaders and unloaders
- Maintenance specialties.

Off-site specialist employees at the B level need only meet those competencies required in their area of specialization.

6-3.1.3   Goal. The goal of these competencies is to ensure that the off-site specialist employee B has the knowledge and skills to safely perform the duties and responsibilities assigned in his or her organization's emergency response plan and standard operating procedures. Therefore, within his or her individual area of specialization, the off-site specialist employee B shall be able to:

(a) Assist the incident commander in analyzing the magnitude of an incident involving chemicals and/or containers for chemicals within his or her individual area of specialization by completing the following tasks:

1. Provide and interpret information on the hazards and harmful effects of specific chemicals within his or her individual area of specialization.

2. Provide and interpret information on the characteristics of specific containers for chemicals within his or her individual area of specialization.

3. Provide information on concentrations of chemicals within his or her individual area of specialization from exposure monitoring, dispersion modeling, or any other predictive method.

(b) Assist the incident commander in planning a response to an incident involving chemicals and/or containers for chemicals within his or her individual area of specialization by completing the following tasks:

1. Provide information on the potential response options and their consequences for specific chemicals and/or containers for chemicals within his or her individual area of specialization.

2. Provide information on the personal protective equipment requirements for a specific chemical within his or her individual area of specialization.

3. Provide information on the decontamination methods for a specific chemical within his or her individual area of specialization.

4. Provide information on the federal/provincial regulations that relate to the handling and/or disposal of a specific chemical within his or her individual area of specialization.

5. Develop a plan of action (within the capabilities of the available resources), including safety considerations, for handling chemicals and/or containers for chemicals within his or her individual area of specialization consistent with his or her organization's emergency response plan and standard operating procedures.

(c) Implement the planned response, as developed with the incident commander, for chemicals and/or containers for chemicals within his or her individual area of specialization, consistent with his or her organization's emergency response plan and standard operating procedures and within the capabilities of the available resources, by completing the following tasks:

1. Perform response options within his or her individual area of specialization, specified in the plan of action, as agreed upon with the incident commander and consistent with his or her organization's emergency response plan and standard operating procedures (within the capabilities of the available resources).
2. Don, work in, and doff personal protective equipment needed to implement the response options consistent with his or her organization's emergency response plan and standard operating procedures.

(d) Assist the incident commander to evaluate the results of implementing the planned response by completing the following tasks:

1. Provide feedback on the effectiveness of the response options taken within his or her individual area of specialization.
2. Provide reporting and subsequent documentation of the incident involving chemicals as required in his or her organization's emergency response plan and standard operating procedures.

6-3.2 Competencies — Analyzing the Problem.

6-3.2.1 Providing and Interpreting Information on Hazards of Specific Chemicals. Given a specific chemical within his or her individual area of specialization and an appropriate material safety data sheet (MSDS) or other appropriate resource, the off-site specialist employee B shall advise the incident commander of the chemical's hazards and harmful effects and the potential consequences based on the incident. The off-site specialist employee B shall be able to:

6-3.2.1.1 Given a specific chemical within his or her individual area of specialization and an appropriate material safety data sheet (MSDS), identify and interpret the following hazard information:

(a) Physical and chemical characteristics;
(b) Physical hazards of the chemical (including fire and explosion hazards);
(c) Health hazards of the chemical;
(d) Signs and symptoms of exposure;
(e) Routes of entry;
(f) Permissible exposure limits;
(g) Reactivity hazards; and
(h) Environmental concerns.

All off-site specialist employees at Level B must be able to locate the above hazard information in their organization's MSDSs and understand it. The product specialist described in 6-3.1.2 must be able to interpret the above information and advise the incident commander about it. If the Level B responder cannot interpret this information, he or she must be able to contact the individuals in his or her organization who can interpret it.

6-3.2.1.2 Given examples of specific chemicals used in his or her individual area of specialization and the appropriate resources (as identified in his or her organization's emergency response plan and standard operating procedures), predict the potential behavior of the chemicals based on the damage found, including the consequences of that behavior.

Off-site specialist employees at Level B should be able to predict what will happen when a liquid, solid, or gas is released from its container. What will occur, for example, if the chemical is exposed to air? Will it react, vaporize, or ignite?

6-3.2.1.3 Identify the general types of hazard information available from the other resources identified in his or her organization's emergency response plan and standard operating procedures.

Among the types of hazard information available to the off-site specialist employee from sources identified in the organization's E/R plan and SOPs are containment techniques, medical treatment protocols, container design, reactivity data, and decontamination and remediation procedures.

6-3.2.2 **Providing Information on Characteristics of Specific Containers.** Given a container for specific chemicals within his or her individual area of specialization, the off-site specialist employee B shall advise the incident commander of the characteristics and potential behavior of that container. The off-site specialist employee B shall be able to:

All off-site specialist employees at Level B should be able to identify the person or persons in their organizations they can contact for information on such things as damage assessment, fitting arrangement, and the probability of container failure.

6-3.2.2.1 Given examples of containers for specific chemicals used in his or her individual area of specialization, identify the purpose and operation of the closures found on those containers.

For the containers to which he or she responds, the container specialist must be able to identify the following:

(a) Nonbulk Containers:

   1. Drums
     a. type of bungs and seal caps
     b. closed and open
     c. head drums
     d. closure rings
     e. safety relief devices

   2. Cylinders
     a. valves
     b. valve caps
     c. relief devices
     d. purge devices

(b) Bulk Containers:

   1. Tote bins
     a. valves
     b. bungs, safety and relief devices
     c. spouts
     d. secondary closures
     e. purge devices
     f. couplings

   2. Cargo tanks
     a. venting devices
     b. valves
     c. relief purge devices
     d. connections
     e. manways
     f. emergency shutoff valves

g. excess flow valves

h. gauging devices

i. sampling devices

j. secondary closures

k. safety relief devices

3. Tank cars

a. valves

b. safety and relief devices

c. connections

d. sampling valves

e. vapor and liquid valves

f. heater coils

g. caps

h. bottom outlet

i. dome cover

j. dome gasket, etc.

4. Hopper cars and hopper trucks

a. gates

b. manways

5. ISO containers

a. venting devices

b. valves

c. relief purge devices

d. connections

e. manways

f. emergency shutoff valves

g. excess flow valves

h. gauging devices

i. sampling devices

j. secondary closures

k. safety and relief devices.

This list should also include any other types of containers, such as sample containers and supersacks, that the responder's organization uses. The off-site specialist employee B should know how to contact his or her organization's container specialist, who can provide this type of information.

6-3.2.2.2 Given a chemical container within his or her individual area of specialization, list the types of damage that could occur.

Level B off-site specialist employees who are expected to respond as container specialists should be able to identify the type of damage the containers their organizations use could incur. Such damage might include punctures, scores, gouges, blown rupture disks, damaged gaskets, corrosion, damaged "O" rings, liner failure, weld seam failure, cracked bungs, and frictional damage on drums, among other things.

The off-site specialist employee B should know how to contact his or her organization's container specialist, who can provide this type of information.

6-3.2.2.3 Given examples of containers for specific chemicals within his or her individual area of specialization and the appropriate resources (as identified in his or her organization's emergency response plan and standard operating procedures), predict the potential behavior of the containers and the consequences, based on the damage found.

Based on the types of damage listed in 6-3.2.2.2, all Level B off-site specialist employees who respond as container specialists should be able to assess potential container failures. They should know, for example, that a scored container may BLEVE, that a sheared valve on a cylinder may cause the container to rocket, that a bulging drum may rupture, and that damaged or stripped threads may cause a product release.

If the off-site specialist employee B cannot fulfill this requirement, he or she should know how to contact a container specialist.

6-3.2.2.4 Given his or her organization's emergency response plan and standard operating procedures, identify resources (including a method of contact) knowledgeable in the design, construction, and damage assessment of containers for chemicals within his or her individual area of specialization.

These resources may include, but are not limited to, chemical and mechanical engineers, packaging specialists, and container manufacturers. The carrier, especially the cargo tank carrier, may be another resource.

6-3.2.3 Providing Information on Concentrations of Chemicals. Given a chemical used in his or her individual area of specialization and the applicable monitoring equipment provided by his or her organization for that chemical or the available predictive capabilities (e.g., dispersion modeling, exposure modeling), the off-site specialist employee B shall advise the incident commander of the concentrations of the released chemical and the implications of that information to the incident. The off-site specialist employee B shall be able to:

6-3.2.3.1 Identify the appropriate monitoring equipment for a chemical used in his or her individual area of specialization.

Off-site specialist employees at Level B who are trained to select and use monitoring equipment as part of their regular duties—industrial hygienists, for example—should be able to identify the equipment appropriate for monitoring a chemical used in their areas of specialization. These include:

Flammables: Combustible gas indicators
Corrosives: pH devices
Toxics: Colorimetric tubes, photoionization detectors, flame ionization detectors, other monitoring devices
Radioactives: Geiger counter

6-3.2.3.2 Use the appropriate monitoring equipment provided by his or her organization to determine the actual concentrations of a specific chemical used in his or her individual area of specialization.

All off-site specialist employees at the B level who are trained to use monitoring equipment as part of their normal duties—industrial hygienists, for example—should be able to use and demonstrate the monitoring devices appropriate for the chemicals to which they respond. The equipment should be operated in accordance with the manufacturer's instructions.

6-3.2.3.3 Given information on the concentrations of a chemical used in his or her individual area of specialization, interpret the significance of that concentration information relative to the hazards and harmful effects of the chemical to responders.

The off-site specialist employee B should be able to interpret the results of monitoring in terms of known hazards. If the specialist employee cannot, he or she should be able to contact the appropriate person in his or her organization who can.

6-3.2.3.4 Demonstrate field calibration and testing procedures, as necessary, for the monitoring equipment provided by his or her organization.

Off-site specialist employees at the B level who are trained in the use of monitoring equipment as part of their regular duties—industrial hygienists, for example—should be able to demonstrate field calibration and testing procedures for these instruments.

6-3.2.3.5 Given his or her organization's emergency response plan and standard operating procedures, identify the resources (including a method of contact) capable of providing monitoring equipment, dispersion modeling, and/or monitoring services for chemicals within his or her organization's area of specialization.

Level B off-site specialist employees should be familiar with their organizations' E/R plan, SOPs, and other resources so that they can identify the industrial hygienist, the site safety officer, the equipment supplier, and any other source who can provide this information.

6-3.3 Competencies — Planning the Response.

6-3.3.1 Providing Information on Potential Response Options and Consequences for Specific Chemicals. Given specific chemicals or containers within his or her individual area of specialization and the appropriate resources, the off-site specialist employee B shall advise the incident commander of the potential response options and their consequences. The off-site specialist employee B shall be able to:

6-3.3.1.1 Given a specific chemical within his or her individual area of specialization and an appropriate material safety data sheet (MSDS), identify and interpret the following response information:

(a) Precautions for safe handling, including hygiene practices, protective measures, and procedures for cleanup of spills/leaks;

Using their organizations' MSDSs, off-site specialist employees at Level B should be able to determine what personal protective equipment is needed to deal with spills and decontamination procedures, taking into consideration the class and state of the material, the external hazards, and the possible secondary hazards.

When dealing with flammables, for example, the Level B specialist employee should eliminate ignition sources, use sparkproof tools, and consider using foam to suppress vapors. When dealing with corrosives, the Level B specialist employee should ensure that the equipment and tools specific to the materials involved in the incident are compatible with the material and type of neutralizing materials being used. For example, certain chemicals may react with some metal shovels, pumps, and hoses. When dealing with poison, the Level B specialist employee should consider evacuation or protection in place; should not allow eating, smoking, or gum chewing in hazardous areas; and should ensure that any personal protective equipment and tools that have been used are properly decontaminated.

(b) Applicable control measures, including personal protective equipment; and

Level B off-site specialist employees should be able to locate and identify the above information from an MSDS. Interpreting specific control and remediation procedures may require the expertise of a product specialist, an industrial hygienist, or some other appropriate person in the organization, whom the off-site specialist employee at the B level must know how to contact.

(c) Emergency and first aid procedures.

The first aid section of an MSDS is generally self-explanatory. The off-site specialist employee B should pay particular attention to antidotes, if there are any.

6-3.3.1.2 Given his or her organization's emergency response plan and standard operating procedures, identify additional resources for interpreting response information for a chemical within his or her organization's area of specialization.

Off-site specialist employees at Level B should be able to demonstrate how to access additional sources of assistance from their organization's resources, such as the E/R plan and SOPs.

6-3.3.1.3 Describe the advantages and limitations of the potential response options for a specific chemical within his or her individual area of specialization.

The off-site specialist employee should be able to associate the appropriate response option to the chemicals for which he or she is expected to respond.

Table 6.1 lists three possible options.

Table 6.1

Event	Response	Advantages	Disadvantages
Pesticide fire	Allow to burn	• Minimal run-off • Minimal containment	• Products of combustion • Demand on fire department
Acid spill	Neutralize run-off	• Reduces hazard • Lessens corrosivity	• Exposure from process • Vapor release
Poisonous liquid releasing vapor	Apply foam blanket	• Suppresses vapor	• Contributes to contamination, increases cleanup

6-3.3.1.4 Given his or her organization's emergency response plan and standard operating procedures, identify resources (including a method of contact) capable of:

(a) Repairing containers for chemicals within his or her individual area of specialization;

(b) Removing the contents of containers for chemicals within his or her individual area of specialization; and

(c) Cleanup and disposal of chemicals and/or containers for chemicals within his or her individual area of specialization.

The off-site specialist employee B should be able to locate individuals, identified by such organizational resources as the E/R plan and SOPs, who can accomplish the following tasks, tasks they normally perform in their day-to-day activities:

Normal Job Function	Task(s)
Material handler Container specialist	Repair containers
Material handler Loader/unloader Container specialist	Removing container contents
Environmental specialist	Cleaning up site Disposing of chemicals

6-3.3.2 **Providing Information on Personal Protective Equipment Requirements.** Given specific chemicals and/or containers for chemicals within his or her individual area of specialization and the appropriate resources, the off-site specialist employee B shall advise the incident commander of the appropriate personal protective equipment necessary for various response options. The off-site specialist employee B shall be able to:

6-3.3.2.1 Given a specific chemical within his or her individual area of specialization and an appropriate material safety data sheet (MSDS), identify personal protective equipment, including the materials of construction, that will be compatible with that chemical.

6-3.3.2.2 Given his or her organization's emergency response plan and standard operating procedures, identify other appropriate resources (including a method of contact) capable of identifying the personal protective equipment that is compatible with a specific chemical.

Among the resources a Level B off-site specialist employee can use to identify personal protective equipment compatible with a specific chemical are an organization's safety and product specialists and such references as compatibility charts, data bases, and PPE manufacturers.

6-3.3.2.3 Given an incident involving a specific chemical used in his or her individual area of specialization and the response options for that problem, determine whether the personal protective equipment provided by the organization is appropriate for the options presented.

All off-site specialist employees at Level B who are trained to select and/or use PPE should be able to identify, interpret, and apply compatibility data from compatibility charts and consider the problems associated with the use of Level A equipment, especially if there is potential for fire.

6-3.3.3 Providing Information on Decontamination Methods. Given a specific chemical within his or her individual area of specialization and the available resources, the off-site specialist employee B shall identify appropriate decontamination methods for various response options. The off-site specialist employee B shall be able to:

6-3.3.3.1 Given a specific chemical within his or her individual area of specialization and a material safety data sheet (MSDS) or other resource, obtain the potential methods for removing or neutralizing that chemical.

The off-site specialist employee B should be able to point out the sections on an MSDS that include decontamination and neutralization information or be able to contact an individual in the organization who can provide this information.

6-3.3.3.2 Given a specific chemical within his or her individual area of specialization and a material safety data sheet (MSDS) or other resource, identify the circumstances under which disposal of contaminated equipment would be necessary.

Materials that cannot be decontaminated, such as porous materials like leather or wood; limited-use equipment, such as one-time-use protective clothing; or equipment for which decontamination procedures are unknown may have to be disposed of.

6-3.3.3.3 Given his or her organization's emergency response plan and standard operating procedures, identify resources (including a method of contact) capable of identifying potential decontamination methods for chemicals within his or her individual area of specialization.

See 6-3.3.3.1 of NFPA 472 and Chapter 7 of NFPA 471.

6-3.3.4 **Providing Information on Handling and Disposal Regulations.** Given a specific chemical within his or her area of specialization and the available resources, the off-site specialist employee B shall advise the incident commander of the federal or provincial regulations that relate to the handling, transportation, and/or disposal of that chemical. The off-site specialist employee B shall be able to:

6-3.3.4.1 Given a specific chemical within his or her individual area of specialization and a material safety data sheet (MSDS) or other resource, identify federal or provincial regulations that apply to the handling, transportation, and/or disposal of that chemical.

Among the regulations that apply to the handling, transportation, and disposal of chemicals are Title 49 of CFR-DOT, Title 40 of CFR-EPA, and, in Canada, TDG and/or MOE. The agencies that handle these regulations should be contacted by someone identified in the organization's E/R plan or SOPs as the person responsible for making notifications according to regulations.

6-3.3.4.2 Given a specific chemical within his or her individual area of specialization and a material safety data sheet (MSDS) or other resource, identify the agencies (including a method of contact) responsible for compliance with the federal or provincial regulations that apply to the handling, transportation, and/or disposal of a specific chemical.

The agencies responsible for compliance with regulations applying to the handling, transportation, and disposal of a specific chemical include the Occupational Safety and Health Administration (OSHA), which covers emergency response activities; the U.S. Department of Transportation (DOT), which deals with transportation; and the Environmental Protection Agency (EPA), which handles hazardous materials disposal in the U.S. Transport Canada, Environment Canada, and Labor Canada cover the same issues for the provinces. These agencies should be contacted by someone identified in the organization's E/R plan or SOPs as the person responsible for making the necessary notifications according to regulations. See next item.

6-3.3.4.3 Given his or her organization's emergency response plan and standard operating procedures, identify resources for information pertaining to federal or provincial regulations relative to the handling and/or disposal of a specific chemical.

The off-site specialist employee B should be able to contact a knowledgeable person in his or her organization, such as an environmental specialist, who will know where in the organization's E/R plan or SOPs to locate the needed information. See 6-3.3.4.1 and 6-3.3.4.2.

6-3.3.5 **Developing a Plan of Action.** Given a simulated incident involving chemicals and/or containers used in his or her individual area of specialization, the off-site specialist employee B shall (in conjunction with the incident commander) develop a plan of action, consistent with his or her organization's emergency response plan and standard operating procedures, for handling chemicals and/or containers in that incident. The plan of action developed shall be within the capabilities of the available resources, and shall include safety considerations. The off-site specialist employee B shall be able to:

6-3.3.5.1 Given his or her organization's emergency response plan and standard operating procedures, identify the process for development of a plan of action, including safety considerations.

Each hazardous materials incident is required to have an overall plan of action. However, off-site specialist employees at Level B are expected to develop an action plan for only those tasks and procedures associated with their areas of expertise and regular job duties. They are also expected to define the processes needed to execute that plan.

For example, a loader/unloader who transloads material as part of his or her normal duties may be called upon to develop an action plan for performing the same procedure at the scene of a hazardous materials incident and to define the steps and safety considerations required to execute that plan.

6-3.4 Competencies — Implementing the Planned Response.

6-3.4.1 Performing Response Options Specified in the Plan of Action. Given an assignment by the incident commander within his or her individual area of specialization, the off-site specialist employee B shall perform the assigned actions consistent with his or her organization's emergency response plan and standard operating procedures. The off-site specialist employee B shall be able to:

6-3.4.1.1 Perform assigned tasks consistent with his or her organization's emergency response plan and standard operating procedures and the available personnel, tools, and equipment (including personal protective equipment), including:

(a) Confinement activities;

If the off-site specialist employee B specializes in an area that includes confinement activities, he or she should be able to perform such operations as diking, damming, absorption, and appropriate vapor dispersion and suppression techniques.

(b) Containment activities; and

If the off-site specialist employee B specializes in an area that includes containment activities, he or she should be able to perform such operations as patching, plugging, using capping devices, and repairing and replacing closures.

(c) Product removal activities.

If the off-site specialist employee B specializes in an area that includes product removal techniques, he or she should be able to transfer, vacuum, vent, flare, and scrub hazardous materials or to identify the appropriate resources and equipment for performing such activities.

6-3.4.1.2* Identify factors that may affect his or her ability to perform the assigned tasks.

A-6-3.4.1.2 Factors include heat, cold, working in confined space, working in personal protective equipment, working in a flammable or toxic atmosphere, and preexisting health conditions.

6-3.4.2 Using Personal Protective Equipment. Given an assignment by the incident commander within his or her individual area of specialization, the off-site specialist employee B shall don, work in, and doff the appropriate personal protective equipment needed to implement the assigned response options, consistent with his or her organization's emergency response plan and standard operating procedures. The off-site specialist employee B shall be able to:

6-3.4.2.1 Don, work in, and doff the appropriate respiratory protection and protective clothing for the assigned tasks consistent with his or her organization's emergency response plan and standard operating procedures.

6-3.4.2.2 Identify the safety procedures for personnel wearing personal protective equipment, including:

(a) Buddy system;

All responders should be dressed in the same level of protection so that they can observe and assist each other as required.

(b) Back-up personnel;

Safety procedures require that back-up responders dress in the same level of protection as those on the entry team so that they can perform rescue if they must.

(c) Symptoms of heat and cold stress;

Safety procedures require that personnel be monitored for symptoms such as abnormal pulse, temperature, respiration, changes in skin color, and decreased mental alertness.

(d) Limitations of personnel working in personal protective equipment;

Safety procedures establish time limits for wearing personal protective clothing during an incident. Other limiting factors include changes in the physiological and psychological condition of the individual.

(e) Indications of material degradation of chemical-protective clothing;

Safety procedures require that responders evaluate signs of material degradation, which include discoloration, the loss of integrity and flexibility, the formation of blisters, and melting or stretching of material.

(f) Physical and psychological stresses on the wearer; and

Safety procedures required to maintain the overall well-being of the responder include monitoring for signs of stress. Stress may cause a rise in body temperature, loss of body fluids, elevated pulse and respiration rates, vertigo, nausea, changes in skin color, disorientation, anxiety, and incoherence.

(g) Emergency procedures and hand signals.

Safety procedures require that responders maintain visual contact and that they demonstrate hand signals for loss of air and emergency escape. See also 3-4.3 and 4-3.3 of NFPA 472 for additional information on PPE.

6-3.4.2.3 Identify the procedures for cleaning, sanitizing, and inspecting personal protective equipment provided by the employer.

These procedures should be performed in accordance with the organization's SOPs concerning the use and maintenance of personal protective equipment. The tasks must be performed in accordance with existing regulations such as OSHA 29 CFR 1910.120, "Hazardous Waste Operations and Emergency Response,"[4] and OSHA 29 CFR 1910.134, "Respiratory Protection."[5]

6-3.5 Competencies — Evaluating Progress.

6-3.5.1 Providing an Evaluation of the Effectiveness of Selected Response Options. Given an incident involving specific chemicals and/or containers for chemicals within his or her individual area of specialization, the off-site specialist employee B shall advise the incident commander of the effectiveness of the selected response options. The off-site specialist employee B shall be able to:

6-3.5.1.1 Identify the criteria for evaluating whether or not the selected response options are effective in accomplishing the objectives.

These criteria must be based on the desired outcome, and the responder must determine how the hazards to personnel, the environment, and property change. Effective options result in diminished hazards.

6-3.5.1.2 Identify the circumstances under which it would be prudent to pull back from a chemical incident.

The responder should withdraw from a chemical incident when:

(a) Intervention will not or cannot produce a favorable outcome;
(b) The immediate hazard level is unacceptable; or
(c) The incident is worsening as a result of the option selected.

6-3.5.2 **Reporting and Documenting the Incident.** Given a simulated incident involving chemicals and/or containers for chemicals used in his or her individual area of specialization, the off-site specialist employee B shall complete the reporting and subsequent documentation requirements consistent with his or her organization's emergency response plan and standard operating procedures. The off-site specialist employee B shall be able to:

6-3.5.2.1 Identify the importance of documentation (including training records, exposure records, incident reports, and critique reports) for an incident involving chemicals.

Organizations require documentation in order to help them establish preventive and corrective actions. Documentation is also mandated by federal and usually state and local law. For example, it is required for emergency response training, medical monitoring, and PPE certification. Regulatory organizations require documentation for similar reasons, and this documentation often serves as the basis for establishing penalties. Finally, documentation is frequently used for trend analysis.

6-3.5.2.2 Identify the steps used in keeping an activity log and exposure records as described in his or her organization's emergency response plan and standard operating procedures.

Most E/R plans require that an employee's name and identifying code (usually a social security number) be recorded against the functions he or she performs at a hazardous materials incident and that the time and duration of each activity be accurately documented. Exposure records are compiled in a similar manner, listing the materials to which the employee may be exposed, the duration of exposure, the manner in which the contaminant was determined, and the name of the person who made that determination.

6-3.5.2.3 Identify the requirements for compiling incident reports from his or her organization's emergency response plan and standard operating procedures.

These requirements are a means of providing a factual, objective format for defining the how, why, what, where, when, and who of an incident.

6-3.5.2.4 Identify the requirements for compiling hot zone entry and exit logs from his or her organization's emergency response plan and standard operating procedures.

This information is required so that those in charge of an incident can monitor an employee's health and keep track of the work he or she has performed and the time he or she has spent in the hot zone. See 6-3.5.2.2.

6-3.5.2.5 Identify the requirements for compiling personal protective equipment logs from his or her organization's emergency response plan and standard operating procedures.

Personal protective equipment logs should include information about use time, inspections, testing, and the results of inspection and decontamination procedures. This documentation should also include a list of the contaminants to which the equipment has been exposed, as well as the duration of the exposure.

6-3.5.2.6 Identify the requirements for filing documents and maintaining records as prescribed in his or her organization's emergency response plan and standard operating procedures.

Most organizations keep records to protect themselves and their employees and to provide a written account of the incident. Appropriate sources within the organization, such as regulatory specialists, should be consulted for both internal and external reporting.

6-4 Off-site Specialist Employee A.

6-4.1 General.

6-4.1.1 Introduction. The off-site specialist employee A shall be trained to meet the competencies at the off-site specialist employee C (Section 6-2 in this chapter), off-site specialist employee B (Section 6-3 in this chapter), and hazardous materials technician (Chapter 4) levels relative to the chemicals and containers used in his or her organization's area of specialization. In addition, the off-site specialist employee A shall receive training to meet any applicable federal Department of Transportation (DOT), Occupational Safety and Health Administration (OSHA), Environmental Protection Agency (EPA), or local occupational health and safety regulatory agency requirements.

Off-site specialist employees at Level A can provide the incident commander with considerable assistance since their level of competence when dealing with the range of products and containers their organizations produce, use, or handle would be equivalent to that of the hazardous materials technician.

Level A off-site specialist employees need only demonstrate those technician-level competencies that apply to the materials and containers for which they are expected to respond.

6-4.1.2 Definition. Level A off-site specialist employees are those persons who are specifically trained to handle incidents involving chemicals and/or containers for chemicals used in their organization's area of specialization. Consistent with his or her organization's emergency response plan and standard operating procedures, the off-site specialist employee A shall be able to analyze an incident involving chemicals within his or her organization's area of specialization, plan a response to that incident, implement the planned response within the capabilities of the resources available, and evaluate the progress of the planned response.

6-4.1.3 Goal. The goal of this level of competence is to ensure that the off-site specialist employee A has the knowledge and skills to safely perform the duties and responsibilities assigned in his or her organization's emergency response plan and standard operating procedures. Therefore, in addition to being competent at the off-site specialist employee C and off-site specialist employee B levels, the off-site specialist employee A shall be able to, in conjunction with the incident commander:

(a) Analyze an incident involving chemicals and containers for chemicals used in his or her organization's area of specialization to determine the magnitude of the incident by completing the following tasks:

    1. Survey an incident involving chemicals and containers for chemicals used in his or her organization's area of specialization to:

a. Identify the containers involved;

See 4-2.1.1.

b. Identify or classify unknown materials; and

See 4-2.1.3.

c. Verify the identity of the chemicals.

2. Collect and interpret hazard and response information from printed resources, technical resources, computer data bases, and monitoring equipment for chemicals used in his or her organization's area of specialization.

See 4-2.2.

3. Determine the extent of damage to containers of chemicals used in his or her organization's area of specialization.

See 4-2.3.

4. Predict the likely behavior of the chemicals and containers for chemicals used in his or her organization's area of specialization.

See 4-2.4.

5. Estimate the potential outcomes of an incident involving chemicals and containers for chemicals used in his or her organization's area of specialization.

See 4-2.5.

(b) Plan a response (within the capabilities of available resources) to an incident involving chemicals and containers for chemicals used in his or her organization's area of specialization by completing the following tasks:

1. Identify the response objectives for an incident involving chemicals and containers for chemicals used in his or her organization's area of specialization.

See 4-3.1.1.

2. Identify the potential action options for each response objective for an incident involving chemicals and containers for chemicals used in his or her organization's area of specialization.

See 4-3.2.

3. Select the personal protective equipment required for a given response option for an incident involving chemicals and containers for chemicals used in his or her organization's area of specialization.

See 4-3.3.

4. Select the appropriate decontamination procedures, as necessary, for an incident involving chemicals and containers for chemicals used in his or her organization's area of specialization.

See 4-3.4.

5. Develop a plan of action (within the capabilities of the available resources), including safety considerations, for handling an incident involving chemicals and containers for chemicals used in his or her organization's area of specialization consistent with his or her organization's emergency response plan and standard operating procedures.

See 4-3.5.

(c) Implement the planned response (as developed with the incident commander) to an incident involving chemicals and containers for chemicals used in his or her organization's area of specialization consistent with his or her organization's emergency response plan and standard operating procedures.

1. Don, work in, and doff appropriate personal protective equipment provided by his or her organization for chemicals used in his or her organization's area of specialization, consistent with his or her organization's emergency response plan and standard operating procedures.

See 4-4.2.

2. Perform control functions, as agreed upon with the incident commander, for chemicals and containers for chemicals used in his or her organization's area of specialization consistent with his or her organization's emergency response plan and standard operating procedures.

See 4-4.3.

(d) Evaluate the results of implementing the planned response to an incident involving chemicals and containers for chemicals used in his or her organization's area of specialization.

References Cited in Commentary

1. *Emergency Handling of Hazardous Materials in Surface Transportation* (Washington, DC: Association of American Railroads).

2. OSHA 29 CFR 1910.1200 (Washington, DC: U.S. Government Printing Office).

3. *Packaging for Transporting Hazardous and Non-hazardous Materials* (Washington, DC: Chemical Manufacturers Association, 1989).

4. OSHA 29 CFR 1910.120, "Hazardous Materials Waste Operations and Emergency Response Plan" (Washington, DC: U.S. Government Printing Office).

5. OSHA 29 CFR 1910.134, "Respiratory Protection" (Washington, DC: U.S. Government Printing Office).

7

Referenced Publications

The following documents or portions thereof are referenced within this standard and shall be considered part of the requirements of this document. The edition indicated for each reference is the current edition as of the date of the NFPA issuance of this document.

7-1.1 **NFPA Publications.** National Fire Protection Association, 1 Batterymarch Park, P.O. Box 9101, Quincy, MA 02269-9101.

NFPA 704, *Standard System for the Identification of Fire Hazards of Materials*, 1990 edition
NFPA 1991, *Standard on Vapor-Protective Suits for Hazardous Chemical Emergencies*, 1990 edition
NFPA 1992, *Standard on Liquid Splash-Protective Suits for Hazardous Chemical Emergencies*, 1990 edition.

7-1.2 Other Publications.

7-1.2.1 **U.S. Government Publications.** U.S. Government Printing Office, Superintendent of Documents, Washington, DC 20402.

Title 29 CFR Part 1910.120

Title 29 CFR Part 1910.1200

Emergency Response Guidebook, U.S. Department of Transportation DOT P 5800.4, 1990 edition.

This Appendix is not a part of the requirements of this NFPA document, but is included for information purposes only.

The material contained Appendix A of this standard is included within the text of this handbook, and therefore is not repeated here.

Referenced Publications

B-1

The following documents or portions thereof are referenced within this standard for informational purposes only and thus are not considered part of the requirements of this document. The edition indicated for each reference is the current edition as of the date of the NFPA issuance of this document.

B-1.1 **NFPA Publication.** National Fire Protection Association, 1 Batterymarch Park, P.O. Box 9101, Quincy, MA 02269-9101.

Hazardous Materials Response Handbook.

B-1.2 **Other Publications.**

B-1.2.1 **Chemical Manufacturers Association Publication.** Chemical Manufacturers Association, 2501 M Street NW, Washington, DC 20037.

Packaging for Transporting Hazardous and Non-hazardous Materials, June 1989 edition.

B-1.2.2 **National Fire Academy Publication.** National Fire Academy, Federal Emergency Management Agency, Emmitsburg, MD 21727.

Hazardous Materials Incident Analysis.

B-1.2.3 **National Response Team Publication.** National Response Team, National Oil and Hazardous Substances Contingency Plan, Washington, DC 20593.

NRT-1, *Hazardous Materials Emergency Planning Guide.*

B-1.2.4 **U.S. Government Publications.** U.S. Government Printing Office, Superintendent of Documents, Washington, DC 20402.

Code of Federal Regulations:
Title 40 CFR Part 261.33
Title 40 CFR Part 302
Title 40 CFR Part 355
Title 49 CFR Parts 170–179.
Department of Transportation:
Research and Special Programs Administration, "Performance-Oriented Packaging Standards; Changes to Classification, Hazard Communication, Packaging and Handling Requirements based on UN Standards and Agency Initiative; Final Rule," *Federal Register*, Vol. 55, No. 246, December 21, 1990, pp. 52,402–52,729.

B-1.2.5 **Miscellaneous Publications.**

Benner, Ludwig, Jr., *A Textbook for Use in the Study of Hazardous Materials Emergencies,* 2nd edition, Lufred Industries, Inc., Oakton, VA, 1978

Noll, Gregory G., et al., *Hazardous Materials: Managing the Incident*, Fire Protection Publications, Stillwater, OK, 1988

NIOSH/OSHA/USCG/EPA *Occupational Safety and Health Guidance Manual for Hazardous Waste Site Activities*, October 1985

Wright, Charles J., "Managing the Hazardous Materials Incident," *Fire Protection Handbook*, 17th edition, National Fire Protection Association, Quincy, MA, 1991.

Standard and Commentary for NFPA 473, *Standard for Competencies for EMS Personnel Responding to Hazardous Materials Incidents*

1992 Edition

The text and illustrations that make up the commentary on the various sections of NFPA 473 are printed in black. The text of the standard itself is printed in blue.

Paragraphs that begin with the letter "A" are extracted from Appendix A of the standard. Appendix A material is not mandatory. It is designed to help users apply the provisions of the standard. In this handbook, material from Appendix A is integrated with the text, so that it follows the paragraph it explains. An asterisk (*) following a paragraph number indicates that explanatory material from Appendix A will follow.

1

Administration

1-1 Scope.

This standard identifies the levels of competence required of emergency medical services (EMS) personnel who respond to hazardous materials incidents. It specifically covers the requirements for basic life support and advanced life support personnel in the prehospital setting.

In developing this standard, every effort was made to ensure that the needs and operations of the various types of emergency medical services (EMS) systems were considered. Thus, this document can be applied to emergency medical services associated with fire departments, third services, private ambulance services, and industrial emergency care services. The competencies can be achieved by both career and volunteer personnel.

The standard is based on an analysis of the tasks EMS personnel perform at hazardous materials incidents. After these tasks were identified, the competencies necessary for their performance were developed. Since proficiency in these competencies should be the final result of any training program intended to prepare EMS personnel to respond to hazardous materials incidents, these competencies can serve as the basis for developing training programs in this area.

After many revisions, it became apparent that EMS responders to hazardous materials incidents should be classified on two levels. To avoid confusion with names attached to other hazardous materials responders, these levels will be referred to as Level I and Level II.

1-2 Purpose.

The purpose of this standard is to specify minimum requirements of competence and to enhance the safety and protection of response personnel and all components of the emergency medical services system. It is not the intent of this standard to restrict any jurisdiction from exceeding these minimum requirements. (*See Appendix B.*)

The primary objective of response to hazardous materials incidents is "to save lives and prevent injuries." Since many hazardous materials incidents result in victims who need treatment and transportation to medical facilities, the services of EMS personnel are vital. Management of EMS operations and hazardous materials patients differs from the management of other medical operations and patients, and this standard provides information on the knowledge and skills necessary to conduct safe and effective EMS operations.

1-2.1 The competency requirements for EMS personnel contained herein have been prepared to reduce the numbers of accidents, exposures, and injuries resulting from hazardous materials incidents.

Prevention, as well as treatment, was taken into account when the competencies for operations and emergency medical care were developed. By performing their duties effectively, EMS responders can prevent accidents, exposures, and injuries to themselves, individuals caught in the incident, and other emergency responders.

1-3 Definitions.

Advanced Life Support (ALS).

Emergency Medical Technician-Paramedic (EMT-P). An individual who has successfully completed a course of instruction that meets or exceeds the requirements of the U.S. Department of Transportation National Standard EMT-Paramedic Curriculum and who holds an EMT-P certification from the authority having jurisdiction.

Emergency Medical Technician-Intermediate (EMT-I). (This category may include EMT-Cardiac.) An individual who has completed a course of instruction that includes selected modules of the U.S. Department of Transportation National Standard EMT-Paramedic Curriculum and who holds an intermediate level EMT-I or EMT-C certification from the authority having jurisdiction.

EMT-P and EMT-I are nationally recognized classifications of advanced life support providers. The length of their training and the procedures they perform at each level vary from state to state and sometimes from jurisdiction to jurisdiction within states. The U.S. Department of Transportation standard curricula serve as the basis for the majority of the programs.

Authority Having Jurisdiction. The "authority having jurisdiction" is the organization, office or individual responsible for "approving" equipment, an installation or a procedure.

NOTE: The phrase "authority having jurisdiction" is used in NFPA documents in a broad manner since jurisdictions and "approval" agencies vary as do their responsibilities. Where public safety is primary, the "authority having jurisdiction" may be a federal, state, local or other regional department or individual such as a fire chief, fire marshal, chief of a fire prevention bureau, labor department, health department, building official, electrical inspector, or others having statutory authority. For insurance purposes, an insurance inspection department, rating bureau, or other insurance company representative may be the "authority having jurisdiction." In many circumstances the property owner or his designated agent assumes the role of the "authority having jurisdiction"; at government installations, the commanding officer or departmental official may be the "authority having jurisdiction."

In a document that addresses the broad concept of hazardous materials, there will be many authorities exercising jurisdiction. These will include, at least in some cases and to some extent, the manufacturer, the shipper, and the local community. Every level of government will also play a role.

In this document, the authority having jurisdiction (AHJ) also includes the local medical control authority in the area. For purposes of patient care and protocols, the local medical control authority, rather than a regulatory agency, may be the designated AHJ in certain instances.

Basic Life Support (BLS).

Emergency Medical Technician-Ambulance (EMT-A). An individual who has completed a specified EMT-A course developed by the U.S. Department of Transportation and who holds an EMT-A certification from the authority having jurisdiction.

NOTE: This level in some jurisdictions may be recognized as EMT-Basic (EMT-B).

Emergency Care First Responder (ECFR). An individual who has successfully completed the specified Emergency Care First Responder course developed by U.S. Department of Transportation and who holds an ECFR certification from the authority having jurisdiction.

NOTE: In Canada the terminology used is: Emergency Medical Assistant-1 (EMA-1), Emergency Medical Assistant-2 (EMA-2), and Emergency Medical Assistant-3 (EMA-3).

Like advanced life support, basic life support is divided into two levels, and the length of training and the skills required by those certified at these levels differ. The U.S. Department of Transportation courses serve as models for most programs.

Cold Zone.   This area contains the command post and such other support functions as are deemed necessary to control the incident. This is also referred to as the clean zone or support zone in other documents.

It might appear that there is no outer boundary to the cold zone, but this is not the case. One might equate the outer boundary at a hazardous materials incident with the fire lines that are often established at a major fire or emergency and that are usually controlled by the police department. The public at large would not have access to the cold zone under most circumstances.

Competence.   Possessing knowledge, skills, and judgment needed to perform indicated objectives satisfactorily.

While knowledge and skills can be measured, judgment is not as easily evaluated. Under varying circumstances, the proper exercise of judgment or decision-making skills can change substantially. Nonetheless, training and experience can effectively improve the emergency decision-making process. Since the outcome of an incident is determined by the decisions made in handling it, the need for proper and adequate training becomes paramount.

Components of EMS System.   The parts of a comprehensive plan to treat an individual in need of emergency medical care following an illness or injury. These parts include:

(a) First responders,
(b) Emergency dispatching,
(c) EMS agency response,
(d) Hospital emergency departments,
(e) Specialized care facilities.

The provision of emergency medical care is generally a coordinated effort among personnel from several different agencies, each with a particular role to play in the prehospital care, transportation, and hospital management of sick or injured persons who have been exposed to hazardous materials.

Contaminant/Contamination.   A substance or process that poses a threat to life, health, or the environment.

Inherent in the definition of contaminant is the concept that the offending material is present where it should not be and does not belong. A second aspect essential to the meaning is that the offending material is somehow toxic or harmful.

Control.   The procedures, techniques, and methods used in the mitigation of a hazardous materials incident, including containment, extinguishment, and confinement.

The word *control* can be used interchangeably with the word *mitigation*. Every measure taken to control a hazardous materials incident is part of the mitigation process. Limiting the degree of contamination, by whatever measure, is also part of the control or mitigation process.

Control Zones.   The designation of areas at a hazardous materials incident based on safety and the degree of hazard. Many terms are used to describe the zones involved in a hazardous materials incident. For purposes of this standard, these zones shall be defined as the hot, warm, and cold zones.

Other terms commonly used include *site control zones* and *work zones*. The committee chose to use basic terms such as *hot*, *warm*, and *cold* because the words are simple and easily understood and they clearly suggest the nature of the situation one would expect to encounter in an area so designated.
See NFPA 471 and 472 for additional information.

Decontamination (Contamination Reduction).   The physical and/or chemical process of reducing and preventing the spread of contamination from persons and equipment involved in a hazardous materials incident.

Broadly speaking, decontamination includes actions taken at an incident that serve to reduce or even prevent contamination. The narrower meaning of the word suggests those steps taken to remove contaminants that may have accumulated on equipment or personnel at an incident. Good decontamination practices should incorporate and emphasize the notion of prevention as a first step.

Decontamination Area.   The area, usually located within the warm zone, where decontamination takes place.

The purpose of a decontamination area is to reduce the likelihood that contamination will spread to the cold zone. It is usually located in the warm zone because that is the transition area between the clean, or uncontaminated, area and the hot zone.

Demonstrate.   To show by actual use. This may be supplemented by simulation, explanation, illustration, or a combination of these.

Describe.   To explain verbally or in writing using standard terms recognized in the hazardous materials response community.

Gross Decontamination.   The initial phase of the decontamination process during which the amount of surface contaminant is significantly reduced. This phase may include mechanical removal and initial rinsing.

Decontamination is a multiphase process. In certain situations, only partial removal of contaminants is achieved. Gross decontamination, the first phase of the process, usually removes 80 to 90 percent of the contaminants.

Hazard/Hazardous.   Capable of posing an unreasonable risk to health, safety, or the environment; capable of doing harm.

Hazardous Materials.*   A substance (solid, liquid, or gas) capable of creating harm to people, property, and the environment.

Class/Division.   The general category of hazard assigned to a hazardous material under the DOT regulations. The division is a subdivision of a hazard class.

Class 1 (Explosives)

Division 1.1 — Explosives with a mass explosion hazard

Division 1.2 — Explosives with a projection hazard

Division 1.3 — Explosives with predominantly a fire hazard

Division 1.4 — Explosives with no significant blast hazard

Division 1.5 — Very insensitive explosives

Division 1.6 — Extremely insensitive explosive articles

Class 2

Division 2.1 — Flammable gas

Division 2.2 — Nonflammable, non-poisonous compressed gas

Division 2.3 — Poison gas

Division 2.4 — Corrosive gas (Canadian)

Class 3 (Flammable Liquid)

Division 3.1 — Flammable liquids, flash point $<0°F$

Division 3.2 — Flammable liquids, flash point $0°F$ & above but $<73°F$

Division 3.3 — Flammable liquids, flash point $73°F$ & up to $141°F$

Combustible Liquid

Class 4

Division 4.1 — Flammable solid

Division 4.2 — Spontaneously combustible material

Division 4.3 — Dangerous when wet material

Class 5

Division 5.1 — Oxidizer

Division 5.2 — Organic peroxide

Class 6

Division 6.1 — Poisonous material

Division 6.2 — Infectious material

Class 7 (Radioactive material)

Class 8 (Corrosive material)

Class 9 (Miscellaneous hazardous material)

ORM-D material

Refer to NFPA 472, Section A-2-2.1.2.

A-1-3 Hazardous Materials. There are many definitions and descriptive names being used for the term hazardous materials, each of which depends on the nature of the problem being addressed.

Unfortunately, there is no one list or definition that covers everything. The United States agencies involved, as well as state and local governments, have different purposes for regulating hazardous materials that, under certain circumstances, pose a risk to the public or the environment.

(a) *Hazardous Materials.* The United States Department of Transportation (DOT) uses the term hazardous materials, which covers eight hazard classes, some of which have subcategories called classification, and a ninth class covering other regulated materials (ORM). DOT includes in its regulations hazardous substances and hazardous wastes as an ORM-E, both of which are regulated by the Environmental Protection Agency (EPA), if their inherent properties would not otherwise be covered.

(b) *Hazardous Substances.* EPA uses the term hazardous substance for the chemicals that, if released into the environment above a certain amount, must be reported and, depending on the threat to the environment, for which federal assistance in handling the incident can be authorized. A list of the hazardous substances is published in 40 CFR Part 302, Table 302.4.

(c) *Extremely Hazardous Substances.* EPA uses the term extremely hazardous substance for chemicals that must be reported to the appropriate authorities if released above the threshold reporting quantity. Each substance has a threshold reporting quantity. The list of extremely hazardous substances is identified in Title III of Superfund Amendments and Reauthorization Act (SARA) of 1986 (40 CFR Part 355).

(d) *Toxic Chemicals.* EPA uses the term toxic chemical for chemicals whose total emissions or releases must be reported annually by owners and operators of certain facilities that manufacture, process, or otherwise use a listed toxic chemical. The list of toxic chemicals is identified in Title III of SARA.

(e) *Hazardous Wastes.* EPA uses the term hazardous wastes for chemicals that are regulated under the Resource, Conservation and Recovery Act (40 CFR Part 261.33). Hazardous wastes in transportation are regulated by DOT (49 CFR Parts 170-179).

(f) *Hazardous Chemicals.* The United States Occupational Safety and Health Administration (OSHA) uses the term hazardous chemical to denote any chemical that would be a risk to employees if exposed in the work place. Hazardous chemicals cover a broader group of chemicals than the other chemical lists.

(g) *Hazardous Substances.* OSHA uses the term hazardous substances in 29 CFR Part 1910.120, which resulted from Title I of SARA and covers emergency response. OSHA uses the term differently than EPA. Hazardous substances, as used by OSHA, cover every chemical regulated by both DOT and EPA.

Hazardous Materials Response Team. A group of trained response personnel operating under an emergency response plan and appropriate standard operating procedures to control or otherwise minimize or eliminate the hazards to people, property, or the environment from a released hazardous material.

In most cases, the duties and responsibilities of a constituted response team are clearly listed and described in a standard operating procedure manual. A few fire departments have "dedicated" response teams whose only function is to handle hazardous materials incidents. In a typical fire department, however, designated members of a hazardous materials response unit perform in other capacities, such as engine company, ladder company, or rescue squad, and respond to hazardous materials incidents when specially called upon to do so. The members so designated have been adequately trained for the hazardous materials response.

Fire departments are not required to have a designated hazardous materials response team. However, all members should be trained to the operational level to respond to incidents that might involve hazardous materials.

High Temperature-Protective Clothing. Protective clothing designed to protect the wearer from short-term high temperature exposures. This type of clothing is usually of limited use in dealing with chemical commodities.

This definition of high temperature-protective clothing comes from the Chemical Manufacturers Association (CMA); NFPA does not have a standard on this category of protective clothing or equipment. This type of protective clothing is also referred to as entry fire fighting protective clothing. The user should be aware of the stated limitations in the CMA definition, namely "short-term" and "limited use."

Hot Zone. Area immediately surrounding a hazardous materials incident, which extends far enough to prevent adverse effects from hazardous materials releases to personnel outside the zone. This zone is also referred to as the exclusion zone or restricted zone in other documents.

The hot zone is the area in which contamination does or could take place. It is also the area in which cleanup operations are performed. The boundary between the hot zone and the warm zone should be clearly indicated by some physical means, such as lines, hazard tape, equipment barriers, or the like. Movement of personnel from one zone to another must be tightly regulated and supervised to minimize contamination. This will allow for greater control of the operations within the zone.

Identify.   To select or indicate verbally or in writing using standard terms to establish the identity of; the fact of being the same as the one described.

Incident.   An event involving a hazardous material or a release or potential release of a hazardous material.

There is a wide difference of opinion as to what constitutes a hazardous materials incident when fire is involved. At one extreme are those who would like to categorize every fire as a hazardous materials incident because every fire produces toxic products of combustion that are clearly hazardous. A more reasonable approach is to limit the definition to the actual involvement in fire of an acknowledged hazardous material.

Frequently, however, the line between the two positions is blurred. Is a fire in a fuel storage tank a fire, or should it be classified as a hazardous materials incident? The answer may vary from one agency to another. A fire department might list it as a fire, for example, while an environmental agency might see it as a hazardous materials incident. Several other examples can be used to illustrate the problem, which should not be taken lightly. There are many more complicating factors involved in operating at a designated hazardous materials incident than there are in operating at a fire, if only from the standpoint of personnel.

Incident Commander.   The person responsible for the overall coordination and direction of all activities at the incident scene, as specified in NFPA 1561, *Standard on Fire Department Incident Management System*.

The incident commander can be compared to a fireground commander, who is responsible for overall control of operations. It is important that one person be in command of an incident, although the actual command may move from one person to another as an incident develops and intensifies. See NFPA 472, Chapter 5.

Incident Management System.   An organized system of roles, responsibilities, and standard operating procedures used to manage emergency operations, as described in NFPA 1561, *Standard on Fire Department Incident Management System*. Such systems are often referred to as "Incident Command Systems."

See Supplement 7 for an example of a hazardous materials incident management system.

Local Area. A geographic area that includes the defined response area and receiving facilities for an EMS agency.

An EMS system is an organized group of agencies that provide services within a defined geographic area. This area may be a municipality or several municipalities. For the purposes of this document, the term *local area* can be used to refer to recognized response and mutual assistance areas.

Local Emergency Planning Committee (LEPC). (As mandated by SARA Title III.) The LEPC must include elected state and local officials, police, fire, civil defense, public health professionals, environmental, hospital, and transportation officials as well as representatives of facilities, community groups, and the media.

The LEPC must develop and maintain a current hazardous materials response plan for the area it covers. This plan designates the responsibilities that EMS agencies and personnel are to undertake during hazardous materials emergencies.

Medical Control. The physician providing direction for patient care activities in the prehospital setting.

Medical control consists of both on-line and off-line medical control. This document does not specify whether the direction is given through a communication system or by written protocol.

Medical Surveillance. The ongoing process of medical evaluation of hazardous materials response team members and public safety personnel who may respond to a hazardous materials incident.

Objective. A goal that is achieved through the attainment of a skill, knowledge, or both that can be observed or measured.

Personal Protective Equipment. The equipment provided to shield or isolate a person from the chemical, physical, and thermal hazards that may be encountered at a hazardous materials incident. Adequate personal protective equipment should protect the respiratory system, skin, eyes, face, hands, feet, head, body, and hearing. Personal protective equipment includes both personal protective clothing and respiratory protection.

The purpose of personal protective equipment is to protect the responder against anticipated hazards. Protective equipment has improved dramatically in recent years, but the nature of the hazards to which emergency personnel must respond has also become exceedingly complex. Protective equipment will not prevent injuries or illnesses if it is improperly selected, if it is not used, or if it is not used properly. OSHA and EPA regulations require the use of personal protective equipment.

No combination of protective equipment will protect the responder against every conceivable hazard. And the equipment itself can cause injuries if it is not used appropriately and intelligently. The more sophisticated the equipment, the greater the need for training and experience.

See NFPA 471 and 472 and Supplement 10.

Protective Clothing.   Equipment designed to protect the wearer from heat and/or hazardous materials contacting the skin or eyes. Protective clothing is divided into three types:

(a) Structural fire fighting protective clothing,
(b) Chemical-protective clothing, and
(c) High temperature-protective clothing.

See NFPA 471 and 472 for additional information.

Protocol.   A series of sequential steps describing the precise patient treatment.

Region.   A geographic area that includes the local and neighboring jurisdiction for an EMS agency.

Respiratory Protection.   Equipment designed to protect the wearer from the inhalation of contaminants. Respiratory protection is divided into three types:

(a) Positive pressure self-contained breathing apparatus,
(b) Positive pressure airline respirators, and
(c) Air purifying respirators.

Inhaling toxics is a principal cause of serious injury among responders, so respiratory protection is of the utmost importance.

A positive pressure self-contained breathing apparatus (SCBA) is one in which the pressure inside the facepiece is positive in relation to the immediate environment during both inhalation and exhalation when tested in accordance with 30 CFR, Part II, Subpart H[1] by NIOSH using NIOSH test equipment. SCBA is a respirator worn by the user that supplies a breathable atmosphere that is either carried in or generated by the apparatus and is independent of the ambient environment. See NFPA 1981, *Standard on Open-Circuit Self-Contained Breathing Apparatus for Fire Fighters²*, for additional information on the subject.

A positive pressure self-contained air respirator supplies air in positive pressure, during both inhalation and exhalation, from a source located at some distance from the user and connected to the user by an air line hose.

An air purifying respirator does not have a separate air supply; rather, it passes ambient air through a filtering device before the user inhales it. Such a respirator can be of the demand (or negative pressure) type, which depends on the user's inhalation to bring air into the facepiece, or it may be a powered respirator that delivers a continuous flow of air into the facepiece in a positive pressure configuration. In the latter type, the user can create a negative pressure in the facepiece if he or she is breathing at a maximal rate. See NFPA 472 for additional information.

Safely. To perform the objective without injury to self or others, property, or the environment.

Secondary Contamination.* The transfer of contaminants to personnel or equipment outside the hot zone.

A-1-3 **Secondary Contamination.** A substance is considered to pose a serious risk of secondary contamination if it is likely to be carried on equipment, clothing, skin, or hair in sufficient quantities to be capable of harming personnel outside of the hot zone.

Shall. Indicates a mandatory requirement.

Should. Indicates a recommendation or that which is advised but not required.

Termination. That portion of incident management in which personnel are involved in documenting safety procedures, site operations, hazards faced, and lessons learned from the incident. Termination is divided into three phases: debriefing the incident, post-incident analysis, and critiquing the incident.

This definition differs from what one would expect it to mean at a fire, where termination usually is marked by the departure of the fire department from the scene.

Debriefing the incident involves collecting all the pertinent information about the nature of the incident and about all the emergency actions and operations that had to be utilized to mitigate and control the incident. The collected information is then evaluated in a post-incident analysis. Critiquing an incident brings an operation to a fitting conclusion. It also affords all responders valuable "lessons learned" to enhance future operations at hazardous materials incidents.

Understanding.   The process of gaining or developing the meaning of various types of materials or knowledge.

Warm Zone.   The area where personnel and equipment decontamination and hot zone support takes place. It includes control points for access corridor and thus assists in reducing the spread of contamination. This is also referred to as the decontamination, contamination reduction, or limited access zone in other documents.

One purpose of the warm zone is to reduce the likelihood of contaminating the cold zone. To some extent, then, the warm zone serves as a buffer between the hot and cold zones. The intensity of contamination in the warm zone should decrease the closer one gets to the cold zone, because decontamination procedures are taking place and because the warm zone affords an intervening space between the other two zones.

An *access corridor* is a defined path between the hot and cold zones where personnel and equipment are decontaminated. At very large incidents, several access corridors may be needed. The access corridors must be tightly controlled and supervised to regulate movement between the zones. Persons entering the warm zone from the cold zone must wear appropriate personal protective equipment.

References Cited in Commentary

1. Title 30 CFR, Part II, Subpart 4 (Washington, DC: U.S. Government Printing Office).
2. NFPA 1981, *Standard on Open-Circuit Self-Contained Breathing Apparatus for Fire Fighters*, 1992 edition.

2

Competencies for EMS/HM Level I Responders

2-1 General.

2-1.1 Introduction. All EMS personnel at EMS/HM Level I, in addition to their BLS or ALS certification, shall be trained to meet at least the First Responder Awareness level as defined in NFPA 472, *Standard for Professional Competence of Responders to Hazardous Materials Incidents*, and all competencies of this chapter.

EMS personnel who respond to hazardous materials incidents must be versed in several areas, including emergency medical care, hazardous materials, and EMS aspects of hazardous materials response. Deciding the appropriate level for each of these areas was difficult. The committee and subcommittee sought input from numerous sources and made universal application the basis of its decision.

Initially, Level I was divided into two categories: basic life support and advanced life support. In examining the competencies developed for these two levels, however, the subcommittee determined that there was very little difference between the two, certainly not enough to justify two separate listings. Therefore, EMS/HM Level I competencies apply to both basic life support (BLS) and advanced life support (ALS) personnel. BLS personnel may be either EMS/HM First Responders or EMTs. While many of the EMS personnel who reviewed the document felt the minimum level for BLS personnel should be EMT, the committee and subcommittee recognized that EMS/HM First Responders may be the only EMS personnel available in many areas.

The level of hazardous materials training for the EMS/HM Level I Responder was also difficult to determine. By definition, EMS personnel fall into the categories of HM First Responder, Awareness, and HM First Responder, Operations. In reviewing the competencies for these two levels, the subcommittee realized that the EMS/HM Level I Responder had to be versed in all the competencies of the awareness level and some, but not all, of the competencies at the operational level. Thus, the subcommittee decided to identify first responder awareness as the minimum level and to include in this standard those competencies from the operational level that apply.

This chapter contains competencies that are specific to EMS response to hazardous materials incidents.

2-1.2*   Definition. EMS personnel at EMS/HM Level I are those persons who, in the course of their normal duties, may be called on to perform patient care activities in the cold zone at a hazardous materials incident. EMS/HM Level I Responders shall provide care only to those individuals who no longer pose a significant risk of secondary contamination.

While this standard was being developed, the committee and subcommittee considered several terms for classifying EMS personnel responding to hazardous materials incidents. Initially, they used levels comparable to those defined in NFPA 472. However, it was decided that using the same terms would be confusing. After analyzing hazardous materials scene operations and considering comments from reviewers, it was determined that EMS functions fell into two main categories: patient care and emergency activity coordination. That is, certain individuals were responsible for providing patient care and others could, with additional training, coordinate EMS activities and support other emergency responders. To keep the terms simple and prevent confusion with other responding personnel, the titles selected for these two categories were Level I and Level II.

To help clarify the role of the EMS/HM Level I Responder, the committee and subcommittee identified where this individual would work during the hazardous materials incident: Because of their basic hazardous materials training, responders at this level should be allowed to function in the cold zone, where only minimal protective clothing would be needed. Since EMS personnel at this level would not don chemical-protective clothing and respiratory equipment, subcommittee members felt it was imperative that they be permitted to care only for those patients who have undergone "field decontamination" and have a reduced risk of passing contaminants on to others.

A-2-1.2 See Appendix D.

2-1.3 **Goal.** The goal of the competencies at EMS/HM Level I shall be to provide the individual with the knowledge and skills necessary to safely deliver emergency medical care in the cold zone. Therefore the EMS/HM Level I Responder shall be able to:

The goal of the responder at this level is in keeping with the definition of the position. The individual must have the knowledge and skills necessary to deliver patient care in such a way that both the responder and the patient are safe.

This knowledge and these skills are specific to caring for patients at haz mat incidents. Hazardous materials incidents differ from other situations in which EMS personnel provide patient care in that:

(a) they require more coordination with other response personnel,

(b) the patient is not initially under the care of the EMS provider,

(c) the EMS provider does not effect rescue, and

(d) the incident generally takes considerably more time than most incidents to which EMS personnel respond.

All these factors are contrary to the usual practice of EMS providers. For example, in most situations rapid transport is stressed. In hazardous material incidents, it takes time to prepare personnel to rescue the patient from the hot zone, to decontaminate the patient, and to transfer the patient from the rescuer to the EMS provider. EMS providers may find the waiting or their inability to effect rescue personally stressful.

(a) Analyze a hazardous materials emergency to determine what risks are present to the provider and the patient by completing the following tasks:

Since the goal for this level of responder is to perform in a safe manner, the responder must be able to determine when to respond and what risks are present for both the responder and the patient. Recognizing a hazard is the basis for determining the actions and flow of the entire incident.

1. Determine the hazards present to the Level I Responder and the patient in a hazardous materials incident, and

The responder must be able to determine from clues presented during dispatch, response, and approach whether a hazardous material is present at the scene and whether that product poses a risk to the patient and, in turn, the responder.

2. Assess the patient to determine the risk of secondary contamination.

Patients exposed to a hazardous material may pose a risk to others who come in contact with them. Responders must be able to determine from their knowledge of toxic exposure, patient assessment and treatment, and decontamination procedures what actions are necessary to prepare patients to be treated and transported safely.

(b) Plan a response to provide the appropriate level of emergency medical care to persons involved in hazardous materials incidents by completing the following tasks:

Once the responder has determined that there is a risk, he or she must be able to plan the actions necessary to obtain the help needed to rescue, decontaminate, care for, and transport the hazardous materials victim. The plan should also identify level of protection for EMS personnel, preparation of supplies and vehicles for transport, and level of prehospital treatment.

1. Describe the role of the Level I Responder in a hazardous materials incident,

Level I Responders must clearly understand their duties and responsibilities as part of a planned response to a hazardous materials incident in the community. They should know what their limitations and capabilities are so that they can be sure the proper resources are available to care for victims of exposure in their response area.

2. Plan a response to provide the appropriate level of emergency medical care in a hazardous materials incident,

To ensure that patients exposed to hazardous materials receive the appropriate care, the Level I Responder must know the procedures for treating patients with specific exposures. The responder who is familiar with treatment procedures will have all the necessary supplies and equipment ready to safely assess, care for, and transport the patient.

3. Determine if the personal protective equipment provided is appropriate, and

Responders at this level have a very limited ability to protect themselves from the risks of exposure, so it is important that they understand the limits of their personal protective equipment. At this level, the clothing will provide only splash protection to the skin and offer very little respiratory protection. This knowledge should convince responders to make sure they know what decontamination procedures must be performed so that their personal protective equipment is adequate.

4. Determine if the equipment and supplies provided are adequate.

Certain supplies and equipment must be available on the ambulance to care for the victim of a hazardous materials exposure. The Level I Responder must be able to evaluate the situation and the patient's condition and determine specifically what he or she will need to care for the patient on the scene and during transport. If the responder determines that the supplies and equipment are inadequate, he or she must know how to secure additional resources quickly.

(c) Implement the planned response by completing the following:

The responder's ability to analyze and plan the response will determine how effectively he or she will prepare for, treat, and transport the patient. It is vital that care be taken in preparing the patient, not only to minimize the risk to the responder while he or she is in contact with the patient, but also to minimize the amount of effort needed to return personnel and equipment to service after the incident.

1. Perform the necessary preparations for receiving the hazardous materials patient and preventing secondary contamination,

Preparing to receive a hazardous materials patient is quite different from actions routinely taken to prepare to receive a patient. These preparations are time-consuming, but they are done while other personnel on the scene prepare to rescue, evacuate, and decontaminate the patient. Adequate preparation of the personnel and the ambulance or transporting vehicle generally takes 20 to 30 minutes.

2. Treat the hazardous materials patient, and

Assessment and treatment of a hazardous materials victim will depend on numerous factors associated with the incident, the exposure, and the patient's condition. Most basic patient assessment techniques rely on the responder's

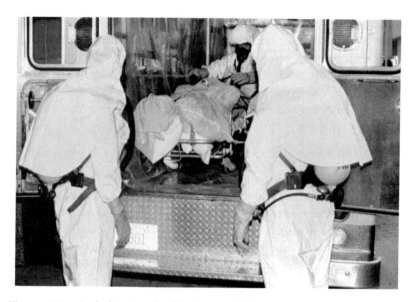

Figure 2.1 Ambulances should be specially prepared for transporting contaminated patients. Otherwise, expensive medical equipment that becomes contaminated may have to be disposed of.

senses of sight, touch, hearing, and smell. Unfortunately, personal protective equipment worn by the Level I EMS/HM Responder dulls these senses, hindering the responder's ability to assess a patient's condition. Because contaminants may remain on the patient, responders may also have to alter their treatment procedures. Care during transport will primarily be supportive according to the patient's symptoms.

3. Transport the patient as appropriate.

Selecting the appropriate vehicle in which to transport the patient or patients is important not only to the victim, but also to the emergency services in the response area. Communities in many parts of the country have a limited number of ambulances. Since a hazardous materials incident will require a lengthy time commitment, the responder must be aware of any alternate means of transportation that can be safely used for patients. Alternatives may include other emergency response units, such as command vehicles or reserve ambulances, and buses for multiple victims.

Equally important is determining which facility will receive the patient. The community plan should designate the hospital. In some jurisdictions, specialty facilities will be assigned to handle hazardous materials victims, while other jurisdictions will allow all hospitals and medical care facilities to receive and treat these patients. The Joint Commission on Accreditation of Healthcare Organizations (JCAHO) requires hospitals to have the equipment necessary to care for victims of chemical exposures. The JCAHO standards have been in place for a number of years; however, their existence does not ensure that the hospital's equipment or treatment will be adequate. Prehospital providers must get to know the medical facilities in their communities, and the best way to do this is to communicate with personnel at the facility before an incident occurs and to test the hospital's capabilities along with their own.

(d) Terminate the incident.

Post-incident, or terminating, activities are especially important in hazardous materials incidents because each incident is both stressful and unique. Responders may have a tendency to ignore post-incident activities because they are tired. However, they must recognize that such activities are important not only to the responder but to patient follow-up and successful management of future incidents, as well.

2-2 Competencies — Analyzing the Hazardous Materials Incident.

2-2.1 Determine the hazards present to the Level I Responder and the patient in a hazardous materials incident. The Level I Responder shall, given an emergency involving hazardous materials, determine the hazards to the responder and the patient in that situation. The Level I Responder shall be able to:

Once the EMS/HM Level I Responder realizes that an incident involves a hazardous material, he or she must be able to evaluate the variable conditions of the incident and the environment, know what resources are available to help assess and evaluate the risk, and interpret the information obtained from these sources.

(a) Assess the nature and severity of the incident (size-up) as they pertain to EMS responsibilities at a hazardous materials incident with evaluation of available resources and a request for any needed assistance.

For EMS responders, size-up consists not only of the observations they make that will help them determine the nature and severity of the incident, but also the factors pertinent to EMS operations.

Assessing the severity of the incident begins at dispatch. It is vital that dispatchers screen all calls carefully to obtain any information that may indicate whether a hazardous material is involved in an incident. Dispatchers should follow protocols that include questions that will allow them to gather the appropriate information.

Some incidents are actually dispatched as hazardous materials incidents. When this occurs, the EMS responder can research the product en route to assess the risk it presents. More often, however, responders are sent without notification that a hazardous material is involved.

Certain types of dispatches should alert the EMS responder to the possibility that a hazardous material may be the cause of a patient's distress. These include "unconscious person" and "trouble breathing" calls and calls that involve several patients with the same problem. These situations are frequently caused by a product in the environment that may also pose a danger to the responder. Calls to high-risk occupancies or to locations such as garden nurseries, pool equipment stores, and chemical plants should also arouse the responder's suspicions.

EMS responders should carry in their ambulances the U.S. DOT *Emergency Response Guidebook* (ERG) or the Canadian *Dangerous Goods Initial Emergency Response Guide*, and a pair of binoculars, at the very least. When the responder suspects a hazardous material, he or she can use the binoculars to make observations from a safe distance and refer to the ERG to identify the product involved and assess the danger it presents.

(b) Evaluate the environmental factors as they affect patient care.

Environmental factors must be considered for two reasons. First, the physical environment of the area in which the incident occurs may determine to some extent the severity of the exposure. If the exposure occurs in an open area, for example, the product will disperse and its effects on the patient will not be as severe as they would have been if the exposure had occurred in an enclosed area. Second, the weather may have an impact on the location and the amount of decontamination that can be performed without compromising a victim's condition. When it is cold and a protected decontamination area is not available, patient decontamination will necessarily be limited because of the danger of hypothermia.

(c) Identify the information resources available and how to access the following:

1. Poison Control Center

Responders usually call poison control centers directly from the scene or go through medical control. Poison control centers can provide detailed information regarding a patient's symptoms and the prescribed treatment. Often, poison control centers will coordinate the flow of medical information between the field and the receiving hospital.

Not all poison control centers have the same capabilities. EMS personnel are advised to coordinate with their local regional poison control center in the planning stage so that they will be able to support the field personnel during an incident.

2. Medical Control

The responder usually reaches medical control using the EMS radio communications systems or telephone. Medical control can also provide information regarding the symptoms of exposure, treatment, and recommendations for decontamination and personal protection.

3. Material safety data sheets

This information should be available at the facility at which an incident has occurred. Material safety data sheets have not been standardized, but most sheets detail exposure limits, the effects of exposure, types of exposure, and the appropriate emergency and first-aid procedures.

4. Reference guide books

Not only should ambulances carry the latest edition of the U.S. DOT *Emergency Response Guidebook* or the Canadian *Dangerous Goods Initial Emergency Response Guide*, they should also carry EMS hazardous materials guides such as the following:

(a) *Emergency Care for Hazardous Materials Exposure* by Bronstein and Currance[1]

(b) *Hazardous Material Injuries* by Stutz et al.[2]

5. Hazardous materials data bases

Several hazardous materials data bases currently available on computer provide detailed health and treatment information. If these data bases are not available at the scene, responders may call poison control and other such centers for the information. Examples of hazardous materials data bases include MEDITEXT, OHMTATDS, TOMAS, and TOXNET.

6. Technical information centers (CHEMTREC, NRC, etc.)

Responders throughout the U.S. and Canada can access CHEMTREC 24 hours a day by dialing (800) 424-9300. CHEMTREC will provide all the available information on a specific chemical and will contact the shipper involved in the incident for detailed assistance and response follow-up. CHEMTREC also notifies the National Response Center of significant incidents and will connect a caller to the National Response Center to report a spill. In Canada, responders can contact CANUTEC by calling collect (613) 996-6666. CANUTEC will provide emergency response assistance and information.

7. Technical specialists

EMS personnel should be able to consult, either directly or through their communications center, a chemist, an occupational health specialist, or some other individual who can give them the information they need to care for patients on the scene. These technical specialists may also prove valuable at the hospital, where the patient may have to undergo more definitive treatment. Lists of technical specialists should be developed during the planning of hazardous materials responses and be accessible during emergency incidents.

8. Agency for Toxic Substances and Disease Registry (ATSDR).

Responders can reach the Agency for Toxic Substances and Disease Registry (ATSDR) 24 hours a day by calling (404) 639-0615. ATSDR will provide health-related support, including on-site assistance, if necessary, in hazardous materials emergencies.

(d) Given a pesticide label, identify and explain the significance of the following:

1. Name of pesticide
2. Signal word

3. EPA registration number
4. Precautionary statement
5. Hazard statement
6. Active ingredient.

Because containers at an incident site will often alert responders to the type of contamination to which a patient has been exposed, EMS personnel must understand the significance of the various parts of a container label. An example is a pesticide label.

The name of the pesticide is usually displayed prominently on the label. Also displayed are signal words. The primary signal words used are "DANGER" and "WARNING." "DANGER," accompanied by a skull and crossbones and the word "POISON" printed in red, indicates highly toxic pesticides. "WARNING" indicates moderately toxic pesticides. The word "CAUTION" is used to identify low-toxicity materials.

The product's EPA registration number is usually printed in smaller letters on the front of the package label. Like the product name, the number ensures positive identification of the pesticide. When seeking additional expert advice, the responder must know the product name, its chemical or common name, or the EPA registration number.

Precautionary statements indicating the dangers the product presents to humans, animals, and the environment also appear on the label. The hazards statement lists the physical and chemical hazards of the substance and its ingredients, which are described as active or inert. Active ingredients must be listed by their chemical names; a common name is sometimes also listed.

Whenever possible, the label and the container should be properly wrapped and taken to the hospital or to an isolated area where the responder can access the information displayed on them.

2-2.2 Assess the patient to determine the risk of secondary contamination. The Level I Responder shall, given a hazardous materials incident with a patient(s), determine the risk of secondary contamination. The responder shall be able to:

Assessing a patient to determine the risk of secondary contamination will involve analysis of his or her exposure to the product. The responder will have to determine the type of exposure, how much of the product the patient was exposed to, how many times the exposure occurred, and how long ago the exposure occurred.

Assessment will also include determining how much decontamination must be done to lower the risk of secondary contamination.

(a) Explain the basic toxicological principles relative to assessment and treatment of victims exposed to hazardous materials, including:

To determine the risk to both patient and rescuer, the EMS care provider must understand some basic toxicological principles. These include:

1. Acute and delayed toxicity

Acute toxicity refers to the sudden, severe onset of symptoms due to an exposure. The effects of delayed toxicity may not develop for hours or longer after an exposure; in some instances, symptoms may not appear until 72 hours after the exposure.

2. Routes of exposure to toxic materials

The routes of exposure, sometimes referred to as routes of entry, include:
(a) Inhalation—the process by which irritants or toxins enter the body through the lungs as a result of the respiratory process.
(b) Ingestion—the process of consuming contaminated food or water.
(c) Absorption—the process by which hazardous materials are absorbed into the body through the skin or other external tissues.
(d) Injection—the process by which a toxic substance is introduced directly into the blood by a needle cannula or some other mechanical means. Contaminants entering the blood stream through an open wound can be considered to be injected.

3. Local and systemic effects

Local effects are those in which a toxic substance comes in direct contact with the skin or other tissue. Systemic effects are the effects a toxic product has on either the entire body or on a specific system or organ.

4. Dose-response as it relates to risk assessment

The chemical dose relationship refers to the response a chemical produces in the human body. It is a cause/effect relationship. The magnitude of the body's response will depend on the concentration of chemical at the site, the concentration of the chemical itself, and the dose administered.

5.* Synergistic effects

A-2-2.2(a)5   As defined in Webster's Dictionary, the word synergism means "a cooperative action of discrete agencies such that the total effect is greater than the sum of the effects taken independently." In the context of hazardous materials, it is important to remember that the signs and symptoms of a given chemical are generally standard for that particular chemical. But when two or more chemicals are involved, the resultant signs and symptoms from an exposure may be dramatically different than what the EMS provider anticipates.

  6. Health hazard as determined by assessing toxicity, exposure, and dose.

Toxicity, exposure, and dose must all be analyzed to understand the risk a product will have on both the patient and the rescuer.

(b) Describe how the chemical contamination of patients alters the principles of triage in hazardous materials incidents.

The hazardous materials entry team begins the initial triage of patients in the hot zone. Because time and resources are generally limited, patients are not triaged in the normal method. Ambulatory patients usually evacuate themselves, while nonambulatory patients are removed from the hot zone based on their location. The first patients removed are usually the first encountered. Once the patients are in the contamination reduction area, the amount of chemical contamination they have suffered and their condition will determine the extent of decontamination performed. Actual triage is not performed until the patients have been transferred from the decontamination area to the cold zone.

(c) Explain the need for patient decontamination procedures at hazardous materials incidents.

Patients must be decontaminated at the site of the hazardous materials incident to reduce their exposure to the hazardous chemical and to reduce the chance of secondary contamination of emergency response personnel, including the EMS provider. It must also be done to prevent the spread of contaminants to other areas and personnel. The location of the decontamination procedures and the amount performed depends on various factors, among them the incident, the chemical, and the patient's condition.

(d) Describe how the potential for secondary contamination determines the extent of patient decontamination required.

A great many chemicals are extremely toxic only in the high concentrations found in the immediate exposure area, but pose little or no risk to those outside the hot zone. Small amounts of some chemicals may produce relatively little acute toxicity, but are suspected of causing cancer or other chronic diseases and are thus considered to pose a greater risk of secondary contamination. These factors must be considered when analyzing the potential for secondary contamination and the extent to which a patient must be decontaminated. Patients exposed to substances that pose little or no risk to persons outside the hot zone will require decontamination only to remove the substance. Those who have been exposed to substances that pose a risk of secondary contamination will have to undergo a more thorough decontamination process. Contaminants that cannot be removed must be contained to protect the rescuers.

(e) Describe the way that personnel, personal protective clothing, apparatus, tools, and equipment become contaminated and the importance and limitations of decontamination procedures.

Personnel, protective clothing, apparatus, tools, and equipment may be contaminated if they come in direct contact with hazardous substances, or with the run-off, smoke, or vapors associated with hazardous materials incidents. These contaminants must be removed as soon as possible to keep them from injuring those who use these items. Decontamination is the time-consuming, labor-intensive process of removing such contaminants using various methods and solutions. One of the problems with decontamination is that the contaminating substances often cannot be seen, making it difficult to determine when they have been removed.

(f) Explain the decontamination procedures as defined by the authority having jurisdiction for patients, personnel, personal protective equipment, and apparatus at hazardous materials incidents.

Each jurisdiction should develop separate procedures for decontaminating personnel, equipment, apparatus, and patients. It is important that EMS/HM Level I Responders be familiar with all these procedures so that they understand the decontamination process their patients will undergo and the process they themselves will have to undergo should they become contaminated.

See NFPA 471 for more information.

2-3 Competencies — Planning the Response.

2-3.1   Describe the role of the Level I Responder in a hazardous materials incident. The Level I Responder shall, given a plan of action by the incident commander, describe their role in a hazardous materials incident as identified in the local emergency response plan or organization's standard operating procedures, including:

Level I Responders must understand what the local emergency response plan and their own organization's standard operating procedures require of them at a hazardous materials incident. It is vital that they understand their roles and responsibilities so that there is no confusion when they are given a plan of action by an incident commander. They must be familiar with their responsibilities and with the limitations of their performance, based on their training and knowledge. See Supplement 14.

(a)   Describe the emergency medical component for the hazardous materials incident response plan as developed by the authority having jurisdiction.

The hazardous incident response plan developed by the authority having jurisdiction should:

• Detail the responsibilities of the EMS response agencies
• Detail the role of the various levels of EMS responders
• Identify where EMS falls in the incident command structure
• List the interactions required of EMS with other agencies on the scene
• Describe the relationship between the prehospital and the hospital component in caring for chemically contaminated patients.

(b)   State the Level I Responder's role within the hazardous materials response plan as developed by the authority having jurisdiction.

The local hazardous materials response plan developed by the jurisdictions having authority should define the EMS Level I Responder's role as being patient care provider in the cold zone.

(c)   State the Level I Responder's role within the hazardous materials Incident Management System.

Level I Responders should be assigned a role within the treatment, triage, or transportation sectors of the hazardous materials incident management system. They will function in these sectors in the cold zone, providing patient care.

2-3.2 Plan a response to provide the appropriate level of emergency medical care to patients in a hazardous materials incident. The Level I Responder shall, given a hazardous materials incident, be able to plan the appropriate emergency medical care, including:

If they understand their roles in a hazardous materials response, Level I Responders working at an incident or participating in a simulation should be able to develop a plan for patient care specific to that incident.

(a) Describe the standard operating procedures for the medical management of persons exposed to hazardous materials, as specified by the authority having jurisdiction.

Local jurisdictions should develop standard operating procedures for the medical management of persons exposed to hazardous materials. These procedures should include:

• Assessment of the hazardous materials victim,
• Treatment,
• Communications, and
• Transportation (including mode of transport and hospital determination).
See Supplement 11 for more information.

2-3.3 Determine if the personal protective equipment provided is appropriate. The Level I Responder shall, given the name of the hazardous material and the type, duration, and extent of exposure and decontamination process, determine if available personal protective clothing and equipment is appropriate to implement the planned response. The Level I Responder shall be able to:

Using the information available on the hazardous material involved in an incident, EMS/HM Level I Responders should be able to determine whether their personal protective clothing will provide them with the protection required.

(a) Describe the application, use, and limitations of the following:

1. Street clothing and work uniforms

Street clothing and work uniforms provide the least, if any, protection against chemical exposures.

2. Structural fire fighter protective clothing

Structural fire fighter clothing is designed to protect fire fighters against adverse environmental conditions during structural fire fighting. These conditions include high temperatures, steam, hot water, and hard particles. This type of protective clothing is not designed for chemical exposures and may provide only limited protection.

3. Respiratory protective equipment

The most commonly available types of respiratory protection equipment include:

(a) Positive pressure SCBA, which can be used in a toxic atmosphere if the air supply lasts long enough to complete the assignment. Its disadvantages include its weight and bulk, which may interfere with performance.

(b) Positive pressure supplied air line respirators, provided that the hose does not impair the fire fighter's mobility and that the air line is not damaged or obstructed. Airborne contaminants will not enter this system, and the operations of the air line will not impede other workers.

(c) Air purifying respirators, which can be used with certain gases and vapors if the specific product is known. They should not be the respirator of choice where the following conditions exist:

- Oxygen deficiency
- IDLH concentrations of a specific substance
- Entry into unventilated or confined spaces
- Presence or potential presence of identified contaminants
- Contaminant concentrations are unknown or exceed designated maximum use concentrations
- Identified gases or vapors have inadequate warning properties.

4. Chemical-protective clothing.

Chemical-protective clothing is designed to protect the wearer's skin and eyes from direct chemical contact; most chemical-protective clothing will not provide thermal protection. It is made from materials compatible with specific chemicals and groups of chemicals. No single material will protect against all chemicals, and mixtures of chemicals make the problem more complex.

See NFPA 471 and NFPA 472 for additional information.

2-3.4 Determine if the equipment and supplies provided are adequate for implementing the planned response. The Level I Responder shall, given a simulated hazardous materials incident, determine if available equipment and supplies are appropriate to implement the planned response. The Level I Responder shall be able to:

To carry out their responsibilities at an incident, EMS/HM Level I responders require certain supplies and equipment for their own protection and to care for the patient.

(a) Describe the equipment and supplies available to the Level I Responder for the care and transportation of the hazardous materials incident patient.

The following equipment and supplies are recommended as the minimum necessary for the EMS/HM Level I responder at the scene of a hazardous materials incident.

General ambulance and patient care supplies:

- Binoculars
- Hazardous materials EMS references
- Disposable patient care equipment, such as bag mask units, blood pressure cuffs, stokes basket, backboard with securing devices, cleaning supplies, irrigating water, patient washing supplies, yellow waste bags, patient-covering disposable sheets, Tyvek® suits, body bags, and disaster pouches
- Clean plastic bags and duct tape to cover equipment, floors, and walls.

Personal protection supplies:

- Splash suit
- Gloves, both inner and outer
- Boots
- Hard hat
- Goggles
- Mask
- Duct tape
- Valuables bag
- Clean uniform, scrubs.

2-4 Competencies — Implementing the Planned Response.

2-4.1 Perform the preparations necessary to receive the patient for treatment and transport. The Level I Responder shall, given a plan for providing patient care at a hazardous materials incident, be able to:

While efforts are being made to evacuate patients from the hot zone, EMS personnel must prepare themselves and their vehicles to receive them. This is an important part of their activities on the scene because it is vital to their own safety, as well as the safety of others. Supplies, equipment, and the ambulance itself must be prepared so that exposure of the materials on the ambulance is limited and the ambulance can be readily cleaned and returned to service.

(a) List the information that should be communicated to the Medical Control/ Receiving facility regarding the hazardous materials incident, including:

As soon as a hazardous materials incident is known to exist and patients are known to be involved, the medical control receiving facility should be notified so that facility personnel can initiate the hospital hazardous materials response plans. At a minimum, the following information should be given.

1. Type and nature of the incident

Did the incident occur at a fixed facility or was it a transportation incident? Was there a fire and explosion or merely a spill or leak? What is the extent of the incident?

2. Chemical involved and its physical state

This will allow the hospital time to discover the potential health care risks the chemical presents.

3. Number of potential patients.

The hospital needs this information to determine whether additional staffing will be needed and whether the facility can handle the potential increase in patients. In many situations, the hospital may have to activate its disaster plan to accommodate a large influx of patients.

(b) Describe the procedure for preparing the vehicle and equipment for the patient.

Several steps are involved in preparing a vehicle and equipment for patients:

(a) All supplies must be removed from cabinets and kits. Items needed to prepare the ambulance and personnel should be removed from storage areas. Any dressings, medications, and irrigation supplies that will be needed should also be removed. Supplies should be removed from kits, and kits should be secured.
(b) The inside walls and floor of the vehicle should be draped with a plastic covering to protect the inside of the ambulance, and all cabinets should be sealed. The ambulance floor should also have some type of covering that will prevent skidding if water gets on the surface. Various commercial kits or large plastic bags can be used to prepare the inside of the vehicle.
(c) Equipment that is not disposable should be wrapped in clear plastic.
(d) The mattress should be removed from the stretcher so that the backboard can be placed on it. The backboard should be covered with several disposable sheets or with a Tyvek® bag to receive the patient. See Supplement 12.

(c) Demonstrate the proper donning, doffing, usage, and limitations of all personal protective equipment provided to the Level I Responder by the authority having jurisdiction for use in their hazardous materials response activities.

EMS personnel should be able to demonstrate the preparations necessary for properly donning, doffing, and using their personal protective equipment. Before they don this equipment, each member of the team should be assessed for fitness. In addition, their vital signs should be taken and hydrating fluids should be given.

The steps for donning the protective clothing consist of:

(a) Removing jewelry and leather goods and placing them in a property bag in a secured area on the ambulance
(b) Applying the first layer of gloves, followed by the splash suit
(c) Putting on the boots, securing the seam with duct tape, and putting on a mask and goggles, making sure that none of the skin around the face is exposed
(d) Putting on the second layer of gloves and using duct tape to secure the seam between the glove and the suit.

This is the minimum protection for the EMS/HM Level I Responder who will be transporting patients who have undergone gross decontamination in the field.

(d) Describe the concept of patient transfer from the incident site to the decontamination area and then to the treatment area.

Once the patient has completed decontamination, EMS personnel should bring their stretcher to the exit from the contamination reduction corridor. The patient will be transferred from the backboard or stretcher used for decontamination to the clean backboard on the ambulance stretcher.

2-4.2 Treat the hazardous materials patient. The Level I Responder shall, given a patient from a hazardous materials incident, provide patient care consistent with the planned response and the organization's standard operating procedures. The Level I Responder shall be able to:

Each organization must ensure that it has adequate standard operating procedures, including medical procedures, to care for patients who may be involved in incidents anticipated in its response area. See Supplement 11.

(a) Describe how chemical contamination alters the assessment and care of the hazardous materials patient.

Chemical contamination alters the assessment of patients in that:

(a) The history the patient gives may not be as reliable as it is for most patients because hazardous materials may induce confusion, amnesia, delirium, seizures, or coma.
(b) The history of the patient's exposure must also be obtained:
    What was the product to which the patient was exposed?
    What was the type of exposure?
    What was the concentration, quantity, and duration of the exposure?
    How many times did it occur?
    Was it acute or chronic?
    How long ago did it occur?
    What has the patient done since being exposed?
(c) Associated injuries become very important, since they may affect the toxic substance's interaction with the body.

Physical assessment is important, especially if the patient's history is unreliable. It is initially done at a distance from the cold zone. The first close assessment is performed by the entry team in the hot zone. The next is performed by the decontamination team, and the last is done by EMS personnel in the cold zone.

Assessment is often limited due to the amount of protection the EMS responder must wear: The protective clothing will dull the senses he or she would normally use for patient assessment. The rescuer's protective clothing will hinder observation, auscultation, and palpation.

Treatment of the patient is primarily symptomatic. Antidotes are given when the specific product is known and an antidote exists. Special care must be taken with wounds to prevent external contaminants from becoming internal contaminants.

(b) List the common signs and symptoms and describe the EMS treatment protocols for the following:

   1. Corrosives (e.g., acid, alkali)
   2. Pulmonary irritants (e.g., ammonia, chlorine)
   3. Pesticides (e.g., organophosphates, carbamates)
   4. Chemical asphyxiants (e.g., cyanide, carbon monoxide)
   5. Hydrocarbon solvents (e.g., xylene, methlyene chloride).

The following are examples of treatment protocols.

Corrosives

Common Signs and Symptoms

Target organs are the eyes, skin, and the respiratory and gastrointestinal systems.

Eyes:

- pain
- tearing
- decreased visual acuity
- photophobia
- blepharospasm

Skin: Generally resemble thermal burns

- pain
- edema
- sloughing
- tenderness
- erythema
- discoloration of skin

Respiratory: Pain involving nose, mouth, throat, and chest

- dysphagia
- coughing
- weakness
- syncope

- hoarseness
- dyspnea

- dizziness
- wheezing, rales, or rhonchi

Associated findings include tachycardia, tachypnea, and cyanosis.

Gastrointestinal:

- chest pain
- drooling
- dysphagia
- abdominal tenderness
- vomiting, sometimes with blood

- pain
- muffled or slurred speech
- oropharyngeal burns
- abdominal pain

Systemic toxicity: Symptoms of metabolic acidosis with injection of acids coma, bradycardia, hypotension, weakness, excited behavior, tremor, lethargy, ataxia, and nystagmus.

EMS Treatment Protocols

1. Decontaminate patient.
2. Evaluate and support ABCs (airway, breathing, and circulation).
3. Provide hi-flow oxygen.
4. Attach to cardiac monitor.
5. Aerosolized bronchodilators for wheezing.
6. Continuous flushing of affected skin and eyes with copious amount of water or saline.
7. DO NOT induce vomiting with ingestion.

Pulmonary Irritants

Common Signs and Symptoms

- cough
- bronchoconstriction
- hoarseness/stridor
- pulmonary edema
- cyanosis
- increased mucus secretions
- chest pain

- airway edema
- difficulty breathing
- fatigue
- respiratory arrest
- changes in or decreased level of consciousness

EMS Treatment Protocols

1. Decontaminate patient.
2. Evaluate and support ABCs.
3. Administer oxygen as per local protocol.
4. Attach cardiac monitor if clinically indicated.
5. Continuous flushing of affected skin and eye areas with copious amounts of water or saline.
6. DO NOT induce vomiting with ingestion. Immediately dilute with 1 glass of water or milk.

Pesticides

Common Signs and Symptoms

Muscarinic Effects (easy-to-remember mnemonics):

- Salivation
- Lacrimation
- Urination
- Defecation
- Gastrointestinal
- Emesis

- Salivation
- Tremor
- Urination
- Miosis
- Bradycardia
- Lacrimation
- Emesis
- Diarrhea

Central Nervous System Effects:

- headache
- respiratory depression

- anxiety
- seizures

EMS Treatment Protocols

1. Decontaminate patient.
2. Evaluate and support ABCs.
3. Attach to cardiac monitor.
4. If patient is symptomatic due to organophosphate poisoning, administer atropine according to local medical protocol.
5. DO NOT induce vomiting with ingestion. If available, administer activated charcoal.
6. Continuous flushing of affected parts and eyes with copious amounts of water or saline.

Chemical Asphyxiants

Common Signs and Symptoms

These depend on the product and may be immediate or delayed, depending on product.

- anxiety
- hyperventilation
- muscle weakness
- dyspnea
- seizures

- headache
- dizziness
- chest pain
- syncope
- coma

EMS Treatment Protocols

1. Decontaminate patient.
2. Evaluate and support ABCs.
3. Administer oxygen according to local protocol.
4. Attach to cardiac monitor.
5. Support blood pressure as needed according to local protocol.
6. Continuous flushing of affected skin and eye with copious amounts of water or saline.

Hydrocarbon Solvents

Common Signs and Symptoms

- skin irritation
- chemical conjunctivitis
- headache
- drowsiness
- cough
- stupor
- pulmonary edema

- dizziness
- weakness
- coma
- chest pain
- tachyarrythmias
- respiratory depression

EMS Treatment Protocols

1. Decontaminate patient.
2. Evaluate and support ABCs.
3. Administer oxygen as indicated by local medical protocols.
4. Attach to cardiac monitor.
5. If complaining of skin or eye irritation, continuously flush skin or eyes with copious amounts of water or saline.

6. DO NOT induce vomiting with ingestion. Dilute with 1 glass of water and give activated charcoal if available.

7. Avoid epinephrine, bronchodilators, terbutaline, and other beta-adrenergic agents.

(c) Explain the potential risk with invasive procedures for hazardous materials patients.

Invasive procedures are performed only in life-threatening situations or when the patient is extremely unstable. The danger with invasive procedures is the risk of external contaminants entering the patient's bloodstream and spreading rapidly throughout the body.

(d) Demonstrate the ability to perform the following EMS functions within the Incident Management System on incidents involving multiple hazardous materials patients:

1.* EMS control

A-2-4.2(d)(1) EMS control activities at a hazardous materials incident include but are not limited to:

(a) Identification of EMS needs including appropriate level of protection for EMS personnel and equipment, resources for patient care, and decontamination of patient and EMS personnel.

(b) Securing of resources to meet EMS needs.

(c) Assignment of personnel, in the cold zone, to coordinate triage, treatment, disposition, and transport as required.

(d) Assignment of appropriately trained personnel to perform medical monitoring and other EMS support functions for hazardous materials response personnel in the cold zone.

(e) Assignment of appropriately trained personnel to provide patient care, assist with patient decontamination, and any other EMS support functions, as may be required in the warm zone.

2. Triage

3. Treatment

4. Disposition and transportation.

The following are some of the duties and responsibilities of personnel serving in the EMS functions in the incident management system.

1. EMS control (see Appendix A)
2. Triage

   • Establish triage area
   • Sort and prioritize patients
   • Reevaluate patients as necessary
   • Deliver patients from decontamination area to treatment area.

3. Treatment

   • Communicate with triage personnel
   • Determine needs and locate treatment area
   • Request personnel and supplies
   • Process and treat patients.

4. Disposition and transportation

   • Establish work area and communications
   • Coordinate activities between treatment and transportation
   • Contact hospital and advise of patients and conditions
   • Coordinate with medical control for distributing patients
   • Establish evacuation plan
   • Request resources needed.

2-4.3 Transport the patient. The Level I Responder shall, given a patient from a hazardous materials incident, transport the patient as specified in the local emergency response plan and the organization's standard operating procedures. The Level I Responder shall be able to:

The transportation of patients from hazardous materials incidents will differ from the normal transportation process because the amount of interaction with the patient will be limited. This is so because EMS personnel are dressed in protective clothing and the patient is wrapped to contain any residual contaminants. Hazardous materials patients are generally not given a secondary assessment or secondary treatment (splinting, etc.) during transport.

(a) Identify the capabilities of the medical facilities available in the local area to receive hazardous materials patients.

The local response plan should indicate the capabilities of all the medical facilities in the given area. In consultation with medical control, the responder will decide which facility should receive the patient from the incident. If there are a number of patients, the responder should determine whether they should be transported to one facility or distributed among several.

(b) Identify the acceptable vehicles available to transport hazardous materials patients from the treatment area to a receiving facility.

Based on the local operating procedures, the EMS provider, in coordination with his or her supervisors, will determine whether ambulances are the appropriate mode of transportation for patients from a specific incident. Alternate means of transportation include command vehicles and buses, if there are several patients. In determining which vehicles to use, the responder must consider the local and state laws addressing patient transportation.

(c) List the pertinent patient information that should be communicated to the receiving facility, including:

1. Estimated time of arrival
2. Age/sex
3. Patient condition/chief complaint
4. Associated injuries
5. Routes, extent, and duration of chemical exposure
6. Pertinent medical history
7. Signs and symptoms
8. Vital signs
9. Treatment, including decontamination and patient response
10. Pertinent chemical characteristics.

Once the patient is in the care of the Level I Responder, it is important that the responder obtain the information listed and relay it to the hospital in sequence. This information will help the hospital staff prepare to treat the patient.

(d) Describe the actions necessary for the coordinated delivery of hazardous materials incidents patients to a receiving facility.

In communicating with the hospital, EMS personnel should determine where in the hospital the patient is to be delivered. Hospital personnel should meet the ambulance at the designated site and transfer the patient from the ambulance stretcher to the hospital litter.

(e) Explain the special hazards associated with air transportation of patients exposed to hazardous materials.

Any decision to transport the patient by helicopter or fixed-wing aircraft should be carefully weighed against the risk to the air crew. Contaminants may still be on the patient, and these contaminants could cause the flight crew to become ill or crash the aircraft.

When definitive decontamination has been completed, air transportation may be an option. Precautions must be taken to ensure that no contaminants from the scene are carried aboard the aircraft. This includes the patient's clothing and valuables, and gastrointestinal specimens.

2-5 Competencies — Terminating the Incident.

2-5.1 Perform the reporting, documentation, and follow-up required of the EMS component of the hazardous materials incident. The Level I Responder shall, upon termination of the hazardous materials incident, complete the reporting, documentation, and EMS termination activities as required by the local emergency response plan or the organization's standard operating procedures. The Level I Responder shall be able to:

When the patient has been transferred to the care of the hospital staff, the EMS personnel must clean and decontaminate the transporting vehicle and complete the appropriate personal decontamination. They must then complete the appropriate patient records, including incident and exposure reports.

(a) List the information to be gathered regarding the exposure of the patient and the EMS provider and describe the proper reporting procedures, including:

1. Product information
2. Routes, extent, and duration of exposure
3. Actions taken to limit exposure and contamination
4. Treatment rendered
5. Patient condition and disposition.

Each EMS organization and program has a standard incident report that requires specific patient information. These reports generally have a comment or narrative area in which to list information about the product, the exposure, and the incident activities. Patient care information should also be completed, as it will be needed to develop an appropriate patient care plan.

(b) Identify situations that may necessitate critical incident stress debriefing intervention.

Because hazardous materials incidents may last a long time and result in personal injury and mass casualties, they produce additional stress not encountered in the "usual" emergency incident. The increased likelihood of psychological stress may warrant a critical incident defusing and possibly a debriefing. Factors that may indicate the need for a debriefing include the injury or death of emergency responders; major trauma, especially dismemberment; and mass casualties.

(c) Describe the EMS provider's role in the post-incident critique.

Since significant hazardous materials incidents do not occur frequently, it is important to critique each incident. The critique should include operational, as well as hospital, personnel. The EMS provider should participate, explaining what worked well and what didn't, what supplies were needed and not available, and what procedures may have to be revised based on the experience.

References Cited in Commentary

1. Alvin C. Bronstein, M.D., and Phillip L. Currance, EMT-P, *Emergency Care for Hazardous Materials Exposure* (St. Louis: The C.V. Mosby Company, 1988).
2. Douglas R. Stutz, Robert C. Ricks, and Michael F. Olsen, *Hazardous Materials Injuries: A Handbook for Pre-Hospital Care* (Greenbelt, MD; Bradford Communications Corp., 1982).

3

Competencies for EMS/HM Level II Responders

3-1 General.

3-1.1 **Introduction.** All personnel at EMS/HM Level II shall be certified to the EMT-A level or higher and shall meet all competencies for EMS/HM Level I in addition to all the competencies of this chapter.

Personnel at EMS/HM Level II require a higher level of EMS certification: the minimum needed is EMT-A. Based on the fact that this is a national standard and that some areas of the country do not have EMS personnel above the level of EMT-A, the committee and subcommittee realized that they had to provide for the functions required by this level. Nonetheless, EMT-Intermediate or EMT-Paramedic are preferred if available.

This section also includes additional hazardous materials competencies from NFPA 472. It was felt these were needed because the responder must be able to make an in-depth analysis to formulate plans that are to be carried out by other EMS personnel.

3-1.2 **Definition.** Personnel at EMS/HM Level II are those persons who, in the course of their normal activities, may be called upon to perform patient care activities in the warm zone at hazardous materials incidents. EMS/HM Level II Responder personnel may provide care to those individuals who still pose a significant risk of secondary contamination. In addition, personnel at this level shall be able to coordinate EMS activities at a hazardous materials incident and provide medical support for hazardous materials response personnel.

Personnel at this level are often more involved in coordination, supervision, and support activities than in patient care. Their contact with patients is usually limited to the warm zone. Because they may work in the warm zone, they require additional knowledge of, and skill in, personal protection. They may also work with response personnel other than EMS personnel.

Each jurisdiction should determine how many and which of their personnel will be trained to function at Level II. This will depend on the community's level of response to hazardous materials incidents. Where there are hazardous materials teams, there should be personnel trained to this level to provide the medical support required for the health and safety of entry personnel.

3-1.3 **Goal.** The goal of the competencies at EMS/HM Level II shall be to provide the Level II Responder with the knowledge and skills necessary to perform and/or coordinate patient care activities and medical support of hazardous materials response personnel in the warm zone. Therefore the Level II Responder shall be able to:

The committee and subcommittee felt the Level II Responder needed additional knowledge and skills to coordinate patient care and the flow of patients from the hot zone to the Level I Responder in the cold zone. This individual serves as the link between the emergency medical services and hazardous materials operations.

(a) Analyze a hazardous materials incident to determine the magnitude of the problem in terms of outcomes by completing the following tasks:

Since the Level II Responder knows more about the hazardous material involved in the incident and about the exposure, he or she can better analyze the magnitude of the problem and possible outcomes.

1. Determine the hazards present to the Level II Responder and the patient in a hazardous materials incident, and

EMS responders working in the warm zone will encounter additional hazards. Because they are closer to the hot zone and are more likely to be exposed to contamination, Level II Responders must know what additional precautions to take when working in the warm zone. These include wearing a higher level of personal protection and taking care where they walk and what items they touch.

Being in the warm zone, Level II Responders have access to more information and can better observe the patients. This puts them in a better position to determine what risks the patients face.

2. Assess the patient to determine the patient care needs and the risk of secondary contamination.

The Level II Responder is the first EMS person to assess patients at a hazardous materials incident and to determine their needs. He or she uses this information to decide whether other rescuers run the risk of secondary contamination.

(b) Plan a response to provide the appropriate level of emergency medical care to persons involved in hazardous materials incidents and to provide medical support to hazardous materials response personnel by completing the following tasks:

Once the Level II Responder is able to analyze the incident and project possible outcomes, he or she must develop a plan to coordinate the EMS aspects of the incident. This plan must include provisions for medical support of the hazardous materials responders.

1. Describe the role of the Level II Responder in a hazardous materials incident,

The Level II Responder may be assigned to one or more functions at a hazardous materials incident, including EMS control, patient decontamination, and medical surveillance.

2. Plan a response to provide the appropriate level of emergency medical care in a hazardous materials incident, and

The Level II Responder must develop the emergency care plan to be implemented by Level I Responders. In some instances, the Level II Responder may only be certified to the EMT-A level, and he or she may have to consult with ALS-certified Level I Responders to determine the care required.

3. Determine if the personal protective equipment provided EMS personnel is appropriate.

After analyzing the pertinent information, the responder should consult with hazardous materials response personnel to determine whether the personal protection available is adequate. If it is not, personnel with a higher level of training will be needed to provide patient care, or definitive decontamination will have to be performed in the field before the patient is transported to a medical facility.

(c) Implement the planned response by completing the following:

Certain components of the EMS plan for the incident must be completed by the Level II Responder, who must have the knowledge and skills necessary to do this.

1. Perform the necessary preparations for receiving the patient,

Depending on the local emergency plan, Level II Responders working in the warm zone may be required to wear a higher level of protection. In these situations, they must be able to provide their own personal protection and to protect their equipment.

2. Perform necessary treatment to the hazardous materials patient,

In certain situations, it may be necessary to begin caring for patients before decontamination efforts begin or while they are in progress. Actually giving care or directly supervising the care is the duty of the Level II Responder.

3. Coordinate and manage the EMS component of the hazardous materials incident, and

This entails assigning Level I personnel to perform functions required by the incident emergency care plan that the Level II Responder has developed.

4. Perform medical support of hazardous materials incident response personnel.

A Level II Responder must also provide medical support to the hazardous materials response personnel. Because hazardous materials personnel are at significant risk of becoming patients themselves, it is necessary to ensure their health and safety.

See Supplement 13.

(d) Terminate the incident.

Since the Level II Responder serves in a supervisory and coordinating role, he or she has more responsibilities during the termination phase. These include providing critical incident stress debriefings, defusing, and critiques, and making appropriate recommendations to the incident commander.

3-2 Competencies — Analyzing the Hazardous Materials Incident.

3-2.1 Determine the hazards present to the Level II Responder and the patient in a hazardous materials incident. The Level II Responder shall, given an emergency involving hazardous materials, determine the hazards to the responders and the patient in that situation. The Level II Responder shall be able to:

Since the Level II Responder is responsible for developing the EMS activity plan on which the outcome for both patients and responders depends, it is important that he or she know enough about hazardous materials to analyze the incident and effectively estimate the risks it presents.

(a) Define the following chemical and physical properties and describe their importance in the risk assessment process:

1. Boiling point

Boiling point is the temperature at which the transition from liquid to gaseous occurs. At this temperature, the vapor pressure of a liquid equals the surrounding atmospheric pressure so that the liquid rapidly becomes a vapor. Flammable materials with low boiling points generally present greater problems than those with high boiling points. For example, the boiling point of acetone is 133°F (56°C), and the boiling points for jet fuels range from 400 to 550°F (204 to 288°C).

2. Flammable (explosive) limits

A material's flammable or explosive range is the difference between its upper and lower flammable limits. The lower flammable limit (LFL) is the minimum concentration of vapor to air below which a flame will not propagate in the presence of an ignition source. The upper flammable limit (UFL) is the maximum vapor-to-air concentration above which a flame will not propagate. If a vapor-to-air mixture is below the LFL, it is described as being "too lean" to burn; if it is above the UFL, it is "too rich" to burn. When the vapor-to-air ratio is somewhere between the LFL and the UFL, fires and explosions can occur, and the mixture is said to be in the flammable range. The flammable range for gasoline is 1.4 percent to 7.6 percent and the flammable range for carbon monoxide is 12.5 percent to 74 percent. It is important that the responder be aware of the range as well as the LFL.

If the responder suspects or knows that flammable vapors are present, it is important that he or she determine the concentration of vapor in air. Combustible gas instruments are used for this purpose.

3. Flash point

The flash point of a liquid is the minimum temperature at which it gives off vapor in sufficient concentration to form an ignitable mixture with air. A liquid's flash point is the primary property or characteristic used to determine its relative degree of flammability. Since it is the vapors of flammable liquids that burn, vapor generation is a primary factor in determining the liquid's fire hazard.

4. Ignition temperature

Ignition temperature and *autoignition temperature* are interchangeable terms. The ignition temperature of a substance, whether solid, liquid, or gaseous, is the minimum temperature required to cause self-sustained combustion in the absence of any source of ignition. The responder should regard assigned ignition temperatures as approximations.

Ignition temperatures can be quite high, especially in relation to a liquid's flash point. For example, the flash point of gasoline is −45°F (−43°C), while its ignition temperature is well over 500°F (260°C).

5. Specific gravity

Specific gravity is the ratio of the weight of a volume of liquid or solid to the weight of an equal volume of water, with the gravity of water being 1.0. A substance with a specific gravity of less than 1.0 will float on water.

6. Vapor density

Vapor density measures the weight of a given vapor as compared with an equal volume of air, with air having a value of 1.0. A vapor density greater than 1.0 indicates it is heavier than air; a value less than 1.0 indicates it is lighter.

Vapor density can be important to the responder, since it will determine the behavior of free vapor at the scene of a liquid spill or gas release.

7. Vapor pressure

Vapor pressure is the pressure exerted on the inside of a closed container by the vapor in the space above the liquid in the container. Products with high vapor pressures have a greater potential to breach their containers when heated, since the pressure increases as the temperature rises. Products with high vapor pressures are more volatile.

Vapor pressure is measured in mm of mercury. The vapor pressure of water is 21 mm mercury, and that of chlorine is 4,800 mm mercury.

8. Water solubility.

Water solubility, or the degree to which a substance is soluble in water, can be useful in determining effective extinguishing agents and methods. Response personnel should consider the property of solubility along with that of specific gravity.

(b) Define the following terms:

1. Alpha radiation

Alpha radiation involves the alpha particle, a positively charged particle emitted by some radioactive materials. It is less penetrating than beta and gamma radiation and is not considered dangerous unless ingested. If ingested, alpha radiation will attack internal organs.

2. Beta radiation

Beta radiation involves the beta particle, which is much smaller but more penetrating than the alpha particle. Beta particles can damage skin tissue, and they can damage internal organs if they enter the body. Full protective clothing, including positive pressure self-contained breathing apparatus, will protect against most beta radiation.

3. Gamma radiation.

Gamma radiation is especially harmful since it has great penetrating power. Gamma rays are a form of ionizing radiation with high energy that travels at the speed of light. It can cause skin burns and can severely injure internal organs. Protective clothing is inadequate in preventing gamma radiation from harming the body.

(c) Define the following toxicological terms and explain their use in the risk assessment process:

1. Threshold limit value (TLV-TWA)

This is the maximum concentration, averaged over 8 hours, to which a healthy adult can be repeatedly exposed for 8 hours per day, 40 hours per week.

2. Lethal concentration and doses ($LD_{50/100}$)

The LC is the median lethal concentration of a hazardous material. It is defined as the concentration of a material in air that, on the basis of laboratory tests (inhalation route), is expected to kill 50 percent of a group of test animals when administered in a specific time period.

The LD of a substance is a single dose that will cause the death of 50 percent of a group of test animals exposed to it by any route other than inhalation.

3. Parts per million/billion (ppm/ppb)

The values used to establish the exposure limits above are quantified in parts per million or parts per billion. A good reference to remember is that 1 percent equals 10,000 ppm, 1 percent equals 1,000 ppb. So if you obtain a reading from a sampling instrument of 0.5 percent, that is equivalent to 5,000 ppm, or 500 ppb. If you then determine the TLV is 7,500 ppm, you can relate the reading from the instrument to determine the degree of hazard.

4. Immediately dangerous to life and health (IDLH)

This is the maximum level to which a healthy worker can be exposed for 30 minutes and escape without suffering irreversible health effects or impairment. If at all possible, exposure to this level should be avoided. If that is not possible, responders should wear Level A or Level B protection with positive pressure self-contained breathing apparatus or a positive pressure supplied-air respirator with an auxiliary escape system. This limit is established by OSHA and NIOSH.

5. Permissible exposure limit (PEL)

This is a term OSHA uses in its health standards covering exposures to hazardous chemicals. It is similar to the TLV/TWA established by the ACGIH. PEL, which generally relates to legally enforceable TLV limits, is the maximum concentration, averaged over 8 hours, to which 95 percent of healthy adults can be repeatedly exposed for 8 hours per day, 40 hours per week.

6. Short term exposure limit (TLV-STEL)

This is the maximum average concentration, averaged over a 15-minute period, to which healthy adults can be safely exposed for up to 15 minutes continuously. Exposure should not occur more than four times a day with at least 1 hour between exposures.

7. Ceiling level (TLV-C).

This is the maximum concentration to which a healthy adult can be exposed without risk of injury. It is comparable to the IDLH, and exposures to higher concentrations should not occur.

(d) Given a specific hazardous material and using the information sources available to the Level II Responder, demonstrate extracting appropriate information about the physical characteristics and chemical properties, hazards, and suggested medical response considerations for that material.

Responders must know which references provide the information necessary to determine the health effects of the hazardous materials involved in an incident and to establish the medical treatment that can be used to combat them. They must also understand which properties have an impact on a patient's reactions and on the care the patient must receive.

3-2.2 Assess the patient and conditions to determine the risk of secondary contamination. The Level II Responder shall, given a hazardous materials incident with a patient(s), determine the risk of secondary contamination. The Level II Responder shall be able to:

The Level II Responder can apply the information on the product and its properties to the information he or she obtains while assessing the patient to determine the risk of secondary contamination to others.

(a) Identify sources of technical information for the performance of patient decontamination.

Sources of information include EMS reference books, the poison control center, EMS/HM data systems, and the ATSDR.

(b) Identify the factors that influence the decision of when and where to treat the patient and the extent of patient care, including:

These factors may indicate that treatment is required as soon as the patient enters the warm zone.

1. Hazardous material toxicity

The more toxic the product, the faster it should be removed from the patient and the more severe the patient's symptoms may be.

2. Patient condition

The patient may be in critical condition due to factors other than the hazardous materials exposure; exposures are often associated with other medical problems or trauma. If the patient's condition is unstable, the time between decontamination and the implementation of life-saving treatment may have to be as short as possible.

3. Availability of decontamination.

Some jurisdictions can provide complete patient decontamination in the field year round, but this is more the exception than the rule. In most areas, resources are only available to deal with gross contaminants by mechanical removal, disrobing, and a thorough rinsing. Definitive decontamination, which consists of two thorough soapings and rinsings, generally must be performed in a sheltered environment or at a medical facility.

3-3 Competencies — Planning the Response.

3-3.1 Describe the role of the Level II Responder at a hazardous materials incident. The Level II Responder shall, given a plan of action by the incident commander, describe his or her role in hazardous materials incident as identified in the local emergency response plan or the organization's standard operating procedures. The Level II Responder shall be able to:

The local emergency response plan should define the duties and responsibilities of EMS/HM responders at each level. Jurisdictions can use this standard as a guide to updating their own plans if these plans do not currently contain this information.

The Level II EMS/HM Responder should perform the following functions:

- Coordinate with the incident commander the delivery of emergency medical services
- Maintain communications with the incident commander
- Establish and maintain communications with EMS personnel
- Establish EMS control operations and direct the administration of all EMS operations, in accordance with the local SOPs
- Make provisions for the appropriate level of personal protection for the EMS responders
- Advise the incident commander whether additional personnel or resources are needed to implement the EMS medical care plan
- Establish and coordinate patient contamination control operations, in accordance with the local SOPs
- Ensure that patients receive the appropriate emergency medical care according to the incident's emergency medical plan
- Ensure the efficient transport and transfer of patients to a designated medical facility
- Make provisions for the ongoing medical surveillance of hazardous materials response personnel.

(a) Describe the importance of coordination between various agencies at the scene of hazardous materials incidents.

To perform their duties and responsibilities, Level II Responders must coordinate with personnel representing the numerous agencies on the scene of a hazardous materials incident. These include both emergency response agencies and support agencies, such as the health department, environmental agencies, emergency management agencies, the police and fire departments, and disaster response teams.

By coordinating with these agencies, EMS personnel will have the resources they need to perform their functions safely. It will also prevent duplication of effort and ensure effective on-scene operations.

3-3.2 Plan a response to provide the appropriate level of emergency medical care to persons involved in hazardous materials incidents and to provide medical support to hazardous materials response personnel. The Level II Responder shall, given a hazardous materials incident, be able to develop the plan for EMS activities. The Level II Responder shall be able to:

Developing a plan of response for the EMS component is one of the most important functions of the Level II Responder, since the flow of EMS operations depends on this plan. To assist the Level II Responder, each organization should have a standard format outlining the plan to serve as a guide at each incident. This will ensure that important items are not missed.

(a) Given a simulated hazardous materials incident, assess the problem and formulate and implement a plan including:

1. EMS control activities
2. EMS component of an incident management system
3. Medical monitoring of personnel utilizing chemical-protective and high temperature-protective clothing
4. Triage of hazardous materials victims
5. Medical treatment for chemically contaminated individuals
6. Product and exposure information gathering and documentation.

The functions listed are an integral part of the EMS response at a hazardous materials incident. Each can and should be broken down into components and itemized on a checklist so they can be addressed for each incident. At any major hazardous materials incident, at least one EMS responder should be assigned to supervise each of these functions.
See Supplement 14.

(b) Describe the importance of pre-emergency planning relating to specific sites.

The response plans for any hazardous materials incident should specify in general terms the placement of the following sites on and off scene for the EMS responders:

• Staging of ambulances
• Triage and treatment sites
• Patient and personnel decontamination areas
• Medical surveillance.

(c) Describe the hazards and precautions to be observed when approaching a hazardous materials incident.

When approaching a hazardous materials incident, EMS responders should take the following precautions:

- Look for environmental clues that indicate the presence of a hazardous material
- Approach the scene from uphill and upwind
- Secure a vehicle that can be used for a rapid evacuation, if necessary
- Be careful not to drive through any run-off or spill.

(d) Describe the considerations associated with the placement, location, and setup of the patient decontamination site.

The patient decontamination site should be:

- Near the exit from the hot zone to minimize the spread of contaminants
- Located in an area that will contain run-off
- Protected from the environment and provide privacy, if possible
- Located with the technical decontamination area, if it does not interfere with the decontamination of response personnel.

(e) Explain the advantages and limitations of the following techniques of decontamination and how they may or may not be applicable to patient decontamination:

  1. Absorption

Advantages:

- Easily applied
- Nonreactive with the product
- Best for water-reactive products
- Makes liquids easy to pick up.

Disadvantages:

- Generally expensive
- Materials are not rendered harmless
- Must be bagged and treated as hazardous waste
- Generally effective only for flat surfaces.

Patient Application:

- Limited, although certain pads can be used to absorb product from the skin if water cannot be used for flushing.

2. Chemical degradation

Advantages:

- Renders the product neutral or harmless.

Disadvantages:

- Generally produces exothermic reaction, which may burn skin
- Exact product may not be known, or there may be a mixture of chemicals.

Patient Application:

- Is impractical for patient use because it produces heat in neutralizing the product

3. Dilution

Advantages:

- Water is readily available
- Can be applied rapidly
- Lowers the concentration of the product and may render it harmless.

Disadvantages:

- Run-off must be contained
- May be water-reactive.

Patient Application:

- Most common and generally effective means of patient decontamination.

4. Isolation.

Advantages:

- Does not require supplies other than disposal containers.

Disadvantages:

- Produces hazardous waste and must be disposed of properly.

Patient Application:

* Can only be applied to the management of patient wastes.

(f) Describe when it may be prudent to pull back from a hazardous materials incident.

When an incident threatens to escalate or changing conditions increase the risk to personnel, responders should consider pulling back from the incident and relocating operations.

3-3.3 Determine if personal protective equipment provided is appropriate. The Level II Responder shall, given the name of exposure, determine if the protective clothing and equipment available to EMS personnel is appropriate to implement the planned response.

(a) Identify the advantages and dangers of search and rescue missions at hazardous materials incidents.

A search and rescue mission may disclose the existence and location of victims and result in their rescue. However, it may also place the rescuer in a risky position.

(b) Identify the advantages and hazards associated with the rescue, extrication, and removal of a victim from a hazardous materials incident.

The advantage of rescuing, extricating, and removing a patient from a hazardous materials incident is that it may save the patient's life. The risks to rescuers are significant, however. The time available for a rescue is limited by the responders' air supply needs, and it is often difficult for an entry team, which usually consists of only two or three persons, to immobilize and extricate a patient adequately.

(c) Describe the types, application, use, and limitations of protective clothing used by EMS personnel at hazardous materials incidents.

The personal protection available to EMS personnel generally provides the minimal chemical protection. Because Level I Responders work in the cold zone, they should have splash protection for exposed skin surfaces. Generally, they have no respiratory protection but a face mask.

Level II Responders should be familiar with the level of protection used by other responders who work in the warm zone, including Levels C and B personal protective equipment (PPE).

(d) Demonstrate how to interpret a chemical compatibility chart for chemical-protective clothing.

Because the competency calls for a demonstration of ability, the Level II Responder must be familiar with the different charts available locally and be able to interpret the information on the charts as it applies to the selection of PPE.

3-4 Competencies — Implementing the Planned Response.

3-4.1 Perform the preparations necessary to receive the patient for treatment and transport. The Level II Responder shall, given a plan for providing patient care at a hazardous materials incident, be able to:

(a) Demonstrate the proper donning, doffing, and usage of all personal protective equipment provided to the Level II Responder by the authority having jurisdiction.

Since the Level II Responder works in the warm zone, he or she must be able to select the appropriate level of PPE and be able to don and doff it efficiently.

3-4.2 Perform the necessary treatment to hazardous materials patients. At the scene of a hazardous materials incident, the Level II Responder shall be able to provide or coordinate the patient care. The Level II Responder shall be able to:

Because the Level II Responder may be the first EMS person to reach the patient, he or she must be able to implement the necessary care-for-life support in coordination with other activities taking place in the warm zone.

(a) Given a simulated hazardous materials incident and using local available resources, demonstrate the implementation of the patient decontamination procedure. (*See Appendix E.*)

(b) Explain the principles of emergency decontamination and its application for critically ill patients.

There are two situations to consider. The first is that in which the severity of the patient's condition is due to exposure to a hazardous material, and the second is that in which the critical illness or injury is not a result of the exposure.

Where the patient's condition is due to the hazardous material exposure, it is vital that as much of the product as possible be removed from the victim as rapidly as possible. Where the condition is not a result of the exposure, life support can be applied while the product is being removed or removal may be delayed while responders attempt to stabilize the patient.

(c) Demonstrate the ability to coordinate patient care activities including treatment, disposition, and transportation of patients.

In a simulated situation, the responder must be able to demonstrate an ability to coordinate the activities necessary for effective patient care throughout the various phases of the incident.

3-4.3 Coordinate and manage the EMS component of the hazardous materials incident. The Level II Responder shall be able to:

(a) Given a simulated hazardous materials incident, the Level II Responder shall be able to demonstrate the ability to establish and manage the EMS component of an incident management system.

During a simulation, the Level II Responder must demonstrate the ability to coordinate the functions associated with any of the EMS components in the incident management system. This includes EMS control and the duties of triage officer, treatment officer, disposition and transportation officer, and safety officer.

3-4.4 Perform medical support of hazardous materials incident response personnel. The Level II Responder shall be able to:

The Level II Responder will either perform or supervise the medical support for hazardous materials response personnel.

(a) Explain the components of pre-entry and post-entry assessment, including:

1. Vital signs

These include temperature, pulse, respiration (rate and character), and blood pressure.

2. Body weight

Fluid loss is best measured in the field by measuring the responder's pre- and post-entry body weights.

3. General health

An idea of the responder's general health can be obtained by observing his or her physical appearance and asking specific questions about his or her well-being. Those who are recovering from an illness or feel "under the weather" should not be allowed to serve in any level of personal protective equipment but a work uniform.

4. Neurological status

A responder's neurological status must be assessed to determine baseline functioning. Changes in neurological status are early indicators of stress, exposure to toxic products, or both.

5. Electrocardiographic rhythm strip, if available.

This provides a baseline reading.

(b) Explain the following factors and how they influence heat stress for hazardous materials response personnel, including:

1. Hydration

This is probably the most important factor for working effectively in the heat produced by protective clothing. Response personnel must be prehydrated, and replacement fluids must be administered after entry to effect a rapid recovery.

2. Physical fitness

The degree of a responder's physical fitness is important to his or her ability to work for any length of time in heat stress conditions.

3. Environmental factors

Extreme heat or cold will make it more difficult for personnel to function in PPE. Heat increases the heat stress inherent in wearing PPE, and extreme cold exacerbates the difficulty of maneuvering in PPE.

4. Activity levels

The amount and type of work to be performed can produce various levels of heat exhaustion and must be considered when determining the entry times.

5. Level of PPE

The higher the level of PPE, the more heat stress the responder will suffer and the more difficult he or she will find it to move.

6. Duration of entry.

Entry must be timed to ensure an adequate air supply for decontamination. The entry time may have to be further limited during extremely hot weather because of heat stress.

(c) Explain the medical monitoring protocols and demonstrate medical monitoring procedures for personnel at the scene of a hazardous materials incident.

Each organization should develop medical monitoring protocols that not only define the procedures to be performed but also establish normal values and exclusionary limits for personnel. The Level II Responder must be able to use the monitor and must know the local acceptable values. If the Level II Responder has not been trained to the ALS level, an ALS provider should apply the EKG monitor and interpret the reading.
See Supplement 13.

(d) Describe the criteria for site selection of a medical monitoring station.

The medical monitoring site should be near the dressing area but away from noise and commotion, if possible. If the baseline readings are to be accurate, personnel must be assessed in a quiet environment.

(e) Demonstrate the ability to set up and operate a medical monitoring station.

In a simulated situation, the Level II Responder must be able to select an appropriate site and establish a monitoring station with the supplies and equipment necessary to perform assessments and keep accurate records. The responder must also demonstrate an ability to supervise the station so that medical support is accomplished in a timely manner and does not delay dressing and entry.

(f) Demonstrate the ability to interpret and analyze data obtained from medical monitoring of hazardous materials response personnel.

The Level II Responder must be able to review a reading and determine whether it is within the normal or safe limits for entry.

(g) Given a simulated hazardous materials incident, demonstrate proper documentation of medical monitoring.

Each organization must have an established recordkeeping system to document pre- and post-entry physical findings. The Level II Responder is responsible for ensuring that these records are properly completed.

3-5 Competencies — Terminating the Incident.

3-5.1 Perform the reporting, documentation, and follow-up required of the EMS component of the hazardous materials incident. The Level II Responder shall, upon termination of the hazardous materials incident, complete the reporting, documentation, and EMS termination activities as required by the local emergency response plan or the organization's standard operating procedures. The Level II Responder shall be able to:

(a)* Describe the information regarding incident EMS activities that should be relayed through the chain of command to the incident commander.

A-3-5.1(a) The type of information that should be made available to the incident commander would include but not necessarily be limited to the following:

(a) Patients.

1. Number
2. Condition
3. Disposition.

(b) Hazardous materials response personnel.

1. Number of personnel screened
2. Adverse reactions noted
3. Personnel transported for further treatment
4. Completed records
5. Recommended medical, physical, and psychological needs for immediate rehabilitation
6. Recommended medical surveillance followup.

(c) Availability of EMS personnel and equipment.

The Level II Responder should check to see that all personnel and equipment have been adequately decontaminated and readied for service. The responder also should see that someone contacts the medical facilities that received patients to determine whether they need any additional information about the patients they received. In addition, the responder must collect the information required to complete the incident and exposure records and determine whether a critical incident stress defusing or debriefing is needed, based on his or her evaluation of the incident.

(b) Describe the activities required in terminating the EMS component of a hazardous materials incident.
(c) Describe the process and demonstrate the ability to conduct the EMS portion of an incident critique.

Where hazardous materials incidents occur infrequently, each incident should be critiqued. Critiques often carry negative connotations because they may become finger-pointing sessions if not conducted properly. To be successful, the critique facilitator should take the following steps:

- Assign specific presentations on each aspect of the incident.
- Prepare maps, tapes, and drawings of the site so those involved can visualize the incident. If pictures and videos are available, they should be used.
- Develop and follow an agenda to ensure that all the facts are presented and that recommendations can be developed for revising procedures or protocols.
- Prepare and distribute a written report so that everyone can benefit from the "lessons learned."

(d) Explain the process of making revisions to EMS operating procedures and response capabilities as a result of information learned.

Each organization and jurisdiction should have a process by which the recommendations resulting from incident critiques can be routed. Level II Responders should be familiar with and use the mechanism in their jurisdiction.

4

Referenced Publications

The following documents or portions thereof are referenced within this standard and shall be considered part of the requirements of this document. The edition indicated for each reference is the current edition as of the date of the NFPA issuance of this document.

4-1.1 **NFPA Publications.** National Fire Protection Association, 1 Batterymarch Park, P.O. Box 9101, Quincy, MA 02269-9101.

NFPA 472, *Standard for Professional Competence of Responders to Hazardous Materials Incidents*, 1989 edition
NFPA 1561, *Standard on Fire Department Incident Management System*, 1990 edition.

The material contained Appendix A of this standard is included within the text of this handbook, and therefore is not repeated here.

Training

This Appendix is not a part of the requirements of this NFPA document, but is included for information purposes only.

B-1 General.

The Emergency Medical Services (EMS) personnel responding to hazardous materials incidents should be trained and should receive regular continuing education to maintain competency in three areas: emergency medical technology, hazardous materials, and specialized topics approved by the authority having jurisdiction.

B-1.1 **EMS Training.** Recognized US DOT, state, regional, or local training curricula should constitute the entry level EMS preparation for continuing hazardous materials training. At a hazardous materials incident it is desirable that all EMS BLS Provider personnel be trained to the US DOT EMT-A level or equivalent.

B-1.2 **Hazardous Materials Training.** The foundation for EMS response to a hazardous materials incident should be the competencies described in NFPA 472, *Standard for Professional Competence of Responders to Hazardous Materials Incidents.*

B-1.3 Specialized Training. Following completion of approved EMS training and appropriate level of hazardous materials instruction described in NFPA 473, the authority having jurisdiction should stipulate additional specialized instruction that the EMS personnel responding to hazardous materials incidents must complete.

B-2 Training Plan.

B-2.1 The authority having jurisdiction should develop a formal training plan and provide a program to train EMS personnel to the level being utilized.

B-2.2 A training plan should be developed and contain guidelines for the following functional categories:

(a) Program management
(b) Content development
(c) Instructor competencies
(d) Technical specialist competencies.

B-2.3 The training plan should be criteria-based to maintain a consistent quality of curriculum and instruction.

B-2.4 The training plan should specify entry knowledge and skill levels, training, and refresher training for both students and instructors.

B-2.5 The training plan should define evaluation criteria for successful completion of knowledge and skill objectives of the training program.

B-2.6 The training plan should provide for supervised field experience for EMS HAZMAT Responder and EMS HAZMAT Coordinator training levels.

B-3 Training Program.

The training program should be a comprehensive competency-based guideline of the implementation and presentation of the required subject material. As a minimum it should address the following areas.

B-3.1 Program Manager.

B-3.1.1 The Program Manager should have the authority and responsibility for the overall implementation of the program.

B-3.1.2 The Program Manager should be able to demonstrate knowledge of the following:

(a) The content of NFPA 472, *Standard for Professional Competence of Responders to Hazardous Materials Incidents*; NFPA 471, *Recommended Practice for Responding to Hazardous Materials Incidents*; and NFPA 473, *Standard for Competencies for EMS Responders to Hazardous Materials Incidents,*
(b) EMS delivery systems,
(c) Budgeting and financial planning, and
(d) Processes used to develop instructional materials.

B-3.1.3 The Program Manager should demonstrate the skill and ability to:

(a) Coordinate the training program,
(b) Evaluate program effectiveness, and
(c) Identify instructors and technical specialists.

B-3.2 Content. The content of the training program should include the competencies of NFPA 473 as a minimum.

B-3.3 Evaluation. Recognizing the need for technically sound curricula and instruction to meet the competencies outlined in this standard, careful evaluation of all instructors' training, background, and experience should be made.

B-3.3.1 The authority having jurisdiction should ensure that the training program meets the needs of the local area.

B-3.3.2 The Program Manager should ensure that the training program meets the needs of the hazardous materials response team and the EMS providers.

B-4 Instruction.

The need exists for technically sound curricula and delivery to meet the competencies outlined in this standard.

B-4.1 Instructors. The instructor should:

(a) Have mastery of the material he/she presents,

(b) Have an understanding of the training program objectives, and

(c) Have the ability to teach and evaluate.

B-4.2 Technical Specialist. The technical specialist is a person who has technical expertise and practical knowledge in a specific area. This category is intended to support training activities by allowing individuals not otherwise qualified at the instructor level to present an essential segment for which they do have expertise.

B-4.3 Final Evaluation. Upon completion of the training program the student should demonstrate competency in all prescribed content areas. This evaluation should include written and practical testing as specified by the program manager and instructors.

Recommended Support Resources

This Appendix is not a part of the requirements of this NFPA document, but is included for information purposes only.

C-1 General.

Emergency medical service personnel who respond to hazardous materials incidents must operate within a network of support resources. This appendix addresses the general classes of these resources and presents a recommended minimum level of support necessary for adequate emergency medical response.

C-2 Poison Control Centers (PCC).

C-2.1 **Goal.** In addition to providing support to the general hazardous materials response, the goal of the Poison Control Center is to provide the emergency medical personnel who respond to hazardous materials incidents with medical guidance, information, and advice during incidents involving toxic chemical releases and associated injuries. The PCC should regularly participate in the following activities together with the EMS component of the hazardous materials incident response:

C-2.2.1 Preplanning Assistance.

(a) Training,
(b) EMS HAZMAT standard operating procedures review,
(c) EMS reference materials.

C-2.2.2 Technical Advice. The ability to coordinate decontamination, treatment, and transportation of injured persons. The PCC should be available to the EMS personnel who respond to hazardous materials incidents for emergency consultation around the clock and during the normal working hours for nonemergency consultation. Poison Control Centers should be capable of providing advice regarding:

(a) Identify of ingredients,
(b) Toxicity of substances involved and symptoms and signs of exposure,
(c) Level of protective clothing recommended,
(d) Potential for secondary contamination,
(e) Recommended decontamination procedures, and
(f) Specific treatment and or antidotes.

C-2.2.3 Data Bases. The PCC should supervise and review the EMS data bases used during hazardous materials incident response.

C-2.2.4 Medical Surveillance. The PCC should provide support for:

(a) Surveillance quality assurance program design
(b) Surveillance Q/A program review
(c) Medical followup activities.

C-3 Chemical Injury Treatment Centers.

C-3.1 Goal. The emergency medical responders to hazardous materials incidents should transfer chemically injured patients to facilities having adequate chemical injury treatment capability. All such facilities should have a minimum level of competency to receive those patients including:

C-3.1.1 Patient Decontamination Capabilities.

(a) Decontamination area
(b) Proper ventilation system

(c) Restricted access

(d) Runoff containment.

C-3.1.2 A cadre of trained in-house hazardous materials incident injury treatment personnel.

C-3.1.3 Personal protective clothing for hospital personnel that may treat HAZMAT patients.

C-3.1.4 Formal hazardous materials incident response procedures directed to EMS providers and hospital personnel.

C-4 Communications.

The network of emergency medical response resources to hazardous materials incidents should be linked by an adequate communication system within the incident command post. The following components are suggested as a minimum.

C-4.1 **Radiotelephone.** All mobile and fixed EMS components should be able to coordinate EMS hazardous materials incident response via at least one dedicated frequency. All fixed facilities shall have r-f emergency power capability for at least one radio channel.

C-4.2 **Telephone Service.** There should be telephone service within the Medical Section/Division; preferably a cellular telephone.

C-4.3 **Computer.** All components of the EMS hazardous materials incident response system should have an orientation to and direct or indirect access to computerized chemical databases, computerized preplans, and computerized operational command and control.

C-4.3.1 **Fixed Installation.** Computer generated information should be readily available to field and clinical EMS hazardous materials response personnel via at least two of the following:

(a) Verbal transmission

(b) Fax transmission

(c) Modem transmission.

C-4.3.2 **Mobile.** On scene EMS response personnel should have immediate direct access to a field computerized highly toxic hazardous materials database and computerized command and control information.

C-4.4 **Other Resources.**

(a) CHEMTREC (CMA)

(b) ATSDR (HHS)

(c) Private resources.

D

Medical Treatment Considerations

D-1

The assessment and prehospital care of patients involved in hazardous materials incidents, and who are potentially chemically contaminated, should include the following steps:

(a) Provide for the safety of the EMS provider by securing the scene, ensuring appropriate decontamination of the patient, and protecting against exposure to communicable diseases and hazardous materials.

(b) The patient's airway should be secure and regularly monitored.

(c) The patient's breathing should be monitored and assisted when necessary.

(d) Supplemental oxygen should be administered if the surrounding environment safely permits.

(e) Bleeding should be controlled. This may be accomplished by the application of pressure bandages. Lower extremity bleeding may be controlled through the use of pneumatic anti-shock garments.

(f) When trauma may have involved cervical spine injury, an appropriate stabilization, immobilization collar should be applied.

(g) Cardiopulmonary resuscitation if indicated.

(h) In general, avoid all prophylactic invasive procedures unless required by life-threatening conditions. This includes the establishment of intravenous lines.

(i) Direct medical control should be established.

The authority having jurisdiction should ensure that a written prehospital medical standard operating procedures protocol is in place to provide direction to EMS personnel who respond to hazardous materials incidents.

This Appendix is not a part of the requirements of this NFPA document, but is included for information purposes only.

E-1

Patient decontamination, if required, should be carried out in the warm zone by properly trained personnel wearing appropriate chemical protective clothing and respiratory equipment.

Protocol(s) should be written to address the following:

(a) Determination of the potential for secondary contamination and the necessity for and extent of decontamination.

(b) Selection of appropriate personal protective equipment to be worn by personnel in the warm zone who are assisting with or performing decontamination.

(c) Decontamination of patients when the exposure is to an unidentified gas, liquid, or solid material.

(d) Emergency decontamination of patients with critical injuries and illness requiring immediate patient care or transport.

Referenced Publications

F-1

The following documents or portions thereof are referenced within this standard for informational purposes only and thus are not considered part of the requirements of this document. The edition indicated for each reference is the current edition as of the date of the NFPA issuance of this document.

F-1.1 **NFPA Publications.** National Fire Protection Association, 1 Batterymarch Park, P.O. Box 9101, Quincy, MA 02269-9101.

NFPA 471, *Recommended Practice for Responding to Hazardous Materials Incidents*, 1989 edition

NFPA 472, *Standard for Professional Competence of Responders to Hazardous Materials Incidents*, 1989 edition.

F-1.2 Other Publications.

F-1.2.1 **US Government Publications.** US Government Printing Office, Superintendent of Documents, Washington, DC 20402.

Title 29 CFR Part 1910.120
Title 40 CFR Part 261.33

Title 40 CFR Part 302
Title 40 CFR Part 355
Title 49 CFR Parts 170-179

F-2

.The following documents are not referenced within this standard, but may be useful to the reader.

Poisoning and Drug Overdose, Kent R. Olson, M.D., ed, Appleton & Lange, Norwolk, CT, 1990.

Hazardous Materials Exposure — Emergency Response and Patient Care, Jonathan Borak, M.D., Michael Callan, William and Abbot, Brady, Englewood Cliffs, NJ, 1991.

Emergency Care for Hazardous Materials Exposure, Alvin C. Bronstein, M.D., FACEP, and Phillip L. Currance, EMT-P, The C. V. Mosby Co., St. Louis, MO, 1988.

Supplements

The following supplements are included in this handbook to provide additional information for responders
to hazardous materials incidents. Since they are not
part of the standard, they are printed in black.

Supplement 1: Hazardous Materials Information Exchange (HMIX)

This supplement discusses the Hazardous Materials Information Exchange (HMIX), a service that allows the user to exchange information about hazardous materials management, including training programs, emergency management, technical assistance, and regulations. It is available free of charge from the Federal Emergency Management Agency.

Background

The Hazardous Materials Information Exchange (HMIX) is a computerized bulletin board designed specially for the distribution and exchange of hazardous materials information.

The HMIX provides a centralized database for sharing information pertaining to hazardous materials emergency management, training, resources, technical assistance, and regulations. With the HMIX, you can retrieve information, provide information to other users, or interact with peers.

To serve the hazardous materials community, the HMIX is available 24 hours a day, 7 days a week. Each user is allowed 30 minutes of access time per session on the system.

The HMIX is not intended to provide assistance during an actual emergency! The two primary features of the HMIX are the "Bulletin Board" and the "Message Exchange." The bulletin board feature allows the user to view:

- **Main Board Bulletins:** A listing of current HAZMAT news items.
- **Topic Listings:** A listing of subjects within the HMIX.

The message exchange feature allows the user to communicate with other users in three ways:

- **Electronic Mail:** Send and receive messages.
- **Electronic File Transfer:** "Upload" information onto the HMIX or "Download" information from the system onto your computer.
- **Chat:** On-line communication with other users.

Accessing the HMIX

The necessary tools: a computer
communications software
a modem capable of transmitting at
2400, 1200, or 300 baud

The modem set up: No parity
8 data bits
1 stop bit
VT-100 or TTY emulation

Specific instructions for some of the more common communications software packages can be provided by the system operators by calling the toll-free number listed. Dial the HMIX through your computer:

COMMERCIAL ACCESS (708) 972-3275
FTS ACCESS 972-3275

From 5:00 p.m. to 8:30 a.m. (Central Time), one node of the system is available on the toll-free assistance line listed. During business hours, that number is a voice line.

Having problems accessing the system? For technical assistance contact the system operator on the toll-free number, Monday through Friday between 8:30 a.m. and 5:00 p.m. Central Time.

1-800-PLAN-FOR or 1-800-752-6367

Illinois residents dial *1-800-367-9592*

SUCCESS!!! This message indicates that you have successfully accessed the HMIX:

HAZARDOUS MATERIALS INFORMATION EXCHANGE
PCBoard (R) Version 14.0/E9

Do you want color (Enter) = no

Do you want color screens? (In order to view the colors, you must have a graphics card and a color monitor.)

If NO, press the <ENTER> key.

If YES, enter "Y" or "Yes", and you will be in the color mode. When in color mode, more characters are being transferred and transmission of data will be somewhat slower.

Registering on the HMIX

WELCOME TO THE
HAZARDOUS MATERIALS INFORMATION EXCHANGE
MANAGED BY THE
FEDERAL EMERGENCY MANAGEMENT AGENCY
TECHNOLOGICAL HAZARDS DIVISION
STATE AND LOCAL PROGRAMS AND SUPPORT DIRECTORATE
500 C Street, S.W.
Washington, D.C. 20472

&
DEPARTMENT OF TRANSPORTATION
RESEARCH AND SPECIAL PROGRAMS ADMINISTRATION
OFFICE OF HAZARDOUS MATERIALS TRANSPORTATION
400 7th Street, S.W.
Washington, D.C. 20472

What is your first name? XXXXXXX
What is your last name? XXXXXXXX
Checking user's file please wait.......

- The system will scan the user's file.
  - If your name is found, you will be prompted to enter your password.
  - If your name is NOT found, this message appears:

[Your name] not found in user's file.
(R) to re-enter your name or (C) to continue to log on as new user?

- If you are a registered user and mistyped your name:
  Type (R)e-enter and retype your name

- If you are a first time user:
  Type (C)ontinue, and you will receive instructions on how to register

---

Would you like to register with us <Enter> = yes?

- If you answer NO to the above question, you will be immediately disconnected.

- To begin the registration questionnaire, type a (Y)es or press the <Enter>.

REGISTRATION QUESTIONNAIRE

PASSWORD (One word please!)? (...........................)

Make up your own password. REMEMBER IT!!! It will be your permanent
password to log onto the HMIX.

> **Re-enter PASSWORD to verify? (...........................)**
>
> Password will not show on screen. Dots will echo.
>
> **City and State calling from? (...........................)**
> **Business or Data phone # is? (...........................)**
> **Home or Voice phone # is? (...........................)**
> **Enter your organization? (...........................)**
> **Please wait. Adding name to Quick Index File........**

• After you have registered as a new user or the system has recognized your password, you will view a news bulletin:

> **[HAZMAT NEWS]**
>
> **If you have any questions, problems, or comments about the operation of the bulletin board, please type (C)omment to Sysop at the main menu prompt, or call our toll-free numbers and leave a message.**
>
> **1-800-PLANFOR (752-6367) AND 1-800-367-9592 (In Illinois)**
>
> **Federal law requires that anyone who releases a reportable quantity of a hazardous substance into the environment must immediately notify the National Response Center (NRC). In the event of an actual emergency, immediately notify the NRC at:**
>
> **1-800-424-8802 OR (202) 267-2675 (In Washington, D.C. area)**
>
> **For immediate advice at the scene of a chemical emergency, call CHEMTREC at 1-800-424-9300 or 202-483-7616 (in Washington, D.C.). Communication with CHEMTREC does NOT constitute compliance with Federal reporting requirements.**
>
> **For agency-specific questions contact the FEMA HMIX Coordinator at (202) 646-2860 or the DOT HMIX Coordinator at (202) 366-4448.**

HMIX Menu

The HMIX is a menu driven system. The menus guide the user through a series of options — paths which the user can follow through the HMIX. This system enables both the novice and experienced user to effectively access its many functions.

Most of the commands are listed in the EXTENDED MENU which follows the "News Bulletin." The "Extended Menu" lists the following options:

```
                          [EXTENDED MENU]

      COMMAND                   DESCRIPTION

      VIEWING THE TOPICS:

      (A)bandon Topic           Places you back at the Main Board
      (B)ulletin Listings       Lists Main Bulletins or Subtopics
      (G)oodbye                 Hang up
      (H)elp Functions          Get on-line help for those functions
      (J)oin a Topic            Lists Topic Menu
      (S)cript Questionnaire     Generates mailing label for user's guide
      (X)pert Mode              Turn on/off menu display

      MESSAGE EXCHANGE:

      (CHAT) between nodes      Talk to other users
      (C)omment to SYSOP        Send questions or comments to system
                                operator
      (E)nter a Message         Send a message to other users
      (O)perator Page           To receive on-line assistance
      (R)ead Messages           Read a message

      TRANSFER DATA:

      (D)ownload a File         Download a file to your computer
      (F)ile Directories        Lists files available for downloading
      (U)pload a File           Send a file for inclusion on the board

      To make a selection, type the first letter of the command.

      Type H M (with a space in between) to view this extended menu listing.
```

The "Extended Menu" will only appear once after the "HAZMAT NEWS" bulletin. (Unless you type H M to view the listing at a command prompt.) The "Topic Menu" will be the only menu displayed thereafter. The "Topic Menu" looks like this:

[TOPIC # MENU]

(B)ulletins/subtopics	(E)nter a message	(F)ile Directories
(J)oin a topic	(R)ead messages	(U)pload a file
(G)oodbye	(C)omment to SYSOP	(O)perator page
(H)elp		

Type H M (with a space between) to view the "Extended Menu."

Contents of HMIX

Outlined below are the subject matter topics and subtopics of the Hazardous Materials Information Exchange (HMIX):

0. MAIN BOARD BULLETINS

1. FEDERAL TRAINING

Subtopics Available
By Agency

    1. Federal Emergency Management Agency (FEMA)
    2. Department of Transportation (DOT)
    3. Occupational Safety and Health Administration (OSHA)
    4. Environmental Protection Agency (EPA/OSWER)
    5. Other Agencies

By Topic

    6. Incident Response Training
    7. Emergency Preparedness/Planning/Mitigation Training
    8. Enforcement Training

9. Motor Carrier Safety Training *EPA/OSWER Training
10. Site Assessment
11. Risk Assessment
12. Ground-water Remediation
13. Treatment Technologies
14. Response and Preparedness
15. Community Relations
16. Health and Safety
17. Quality Assurance
18. Computer Systems and Tools
19. CERCLA Specific Courses/Superfund University Training Institutes (SUTI)
20. RCRA Specific

2. INDUSTRY AND ASSOCIATIONS

1. HMIX Private-Sector Criteria
2. NATaT and the National Center for Small Communities
3. Trade Associations
4. International Association of Fire Fighters

3. CALENDAR OF CONFERENCES

Listing for Six Months
1. Current Month
2. Second Month
3. Third Month
4. Fourth Month
5. Fifth Month
6. Sixth Month

4. INSTRUCTIONAL MATERIAL AND LITERATURE LISTING

Subtopics Available
1. Fire Protection & Prevention
2. Training
3. Transportation
4. Laws and Regulations
5. Emergency Management/Response

6. Emergency Medicine/Decontamination
7. Ind. Hygiene/Env. Health/Worker Safety
8. Federal Publications
9. Waste Management

Media
1. Literature
2. Films/Slides/Videos
3. Software
4. Newsletters/Journals

5. TOLL-FREE NUMBERS AND ON-LINE DATABASES

Subtopics Available
1. Federal and State Toll-Free Technical Assistance Sources
2. Private Sector Toll-Free Technical Assistance
3. Federal and State Agency On-line Databases
4. Commercial and Private On-line Databases

6. LAWS AND REGULATIONS

Subtopics Available
1. Hazardous Materials Transportation Regulations
2. Pending Transportation Legislation
3. Interpretations of the Hazardous Materials Transportation Regulations (New)
4. Interpretations of the Hazardous Materials Transportation Regulations (Archived)
5. RSPA Inconsistency Rulings
6. Environmental Protection Agency (EPA)
7. Occupational Safety and Health Administration (OSHA) Regulations
8. SARA Threshold Planning and Spill Quantities (302 List)
9. Other Regulations
10. Reviews of Regulatory Developments
11 & 12. Federal Register Announcement for National Oil and Hazardous Substance Pollution Contingency Plan Introduction, Summary, and Information

7. CONTACTS

Subtopics Available
1. Federal Agencies
2. Federal Regional Offices
3. Professional Coalitions
4. Trade Associations
5. Research Centers
6. Environmental Groups
7. State and Local Public Interest Groups

8. DEPARTMENT OF TRANSPORTATION

Subtopics Available
1. Cooperative Hazardous Materials Enforcement Development Program (COHMED)
2. Federal Highway Administration, Office of Motor Carrier Safety Field Operations HAZMAT Division

1. Program Description
2. Contacts
3. Newsletter/Technical Advisory Bulletin/On-Guards
4. Calendar
5. Activities

40. RSPA Program Description (May 89)
41. Outline of Title 49 (May 89)
42. DOT News Releases and Information Items
43. RSPA Publications Listing
44. DOT Preferred Routes
45. DOT Exemptions Which Must Be Carried Aboard the Motor Vehicle
46. Hazardous Materials Information Systems (HMIS)
47. DOT Contacts, Libraries, and Public Dockets
48. Information on Drug Law, Films, and Literature
49. Motor Carrier Safety Assistance Program (MCSAP)
50. Commercial Vehicle Safety Alliance (CVSA) News (Feb 89)
51. Downloadable Files

9. FEDERAL EMERGENCY MANAGEMENT AGENCY (PRIVATE TOPIC)

10. SARA LISTING OF CHEMICALS

11-20. STATE SPECIFIC INFORMATION

    (11) Region I (CT, MA, ME, NH, RI, VT)
    (12) Region II (NJ, NY, Puerto Rico, Virgin Islands)
    (13) Region III (DE, DC, MD, PA, VA, WV)
    (14) Region IV (AL, FL, GA, KY, MS, NC, SC, TN)
    (15) Region V (IL, IN, MI, MN, OH, WI)
    (16) Region VI (AR, LA, NM, OK, TX)
    (17) Region VII (IA, KS, MO, NE)
    (18) Region VIII (CO, MT, ND, SD, UT, WY)
    (19) Region IX (AZ, CA, HI, NV, American Samoa, Guam, Western Pacific Islands)
    (20) Region X (AK, ID, OR, WA)

21. ARCHIE COMMUNICATION

HMIX Features

Main Board Bulletins

The Main Board "Bulletin Listing" contains "news" items and bulletins of continuing interest.

- To view this listing you must be at the "Main Board Command" prompt.

- Type (B)ulletin Listing at the prompt.

(# min. left) Main Board Command? B

[BULLETIN LISTINGS]
New Bulletins and Announcements
MM/DD/YR1   **Example: Proposed Rulemaking, Training Course, News Bulletin, etc.**

Bulletins of Continuing Interest

MM/DD/YR13   Public Notice Regarding Privacy and Other Legal Matters

MM/DD/YR14   Rules and Guidelines for Users of the HMIX
Bulletin Board

- You may view any of the bulletins by entering the number desired.

(H)elp, (1-17), Bulletin List Command?

- Type (R)e-list to view the bulletins again or press <Enter> to return to the main menu.

- Main Bulletins are added and purged on a regular basis. Bulletins of continuing interest to the users remain on the board indefinitely, but are updated to advise users of HMIX activities or to update the information in the bulletin.

Topic Listings

- To look at the "Topic Listings," type (J)oin a topic at the Main Board Command prompt.

(# min. left) Main Board Command? J

[TOPIC LISTINGS]

Topics 11-20 contain State-specific information:

(0) Returns you to Main Board	(11) Region I (CT, MA, ME, NH, RI, VT)
(1) Federal Training Courses	(12) Region II (NJ, NY, Puerto Rico,
(2) Industry and Associations	Virgin Islands)
(3) Calendar of Conferences	(13) Region III (DE, DC, MD, PA, VA, WV)
(4) Instructional Material and	(14) Region IV (AL, FL, GA, KY, MS, NC,
	SC, TN)
Literature Listing	(15) Region V (IL, IN, MI, MN, OH, WI)
(5) Toll-free (800) Numbers	(16) Region VI (AR, LA, NM, OK, TX)
and On-line Databases	

---

(6) Laws and Regulations	(17) Region VII (IA, KS, MO, NE)
(7) Contacts	(18) Region VIII (CO, MT, ND, SD,
(8) Department of	UT, WY)
Transportation	(19) Region IX (AZ, CA, HI, NV, American
(9) FEMA Headquarters	Samoa, Guam, Western Pacific
(Private)	Islands)
(10) SARA Listing of	(20) Region X (AK, ID, OR, WA)
Chemicals	(21) ARCHIE Communication

Topic # to join <Enter> = none?

• Once the listing appears, enter the number of the topic desired, or press <Enter> to remain where you are currently on the board.

• Entering a (1) places you in Topic #1.

[Topic1]

This section offers a listing of training courses which are sponsored by the Federal Government. (Note: Some of the courses listed have been designed specifically for use by Federal and/or State employees. Others who are interested should telephone the contact listed to inquire as to course availability.)

• Each topic is further subdivided for easier access of the information. Enter (B)ulletin to list the subtopics. (See section "Contents of HMIX" for topic area subtopics.)

– Entering the corresponding numbers will list the subtopics.

– Type (R)e-list to view subtopics again.

– <Enter> will return you to the topics command prompt and display the topic menu.

– Type (H)elp to receive information on how to display bulletin listings.

– To return to the Main Board Command Prompt, type (A)bandon to leave the topic.

– Type (G)oodbye to conclude your session on the HMIX.

• There are two ways to view another topic:

– Type (J)oin to re-list all the topics and their respective numbers.

– If you know which topic you would like to join, you may "stack" the commands. That is, type (J)oin followed by a space and then the number of the topic. A stacked command looks like this:

(# min. left) Topic Command? J 4

[INSTRUCTIONAL MATERIAL AND LITERATURE LISTING]

This library contains literature and instructional material. Subtopics are viewed by selecting, first a subject matter category, and then the media type. For example, to view a listing of Fire Protection and Prevention literature, type 11 (without a space in between).

Subject	Media
1. Fire Protection & Prevention	1. Literature
2. Training	2. Films/Slides/Videos
3. Transportation	3. Software
4. Laws and Regulations	4. Newsletters/Journals
5. Emergency Management/Response	
6. Emergency Medicine/Decontamination	
7. Ind. Hygiene/Env. Health/Worker Safety	
8. Federal Publications	
9. Waste Management	

• Topic 10 is a unique topic area because it contains a "door" in the system that allows users to search the "SARA Listing of Chemicals" by:

– chemical name or fraction of the name

– synonym

–chemical abstract number (CAS number)

–transportation ID number

• Once you have joined Topic 10, type "OPEN 1" at the command prompt to open the door, or "B" for (B)ulletins.

Electronic Mail

The HMIX offers an electronic mail feature which allows you to leave messages to other users, receive private messages, and read public messages.

To Read Messages

• Type (R)ead messages at the main board or topic command prompt.
  (# min. left) Main Board Command? R

• The system will respond with the following:
  (H)elp, (range of numbers), Message Read Command?

• Enter the number of the message you would like to read or one of the following subcommands:

(#)	**selects a specific message within the range.**
(F)rom	**selects only messages left by you.**
(S)ince	**selects public messages that have been sent since last message read.**
(Y)our	**selects only messages left for you.**
(NEXT) or (+)	**reads next higher message number available.**
(PREV)ious or (-)	**reads next lower message number available.**

• Message numbers and the other commands above can be stacked if desired:

(# min. left) Main Board Command? R F Y S

SAMPLE MESSAGE

DATE: 00-00-00 (time)

TO: ALL

FROM: SYSOP

SUBJ: WELCOME!

Hello! Welcome to the Hazardous Materials Information Exchange (HMIX)! If you have any questions or comments, please send a comment to the system Operator or to one of the HMIX Coordinators.

- At the end of each message, you have the following options:

<ENTER>	read more messages or return to the message read command.
(N)o	stop reading messages.
(NS)	continue reading messages in a non-stop format.
(RE)ply	reply to the message.
***(K)ill	erase a message once it has been read.

****Please use the (K)ill messages command to delete your private messages after having read them. You can also use the (K)ill command to delete messages you have sent to others. Only the sender can delete public messages.*

To Send a Message

To send a message to another user, type (E)nter at the main board command. Messages can be sent to all users, a particular group of users, or one individual user.

(# min. left) Main Board Command? E

- Enter the name of the person to whom the message will be addressed. If the message is to ALL, just press <Enter> at the "TO:" prompt.

- The brackets above the "TO:" and "SUBJ:" indicate the maximum length of the entry allowed (up to 25 characters). Attempting to exceed the maximum entry length will result in a BEEP being returned to you and the system will wait until you either press <Enter> to enter the subject as-is, or backspace over your current entry and modify it.

Message Protection

• Next, you will be asked to enter a protection level for the message.

(G)roup	**allows assigning a password to the message which only other callers who know the common password will be allowed to read.**
(R)eceiver	**makes the message private to all except you and the person to whom it is addressed.**
(S)ender	**allows you to assign a password to the message so that only you can kill it later. This prevents the other person to whom the message is addressed from killing it.**
(N)one	**means that the message will be open for all to read.**

***You cannot assign receiver protection to a message addressed to "ALL."

To Enter a Message

• After completing the above step, you can enter your text.

– Each line will be preceded by its line number.
– Up to 72 characters are allowed per line.
– Beyond the 72 character limit will automatically "wrap" down to the next line.

• To complete entering your text, press <Enter> alone on a blank line, at which time the following commands will be displayed:

(A)bort, (C)ontinue, (D)elete, (E)dit, (H)elp, (I)nsert, (L)ist, (S)ave

(A)bort   **abandons or cancels your message entry.**

(C)ontinue allows you to continue entering text.

(D)elete   allows you to delete a line of text from the message.

(E)dit   allows you to edit a line of text.

(H)elp   displays this Help file.

(I)nsert   allows you to insert a line of text in front of another.

(L)ist   re-lists your text entry to the screen.

(S)ave   writes the message to the disk.

(SC)   allows you to save the same message to another individual.

To Edit a Message

The format to (E)dit a line of text is: old text;new text

old text	the text you wish to replace, followed by a ";"
new text	new wording

- The old text search is case sensitive, so capital letters must match exactly.

- If your new text causes that line to exceed 72 characters, the characters beyond 72 will be truncated from the line.

EDIT EXAMPLE

ENTER YOUR TEXT.<ENTER> ALONE TO END. (72 CHARS/LINE, 99 LINES MAXIMUM)
(—————————————————————————————)
1: HELLO!
2: This is an exmaple of the edit function.
3:
(A)bort, (C)ontinue, (D)elete, (E)dit, (H)elp, (I)nsert, (L)ist, (S)ave
Text Entry Command? E

Edit Line #? 2
(—————————————————————————————)

2:    This is an exmaple of the edit function.

Enter (Oldtext;Newtext) or (Enter) alone for "no change."

?   exmaple;example

***NOTE: ONCE A MESSAGE HAS BEEN SAVED, IT MAY NOT BE EDITED.*

To Send a Group Message

- Before sending a (G)roup message, the group must be informed of the password.

- Once the message is saved, a message number will be assigned. Enter a second message (public) to the group to advise them of the message number to read.

Comment to System Operator

If you wish to leave a comment or question for the system operator, you can do so by typing (C)omment to SYSOP at the main board command.

(# min. left) Main Board Command? C

Leave a comment for the sysop <Enter> = no? yes

- Other than that, it is the same as entering a message.

- Messages left for the sysops are NOT public messages.

Electronic Data Transfer

To transfer data, your communications software must be capable of transferring files from the system.

Become familiar with your software before attempting to download or upload files.

(U)pload   **allows sending a file from your machine to the HMIX**
(D)ownload   **allows transferring a file from the HMIX to your machine**

- The system needs to know what type of transfer protocol your software supports. If you are not sure, choose Xmodem. In most cases, it will be compatible with your software.

- The HMIX SYSOPS are available to answer questions concerning the compatibility of software and transfer techniques.

Uploading a File

Information you would like to have included on the HMIX can be uploaded at the Main Board Command. It will not, however, go directly into the system files. Information sent to the HMIX is first reviewed by the system operator, edited if necessary, and then put into the appropriate topic.

- Type (U)pload at the Main Board Command prompt.

- Enter the filename, transfer protocol, and a short description for use by the system operator.

- Include the topic and subtopic numbers where you would like the information to be placed.

- You have 60 seconds to initiate a transfer.

- If the task is aborted, you will be returned to the Main Board or topic command.

UPLOAD EXAMPLE

(# min. left) Main Board Command? U
Filename to Upload <Enter> = none? SAMPLE.TXT

Checking file transfer request. Please wait...
Before beginning, enter a description of (SAMPLE.TXT)

(————————————————————————)
? Topic #/Subtopic #, (Your Name)....................

(A)   Ascii (Non-Binary)
(X)   Xmodem
(C)   CRC Xmodem (CRC)
(O)   1K-Xmodem (PCBoard Ymodem)
=>(N)   None

Protocol Type for Transfer, <Enter> or (N)=abort? X
Upload Status: Screened Before Posting
Protocol Type: Xmodem (Checksum)
File Selected: SAMPLE.TXT
(Ctrl-X) Abort Transfer

Transfer Successfully Completed. (000 cps avg.)
Thanks for the file!

Downloading a File

All of the information included on the HMIX is considered in the "Public Domain" and can be downloaded by users. Within topics, files are available to download. At the topic command, type (F)ile Directories to view the downloadable files. Listed will be the "filename," date, and file size.

DOWNLOAD EXAMPLE

(# min. left) Main Board Command? J 6
(# min. left) Laws (6) Topic Command? F

[LAWS AND REGULATIONS]

These files are available for downloading. Enter (D) followed by the filename on the left. If you would like to download all of these files at once, please download: LBLTW (This file is updated daily at 1:00 AM CST). The archived filename for this topic is LBLT.EXE Please read main board bulletin "Board Status" for further information.

Filename:	Size:	Updated:	Description:
LBLT1	1234	00-00-00	HazMat Transportation Regulations
LBLT2	1234	00-00-00	Pending Transportation Regulations
"	"	"	"
LBLT9	1234	00-00-00	Other Regulations
LBLTW	9999	00-00-00	ASCII file of all lbit files

• At the command, type (D)ownload. The format for downloading will be similar to that of uploading in the previous example. Return to your communications software package for instructions on receiving a file.

Additional Features

For your convenience, HMIX offers additional features. These include:

Chat (CHAT)

The CHAT option can be used to "talk" on-line to other users.

• After typing CHAT, you will see a list of other nodes and other users who are currently on the system.

• Anyone can enter CHAT, but you can only communicate after the other person responds by also typing CHAT.

• Be sure to allow time for the other person to respond. Lines of text are transmitted only after pressing the <Enter> key.

• Please wait until it is your turn to type! End your text with a "/go" to indicate it is the other person's turn to type, and then wait.

• If you engage in a CHAT with another user, please limit your session to a reasonable length of time to allow others access to system.

• Any "CHAT-ting" user can end the session by typing (Q)uit.

CHAT SUBCOMMANDS

(#)	the number of the node with which you wish to CHAT
(G)roup	places you in Group CHAT with others
(Q)uit	quits Node CHAT
(E)nd	also quits Node CHAT
(U)ser status	displays the status of the other Nodes
(H)elp	gives a quick display of these subcommands

Color Mode (M)

The color (M)ode option turns the color displays on and off. You may choose this option at any topic command prompt.

Expert Mode (X)

The e(X)pert mode option can be used to suppress the menu listings. You can type X at any topic command prompt. Additionally, all of the other command options will be shortened to only the first letter. Only experienced users should invoke this command. To cancel the expert mode, type (X).

Operator Page (O)

By typing (O), this option pages the system operator so a user may receive on-line assistance. When the system operator is available, you will receive the message "SYSOP CHAT active at (time)...Hello, this is (name)." This option can be used if you have a question or problem which can be answered briefly. For other questions, please call the toll-free numbers listed in this publication.

Script Questionnaire (S)

If the user types (S) at the command prompt, the user will be provided instructions for completing the script questionnaire. This feature is used to order additional copies of the User's Guide and creates a mailing label with information entered by the user.

Transfer Protocol (T)

If you transfer data frequently and always use the same protocol desired. This option automatically sets the protocol for future sessions.

Glossary

ASCII	An acronym for "American Standard Code for Information Interchange." ASCII is one of the standard formats for representing characters so that files can be shared between programs.
BAUD RATE	The speed of transmission of data bits per second.
BIT	The smallest unit used to store information in your computer's memory (or on disk).
BYTE	The amount of space needed to store a single character (number, letter, or code).
DOS	The acronym for disk operating system.
FILE	A collection of related information.
FILENAME	The identifier given to a FILE saved by the disk operating system. A filename can be from one to eight characters long and can have an extension of up to three characters separated from the filename by a period(.). When saving a file to be uploaded onto the HMIX, use the extension describing the file: (.TXT, .WPF, .MAC, etc.).
HARDWARE	Any piece of physical electronic equipment. (Example: computer, modem, printer)

MODEM   A device that is connected to a computer (internally or externally) and a phone line. This is the piece of HARDWARE that is needed for your computer to communicate with the HMIX or other bulletin board systems.

NODE   The entry point, accessible through a phone line, into the HMIX. The HMIX has 5 nodes allowing 5 users to access the HMIX simultaneously.

PROTOCOL   A set of procedures or conventions that are used to formalize information transfer and error control between points.

SOFTWARE   Any type of program. (Example: DOS, communications software, word processing software.)

Public Notice Regarding Privacy and Other Legal Matters with Respect to the HMIX Electronic Bulletin Board

IN ORDER TO RETAIN THE INTEGRITY OF THE HAZARDOUS MATERIALS INFORMATION EXCHANGE (HMIX) AND ALLOW THE BROADEST DISSEMINATION OF INFORMATION, THE FOLLOWING GUIDELINES HAVE BEEN ESTABLISHED.

1. PURSUANT TO THE ELECTRONIC AND COMMUNICATIONS PRIVACY ACT OF 1986, 18 USC 2510 et. seq., NOTICE IS HEREBY GIVEN THAT THERE ARE NO FACILITIES PROVIDED BY THIS SYSTEM FOR SENDING AND RECEIVING SENSITIVE OR CONFIDENTIAL ELECTRONIC COMMUNICATIONS. ALL MESSAGES SHALL BE DEEMED TO BE READILY ACCESSIBLE TO THE GENERAL PUBLIC.

DO NOT PLACE SENSITIVE OR PROPRIETARY INFORMATION ON THIS SYSTEM!

NOTE: Subjects of interest to only one individual or a selected few may be left as "Receiver Only" or "Sender Only" messages. These messages must comply with Item #2 of the HMIX Rules and Guidelines Bulletin. They will not normally be viewed by other users of the board, however, they will be readily accessible to the public. Normal message disposition is to archive all correspondence at the end of each month. Any justifiable request for information contained on the system will be granted. Furthermore, ALL messages become a part of the HMIX system and may be viewed at any time by the SYSOP, or the HMIX sponsors. If deemed appropriate, the SYSOP or the HMIX sponsors may transfer a communication from "Receiver Only" to the general notices.

2. Anyone accessing this system and leaving any message, comment, file, or other form of communication grants the System Operator and/or Co-System Operator(s) (hereafter collectively known as the SYSOP) or the HMIX sponsors the specific right to read and dispose of said item in any manner that the SYSOP or the HMIX sponsor determines to be proper disposition (including transferring between "Receiver Only" and the general notices, or deleting).

3. The user understands that there is risk associated with using software downloaded from any bulletin board system (BBS). In particular, any file downloaded from this bulletin board is done so entirely at the user's risk. The operators of this board offer no warranties or guarantees of ANY kind and are not liable for any damage or loss of data incurred by anyone using a file downloaded from this BBS.

4. The SYSOP of this board reserves the right to suspend, remove, or deny access to this bulletin board any person who violates a board policy or for any reason deemed appropriate by the SYSOP.

Rules and Guidelines for Users of the HMIX Bulletin Board

1. DO NOT use an alias on this board. Use only your REAL name. Do NOT register more than once (using variations of your name) for the purpose of exceeding your daily allotted time limit. If you have a special need for more time on the board, please telephone or leave a message to the SYSOP briefly stating your reasons.

2. The following materials are NOT suitable for posting on the HMIX:

- Any copyrighted material, including program listings, source or compiled code, or any portion thereof, without permission of the copyright owner. All items posted on the HMIX are considered in the public domain.

- Classified materials of any kind.

- Any credit card number or other financial access or credit coding information which would allow unauthorized use of any other person's credit.

- Any access code, password, or confidential procedure which would aid others in gaining unauthorized access to any confidential information or file.

- Any defamatory statement or materials, or the use of foul language.

- Any explicitly declared trade secret.

- Any proprietary software.

- Any commercial advertisement which suggests or may suggest endorsement by the Federal Emergency Management Agency, the Department of Transportation, or other Government entity without its expressed written permission.

3. "Receiver Only" or "Sender Only" messages are permitted on the board, subject to the statements regarding privacy and other legal matters as described in the previous section. However, we encourage you to make "Public Messages" the RULE.

Criteria for the Inclusion of Private Sector Information

Any materials submitted for inclusion on the Hazardous Materials Information Exchange (HMIX) will be reviewed based upon the following criteria:

- Only information pertaining to the prevention of, preparation for, or mitigation of hazardous materials incidents or other related hazardous materials topics will be accepted.

- Information received to be included in one of the informational topics must be timely and a certification provided that the materials are available immediately (not developmental).

- Materials provided should be submitted 1) as a text file (ASCII) and upload directly to the HMIX following the procedures outline, or 2) typed in an appropriate format and clearly legible so that the materials can be read by an optical scanner.

    Information should include:

- Description of where the materials should be included (Topic # and Subtopic #).

- Title of the course, conference, or resource document, etc.

- Description or abstract.

- Prerequisites (when applicable).

- Sponsoring organization.

- Contact person (including title, address, voice phone number).

- Cost of goods or services.

- Dates (when applicable).

- Location of this activity.

***NOTE: Dated materials are deleted by the SYSOP at the end of each month.*

COMMERCIAL ACCESS: (708) 972-3275
FTS ACCESS: 972-3275

For Technical Assistance:
1-800-PLANFOR
1-800-752-6367

Illinois residents dial:
1-800-367-9592

Supplement 2: So You Want To Start a Haz Mat Team!

William J. Keffer, EPA Region VII

This supplement identifies major areas of concern for a community forming a new hazardous materials response team. The authors cover team make-up, resources, and protective clothing and present an example of personnel protection measures in the form of heat stress monitoring of responders wearing Level C protective clothing. Special thanks to William J. Keffer, the author of this paper, for permitting its inclusion as a supplement to the Hazardous Materials Response Handbook.

Incident Expectations

Many local government and nonmanufacturing industry groups are becoming increasingly concerned about the risk and expense of dealing with hazardous materials releases. Plans are being made to develop the special capabilities needed to handle these incidents, but little practical material is available to assist groups in forming hazardous materials (haz mat) teams.

The following article is designed to provide some insight into what a haz mat response program would entail, as well as information on how to form a haz mat team.

According to the records collected from larger cities in Region VII over the past decade and discussions with members of various operational teams, it appears that haz mat releases account for 0.5 to 1.0% of the fire calls made by a local jurisdiction. Approximately half of the incidents will involve very small quantities (less than 55 gallons), and 75% occur at fixed facilities. (See Supplement III of this publication for further information on how to determine what hazardous materials are most likely to be released in a given community.)

Haz Mat Team Make-up

Composition of a haz mat team should be multi-disciplinary. Suggested membership for a small local team would have a minimum of 10 persons, possibly distributed as follows:

3 to 6 fire service personnel

A minimum of 2 health authorities; local department, poison control, or medical

A minimum of 1 law enforcement representative—especially for security and evacuation coordination

2 to 4 industry or academic types for special advice and assistance

A minimum of 1 sponsoring agency administration type for legal management of storage and disposal

These team members should have ample opportunity to train, cross-train, and practice haz mat skills in a variety of exercises including hands-on. Development of the training program should rely heavily on the Occupational Safety & Health Administration (OSHA) Haz Mat Worker Protection Rule and NFPA 472 and 471.

Resources

If a community or facility plans a group around the 10-person base, discounts the regular salaries of the individuals, and assumes they already have respiratory gear (SCBAs), a team can be developed and maintained for about

$12,000 to $15,000. This team could present a highly effective Level B (nonencapsulated) response posture that will provide personnel safety for more than 90% of all incidents. This $12,000 to $15,000 will be spent in a variable manner after the first year, but for the first year these monies would break down as follows:

Comprehensive physical exams at about $500 each	$5000
Out-of-service training (2 weeks)	$2000
Communications for non-fire service (pagers)	$2000
References	$1000
Personal protective gear	$3000
Mitigation and screening supplies	$1500

Items such as transportation, in-service training, and premium pay have not been costed. Properly managed programs could generally rely on vehicles readily available within a municipality and in-service training needs still vary.

Protective Clothing

Protective clothing, and the training needed for its proper use, are important considerations when preparing a newly formed haz mat team. Without proper protection, a haz mat team member is as helpless as the general public when dealing with the release of a hazardous material. Even becoming actively involved in effective size-up or simple mitigation procedures for a haz mat incident often requires the responder to wear protective equipment.

The purpose of personal protective gear is to minimize or eliminate the routes of exposure to the individual wearer, namely via inhalation and dermal (skin) exposure. In most hazardous materials incidents, respiratory protection is essential. Without adequate monitoring instruments, the only respiratory protection recommended is the self-contained breathing apparatus (SCBA). The SCBA is expensive (approximately $1400) and requires regular maintenance. It is impractical to assume that every first responder from law enforcement, fire, and health agencies will have a unit always available. A more feasible option would be a cooperative agreement between fire service and other responder agencies. Most fire service organizations that usually work on-scene at a haz mat incident are provided with SCBAs. An SCBA, as with any protective gear, should never be worn without advance formal training, prior experience, and regular fit testing and practice. Local fire departments may be able to provide the opportunity for the necessary training.

To determine the appropriate type of protection, the following must first be determined:

1. The positive identification of the material involved through:
   a. Bill of lading, consist, manifest, or other shipping papers or labels;
   b. Placards, UN or NA ID number;
   c. Talking with the driver or conductor.

2. The hazardous properties of the materials through:
   a. Data available from the various handbooks and guides. Examples: DOT *Emergency Response Guidebook*, NIOSH/OSHA *Pocket Guide to Chemical Hazards*, CHRIS Manual, NFPA *Fire Protection Guide on Hazardous Materials*.
   b. Effects of the materials on bystanders—respiratory problems, skin or mucous membrane irritation, etc.
   c. Environmental indicators—dead birds or other animals, discolored or burned vegetation.

There are more than a dozen types of material used for protective clothing. The quality of assembly can also vary, making acquisition a confusing process. Also, it is essential that sizes of suits, gloves, and boots are appropriate to each individual since a breach in the clothing eliminates the protection offered.

Table 1 is a simple chart to guide clothing selection for common chemicals. This chart is very conservative since personnel are rarely exposed to 100% concentrations of the chemicals encountered.

A basic splash protection kit is small enough to be kept in the trunk of a patrol car. The individual kits can be made up for under $150 and can be packaged and stored in a large, heavy PVC bag, which can also be used for the materials after they are contaminated. Each kit would contain one each of the items listed in Table 2.

Following is information on protective clothing for some of the more commonly released materials.

Anhydrous Ammonia

The equipment should prevent any possibility of skin or eye contact with the spilled product. This may include rubber boots, gloves, face shield, splash-proof goggles, and other impervious and resistant clothing. Fully

Table 1

Chemical	Materials for ≥ 2 Hours	Materials for 1/2 Hour
Gasoline	Viton/Neoprene, Nitrile PVA	Butyl, PE, PVC
Chlorine	Vitron, Neoprene Saranex	PE, PVC
Ammonia		
Methyl chloride	Viton/Neoprene	Butyl, Rubber Neoprene, Nitrile, PE, PVC
Methanol	Butyl, CPE, Viton, Teflon, Saranex	Rubber Neoprene, Nitrile, PVA, PVC
Toluene	Viton, Teflon	Butyl, Rubber, Neoprene, Nitrile, PE, PVA, PVC
Hydrochloric acid	Neoprene, Nitrile Viton, CPE, Rubber	PVC
Sulfuric acid	CPE, Saranex, Viton	PVC, Nitrile
Nitric acid	Viton, Saranex	Nitrile, PVC
Sodium hydroxide	CPE, Nitrile	
PCBs	Neoprene, PVA, Viton	Rubber, Polyethylene

encapsulating suits with self-contained breathing apparatus (SCBA) may be advisable in some cases to prevent contact with vapor or fume concentrations in the air. Compatible materials may include butyl rubber, natural rubber, neoprene, nitrile rubber, and polyvinyl chloride.

Respiratory Protection: For unknown concentrations, fire fighting, or high concentrations (above 500 ppm), an SCBA with full facepiece should be worn. For lesser concentrations, a gas mask with chin-style or front- or back-mounted ammonia canister (500 ppm or less) should be used within the limitations of these devices.

Chlorine

The equipment used should prevent any possibility of skin or eye contact with the spilled product. This may include rubber boots, gloves, face shield, splash-proof safety goggles, and other impervious and resistant clothing. Fully encapsulating suits with SCBA may be necessary to prevent contact with vapor or fume concentrations in the air. Compatible materials may include neoprene, chlorinated polyethylene, polyvinyl chloride, viton, and saranex.

Table 2

Items	Cost	Source
Coverall, Tyvek or Polypropylene	$2.80 each $70/case of 25	Saf-T-Glove, KCMO Kimberly-Clark Co. Grove, IL
Coverall, Saranex, or CPE	$10 to $25 each $225/case of 25	Saf-T-Glove, KCMO Chemron, Buffalo, NY
Acid Suit Coverall PVC/Nylon	$50 each	Sijal, Oreland, PA
Surgical Gloves Vinyl	$3.00/box of 100	Edmont, Coshocton, OH
Vinyl Outer Gloves	$15.72/cs of 12 pair	Edmont, Coshocton, OH
PVC-lined Gloves	$21.00/cs of 12 pair	Edmont, Coshocton, OH
Neoprene-lined Gloves	$31.44/cs of 12 pair	Edmont, Coshocton, OH
Buna-Nitrile coated Gloves	$24.96/cs of 12 pair	Edmont, Coshocton, OH
PVA/Organic Solvent Gloves	$113.16/cs of 12 pair	Edmont, Coshocton, OH
Butyl Gloves	$17.28/cs of 12 pair	Edmont, Coshocton, OH
Rubber Overboots	$19.22/pair	Ranger, Endicott, NY
Safety Goggles	$3.96/pair	Obtain locally
Duct Tape	$1.59/roll	Obtain locally
Large Plastic Bags	$3.99/box	Obtain locally
DOT Emergency Response Guidebook	$4.00 each	DOT, Kansas City, MO

Respiratory Protection: For unknown concentrations, fire fighting, or high concentrations (above 25 ppm), a full facepiece SCBA should be worn. For lesser concentrations, either a gas mask with chin-style or front- or back-mounted chlorine canister (25 ppm or less) or a chlorine cartridge respirator with a full facepiece may be used.

Hydrochloric Acid

The equipment used should prevent any possibility of skin or eye contact with the spilled product. This may include rubber boots, gloves, face shield, splash-proof safety goggles, and other impervious and resistant clothing. Fully encapsulating suits with SCBAs may be necessary to prevent contact with vapor or fume concentrations in the air. Compatible materials may include butyl rubber, natural rubber, neoprene, nitrile rubber, nitrile rubber/polyvinyl chloride, chlorinated polyethylene, polyvinyl chloride, styrene-butadiene rubber, viton, nitrile-butadiene rubber, saranex and polycarbonate.

Respiratory Protection: For unknown concentrations, fire fighting, or high concentrations (above 100 ppm), an SCBA with full facepiece should be worn. For lesser concentrations, a gas mask with chin-style or front- or back-mounted acid gas canister (100 ppm or less) or an acid cartridge respirator with a full facepiece may be used.

Methanol

The equipment should prevent repeated or prolonged contact and any reasonable probability of eye contact with the spilled product. This may include rubber boots, gloves, face shield, splash-proof safety goggles, and other impervious and resistant clothing. Compatible materials may include butyl rubber, natural rubber, neoprene, neoprene/styrene-butadiene rubber, nitrile rubber, nitrile rubber/polyvinyl chloride, polyethylene, chlorinated polyethylene, polyurethane, styrene-butadiene, viton, and nitrile-butadiene rubber.

Respiratory Protection: For unknown concentrations, fire fighting, or high concentrations (above 200 ppm), an SCBA with full facepiece or the equivalent should be worn.

Methyl Chloride

The equipment should prevent skin contact with cold methyl chloride or the cold containers and any reasonable probability of eye contact with the cold product. This may include rubber boots, gloves, face shields, splash-proof goggles, and other impervious and resistant clothing. Compatible materials may include neoprene and nitrile-butadiene rubber.

Respiratory Protection: For unknown concentrations, fire fighting, or high concentrations (above 100 ppm), an SCBA with full facepiece should be worn.

Nitric Acid

The equipment should prevent any possibility of skin or eye contact with the spilled product. This may include rubber boots, gloves, face shields, splash-proof safety goggles, and other impervious and resistant clothing. Fully encapsulated suits with an SCBA may be advisable in some cases to prevent contact with high vapor or fume concentrations in the air. Compatible materials may include natural rubber, neoprene, nitrile rubber, polyethylene, chlorinated polyethylene, polyvinyl chloride, viton, nitrile-butadiene rubber, and saranex for concentrations more than 70% nitric acid.

Respiratory Protection: For unknown concentrations, fire fighting, or high concentrations (above 250 mg/m^3), an SCBA with a full facepiece should be worn. For lesser concentrations, a gas mask with chin-style or front- or back-mounted canister or a chemical cartridge respirator with a full facepiece should be used within the limitations of these devices. The canister or cartridge should provide protection against nitric acid and should not contain oxidizable materials such as activated charcoal.

Sodium Hydroxide

The equipment should prevent any possibility of skin or eye contact with the spilled material. This may include rubber boots, gloves, face shield, safety goggles, and other impervious and resistant clothing for solids or liquids. Fully encapsulating suits with SCBAs may be advisable in some cases to prevent contact with high dust or mist concentrations in the air. Compatible materials may include butyl rubber, natural rubber, neoprene, neoprene/styrene-butadiene rubber, nitrile rubber, polyethylene, chlorinated polyethane, polyurethane, polyvinyl alcohol, and polyvinyl chloride for the solid and its solutions, as well as nitrile rubber/PVC, styrene-butadiene, viton, nitrile-butadiene rubber, and saranex for 30-70% solutions.

Respiratory Protection: For unknown concentrations, fire fighting, or high concentrations (above 100 mg/m^3), an SCBA with full facepiece should be worn. For lesser concentrations, a high efficiency mist and particulate filter respirator with full facepiece should be worn.

Sulfuric Acid

The equipment should prevent any possibility of skin or eye contact with the spilled product. This may include rubber boots, gloves, face shield, splash-proof goggles, and other impervious and resistant clothing. Fully encapsulating suits with SCBA may be advisable in some cases to prevent contact with high vapor or fume concentrations in the air. Compatible materials may include butyl rubber, neoprene, nitrile rubber, chlorinated polyethylene, polyvinyl chloride, styrene-butadiene rubber, viton, and nitrile-butadiene rubber for concentrated (more than 70%) acid as well as less concentrated solutions.

Respiratory Protection: For unknown concentrations, fire fighting, or high concentrations (above 50 mg/m^3), an SCBA with full facepiece should be worn. For lesser concentrations, a gas mask with chin-style or front- or back-mounted acid gas canister and high efficiency particulate filter or a high efficiency particulate filter with a facepiece should be worn.

Toluene

The equipment should prevent repeated or prolonged contact and any reasonable probability of eye contact with the spilled product. This may include rubber boots, gloves, face shield, splash-proof goggles, and other impervious and resistant clothing. Compatible materials may include polyurethane, polyvinyl alcohol, viton, nitrile-butadiene rubber, saranex, and fluorine/chloroprene.

Respiratory Protection: For unknown concentrations, fire fighting, or high concentrations (above 200 ppm), an SCBA with full facepiece should be worn. For lesser concentrations, a gas mask with chin-style or front- or back-mounted organic vapor canister (2000 ppm or less) or an organic vapor cartridge respirator with a full facepiece (1000 ppm or less) should be used within the limitations of these devices.

Polychlorinated Biphenyls

The equipment used should prevent any possibility of skin or eye contact with the spilled product. This may include rubber boots, gloves, face shield, splash-proof goggles, and other impervious and resistant clothing, rubber boots, boot covers, viton gloves, and saranex coated tyvek suits.

Respiratory Protection: For unknown concentrations, SCBA or approved air purifying respirator, full-face, canister-equipped (MSHA/NIOSH-approved), and a knowledge of the limitations of such equipment.

Personal Protective Clothing

1. **Before Donning Protective Clothing**

   a. Determine identification of material and associated hazardous properties.
   b. Decide if it is essential for you to enter the contaminated area.
   c. Determine if the personal protective equipment available is adequate protection for the hazards involved.

2. **How to Determine Appropriate Level of Protection**

   a. If the hazardous material involved has an inhalation hazard (i.e., toxic vapors or toxic combustion products), *respiratory protection is required.*

      (1) Indications:

         (a) Data from literature on material
         (b) Visible fire
         (c) Visible cloud of vapors

    (d) Irritation of eyes, nose, throat; nausea of bystanders

    (e) Environmental indicators — e.g., dead birds, animals, etc.

  b. If the hazardous material involved is a skin contact hazard, *dermal splash protection is required.*

  (1) Indications:

    (a) Data from literature on material

    (b) Chance of contact with material while mitigating incident

    (c) Irritation of skin from material contact by bystanders

3. **Decontamination Procedure—Must be done in contamination reduction area or warm zone**

  a. Open heavy-duty plastic bag.

  b. Remove booties and dispose in bag.

  c. Remove outer gloves and dispose in bag.

  d. Remove respiratory equipment, if utilized.

  e. Remove protective coveralls and dispose in bag.

  f. Remove inner gloves and dispose in bag.

  g. Seal bag with duct tape.

  h. Call State Department of Natural Resources or EPA for disposal or specific decontamination instructions.

WARNING: Attempting to mitigate a hazardous material incident and utilizing personal protective equipment should never be attempted without *formal training* and *prior familiarization with equipment.*

Heat Stress Monitoring in Level C

This section has been prepared to update recent experiences and improve current heat stress monitoring (HSM) at Superfund cleanup activities where EPA protective ensembles known as Level C are in use. It is hoped that a uniform procedure can be adopted that will allow for maximum productivity, while educating workers and monitoring effectively for heat stress.

As the title states, this section applies to Level C protection. Once specific guidelines (e.g., vital signs) are in place, they will apply to all levels of protection. However, some of the methods and equipment may vary.

Below are the issues to be addressed:

1. Temperature at which HSM will be implemented.
2. Vital signs to be measured and their ceiling values.
3. Work to rest ratios.
4. Equipment needs.
5. Fluid replacement.
6. Cooling devices.
7. Acclimation of workers.
8. Recordkeeping.
9. Education of workers in recognition.

Temperature

Following is a formula to be used in the field for the purpose of determining the necessity of heat stress monitoring. Ambient temperature, humidity, and cloud cover (solar load) are the three factors requiring consideration during the estimation of the actual thermal load inflicted on the body's cooling system. Acclimation is also important in considering the body's ability to effectively cool itself. The problem is that all people do not react to heat loads in the same manner or at the same rate.

Adjusted Temperature = Tdb* + 13 (% Cloud cover factor)
        No Clouds   = 1.00
        25% Clouds = 0.75
        50% Clouds = 0.50
        75% Clouds = 0.25
        100% Clouds = 0.00

The Wet Bulb Globe Temperature (WBGT) may also be used to determine the radiant heat load. Both the WBGT and the adjusted temperature should be measured and the higher value utilized. For Level C work, monitoring should begin at the adjusted temperature of 75°F, with work to be halted at a temperature of 110°F, unless effective auxiliary personal cooling devices are in use and only acclimated workers are used.

Example: Tdb = 70°F
Cloud Cover = 50%
WBGT   = 75°F

---

*Tdb = dry bulb temperature

Tdb + 13 (% cloud cover)
70 + 13 (.50) = 70 + 6.5 = 76.5°F
Adjusted Temperature = 76.5°F
WBGT = 77°F

WBGT (77°F) would be used as the actual temperature and heat stress monitoring would be initiated.

Vital Signs

Monitoring of the individual worker's vital signs is the most effective way to prevent heat illnesses. It is important to assign values that will identify the heat illness symptoms before they become serious. Previously, temperature, heart rate, and blood pressure were used as indicators of heat stress. It is now believed that skin temperature and body water loss are the most important vitals to monitor. Heart rate should also be monitored as it reflects the heart's reaction to the added thermal load. It is believed that blood pressure is not affected by heat stress and should no longer be routinely monitored. The following are suggested recommendations for vital signs monitoring:

Body Weight

When wearing impermeable clothing, it is possible to have a sweat rate as high as 3.5 L/hr. These lost fluids must be replaced intermittently throughout the day. Thirst alone is not a good indicator of proper fluid replacement. The amount of weight loss should be replaced by the equivalent weight in replacement fluid. Body Water Loss (BWL) should not be allowed to exceed 1.5% of total body weight. A scale should be kept in the break area and should be accurate to plus or minus ¼ pound. The color of excreted urine is also an indicator of the need for fluid replacement. If fluids are not properly replaced, the urine becomes a deep yellow color. There are some interferences; for instance, taking vitamins will darken the color. These factors can be monitored by field personnel themselves as a general indication of their fluid intake as compared to their body water loss.

Temperature

The deep body core temperature (rectal) is the most representative of actual temperature in the body; however, in the field it has not been practical to use this measurement. As a substitute, the oral temperature has been used. The average oral temperature is 37°C (98.6°F), but is not necessarily a norm for every individual. Current standards indicate that a maximum rise in temperature to 99–99.6°F is all that should be tolerated for civilian workers. A problem with this type of cutoff is that it does not allow for adjustment for workers who have abnormal base temperatures. A baseline should be established for each individual from data collected over a two-week period. As a guideline, the maximum rise in temperature should not exceed 1.5°F, and should return to within 0.5°F before the worker is allowed to return to the hot zone. It is very important that initial temperature readings, as well as other measurements, be made promptly (within a very few minutes) on exit from the work zone for break periods.

Heart Rate

The heart rate is probably the best indicator of overall stress applied to the body. The pulse becomes more rapid as the body tries to cool itself, and exhibits the effects of aerobic exercise. A maximum heart rate should be established using a method called the age adjusted heart rate. This maximum should not be exceeded at any point, but can be maintained during Level C activities. Aerobic exercise is generally performed at 70 to 85% of the maximum attainable heart rate, which is generally considered to be 220 beats per minute. The age adjusted heart rate is figured as follows:

(0.7) (220 − age)

A 20-year-old would have an age adjusted heart rate of 140 beats per minute and is calculated as follows:

(0.7) (220 − 20)

(0.7) (200) = 140 beats per minute

Work to Rest Ratios

A significant change in the vital signs measurements can signify heat stress candidates. Careful monitoring by a qualified person (i.e., nurse, EMT, safety officer) will aid in the reduction of heat stress injuries. The individual assigned these duties should be trained to interpret these readings and advise the OSC and response manager to make the proper adjustments in work and rest periods.

The less severe cases of heat illnesses can be mitigated by rest and cooling of the body. In order to focus attention to the possible victims of heat stress, the following guidelines should be met:

1. If, at any time, the maximum age adjusted heart rate is exceeded, or the heart rate has not returned to within 10% of the baseline measurement at the end of the rest period, the next work period should be shortened by 33% and the affected individual should not return to work until the heart rate is reduced.
2. If, at any time, the oral temperature rises 1.5°F above the baseline, the subsequent work period should be reduced by 33%.
3. If, at any time, the BWL exceeds 1.5% of the total body weight, the worker shall be instructed to increase his/her fluid intake.

Fluid Replacement

Fluid replacement and rest periods are the most effective deterrents of heat stress. Proper fluid replacement cannot be gauged by thirst because the sense of thirst is satisfied before the proper amounts of fluid are ingested. The following are suggestions that should be used to monitor and control the replacement of fluids:

1. All workers should be assigned their own replacement fluids container with their names marked clearly on the outside. This procedure helps to gauge the amount of fluids consumed by each worker.
2. Water is excellent and highly recommended, but workers should be allowed to drink something they like. Alcoholic and caffeinated drinks are not good replacement fluids, as they tend to increase the rate of body fluid loss. Fruit juices and electrolyte replacement drinks (Gatorade) should be diluted 3:1 if they are used. Water replacement is more important than the replacement of electrolytes since sweat contains only $\frac{1}{3}$ of the electrolyte balance of the blood. Whichever drink is used, it should be served cooled to between 50 and 60°F.
3. Salt tablets are not suggested as they are under scrutiny at this time. It is suggested that food be salted generously at meals if workers have been exposed to excessive heat.
4. Fluid intake should begin before the work shift is started after the morning weight is recorded. This provides for body water that will be lost in the first work period.
5. For each $\frac{1}{2}$ pound lost, 8 oz. of replacement fluids should be consumed.
6. One and one half percent body water loss is acceptable, as long as it is replaced before the following morning.

Equipment Needs

Body Weight: As stated before, scales should be used that are accurate to ± ¼ pound. They should also be rugged and easily calibrated in the field.

Heart Rate Monitors: In most cases the workers can take their own heart rate after brief instruction. The heart rate should be measured by the radial pulse for 30 seconds. For more reliable readings there is a rugged, battery-powered commercial instrument that could be mounted in the break area which, when lightly held in the hand, gives the heart rate on a digital read-out after approximately 30 seconds.

Temperature: Oral temperatures may be measured by digital thermometers currently used by the nurses at the Superfund cleanup sites.

Cooling Devices

Cooling devices generally should not be considered for temperatures under 95°F. Available cooling vests (ice) are, in general, bulky and heavy and will contribute more to the stress load than the relief provided. The vests that are hooked to some type of cooling unit or compressed air are impractical for workers not performing tasks in a fixed location. Employees who have been asked to wear the cooling units have complained about the weight and uneven cooling and have frequently expressed a preference to just stay hot and finish the work.

Acclimation

Acclimation to Level C is a very important step in the prevention of heat stress. The body needs time to adjust to the additional demands of wearing impermeable clothing in the heat, just as it does to adjust to cold weather. Impermeable clothing in hot weather prevents cooling by evaporation, which is the most effective mechanism that the body has to cool itself. With the added thermal stress and restriction of the natural cooling system the body uses, the first two weeks of Level C work can be the most dangerous, especially if the ambient temperatures are elevated and the work is physically demanding. Acclimation periods allow the body to adjust gradually to the added stress inflicted by donning impermeable protective gear.

Nonacclimated workers starting in Level C should participate in a program of increased personal monitoring and reduced work periods during the first two weeks of exposure. In the absence of additional monitoring by competent health personnel, the following suggested schedule should provide a safe working buildup for proper acclimation:

Day 1-3   Light work during the morning or late afternoon not to exceed two hours.

Day 4-6   Light work during the morning or late afternoon not to exceed three hours.

Day 6-8   Light work during the morning or late afternoon not to exceed four hours.

Day 8-10   Moderate work during the morning and afternoon, approximately four hours.

Day 10-12 Moderate work during the middle of the day, approximately five hours.

Day 12-14 Moderate work during the middle of the day, approximately six hours.

After day 14 Full days of moderate work.

Recordkeeping

Recordkeeping is vital in determining an individual's reaction to added thermal stress and will assist in job placement. A matrix considered to be a good system for heat stress tracking and recordkeeping is available from the Emergency Response Team of Edison, NJ. Using ERT's system of recordkeeping will produce an abundance of paperwork, but it can be incorporated into data using graphs or stored on computer disks. Any way that it is retained will provide valuable information for future references and make data available for future studies.

Worker Education

Education of workers required to wear personal protective gear is the most important preventative technique. Heat illness occurs more often through ignorance than any other contributing factor. It is imperative that personnel understand heat stress, its signs, symptoms, and contributing factors. The following two references should be mandatory reading for all personnel involved in wearing impermeable protective gear or in directing others who are. Daily safety briefings at job sites where protective gear is being worn are also important as a preventative measure.

1. NIOSH Pub. No. 80-132, *Hot Environments*, U.S. Dept. of Labor.
2. NIOSH/OSHA/USCG/EPA, Occupational Safety and Health Guidance Manual for Hazardous Waste Site Activities, U.S. Dept. of Health and Human Services.

Supplement 3: Beginning the Hazard Analysis Process

William J. Keffer, EPA Region VII

The first step in preparing to deal with hazardous materials incidents is to identify the types and possible sources of accidental releases. This supplement outlines the steps a community should take to analyze those facilities that use hazardous materials and the transportation corridors that bring hazardous materials into the community. Included is an item-by-item review of a material safety data sheet and instructions on using the information presented in it. Our thanks to author William J. Keffer for submitting his paper for inclusion as a supplement to the Hazardous Materials Response Handbook.

Background

Several recent federal studies show that there are between 5 and 6 million known chemicals, with this number growing at the rate of about 6,000 chemicals per month. Furthermore, a recent computer review by the Chemical Abstract Service of the complete list of known chemicals indicates that a first responder might reasonably be expected to encounter any of 1.5

million of these chemicals in an emergency, with 33,000 to 63,000 of them considered hazardous. To complicate matters, these hazardous chemicals are known by 183,000 different names. Fortunately, not all of these chemicals are equally common.

The U.S. Department of Transportation (DOT) and the U.S. Environmental Protection Agency (EPA) have used several measures of toxicity and volume of production to develop a shortened list of chemicals that are considered hazardous when transported in commerce. This list is comprised of about 2,700 chemicals, all of which are listed in 40 CFR 172.101 and the 1990 *Emergency Response Guidebook for Hazardous Materials Incidents.*

The Occupational Safety and Health Administration (OSHA) regulates about 400 chemicals on the basis of occupational exposures. This list is found in the NIOSH (National Institute of Occupational Safety and Health) *Pocket Guide to Chemical Hazards.*

Even this abbreviated list can be intimidating to local response personnel hoping to develop a comprehensive hazard analysis for their community. Further complicating their job is the fact that, according to a recent study by the National Academy of Sciences, National Research Council (NRC), there is so little known about seven-eighths of the 63,000 hazardous chemicals that not even a partial assessment can be made of their health hazards. Some conclusions drawn from the NRC study are as follows:

1. Pesticides—Of the 3,350 pesticides classified as important chemicals, information sufficient to make a partial assessment of the health hazard is only available on about 1,100 to 1,200 (34%) of them.
2. Drugs—Of the 1,815 drugs or drug ingredients noted, about 36% have enough information for a partial assessment.
3. Food Additives—For the 8,627 food additives listed, there is partial information on 19%.
4. Other Chemicals—For the remaining 48,500 industrial chemicals, there is enough information on just 10% to develop a partial assessment.

These two points—the lack of a generally accepted name for chemicals considered hazardous and the lack of data for assessing the risk—create a stumbling block for emergency response personnel and community officials who are charged with developing a viable, effective local hazardous materials management system. Without a contingency plan, however, based on effective and accurate hazard analysis prior to an emergency, it would be difficult and time-consuming to develop the necessary information in the midst of an emergency. In order to bring what appears to be an insurmountable task into perspective, the local response community must get involved in the *hazard analysis* process.

Hazard Analysis is the process of identifying chemicals present in the community—either at fixed facilities or passing through transportation corridors—and evaluating the hazard, vulnerability, and risk they present. A hazard analysis can be performed by any individual or small group that understands the principles of hazard analysis.

The goal of this paper is to provide a summary of the sources of information and methods for use by local agencies and hazardous materials teams to improve their understanding of the hazardous materials problem in the community—information that can be used in both planning and emergency response operations. It will also introduce the Material Safety Data Sheet (MSDS) and its utility in the hazard analysis process.

Hazard Analysis Data Sources

The purpose of hazard analysis is to gather data on the locations, quantities, and health hazards of chemicals most likely to be released in a community. This process may seem monumental from two aspects: (1) the sheer numbers of chemicals out there, and (2) the lack of information in a usable form available on these chemicals. In attempting to pare the list down to a workable size and gather data on the location and identity of the chemicals found in and being transported through the community, a variety of methods may be used, including:

1. Historical records of chemicals having the most frequent instances of release on a national level. For example, all previous studies have shown that the most commonly released hazardous chemical is commercial vehicle fuel (gasoline). The EPA commissioned a national study[1] in 1985 to look at hazardous chemicals and the sources of their release. This survey covered 6,928 incidents nationwide involving chemicals other than fuel. The source of releases study indicated:

   74.8% were fixed facility (in-plant) incidents
   25.2% were in-transit incidents

---

1. Acute Hazardous Events, Data Base Industrial Economics, Inc., Cambridge, MA 02140, December 1985.

The fixed facility incidents were distributed as follows:

20.7% storage
19.4% valves and pipes
14.1% process
17.9% unknown
27.8% other

The in-transit incidents were distributed by mode as follows:

54.5% truck
14.1% rail
3.8% water
3.1% pipeline
2.5% other

Perhaps the most useful data from this national study is the information on the chemicals most commonly involved in the 6,928 incidents—49.5% of the incidents involved only 10 chemicals:

23.0% polychlorinated biphenyls (PCBs)
6.5% sulfuric acid
3.7% anhydrous ammonia
3.5% chlorine
3.1% hydrochloric acid
2.6% sodium hydroxide
1.7% methanol/methyl alcohol
1.7% nitric acid
1.4% toluene
1.4% methyl chloride

Of the 6,928 incidents, 468 involved human injury or death. The same 10 chemicals listed above accounted for 35.7% of the death and injury events, though not at the same rate as they occurred:

9.6% chlorine
6.8% anhydrous ammonia
5.6% hydrochloric acid
4.7% sulfuric acid
2.8% PCBs
2.4% toluene
1.9% sodium hydroxide

1.5% nitric acid

0.4% methyl alcohol

0.1% methyl chloride

When the data for occurrence and injury is viewed for chemicals like chlorine and PCBs, it is apparent that the release and injury data are different. That is, although PCBs were involved in more incidents, chlorine, when released, posed a greater threat to humans. Determining this potential for causing injury to humans is the area of hazard assessment that takes the most effort on the part of the local response community. Gathering this information in a systematic manner cannot be done in the midst of an incident.

2. Summaries of previous incidents from emergency management and environmental response organizations at the local, state, regional, and federal levels. For example, EPA Region VII and states in the region have comprehensive, computerized records of all reported incidents by county since 1977. These records are available to any jurisdiction on request.
3. Local fire and police department records may disclose many incidents involving hazardous materials.
4. Local yellow pages and the state industrial directory will show most local fixed facilities that manufacture, store, or use chemicals. As an aid to this search, the EPA has recently prepared a summary for 14 types of facilities that shows what types of hazardous chemicals may be encountered. Copies of the summaries may be obtained from EPA by calling the national RCRA (Resource Conservation and Recovery Act) Industry Hotline toll free at 800-424-9346.

Once the chemicals in a community have been identified in name and quantity (either for contingency planning or emergency response), there are several national data bases for evaluating the hazard, vulnerability, and risk presented by those chemicals, including:

A. Poison Control Centers. If the chemical is a consumer product, the quickest way to get comprehensive hazard information is through the regional poison control center.
B. Manufacturer's technical medical staff. If the chemical is an industrial bulk chemical, effective assistance is generally available from CHEMTREC (Chemical Transportation Emergency Center) through the technical medical staff of the company that manufactures the chemical. Dial 800-CMA-8200 for nonemergency situations; 800-424-9300 for emergencies.

C. Agency for Toxic Substances and Disease Registry (ATSDR). If the chemical is a mixture or a waste, if a second opinion is required, or if the chemical is unknown, a good source of information is the agency at the Centers for Disease Control called the Agency for Toxic Substances and Disease Registry.

When contacting any of these sources, remember there is a lack of full information on health assessment for many chemicals. Answers and information from any of the sources listed above may be qualified, and each local response group, as part of their contingency planning, should locate a competent medical authority to work with the response community and assist in obtaining and interpreting health effects data.

Material Safety Data Sheets (MSDS)

The recent state and federal legislation on hazard communication, right-to-know, and mandatory local notification for certain hazardous chemicals will assist in developing pre-emergency and on-scene hazard assessments of the chemicals in the community. This legislation is bound to make the MSDS one of the best sources of information on chemical hazards.

Personnel from the Chemical Manufacturers Association, the sponsor of CHEMTREC, have available some 190,000 different MSDS and are receiving more all the time. For most local communities, it will be necessary to work with hard copies of the MSDS and begin with the 10 most commonly released chemicals. Later, one can add to that list based on the results of the local hazard analysis that determined additional unique or peculiar local hazards.

All local response groups have access to MSDS sheets by local industry, as required by SARA, Title III, so it is helpful to become familiar with them. However, remember that MSDS are not a cure-all and require some systematic way to approach their use. In any case, becoming familiar with the type of information presented on the MSDS and how that information will be of assistance in making a hazard assessment is essential—whether for pre-emergency planning or when responding to an emergency.

Minimum content of an MSDS is mandated by OSHA. Each sheet must contain the following sections:

1. The chemical name, chemical formula, common synonyms, chemical family, and manufacturer's name and emergency telephone number.

2. Hazardous ingredients and regulatory exposure limits, if any.
3. Physical data.
4. Fire and explosion hazard data.
5. Health hazard data.
6. Reactivity data.
7. Spill or leak procedures.
8. Special protection information.
9. Special precautions.

Response and/or planning personnel are encouraged to review the MSDS for at least the 10 most commonly released chemicals (previously listed), plus gasoline. They should obtain current MSDS from companies in their community, even though there are several sources of generic MSDS, including some where the information is computerized, to establish and maintain good working relationships with local companies.

In reviewing these 11 MSDS, take note of the different ways information is presented and the lack of uniform presentation. It is not required to use the standardized content in the format suggested by OSHA. From the varying presentations, you will gain some insights on the use of MSDS and factors to be considered in interpreting them. The depth of information furnished in MSDS varies depending on the extent of what is known about the chemical, as well as the management attitude of the company providing the information.

MSDS Section 1—Materials Identification

This section identifies the chemical by name, synonyms, and/or chemical family name. The manufacturer's name and emergency telephone number are listed in order to obtain additional data and assistance. Several preparers choose to emphasize the health hazards, precautionary measures, and emergency contacts at the top of the sheets.

MSDS Section 2—Ingredients and Hazards

Absolute clarity in describing all ingredients of a material and their hazardous components is essential; however, experience indicates that is not always the case. Most of the 11 reviewed MSDS address pure substances or aqueous dilutions, with the notable exception of that for gasoline. When discussing chlorine and its hazards, the MSDS preparer assumed chlorine to be the pure chemical in the gaseous form, whereas we are more likely to encounter chlorine as a solid (HTH—commonly used in swimming pool chemical control) or as commercial bleaches (a liquid that is fairly dilute). Most commercial bleaches for domestic use contain from 30,000 to 50,000 parts per million (ppm) sodium hypochlorite as a source of chlorine.

Upon reviewing the ingredients and hazards of anhydrous ammonia and ammonium hydroxide, we see that anhydrous ammonia is a colorless gas with an extremely pungent odor and that ammonium hydroxide is a clear, colorless liquid. Although their forms are different and their ability to impinge on exposures when released is different, the hazard is the same. Ammonia is intensely corrosive to human tissue, whether it is inhaled, contacts the skin, or is ingested. OSHA regulates workplace exposures of ammonia at 50 ppm (permissible exposure limit, PEL) while the American Conference of Governmental Industrial Hygienists (ACGIH) recommends a level of 25 ppm (threshold limit value, TLV). Additionally, OSHA regulations state that at concentrations of 500 ppm in air, the material becomes immediately dangerous to life and health (IDLH).

The OSHA system is designed to provide working conditions for reasonably healthy adult humans for 8-hour exposures for 40 hours per week for 40 years. This data is not directly applicable to general populations. Obviously, anyone with preexisting respiratory ailments would be expected to be more affected by irritants and by those chemicals that affect the central nervous system (CNS). The limits are not applicable to children, especially those in the first year of life, since their metabolism and nervous system responses are significantly different than adults or older children.

Let's look at the form and hazard information extracted from the 11 MSDS selected:

Name	Form	Exposure Limit		IDLH
Ammonium hydroxide	liquid	50	ppm	500 ppm
Anhydrous ammonia	gas	50	ppm	500 ppm
Chlorine	gas	1	ppm (ceiling)	25 ppm
Gasoline (unleaded)	liquid	300	ppm (ACGIH)	
		500	ppm (OSHA) petroleum distillate	
		10	ppm benzene	
Hydrochloric acid	liquid	5	ppm	100 ppm
Methyl alcohol	liquid	200	ppm	25,000 ppm
Nitric acid	liquid	2	ppm	100 ppm
Polychlorinated biphenyls	liquid	0.5	mg/m^3 (ACGIH)	
Sodium hydroxide	solid	2	mg/m^3	200 mg/m^3
Sulfuric acid	liquid	1	mg/m^3	80 mg/m^3
Toluene	liquid	200	ppm	2,000 ppm

All of the OSHA limits are for airborne concentrations and vary widely among the substances listed. It is important to note that the ratio of exposure limit to IDLH concentration also varies widely. The greater the range between these two numbers, the greater the safety factor for an exposed person to avoid permanent harm or death.

One important area of information not available from this section is how the liquids and the solid on the list above become airborne concentrations and how fast this occurs. For this information, we will have to look elsewhere on the sheets.

MSDS Section 3—Physical Data

In the process of hazard assessment, the ability to evaluate physical data combined with health hazard data is essential. The common physical properties provided on the reviewed MSDS include boiling point, freezing point, specific gravity, vapor pressure, vapor density, solubility, and appearance. Other parameters may be provided at the discretion of the company completing the sheets.

Let's look briefly at the range in characteristics of the chemicals from the 11 sheets reviewed and discuss the use or implications of each for the first responder:

1. *Boiling point*

The temperature at which a liquid turns to a vapor.

Chemical	Boiling Point
Ammonium hydroxide	36°C
Anhydrous ammonia	–33°C
Chlorine	–34°C
Gasoline (unleaded)	38° – 204°C
Hydrochloric acid (37%)	53°C
Hydrochloric acid (35%)	65.6° – 110°C
Methyl alcohol	64.5°C
Nitric acid (60 – 68%)	122°C (67%)
Polychroniated biphenyls	360° – 390°C
Sodium hydroxide	1390°C
Sodium hydroxide solution (50%)	145°C
Sulfuric acid	310°C
Toluene	231° – 232°C

Since the ambient temperature of this planet ranges from around –20°C to 50°C (–10°F to 120°F), any chemical with a boiling point below the ambient temperature will rapidly become a gas when released from its container. This is certainly the case for chlorine and anhydrous ammonia. Other materials with boiling points only slightly above normal ambient temperature will, if confined in a container, rapidly expand and pressurize that container with the potential for a rapid release if heated even slightly. Other materials, such as PCBs and sodium hydroxide pellets, will be unaffected by the heat of normal structural fires, but could be affected by the application of water to that fire. Sodium hydroxide pellets, for example, will dissolve in water to form a corrosive liquid.

2. *Freezing point*

Temperature at which the liquid form of a chemical will turn into the solid form.

3. *Melting point*

Temperature at which the solid form of a chemical will turn into the liquid form.

The two physical parameters above may be of limited use to response personnel for most chemicals. There are several chemicals for which control measures such as freezing are effective and where dry ice, for example, may be used to mitigate a release. Similarly, there are some chemicals where the form change under structural fire temperatures may be significant and may seriously alter the hazard to response personnel. Exposure of low melting point solids and most liquids to fire temperatures may result in production of toxic materials in the smoke plume.

4. *Specific gravity*

Density of a chemical compared to the density of water. If the specific gravity is less than one, the chemical will float on water. If the specific gravity is greater than one, the chemical will sink. In either case, it is important for response personnel to consider the property of solubility concurrently with specific gravity. These properties for the 11 chemicals are listed in the table on the following page.

Toluene, gasoline, and methyl alcohol are all flammable or combustible liquids with similar TLV levels. A glance at their respective solubilities, however, shows that mitigation techniques would have to be substantially different due to their solubility; i.e., methyl alcohol is completely miscible in water whereas the others are relatively insoluble. Not only would fire fighting methods differ, but additional attention would have to be paid to the

Chemical	Exposure Limit	Specific Gravity	Solubility
Ammonium hydroxide	50 ppm	0.9	infinite
Anhydrous ammonia	50 ppm	0.68	soluble
Chlorine	1 ppm	2.4	0.7%
Gasoline (unleaded)	300 ppm	0.7 – 0.8	insoluble
Hydrochloric acid	5 ppm	1.18	infinite
Methyl alcohol	200 ppm	0.8	miscible
Nitric acid	2 ppm	1.41	complete
Polychlorinated biphenyls	0.5 mg/m^3	1.5	0.01 ppm
Sodium hydroxide	2 mg/m^3	2.13	111 gm/100 gm
Sulfuric acid	1 mg/m^3	1.84	infinite
Toluene	200 ppm	0.86	0.05 gm/100 gm

solubility when environmental damage is possible. Many chemicals that are listed as only slightly soluble can still cause significant environmental toxicity to plants or aquatic life. Toxicity of methyl alcohol is 250 ppm and toluene 1,180 ppm; therefore each of them presents a serious environmental hazard if significant runoff is allowed to occur.

Most MSDS do not provide environmental risk information; therefore, this data will have to be sought from other sources. One excellent source for environmental risk information for many common chemicals is the EPA's OHM-TADS (Oil & Hazardous Materials—Technical Assistance Data System). Access to this system can be gained through any EPA regional office.

5. *Vapor density*

Density of a gas compared to the density of air. If the vapor density is less than one, the material will rise in still air and dissipate. If the vapor density is greater than one, the vapor will attempt to sink in still air and potentially collect in low spots and valleys.

6. *Vapor pressure*

Pressure exerted by vapors against the sides of the container. Vapor pressure is very temperature dependent. The lower the boiling point of the liquid, the greater the vapor pressure it will exert at a given temperature. In more common terms, the higher the vapor pressure, the more rapidly the material will change from the liquid form to a vapor when released to the environment and the higher the equilibrium concentration with air will be.

Boiling point, vapor pressure, and vapor density for the compounds of interest are listed in Table 1:

Table 1

Chemical	Boiling Point	Vapor Pressure	Vapor Density
Ammonium hydroxide	36°C	115 mm Hg @ 20°C	1.2
Anhydrous ammonia	–33°C	23 atm @ 20°C	0.6
Chlorine	–34°C	4,800 mm Hg @ 20°C	2.49
Gasoline (unleaded)	38° – 204°C	N/A	N/A
Hydrochloric acid (37%)	53°C	190 mm Hg @ 20°C	1.27
Methyl alcohol	64.5°C	97 mm Hg @ 20°C	1.1
Nitric acid (60–68%)	122°C (67%)	62 mm Hg @ 20°C	2 – 3
Polychlorinated biphenyls	360°C–390°C	<1 mm Hg @ 20°C	N/A
Sodium hydroxide	1390°C	negligible	
Sodium hydroxide solution (50%)	145°C	6.3 mm Hg @ 104°F	N/A
Sulfuric acid (96%)	310°C	22 mm Hg @ 145°C	<0.3 @ 25°C*
Sulfuric acid (93.2%)		<0.3 mm Hg @ 25°C	3.4
Toluene	231°–232°C	22 mm Hg @20°C	3.14

*Some MSDS contain incorrect data. The vapor density of sulfuric acid is 3.4 and not 0.3.

This data was found on the actual MSDS reviewed and, if you were to glance at the MSDS for sulfuric acid, it would be apparent that errors do creep in. Also, the detail of the information furnished varies from rough estimates or general statements for some materials to multiple listings for others.

If you can picture a room in which a release occurs from its container, and then look at the range of vapor pressures for the most commonly released substances, it will be apparent that both chlorine and anhydrous ammonia will present an almost instantaneous vapor (inhalation) hazard. Since both of these chemicals are soluble to some extent, a fog line may be helpful in suppression of volatization or reduction of concentrations, even when the release is continuous. However, when the intent is to reduce vapor production, the water from the hose lines should not enter pooled materials like ammonia or chlorine. For materials like sodium hydroxide and PCBs, a vapor hazard is not likely to exist under real-world conditions.

MSDS Section 4—Fire and Explosion Data

Most of the MSDS reviewed contain specific information for the fire fighter on the physical characteristics of the chemicals when involved in a fire. These characteristics, as summarized in Table 2, should be made familiar to most fire personnel.

Table 2

Chemical	Flash Point	Autoignition Temperature	Flammablity Limits	Extinguishing Media
Ammonium hydroxide	—	—	—	—
Anhydrous ammonia	1208°C	651°C	16 – 27%	shut off gas
Chlorine	N/A	N/A	N/A	N/A
Gasoline (unleaded)	–45°F	536° – 853°F	1.5 – 7.6%	dry chemical water spray
Hydrochloric acid	N/A	N/A	N/A	N/A
Methyl alcohol	52°F	385°F	6 – 36.5%	water spray
Nitric acid	none	none	N/A	N/A
PCBs	N/A	N/A	N/A	N/A
Sodium hydroxide	none	none	N/A	N/A
Sulfuric acid	none	none	N/A	N/A
Toluene	40°F	480°F	1.3 – 7.1%	dry chemical

In addition to these normal fire characteristics, the chemicals have other fire-related hazards, some of which are reported in the fire and explosion section. Examples of these are:

1. Chlorine and anhydrous ammonia are generally stored in pressure containers, and the violent rupture of these containers represents a significant hazard, in addition to their toxicity.
2. Many of the chemicals listed generate toxic vapors or mists when involved in a fire, thus representing an additional hazard.
3. Hydrochloric acid, nitric acid, sodium hydroxide, and sulfuric acid are such vigorous oxidizers or reducers that, although they are not flammable hazards themselves, they react with metals to produce hydrogen gas, which is extremely flammable.

MSDS Section 5—Health Hazard Information

This section of the MSDS presents information on the routes of exposure (inhalation, ingestion, dermal) and, in some cases, the severity of these risks (low, moderate, high). This information is essential for the selection of appropriate personal protective equipment and safety procedures for response actions at incidents. Some of the MSDS reviewed highlighted the major hazards in section one on the sheets, while others give a more detailed formal listing of the hazards in this section. Some sheets list the NFPA 704 rating for the specific chemical (use of the 704 system by local industry should be encouraged as it provides local emergency response personnel a basis for quick judgments about the severity of personal exposure). A brief summary of the hazards for each chemical is listed in Table 3.

Table 3

Chemical	Hazards
Ammonium hydroxide	corrosive—severe eye and skin irritant
Anhydrous ammonia	corrsove—severe eye and skin irritant
Chlorine	corrosive—life threatening toxic effects may occur at concentrations of 25 ppm on short exposures
Gasoline	flammable—irritant—CNS effects—some evidence of carcinogenicity—also numerous chronic effects
Hydrochloric acid	corrosive—may be fatal if ingested
Methyl alcohol	flammable—may be fatal if ingested
Nitric acid	corrosive—strong oxidizer at higher concentrations
PCBs	very long-lasting material—some evidence of liver damage—carcinogenic risk and adverse reproductive effects
Sodium hydroxide	corrosive—may be fatal if swallowed, causing severe burns
Sulfuric acid	corrosive—causes severe burns—may be fatal if swallowed—harmful if inhaled
Toluene	flammable—chronic skin irritant—various systematic effects on CNS, liver, kidneys

The most common hazard of the chemicals listed is their corrosive effect on nearly every part of the human body. The effects of chlorine and sulfuric acid are very similar. What makes chlorine a greater risk is the volatility of the liquid when released compared to the acid, which is already a liquid at ambient temperatures and volatilizes very slowly.

MSDS Section 6—Reactivity Data

Generally, four areas of information are presented in this section, and all are potentially useful to those responding to a hazardous chemical emergency.

1. Stability—Is the material stable at ambient temperature and pressure or at normal storage conditions? Most of the chemicals reviewed are stable and not liable to undergo spontaneous changes.

2. Polymerization—Will the chemical change through polymerization at normal conditions of storage and temperature? For chemicals that spontaneously polymerize, this frequently leads to generation of heat and potential container failure.
3. Decomposition—What new chemicals and what hazards will be created by the thermal decomposition of the chemical? Important information is included in this section for officials concerned about the exposures of response personnel and general population if the chemical is exposed to fire. For example, formaldehyde may be formed from fire involving methyl alcohol, oxides of nitrogen from anhydrous ammonia, and from most of the other chemicals, oxides of carbon that may increase the hazards from simple asphyxiation.
4. Incompatibles—What materials may cause violent reactions with the chemical? Note especially the MSDS for gasoline.

With six million chemicals in the world, many can have a large number of potentially violent reactions. It is important to have some idea of the likelihood of these chemicals coming in contact with each other. Many of the chemicals reviewed are potent acids or bases, and they will certainly be incompatible with chemicals of widely differing pH. For example, the sheet for sulfuric acid lists water as being incompatible. The mixing of sulfuric acid (96%) with water (at pH of 7) releases enough heat to cause a violent reaction.

MSDS Section 7—Spill or Leak Procedures

This section contains suggested steps for handling releases of the chemical in question. The information provided is usually similar to the 1990 DOT *Emergency Response Guide*. It is important to note the order in which the material is presented. If the material is extremely flammable, but not particularly toxic, initial advice will usually be in control ignition sources. If the material is extremely toxic, initial advice will generally be to evacuate.

MSDS Section 8—Special Protection Information

On many MSDS now in use, this section is not very specific. Hopefully, improvements will be made that recommend specific respiratory and clothing information. It is important to know that none of the impervious clothing is suitable for all chemicals. For example, polyethylene protective clothing is not recommended for concentrated sulfuric acid, but is suitable for more dilute solutions.

Special problems may be created for first responders by those materials that adversely affect normal fire fighter protective clothing. These materials (i.e., chlorobenzene, methyl iodide, etc.), for which breakthrough times are less than one tank of air, may not offer any useful protection to the responder. Once a hazard analysis is completed and response organizations are at the point of their contingency planning where they are selecting response equipment, it is suggested that they obtain a copy of "Guidelines for the Selection of Chemical Protective Clothing," Arthur D. Little Co., Cambridge, Massachusetts, February 1987, and check out the recommendations for protective clothing for the chemicals in their community.

MSDS Section 9—Special Precautions

Many MSDS do not contain any information in this area. In some cases, for extremely flammable materials, there is an additional warning about sparks and radiant heat; for chlorine, there is a warning about igniting other combustible materials on contact; and for many other chemicals there is a reiteration of standard storage and handling procedures.

One important area that may be covered on some MSDS is the hazard of the chemical to animal or aquatic life. This information is frequently based on scientific testing of the compound or chemical in controlled laboratory settings. The information from this testing is presented in terminology different than the regulatory TLV and PEL information and will take an additional effort on the responder's part to be able to evaluate the information.

Depending on the potential use of the chemical, a series of tests may be run over a period of time in a manner that resembles successive elimination. Tests are run for a variety of acute and chronic effects as well as exotic effects on more and more complex animals. Initial screening is done on bacteria, which allows the testing of large numbers of individuals and numerous generations in small spaces and in a short time. Subsequent tests may be conducted with any of a wide variety of rodents, pigs, dogs, and, sometimes, primates. The sequence of the tests is shown in the following chart.

Tests are done to evaluate the physical/chemical properties of the substance, note routes of entry into the organisms being tested, and document exposure variables. The tests are used to evaluate the biological fate of the chemicals and develop a dose/response curve for the specific effects being evaluated. A hypothetical dose/response curve is shown in Figure 1. The most common

Chronology of Testing

Early	Acute toxicity—handling hazards
•	1st level screen—mutagen/carcinogen
•	Subchronic toxicity—target organ, toxic dose—rodent
•	Birth defects—teratology
•	2nd level screen—mutagen/carcinogen
•	Absorption/distribution/metabolism/excretion—lab animals (metabolism/pharmacokinetics)
•	Subchronic toxicology—non-rodent species
•	Reproduction study
•	Behavioral tests
•	Synergism/potentiation
•	Residue evaluation
•	Long-term studies—carcinogenesis—rodents
•	Definitive test for mutagenesis—rodents
•	Metabolism/pharmacokinetics—humans
Late	Epidemiology

expression of the results of these tests is the dose or concentration at which 50% are affected, known as the TD_{50}. Toxicologists exhibit their skills by the accuracy with which they can extrapolate animal data to predict effects on man. In general, TD_{50} data is commonly given for pesticides and other chemicals developed for pest and weed control. An example of interpreting this data for responders is shown below:

Relative Index of Toxicology

Toxicity Rating	Probable Oral Dose	Lethal Dose for Average Adult Human
Practically nontoxic	>15g/kg	more than a quart
Slightly toxic	5–150g/kg	between pint and quart
Moderately toxic	0.5–5g/kg	between ounce and quart
Very toxic	50–500mg/kg	between teaspoon and ounce
Extremely toxic	5–50mg/kg	7 drops to teaspoonful
Supertoxic	<5mg/kg	a taste (less than 7 drops)

Dose/response curves deal with acute exposures, but it is important to also consider the potential for repetitive exposures at lower doses, which may accumulate in the body. This situation is called chronic exposure and is shown diagrammed below. The two toluene sheets discussed earlier provide chronic exposure data indicating the potential for brain cell damage from long-term inhalation of toluene vapor.

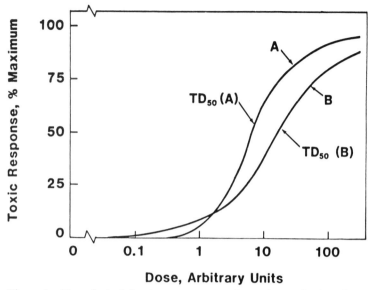

Dose, Arbitrary Units

Figure 1   Hypothetical dose/response curve of two chemicals, A and B.

Summary

Local government emergency response personnel can avoid much confusion if they concentrate on utilizing MSDS in pre-emergency planning. That is, identify in advance the buildings or processes that store or utilize extremely toxic materials and make basic decisions regarding "no attack" before an emergency occurs.

Once response community personnel have completed the hazard analysis and have evaluated the collected MSDS in the manner listed above, numerous additional questions need to be answered to complete the contingency planning portion of the work. Some of the questions that the hazard analysis and contingency planning process will help answer include:

1. What risks will be encountered by first responders?
2. What protective equipment (respiratory and clothing) will be needed by the responders?
3. What type of support resources are likely to be necessary for mitigation supplies or medical assistance?

One useful way to clarify these questions and to provide practical guidance for the response community is to use the highest risk situation from the hazard analysis as a scenario for a hands-on drill/exercise for the participating members of the response community.

Supplement 4: Hazardous Materials Emergency Planning Guide

This supplement is reprinted from a U.S. government publication detailing the organization and planning of a community hazardous materials incident response plan. This plan was developed by the National Response Team of the National Oil and Hazardous Substances Contingency Plan, 2100 2nd Street S.W.,Washington, DC 20593.

The Background of This Guidance

This Haz Mat Emergency Planning Guide has been developed cooperatively by 14 federal agencies. It is being published by the National Response Team in compliance with Section 303(f) of the "Emergency Planning and Community Right-to-Know Act of 1986," Title III of the "Superfund Amendments and Reauthorization Act of 1986" (SARA).

This guide replaces the Federal Emergency Management Agency's (FEMA) Planning Guide and Checklist for Hazardous Materials Contingency Plans (popularly known as FEMA-10).

This guide also incorporates material from the U.S. Environmental Protection Agency's (EPA) interim guidance for its Chemical Emergency Preparedness Program (CEPP) published late in 1985. Included are Chapters 2 ("Organizing the Community"), 4 ("Contingency Plan Development and Content"), and 5 ("Contingency Plan Appraisal and Continuing Planning"). EPA is revising and updating CEPP technical guidance materials that will include site-specific guidance, criteria for identifying extremely hazardous substances, and chemical profiles and a list of such substances. Planners should use this general planning guide in conjunction with the CEPP materials.

In recent years, the U.S. Department of Transportation (DOT) has been active in emergency planning. The Research and Special Programs Administration (RSPA) has published transportation-related reports and guides and has contributed to this general planning guide. The U.S. Coast Guard (USCG) has actively implemented planning and response requirements of the National Contingency Plan (NCP), and has contributed to this general planning guide.

The U.S. Occupational Safety and Health Administration (OSHA) and the U.S. Agency for Toxic Substances and Disease Registry (ATSDR) have assisted in preparing this general planning guide.

In addition to its FEMA-10, FEMA has developed and published a variety of planning-related materials. Of special interest here is *Guide for Development of State and Local Emergency Operations Plans* (known as CPG 1-8) that encourages communities to develop multi-hazard emergency operations plans (EOPs) covering all hazards facing a community (e.g., floods, earthquakes, hurricanes, as well as hazardous materials incidents). This general planning guide complements CPG 1-8 and indicates in Chapter 4 how hazardous materials planners can develop or revise a multi-hazard EOP. Chapter 4 also describes a sample outline for an emergency plan covering only hazardous materials, if a community does not have the resources to develop a multi-hazard EOP.

The terms "contingency plan," "emergency plan," and "emergency operations plan" are often used interchangeably, depending upon whether one is reading the NCP, CPG 1-8, or other planning guides. This guide consistently refers to "emergency plans" and "emergency planning."

This guide will consistently use "hazardous materials" when generally referring to hazardous substances, petroleum, natural gas,[1] synthetic gas, acutely toxic chemicals, and other toxic chemicals. Title III of SARA uses the term "extremely hazardous substances" to indicate those chemicals that could cause serious irreversible health effects from accidental releases.

The major differences between this document and other versions proposed for review are the expansion of the hazards analysis discussion (Chapter 3) and the addition of Appendix A explaining the planning provisions of Title III of SARA.

1. Introduction

1.1 The Need for Hazardous Materials Emergency Planning

Major disasters like that in Bhopal, India, in December 1984, which resulted in 2,000 deaths and over 200,000 injuries, are rare. Reports of hazardous materials spills and releases, however, are increasingly commonplace. Thousands of new chemicals are developed each year. Citizens and officials are concerned about accidents (e.g., highway incidents, warehouse fires, train derailments, industrial incidents) happening in their communities. Recent evidence shows that hazardous materials incidents are considered by many to be the most significant threat facing local jurisdictions. Ninety-three percent of the more than 3,100 localities completing the Federal Emergency Management Agency's (FEMA) Hazard Identification, Capability Assessment, and Multi-Year Development Plan during fiscal year 1985 identified one or more hazardous materials risks (e.g., on highways and railroads, at fixed facilities) as a significant threat to the community. Communities need to prepare themselves to prevent such incidents and to respond to the accidents that do occur.

Because of the risk of hazardous materials incidents, and because local governments will be completely on their own in the first stages of almost any hazardous materials incident, communities need to maintain a continuing

---

1. We recognize that natural gas is under a specific statute, but because this is a general planning guide (and because criteria for the list of extremely hazardous substances under Title III of SARA may be expanded to include flammability), local planners may want to consider natural gas.

preparedness capacity. A specific tangible result of being prepared is an emergency plan. Some communities might have sophisticated and detailed written plans but, if the plans have not recently been tested and revised, these communities might be less prepared than they think for a possible hazardous materials incident.

1.2 Purpose of This Guide

The purpose of this guide is to assist communities in planning for hazardous materials incidents.

"Communities" refers primarily to local jurisdictions. There are other groups of people, however, that can profitably use this guide. Rural areas with limited resources may need to plan at the county or regional level. State officials seeking to develop a state emergency plan that is closely coordinated with local plans can adapt this guidance to their purposes. Likewise, officials of chemical plants, railroad yards, and shipping and trucking companies can use this guidance to coordinate their own hazardous materials emergency planning with that of the local community.

"Hazardous materials" refers generally to hazardous substances, petroleum, natural gas, synthetic gas, acutely toxic chemicals, and other toxic chemicals. "Extremely hazardous substances" is used in Title III of the Superfund Amendments and Reauthorization Act of 1986 to refer to those chemicals that could cause serious health effects following short-term exposure from accidental releases. The U.S. Environmental Protection Agency (EPA) published an initial list of 402 extremely hazardous substances for which emergency planning is required. Because this list may be revised, planners should contact EPA regional offices to obtain information. This guidance deals specifically with response to hazardous materials incidents—both at fixed facilities (manufacturing, processing, storage, and disposal) and during transportation (highways, waterways, rail, and air). Plans for responding to radiological incidents and natural emergencies such as hurricanes, floods, and earthquakes are not the focus of this guidance, although most aspects of plan development and appraisal are common to these emergencies. Communities should see NUREG 0654/FEMA-REP-1 and/or FEMA-REP-5 for assistance in radiological planning. (See Appendix C.) Communities should be prepared, however, for the possibility that natural emergencies, radiological incidents, and hazardous materials incidents will cause or reinforce each other.

The objectives of this guide are to:

- Focus community activity on emergency preparedness and response;
- Provide communities with information useful in organizing the planning task;
- Furnish criteria to determine risk and to help communities decide whether they need to plan for hazardous materials incidents;
- Help communities conduct planning that is consistent with their needs and capabilities; and
- Provide a method for continually updating a community's emergency plan.
- This guide will *not*:
- Give a simple "fill-in-the-blanks" model plan (because each community needs an emergency plan suited to its own unique circumstances);
- Provide details on response techniques; or
- Train personnel to respond to incidents.

Community planners will need to consult other resources in addition to this guide. Related programs and materials are discussed in Section 1.5.

1.3 How to Use This Guide

This guide has been designed so it can be used easily by both those communities with little or no planning experience and those communities with extensive planning experience.

All planners should consult the decision tree in Exhibit 1 for assistance in using this guide.

Chapter 2 describes how communities can organize a planning team. Communities that are beginning the emergency planning process for the first time will need to follow Chapter 2 very closely in order to organize their efforts effectively. Communities with an active planning agency might briefly review Chapter 2, especially to be sure that all of the proper people are included in the planning process, and move on to Chapter 3 for a detailed discussion of tasks for hazardous materials planning. Planners should review existing emergency plans, perform a hazards identification and analysis, assess prevention and response capabilities, and then write or revise an emergency plan.

Chapter 4 discusses two basic approaches to writing an emergency plan: (a) incorporating hazardous materials planning into a multi-hazard emergency operations plan (EOP) (see Section 1.5.1); and (b) developing or revising a plan dealing only with hazardous materials. Incorporating hazardous materials planning into a multi-hazard approach is preferable. Some communities, however, have

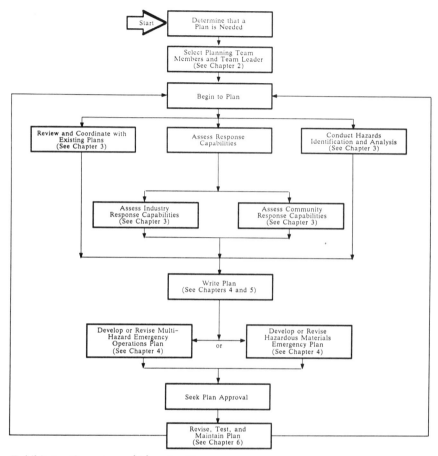

Exhibit 1.   Overview of Planning Process.

neither the capability nor the resources to do this immediately. Communities that choose to develop or revise an EOP should consult FEMA's CPG 1-8 for specific structure requirements for the plan in addition to the discussion in Section 1.5.1. Communities that choose to develop or revise a single-hazard plan for hazardous materials can use the sample outline of an emergency plan in Chapter 4 to organize the various hazardous materials planning elements. (Note: Communities receiving FEMA funds must incorporate hazardous materials planning into a multi-hazard EOP.)

Chapter 5 describes the elements to be considered when planning for potential hazardous materials incidents. All communities (both those preparing an EOP under the multi-hazard approach and those preparing a single-hazard plan) should carefully follow Chapter 5 to ensure that they consider and include the planning elements related to hazardous materials.

Chapter 6 describes how to review and update a plan. Experience shows that many communities mistakenly presume that completing an emergency plan automatically ensures adequate preparedness for emergency response. All communities should follow the recommendations in Chapter 6 to ensure that emergency plans will be helpful during a real incident.

Appendix A is a summary for implementing the "Emergency Planning and Community Right-to-Know Act of 1986." Appendix B is a list of acronyms and abbreviations used in this guidance. Appendix C is a glossary of terms used throughout this guide. (Because this guide necessarily contains many acronyms and technical phrases, local planners should regularly consult Appendices B and C.) Appendix D contains criteria for assessing state and local preparedness. Planners should use this appendix as a checklist to evaluate their hazards analysis, the legal authority for responding, the response organizational structure, communication systems, resources, and the completed emergency plan. Appendix E is a list of references on various topics addressed in this guidance. Appendix F is a listing of addresses of federal agencies at the national and regional levels. Planners should contact the appropriate office for assistance in the planning process.

1.4 Requirements for Planning

Planners should understand federal, state, and local requirements that apply to emergency planning.

1.4.1 Federal Requirements

This section discusses the principal federal planning requirements found in the National Contingency Plan; Title III of SARA; the Resource Conservation and Recovery Act; and FEMA's requirements for Emergency Operations Plans.

A.   National Contingency Plan

The National Contingency Plan (NCP), required by Section 105 of the Comprehensive Environmental Response, Compensation, and Liability Act (CERCLA), calls for extensive preparedness and planning. The National Response Team (NRT), comprised of representatives of various federal government agencies with major environmental, transportation, emergency management, worker safety, and public health responsibilities, is responsible for coordinating federal emergency preparedness and planning on a nationwide basis.

A key element of federal support to local responders during hazardous materials transportation and fixed facility incidents is a response by U.S. Coast Guard (USCG) or Environmental Protection Agency (EPA) On-Scene Coordinators (OSCs). The OSC is the federal official predesignated to coordinate and direct federal responses and removals under the NCP. These OSCs are assisted by federal Regional Response Teams (RRTs) that are available to provide advice and support to the OSC and, through the OSC, to local responders.

Federal responses may be triggered by a report to the National Response Center (NRC), operated by the Coast Guard. Provisions of the federal Water Pollution Control Act (Clean Water Act), CERCLA ("Superfund"), and various other federal laws require persons responsible for a discharge or release to notify the NRC immediately. The NRC Duty Officer promptly relays each report to the appropriate Coast Guard or EPA OSC, depending on the location of an incident. Based on this initial report and any other information that can be obtained, the OSC makes a preliminary assessment of the need for a federal response.

This activity may or may not require the OSC or his/her representative to go to the scene of an incident. If an on-scene response is required, the OSC will go to the scene and monitor the response of the responsible party or state or local government. If the responsible party is unknown or not taking appropriate action, or the response is beyond the capability of state and local governments, the OSC may initiate federal actions. The Coast Guard has OSCs at 48 locations (zones) in 10 districts, and the EPA has OSCs in its 10 regional offices and in certain EPA field offices. (See Appendix F for appropriate addresses.)

Regional Response Teams (RRTs) are composed of representatives from federal agencies and a representative from each state within a federal region. During a response to a major hazardous materials incident involving transportation or a fixed facility, the OSC may request that the RRT be convened to provide advice or recommendations on specific issues requiring resolution.

An enhanced RRT role in preparedness activities includes assistance for local community planning efforts. Local emergency plans should be coordinated with any federal regional contingency plans and OSC contingency plans prepared in compliance with the NCP. Appendix D of this guide contains an adaptation of extensive criteria developed by the NRT Preparedness Committee to assess state and/or local emergency response preparedness programs. These criteria should be used in conjunction with Chapters 3, 4, and 5 of this guide.

B. Title III of SARA ("Superfund Amendments and Reauthorization Act of 1986")

Significant new hazardous materials emergency planning requirements are contained in Title III of SARA (also known as the "Emergency Planning and Community Right-to-Know Act of 1986"). (See Appendix A for a detailed summary on implementing Title III.)

Title III of SARA requires the establishment of state emergency response commissions, emergency planning districts, and local emergency planning committees. The governor of each state appoints a state emergency response commission whose responsibilities include: designating emergency planning districts; appointing local emergency planning committees for each district; supervising and coordinating the activities of planning committees; reviewing emergency plans; receiving chemical release notifications; and establishing procedures for receiving and processing requests from the public for information about and/or copies of emergency response plans, material safety data sheets, the list of extremely hazardous substances prepared as part of EPA's original Chemical Emergency Preparedness Program initiative (see Section 1.5.2), inventory forms, and toxic chemical release forms.

Forming emergency planning districts is intended to facilitate the preparation and implementation of emergency plans. Planning districts may be existing political subdivisions or multijurisdictional planning organizations. The local emergency planning committee for each district must include representatives from each of the following groups or organizations:

- Elected state and local officials;
- Law enforcement, civil defense, fire fighting, health, local environmental, hospital, and transportation personnel;
- Broadcast and print media;
- Community groups; and
- Owners and operators of facilities subject to the requirements of Title III of SARA.

Each emergency planning committee is to establish procedures for receiving and processing requests from the public for information about and/or copies of emergency response plans, material safety data sheets, and chemical inventory forms. The committee must designate an official to serve as coordinator of information.

Facilities are subject to emergency planning and notification requirements if a substance on EPA's list of extremely hazardous substances is present at the facility in an amount in excess of the threshold planning quantity for that substance. (See *Federal Register*, Vol. 51, No. 221, 41570 et seq.) The owner or operator of each facility subject to these requirements must notify the appropriate state emergency response commission that the facility is subject to the requirements.

Each facility must also notify the appropriate emergency planning committee of a facility representative who will participate in the emergency planning process as a facility emergency coordinator. Upon request, facility owners and operators are to provide the appropriate emergency planning committee with information necessary for developing and implementing the emergency plan for the planning district.

Title III provisions help to ensure that adequate information is available for the planning committee to know which facilities to cover in the plan. (See Appendix A for a discussion of how the local planning committee can use information generated by Title III.) Section 303 (d)(3) requires facility owners and operators to provide to the local emergency planning committee whatever information is necessary for developing and implementing the plan.

When there is a release of a chemical identified by Title III of SARA, a facility owner or operator, or a transporter of the chemical, must notify the community emergency coordinator for the emergency planning committee for each area likely to be affected by the release, and the state emergency response commission of any state likely to be affected by the release. (This Title III requirement does not replace the legal requirement to notify the National Response Center for releases of CERCLA Section 103 hazardous substances.)

Each emergency planning committee is to prepare an emergency plan by October 1988 and review it annually. The committee also evaluates the need for resources to develop, implement, and exercise the emergency plan; and makes recommendations with respect to additional needed resources and how to provide them. Each emergency plan must include: facilities and transportation routes related to specific chemicals; response procedures of facilities, and local emergency and medical personnel; the names of community and facility emergency coordinators; procedures for notifying officials and the public in the event of a release; methods for detecting a release and identifying areas and populations at risk; a description of emergency equipment and facilities in the community and at specified fixed facilities; evacuation plans; training programs; and schedules for exercising the emergency plan. (These plan requirements are listed in greater detail in Chapter 5.) The completed plan is to be reviewed by the state emergency response commission and, at the request of the local emergency planning committee, may be reviewed by the federal Regional Response Team.

(Note: Many local jurisdictions already have emergency plans for various types of hazards. These plans may only require modification to meet emergency plan requirements in Title III of SARA.)

Finally, with regard to planning, Title III of SARA requires the NRT to publish guidance for the preparation and implementation of emergency plans. This Hazardous Materials Emergency Planning Guide is intended to fulfill this requirement. Other Title III provisions supporting emergency planning are discussed in Appendix A.

C. Resource Conservation and Recovery Act

The Resource Conservation and Recovery Act (RCRA) established a framework for the proper management and disposal of all wastes. The Hazardous and Solid Waste Amendments of 1984 (HSWA) expanded the scope of the law and placed increased emphasis on waste reduction, corrective action, and treatment of hazardous wastes.

Under Subtitle C of RCRA, EPA identifies hazardous wastes, both generically and by listing specific wastes and industrial process waste streams; develops standards and regulations for proper management of hazardous wastes by the generator and transporter, which include a manifest that accompanies waste shipments; and develops standards for the treatment, storage, and disposal of the wastes. These standards are generally implemented through permits that are issued by EPA or an authorized state. To receive a permit, persons wishing to treat, store, or dispose of hazardous wastes are required to submit permit applications, which must include a characterization of the hazardous wastes to be handled at the facility, demonstration of compliance with standards and regulations that apply to the facility, and a contingency plan. There are required opportunities for public comment on the draft permits, through which local governments and the public may comment on the facility's contingency plan.

It is important that local emergency response authorities be familiar with contingency plans of these facilities. Coordination with local community emergency response agencies is required by regulation (40 CFR 264.37), and EPA strongly encourages active community coordination of local response capabilities with facility plans.

When a community is preparing an emergency plan that includes underground storage tanks (containing either wastes or products), it should coordinate with EPA's regional offices, the states, and local governments. Underground storage tanks are regulated under Subtitle C or I of RCRA.

D. FEMA Emergency Operations Plan Requirements

Planning requirements for jurisdictions receiving FEMA funds are set forth in 44 CFR Part 302, effective May 12, 1986. This regulation calls for states and local governments to prepare an emergency operations plan (EOP) that conforms with the requirements for plan content contained in FEMA's CPG 1-3, CPG 1-8, and CPG 1-8A. These state and local government EOPs must identify the available personnel, equipment, facilities, supplies, and other resources in the jurisdiction, and state the method or scheme for coordinated actions to be taken by individuals and government services in the event of natural, man-made (e.g., hazardous materials), and attack-related disasters.

E. OSHA Regulations

Occupational Safety and Health Administration regulations require employers involved in hazardous waste operations to develop and implement an emergency response plan for employees. The elements of this plan must include: (1) recognition of emergencies; (2) methods or procedures for alerting employees on site; (3) evacuation procedures and routes to places of refuge or safe distances away from the danger area; (4) means and methods for emergency medical treatment and first aid for employees; (5) the line of authority for employees; (6) on-site decontamination procedures; (7) site control means; and (8) methods for evaluating the plan. Employers whose employees will be responding to hazardous materials emergency incidents from their regular work location or duty station (e.g., a fire department, fire brigade, or emergency medical service) must also have an emergency response plan. (See 29 CFR Part 1910.120.)

1.4.2 State and Local Requirements

Many states have adopted individual laws and regulations that address local government involvement in hazardous materials. Local authorities should investigate state requirements and programs before they initiate preparedness and planning activities. Emergency plans should include consideration of any state or local community right-to-know laws. When these laws are more demanding than the federal law, the state and local laws sometimes take precedence over the federal law.

1.5 Related Programs and Materials

Because emergency planning is a complex process involving a variety of issues and concerns, community planners should consult related public and private sector programs and materials. The following are selected examples of planning programs and materials that may be used in conjunction with this guide.

1.5.1 FEMA's Integrated Emergency Management System (CPG 1-8)

FEMA's Guide for Development of State and Local Emergency Operations Plans (CPG 1-8) provides information for emergency management planners and for state and local government officials about FEMA's concept of emergency operations planning under the Integrated Emergency Management System (IEMS). IEMS emphasizes the integration of planning to provide for all hazards discovered in a community's hazards identification process. CPG 1-8 provides extensive guidance in the coordination, development, review, validation, and revision of EOPs (see Section 4.2). (See Appendix F for FEMA's address and telephone number.)

This guide for hazardous materials emergency planning is deliberately meant to complement CPG 1-8. Chapter 4 describes how a community can incorporate hazardous materials planning into an existing multi-hazard EOP, or how it can develop a multi-hazard EOP while addressing possible hazardous materials incidents. In either case, communities should obtain a copy of CPG 1-8 from FEMA and follow its guidance carefully. All communities, even those with sophisticated multi-hazard EOPs, should consult Chapter 5 of this guide to ensure adequate consideration of hazardous materials issues.

1.5.2 EPA's Chemical Emergency Preparedness Program (CEPP)

In June 1985, EPA announced a comprehensive strategy to deal with planning for the problem of toxics released to the air. One section of this strategy, the Chemical Emergency Preparedness Program (CEPP), was designed to address accidental releases of acutely toxic chemicals. This program has two goals: to increase community awareness of chemical hazards and to enhance state and local emergency planning for dealing with chemical accidents. Many of the CEPP goals and objectives are included in Title III of SARA (see Section 1.4.1). EPA's CEPP materials (including technical guidance, criteria for identifying extremely hazardous substances, chemical profiles and list) are designed to complement this guidance and to help communities perform hazards identification and analysis as described in Chapter 3 of this guide. CEPP materials can be obtained by writing EPA. (See Appendix F.)

1.5.3 DOT Materials

The U.S. Department of Transportation's (DOT) *Community Teamwork* is a guide to help local communities develop a cost-effective hazardous materials transportation safety program. It discusses hazards assessment and risk analysis, the development of an emergency plan, enforcement, training, and legal authority for planning. Communities preparing an emergency plan for transportation-related hazards might use *Community Teamwork* in conjunction with this guide.

Lessons Learned is a report on seven hazardous materials safety planning projects funded by DOT. The projects included local plans for Memphis, Indianapolis, New Orleans, and Niagara County (NY); regional plans for Puget Sound and the Oakland/San Francisco Bay Area; and a state plan for Massachusetts. The *Lessons Learned* report synthesizes the actual experiences of these projects during each phase of the planning process. A major conclusion of this study was that local political leadership and support from both the executive and legislative branches are important factors throughout the planning process. Chapter 2 of this guide incorporates portions of the experiences and conclusions from *Lessons Learned*.

DOT's *Emergency Response Guidebook* provides guidance for fire fighters, police, and other emergency services personnel to help them protect themselves and the public during the initial minutes immediately following a hazardous materials incident. This widely used guidebook is keyed to the identification placards required by DOT regulations to be displayed prominently on vehicles transporting hazardous materials. All first responders should have copies of the *Emergency Response Guidebook* and know how to use it.

DOT has also published a four-volume guide for small towns and rural areas writing a hazardous materials emergency plan. DOT's objectives were to alert officials of those communities to the threat to life, property, and the environment from the transportation of hazardous materials, and to provide simplified guidance for those with little or no technical expertise. Titles of the volumes are: Volume I, *A Community Model for Handling Hazardous Materials Transportation Emergencies*; Volume II, *Risk Assessment Users Manual for Small Communities and Rural Areas*; Volume III, *Risk Assessment/ Vulnerability Model Validation*; and Volume IV, *Manual for Small Towns and Rural Areas to Develop a Hazardous Materials Emergency Plan.* (See Appendix F for DOT's address and telephone number.)

1.5.4 Chemical Manufacturers Association's Community Awareness and Emergency Response Program (CMA/CAER)

The Chemical Manufacturers Association's (CMA) Community Awareness and Emergency Response (CAER) program encourages chemical plant managers to take the initiative in cooperating with local communities to develop integrated emergency plans for responding to hazardous materials incidents. Because chemical industry representatives can be especially knowledgeable during the planning process, and because many chemical plant officials are willing and able to share equipment and personnel during response operations, community planners should seek out local CMA/CAER participants. Even if no such local initiative is in place, community planners can approach chemical plant managers or contact CMA and ask for assistance in the spirit of the CAER program.

Users of this general planning guide might also purchase and use the following three CMA/CAER publications: "Community Awareness and Emergency Response Program Handbook," "Site Emergency Response Planning," and "Community Emergency Response Exercise Program." (See Appendix E for CMA's address.)

2. Selecting and Organizing the Planning Team

2.1 Introduction

This chapter discusses the selection and organization of the team members who will coordinate hazardous materials planning. The guidance stresses that successful planning requires community involvement throughout the process. Enlisting the cooperation of all parties directly concerned with hazardous materials will improve planning, make the plan more likely to be used, and maximize the likelihood of an effective response at the time of an emergency. Experience shows that plans are not used if they are prepared by only one person or one agency. Emergency response requires trust, coordination, and cooperation among responders who need to know who is responsible for what activities, and who is capable of performing what activities. This knowledge is gained only through personal interaction. Working together in developing and updating plans is a major opportunity for cooperative interaction among responders.

(As indicated in Section 1.4.1, Title III of SARA requires governors to appoint a state emergency response commission that will designate emergency planning districts and appoint local emergency planning committees for each district. The state commission might follow the guidance in this chapter when appointing planning committees.)

2.2 The Planning Team

Hazardous materials planning should grow out of a process coordinated by a team. The team is the best vehicle for incorporating the expertise of a variety of sources into the planning process and for producing an accurate and complete document. The team approach also encourages a planning process that reflects the consensus of the entire community. Some individual communities and/or areas that include several communities have formed hazardous materials advisory councils (HMACs). HMACs, where they exist, are an excellent resource for the planning team.

2.2.1 Forming the Planning Team

In selecting the members of a team that will bear overall responsibility for hazardous materials planning, four considerations are most important:

- The members of the group must have the ability, commitment, authority, and resources to get the job done;
- The group must possess, or have ready access to, a wide range of expertise relating to the community, its industrial facilities and transportation systems, and the mechanics of emergency response and response planning;
- The members of the group must agree on their purpose and be able to work cooperatively with one another; and
- The group must be representative of all elements of the community with a substantial interest in reducing the risks posed by hazardous materials.

A comprehensive list of potential team members is presented in Exhibit 2.

In those communities receiving FEMA funds, paid staff may already be in place for emergency operations planning and other emergency management tasks. This staff should be an obvious resource for hazardous materials planning. FEMA has two training courses for the person assigned as the planning team leader and for team members—Introduction to Emergency Management, and Emergency Planning. Another course, Hazardous Materials Contingency Planning, is an interagency "train-the-trainer" course presented cooperatively by EPA, FEMA, and other NRT agencies. Course materials and the schedule of offerings are available through state emergency management agencies.

2.2.2 Respect for All Legitimate Interests

While many individuals have a common interest in reducing the risks posed by hazardous materials, their differing economic, political, and social perspectives may cause them to favor different means of promoting safety. For example, people who live near a facility with hazardous materials are likely to be greatly concerned about avoiding any threat to their lives, and are likely to be less intensely concerned about the costs of developing accident prevention and response measures than some of the other groups involved. Others in the community are likely to be more sensitive to the costs involved, and may be anxious to avoid expenditures for unnecessarily elaborate prevention and response measures. Also, facility managers may be reluctant for proprietary reasons to disclose materials and processes beyond what is required by law.

There may also be differing views among the agencies and organizations with emergency response functions about the roles they should play in case of an incident. The local fire department, police department, emergency management agency, and public health agency are all likely to have some responsibilities in responding to an incident. However, each of these organizations might envision a very different set of responsibilities for their respective agencies for planning or for management on scene.

In organizing the community to address the problems associated with hazardous materials, it is important to bear in mind that all affected parties have a legitimate interest in the choices among planning alternatives. Therefore, strong efforts should be made to ensure that all groups with an interest in the planning process are included.

Some interest groups in the community have well-defined political identities and representation, but others may not. Government agencies, private industry, environmental groups, and trade unions at the facilities are all likely to have ready institutional access to an emergency planning process. Nearby residents, however, may lack an effective vehicle for institutional representation. Organizations that may be available to represent the residents' interests include neighborhood associations, church organizations, and ad hoc organizations formed especially to deal with the risks posed by the presence of specific hazardous materials in a neighborhood.

Exhibit 2. Potential Members of an Emergency Planning Team

Part A: Experience shows that the following individuals, groups, and agencies should participate in order for a successful plan to be developed:

- Mayor/city manager (or representative)
- County executive (or representative)/board of supervisors
- State elected officials (or representative)
- Fire department (paid and volunteer)
- Police department
- Emergency management or civil defense agency
- Environmental agency (e.g., air and/or water pollution control agency)
- Health department
- Hospitals, emergency medical service, veterinarians, medical community
- Transportation agency (e.g., DOT, port authority, transit authority, bus company, truck or rail companies)

- Industry (e.g., chemical and transportation)
- Coast Guard/EPA representative (e.g., agency response program personnel)
- Technical experts (e.g., chemist, engineer)
- Community group representative
- Public information representative (e.g., local radio, TV, press)*

Part B: Other groups/agencies that can be included in the planning process, depending on the community's individual priorities:

- Agriculture agency
- Indian tribes within or adjacent to the affected jurisdiction
- Public works (e.g., waste disposal, water, sanitation, and roads)
- Planning department
- Other agencies (e.g., welfare, parks, and utilities)
- Municipal/county legal counsel
- Workers in local facilities
- Labor union representatives (e.g., chemical and transportation, industrial health units)
- Local business community
- Representatives from volunteer organizations (e.g., Red Cross)
- Public interest and citizens groups, environmental organizations, and representatives of affected neighborhoods
- Schools or school districts
- Key representatives from bordering cities and counties
- State representatives (Governor, legislator's office, State agencies)
- Federal agency representatives (e.g., FEMA, DOT/RSPA, ATSDR, OSHA)

2.2.3 Special Importance of Local Governments

For several reasons, local governments have a critical role to play in the development of emergency preparedness. First, local governments bear major responsibilities for protecting public health and safety; local police and fire departments, for example, often have the lead responsibility for the initial response to incidents involving hazardous materials. Second, one of the functions of local government is to mediate and resolve the sometimes competing ideas of different interest groups. Third, local governments have the

---

* Required by Title III of SARA

resources to gather necessary planning data. Finally, local governments generally have the legislative authority to raise funds for equipment and personnel required for emergency response. Support from the executive and legislative branches is essential to successful planning. Appropriate government leaders must give adequate authority to those responsible for emergency planning.

2.2.4 Local Industry Involvement

Because fixed facility owners and operators are concerned about public health and safety in the event of an accidental release of a hazardous material, and because many facility employees have technical expertise that will be helpful to the planning team, the team should include one or more facility representatives. Title III of SARA requires facility owners or operators to notify the emergency planning committee of a facility representative who will participate in the emergency planning process as a facility emergency coordinator. In planning districts that include several fixed facilities, one or more representative facility emergency coordinators could be active members of the planning team. The planning team could consult with the other facility emergency coordinators and/or assign them to task forces or committees (see Section 2.3.2). Title III of SARA also requires facilities to submit to the local emergency planning committee any information needed to develop the plan.

2.2.5 Size of Planning Team

For the planning team to function effectively, its size should be limited to a workable number. In communities with many interested parties, it will be necessary to select from among them carefully so as to ensure fair and comprehensive representation. Some individuals may feel left out of the planning process. This can be offset by providing these individuals access to the process through the various approaches noted in the following sections, such as membership on a task force or advisory council. In addition, all interested parties should have an opportunity for input during the review process.

2.3 Organizing the Planning Process

After the planning team members have been identified, a team leader must be chosen and procedures for managing the planning process must be established.

2.3.1 Selecting a Team Leader

A community initiating a hazardous materials emergency planning process may choose to appoint an individual to facilitate and lead the effort, or may appoint a planning team and have the group decide who will lead the effort. Either approach can be used. It is essential to establish clear responsibility and authority for the project. The chief executive (or whoever initiates the process) should determine which course is better suited to local circumstances. (The emergency planning committee required by Title III of SARA is to select its own chairperson.) Regardless of how the team leader is selected, it is his or her primary responsibility to oversee the team's efforts through the entire planning process. Because the role of leader is so significant, a co-chair or back-up could also be named.

Five factors are of major importance in selecting a team leader:

- The degree of respect held for the person by groups with an interest in hazardous materials;
- Availability of time and resources;
- The person's history of working relationships with concerned community agencies and organizations;
- The person's management and communication skills; and
- The person's existing responsibilities related to emergency planning, prevention, and response.

Logical sources for a team leader include:

- *The chief executive or other elected official.* Leadership by a mayor, city or county council member, or other senior official is likely to contribute substantially to public confidence, encourage commitment of time and resources by other key parties, and expedite the implementation of program initiatives. Discontinuity in the planning process can result, however, if an elected official leaves office.

- *A public safety department.* In most communities, the fire department or police department bears principal responsibility for responding to incidents involving chemical releases and, typically, for inspecting facilities as well. A public safety department, therefore, may have personnel with past experience in emergency planning and present knowledge of existing responsibilities within the community.

- *The emergency management or civil defense agency.* In many communities, officials of such an agency will be knowledgeable and experienced in planning for major disasters from a variety of causes. One of the primary responsibilities of a community's emergency management coordinator is to guide, direct, and participate in the development of a multi-hazard emergency operations plan. In some states, existing laws require that this agency be the lead agency to prepare and distribute emergency plans.

- *The local environmental agency or public health agency.* Persons with expertise and legal responsibility in these areas will have special knowledge about the risks posed by hazardous materials.

- *A planning agency.* Officials in a planning agency will be familiar with the general planning process and with the activities and resources of the community.

- *Others.* Communities should be creative and consider other possible sources for a team leader, such as civic groups, industry, academic institutions, volunteer organizations, and agencies not mentioned above. Experience in leading groups and committees, regardless of their purpose, will prove useful in emergency planning.

Personal considerations as well as institutional ones should be weighed in selecting a team leader. For example, a particular organization may appear to have all the right resources for addressing hazardous materials incidents. But if the person in charge of that organization does not interact well with other local officials, it might be best to look for a different leader.

A response coordinator generally is knowledgeable about emergency plans and is probably a person who gets things done. Be aware, however, that a good response coordinator is not necessarily a good planner. He or she might make a good chief advisor to someone better suited for the team leader job.

2.3.2 Organizing for Planning Team Responsibilities

The planning team must decide who shall conduct the planning tasks and establish the procedures for monitoring and approving the planning tasks.

A. Staffing

There are three basic staffing approaches that may be employed to accomplish the tasks involved in emergency planning:

- *Assign staff.* Previous experience in related planning efforts demonstrates the usefulness of assigning one or more dedicated staff members to coordinate the planning process and perform specific planning tasks. The staff may be assigned within a "lead agency" having related responsibilities and/or expertise, or may be created separately through outside hiring and/or staff loans from government agencies or industry.

• *Assign task forces or committees.* Planning tasks can be performed by task forces or committees composed entirely or in part of members of the planning team. Adding knowledgeable representatives of government agencies, industry, environmental, labor, and other community organizations to the individual task forces or committees not only supplements the planning team expertise and resources, but also provides an opportunity for additional interested parties to participate directly in the process.

• *Hire contractors or consultants.* If the personnel resources available for the formation of a dedicated staff and task forces or committees are limited, and funds can be provided, the planning team may elect to hire contractors or consultants. Work assigned to a contractor can range from a specialized job, such as designing a survey, to performing an entire planning task (e.g., hazards identification and analysis). A disadvantage of hiring contractors or consultants is that it does not help build a community-centered capability or planning infrastructure.

The three approaches presented above are not mutually exclusive. A community may adopt any combination of the approaches that best matches its own circumstances and resources.

B. Managing the Planning Tasks

The monitoring and approval of planning assignments are the central responsibilities of the planning team. In order to have ongoing cooperation in implementing the plan, it is recommended that the planning team operate on a consensus basis, reaching general agreement by all members of the team. Achieving consensus takes more time than majority voting, but it is the best way to ensure that all represented parties have an opportunity to express their views and that the decisions represent and balance competing interests. If it is determined that a consensus method is inappropriate or impossible (e.g., because of the multi-jurisdictional nature of a group), the planning team should formally decide how issues will be resolved.

The team leader should work with the team members to establish clear goals and deadlines for various phases of the planning success. Progress toward these goals and deadlines should be monitored frequently.

Planning meetings, a necessary element of the planning process, often do not make the best use of available time. Meetings can be unnecessarily long and unproductive if planning members get bogged down on inappropriate side issues. Sometimes, when several agencies or groups sit down at one table, the meeting can become a forum for expressing political differences and other grievances fueled by long-standing interagency rivalries. For a team to be effective, a strong team leader will have to make sure that meeting discussions focus solely on emergency planning.

Another point to consider is that the team approach requires the melding of inputs from different individuals, each with a different style and sense of priorities. A team leader must ensure that the final plan is consistent in substance and tone. An editor may be used to make sure that the plan's grammar, style, and content all ultimately fit well together.

On critical decisions, it may be desirable to extend the scope of participation beyond the membership of the planning team. Approaches that might be used to encourage community consensus building through broadened participation in the process include invited reviews by key interest groups, or formation of an advisory council composed of interested parties that can independently review and comment on the planning team's efforts. Chapter 6 contains further guidance on consensus-building approaches.

The procedures to be used for monitoring and approving planning assignments should be carefully thought out at the beginning of the planning process; planning efforts work best when people understand the ground rules and know when and how they will be able to participate. The monitoring and approval process can be adjusted at any time to accommodate variations in local interest.

Planning committees formed according to Title III of SARA are to develop their own rules. These rules include provisions for public notification of committee activities; public meetings to discuss the emergency plan; public comments; response to public comments by the committee; and distribution of the emergency plan.

C. The Use of Computers

Computers are handy tools for both the planning process and for maintaining response preparedness. Because new technology is continually being developed, this guide does not identify specific hardware or software packages that planning teams and/or response personnel might use. Local planners should consult Regional FEMA or EPA offices (see Appendix F) for more detailed descriptions of how some communities are using computers.

The following list summarizes some ways in which computers are useful both in the planning process and for maintaining response preparedness.

- *Word processing.* Preparation and revision of plans is expedited by word processing. Of special interest to planners is the use of word processing to keep an emergency plan up to date on an annual or semiannual basis.

- *Modeling.* Planners might consider applying air dispersion models for chemicals in their community so that, during an emergency, responders can predict the direction, velocity, and concentration of plume movement. Similarly, models can be developed to predict the pathways of plumes in surface water and ground water.

- *Information access.* Responders can use a personal computer on site to learn the identity of the chemical(s) involved in the incident (e.g., when placards are partially covered), the effects of the chemical(s) on human health and the environment, and appropriate countermeasures to contain and clean up the chemical(s). Communities that intend to use computers on scene should also provide a printer on scene.

- *Data storage.* Communities can store information about what chemicals are present in various local facilities, and the availability of equipment and personnel that are needed during responses to incidents involving specific chemical(s). Compliance with Title III will generate large amounts of data (e.g., MSDS forms, data on specific chemicals in specific facilities, data on accidental releases). (See Appendix A.) Such data could be electronically stored and retrieved. These data should be reviewed and updated regularly. Area maps, with information about transportation and evacuation routes, hospital and school locations, and other emergency-related information, can also be stored in computer disks.

State and local planners with personal computer communications capability can access the federally operated National Hazardous Materials Information Exchange (NHMIE) by dialing (312) 972-3275. Users can obtain up-to-date information on hazmat training courses, planning techniques, events and conferences, and emergency response experiences and lessons learned. NHMIE can also be reached through a toll-free telephone call (1-800-752-6367; in Illinois, 1-800-367-9592).

2.4 Beginning to Plan

When the planning team members and their leader have been identified and a process for managing the planning tasks is in place, the team should address several interrelated tasks. These planning tasks are described in the next chapter.

3. Tasks of the Planning Team

3.1 Introduction

The major tasks of the planning team in completing hazardous materials planning are:

- Review of existing plans, which prevents plan overlap and inconsistency, provides useful information and ideas, and facilitates the coordination of the plan with other plans;
- Hazards analysis, which includes hazards identification, vulnerability analysis, and risk analysis;
- Assessment of preparedness, prevention, and response capabilities, which identifies existing prevention measures and response capabilities (including mutual aid agreements) and assesses their adequacy;
- Completion of hazardous materials planning that describes the personnel, equipment, and procedures to be used in case of accidental release of a hazardous material; and
- Development of an ongoing program for plan implementation/maintenance, training, and exercising.

This chapter discusses the planning tasks that are conducted prior to the preparation of the emergency plan. Chapters 4 and 5 provide guidance on plan format and content. Chapter 6 discusses the team's responsibilities for conducting internal and external reviews, exercises, incident reviews, and training. This chapter begins with a discussion of the organizational responsibilities of the planning team.

3.2 Review of Existing Plans

Before undertaking any other work, steps should be taken to search out and review all existing emergency plans. The main reasons for reviewing these plans are (1) to minimize work efforts by building upon or modifying existing emergency planning and response information and (2) to ensure proper coordination with other related plans. To the extent possible, currently used plans should be amended to account for the special problems posed by hazardous materials, thereby avoiding redundant emergency plans. Even plans that are no longer used may provide a useful starting point. More general plans can also be a source of information and ideas. In seeking to identify existing plans, it will be helpful to consult organizations such as:

- State and local emergency management agencies;
- Fire departments;
- Police departments;
- State and local environmental agencies;
- State and local transportation agencies;
- State and local public health agencies;
- Public service agencies;
- Volunteer groups, such as the Red Cross;
- Local industry and industrial associations; and
- Regional offices of federal agencies such as EPA and FEMA.

When reviewing the existing plans of local industry and industrial associations, the planning team should obtain a copy of the CAER program handbook produced by CMA. (See Section 1.5.4.) The handbook provides useful information and encourages industry-community cooperation in emergency planning.

In addition to the above organizations, planning teams should coordinate with the RRTs and OSCs described in Section 1.4.1. Communities can contact or obtain information on the RRT and OSC covering their area through the EPA regional office or USCG district office. (See Appendix F for a list of these contacts.)

3.3 Hazards Analysis: Hazards Identification, Vulnerability Analysis, Risk Analysis

A hazards analysis is a critical component of planning for hazardous materials releases. The information developed in a hazards analysis provides both the factual basis to set priorities for planning and also the necessary documentation for supporting hazardous materials planning and response efforts.

There are several concepts involved in analyzing the dangers posed by hazardous materials. Three terms—hazard, vulnerability, risk—have different technical meanings but are sometimes used interchangeably. This guidance adopts the following definitions:

- *Hazard.* Any situation that has the potential for causing injury to life or damage to property and the environment.

- *Vulnerability.* The susceptibility of life, property, and the environment to injury or damage if a hazard manifests its potential.
- *Risk.* The probability that injury to life or damage to property and the environment will occur.

A hazards analysis may include vulnerability analysis and risk analysis, or it may simply identify the nature and location of hazards in the community. Developing a complete hazards analysis that examines all hazards, vulnerabilities, and risks may be neither possible nor desirable. This may be particularly true for smaller communities that have less expertise and fewer resources to contribute to the task. The planning team must determine the level of thoroughness that is appropriate. In any case, planners should ask local facilities whether they have already completed a facility hazards analysis. Title III requires facility owners or operators to provide to local emergency planning communities information needed for the planning process.

As important as knowing how to perform a hazards analysis is deciding how detailed an analysis to conduct. While a complete analysis of all hazards would be informative, it may not be feasible or practical given resource and time constraints. The value of a limited hazards analysis should not be underestimated. Often the examination of only major hazards is necessary, and these may be studied without undertaking an elaborate risk analysis. Thus, deciding what is really needed and what can be afforded is an important early step in the hazards analysis process. In fact, the screening of hazards and setting analysis priorities is an essential task of the planning team.

The costs of hazards analysis can and often should be reduced by focusing on the hazards posed by only the most common and/or most hazardous substances. A small number of types of hazardous materials account for the vast majority of incidents and risk. The experience from DOT's *Lessons Learned* is that the most prevalent dangers from hazardous materials are posed by common substances, such as gasoline, other flammable materials, and a few additional chemicals. The CEPP technical guidance presents a method that may be used to assist in ranking hazards posed by less prevalent but extremely hazardous substances, such as liquid chlorine, anhydrous ammonia, and hydrochloric and sulfuric acids.

A hazards analysis can be greatly simplified by using qualitative methods (i.e., analysis that is based on judgment rather than measurement of quantities involved). Smaller communities may find that their fire and police chiefs can provide highly accurate assessments of the community's hazardous materials problems. Other, larger communities may have the expertise and resources to utilize quantitative techniques but may decide to substitute qualitative methods in their place should it be cost effective to do so.

Simple or sophisticated, the hazards analysis serves to characterize the nature of the problem posed by hazardous materials. The information that is developed in the hazards analysis should then be used by the planning team to orient planning appropriate to the community's situation. Do not commit valuable resources to plan development until a hazards analysis is performed.

3.3.1 Developing the Hazards Analysis

The procedures that are presented in this section are intended to provide a simplified approach to hazards analysis for both facility and transportation hazards. Communities undertaking a hazards analysis should refer to CEPP technical guidance for fixed facilities and to *Lessons Learned* and *Community Teamwork* for transportation.

The components of a hazards analysis include the concepts of hazard, vulnerability, and risk. The discussion that follows summarizes the basic procedures for conducting each component.

A. Hazards Identification

The hazards identification provides information on the facility and transportation situations that have the potential for causing injury to life or damage to property and the environment due to a hazardous materials spill or release. The hazards identification should indicate:

- The types and quantities of hazardous materials located in or transported through a community;
- The location of hazardous materials facilities and routes; and
- The nature of the hazard (e.g., fire, explosions) most likely to accompany hazardous materials spills or releases.

To develop this information, consider hazardous materials at fixed sites and those that are transported by highway, rail, water, air, and pipeline. Examine hazardous materials at:

- Chemical plants;
- Refineries;
- Industrial facilities;
- Petroleum and natural gas tank farms;
- Storage facilities/warehouses;

- Trucking terminals;
- Railroad yards;
- Hospital, educational, and governmental facilities;
- Waste disposal and treatment facilities;
- Waterfront facilities, particularly commercial marine terminals;
- Vessels in port;
- Airports;
- Nuclear facilities; and
- Major transportation corridors and transfer points.

For individual facilities, consider hazardous materials:

- Production;
- Storage;
- Processing;
- Transportation; and
- Disposal.

Some situations will be obvious. To identify the less obvious ones, interview fire and police chiefs, industry leaders, and reporters; review news releases and fire and police department records of past incidents. Also, consult lists of hazardous chemicals that have been identified as a result of compliance with right-to-know laws. (Title III of SARA requires facility owners and operators to submit to the local emergency planning committee a material safety data sheet for specified chemicals, and emergency and hazardous chemical inventory forms. Section 303 (d)(3) of Title III states that "upon request from the emergency planning committee, the owner or operator of the facility shall promptly provide information . . . necessary for developing and implementing the emergency plan.") Use the CEPP technical guidance for help in evaluating the hazards associated with airborne releases of extremely hazardous substances.

The hazards identification should result in compilation of those situations that pose the most serious threat of damage to the community. Location maps and charts are an excellent means of depicting this information.

B. Vulnerability Analysis

The vulnerability analysis identifies what in the community is susceptible to damage should a hazardous materials release occur. The vulnerability analysis should provide information on:

- The extent of the vulnerable zone (i.e., the significantly affected area) for a spill or release and the conditions that influence the zone of impact (e.g., size of release, wind direction);
- The population, in terms of size and types (e.g., residents, employees, sensitive populations—hospitals, schools, nursing homes, day care centers), that could be expected to be within the vulnerable zone;
- The private and public property (e.g., homes, businesses, offices) that may be damaged, including essential support systems (e.g., water, food, power, medical) and transportation corridors; and
- The environment that may be affected, and the impact on sensitive natural areas and endangered species.

Refer to the CEPP technical guidance or DOT's *Emergency Response Guidebook* to obtain information on the vulnerable zone for a hazardous materials release. For information on the population, property, and environmental resources within the vulnerable zone, consider conducting:

- A windshield survey of the area (i.e., first hand observation by driving through an area);
- Interviews of fire, police, and planning department personnel; and
- A review of planning department documents, and statistics on land use, population, highway usage, and the area's infrastructure.

The vulnerability analysis should summarize information on all hazards determined to be major in the hazards identification.

C. Risk Analysis

The risk analysis assesses the probability of damage (or injury) taking place in the community due to a hazardous materials release and the actual damage (or injury) that might occur, in light of the vulnerability analysis. Some planners may choose to analyze worst-case scenarios. The risk analysis may provide information on:

- The probability that a release will occur and any unusual environmental conditions, such as areas in flood plains, or the possibility of simultaneous emergency incidents (e.g., flooding or fire hazards resulting in release of hazardous materials);
- The type of harm to people (acute, delayed, chronic) and the associated high-risk groups;
- The type of damage to property (temporary, repairable, permanent); and
- The type of damage to the environment (recoverable, permanent).

Use the Chemical Profiles in the CEPP technical guidance or a similar guide to obtain information on the type of risk associated with the accidental airborne release of extremely hazardous substances.

Developing occurrence probability data may not be feasible for all communities. Such analysis can require specialized expertise not available to a community. This is especially true of facility releases that call for detailed analysis by competent safety engineers and others (e.g., industrial hygienists) of the operations and associated risk factors of the plant and engineering system in question (refer to the American Institute of Chemical Engineers' *Guidelines for Hazard Evaluation Procedures*). Transportation release analysis is more straightforward, given the substantial research and established techniques that have been developed in this area (refer to *Community Teamwork* and *Lessons Learned*).

Communities should not be overly concerned with developing elaborate quantitative release probabilities. Instead, occurrence probabilities can be described in relative terms (e.g., low, moderate, high). The emphasis should be on developing reasonable estimates based on the best available expertise.

3.3.2 Obtaining Facility Information

The information that is needed about a facility for hazards analysis may already be assembled as a result of previous efforts. As indicated in Section 1.4.1, industry is required by Title III of SARA to provide inventory and release information to the appropriate emergency planning committee. Local emergency planning committees are specifically entitled to any information from facility owners and operators deemed necessary for developing and implementing the emergency plan. The EPA administrator can order facilities to comply with a local committee's requests for necessary information; local planning committees can bring a civil suit against a facility that refuses to provide requested information. Some state and local governments have adopted community right-to-know legislation. These community right-to-know provisions vary, but they generally require industry and other handlers of hazardous materials to provide information to state or local authorities and/or the public about hazardous materials in the community. Wisconsin, for example, requires all hazardous materials spills to be reported to a state agency. Such requirements provide a data base that the planning team can use to determine the types of releases that have occurred in and around the community.

Requesting information from a facility for a hazards analysis can be an opening for continuing dialogue within the community. The information should be sought in such a way that facilities are encouraged to cooperate and participate actively in the planning process along with governmental agencies and other community groups. Respecting a commercial facility's needs to protect confidential business information (such as sensitive process information) will encourage a facility to be forthcoming with the information necessary for the community's emergency planning. The planning team can learn what the facility is doing and what measures have been put in place to reduce risks, and also identify what additional resources such as personnel, training, and equipment are needed in the community. Because facilities use different kinds of hazard assessments (e.g., HAZOP, Fault-tree analysis), local planners need to indicate specifically what categories of information they are interested in receiving. These categories may include:

- Identification of chemicals of concern;
- Identification of serious events that can lead to releases (e.g., venting or system leaks, runaway chemical reaction);
- Amounts of toxic material or energy (e.g., blast, fire radiation) that could be released;
- Predicted consequences of the release (e.g., population exposure illustrated with plume maps and damage rings) and associated damages (e.g., deaths, injuries);
- Whether the possible consequences are considered acceptable by the facility; and
- Prevention measures in place on site.

The facilities themselves are a useful resource; the community should work with the facility personnel and utilize their expertise. The assistance that a facility can provide includes:

- Technical experts;
- Facility emergency plans;
- Cleanup and recycling capabilities;
- Spill prevention control and countermeasures (SPCC);
- Training and safe handling instructions; and
- Participation in developing the emergency plan, particularly in defining how to handle spills on company property.

Cooperative programs such as CMA's CAER program are also a source for hazard information. One of the major objectives of the CAER program is to improve local emergency plans by combining chemical plant emergency plans with other local planning to achieve an integrated community emergency plan. The planning team should ask the facility if it is participating in the CAER program; this may stimulate non-CMA members to use the CAER approach. If a facility is participating in the CAER program, the emergency plans developed by the facility will serve as a good starting point in information gathering and emergency planning. The CAER program handbook also encourages companies to perform hazards analyses of their operations. Local planners should ask facilities if they have adhered to this recommendation and whether they are willing to share results with the planning team.

3.3.3 Example Hazards Analysis

Exhibit 3 presents an example of a very simple hazards analysis for a hypothetical community. Hazards A, B, and C are identified as three among other major hazards in the community. Information for the exhibit could have been obtained from windshield surveys of the area; the CEPP technical guidance; information gained from facilities under Title III provisions; and/or interviews with fire, police, county planners, and facility representatives. These interviews also could have provided input into the exhibit's qualitative assessments of hazard occurrence.

Once completed, the hazards analysis is an essential tool in the planning process. It assists the planning team to decide:

• The level of detail that is necessary;
• The types of response to emphasize; and
• Priority hazards or areas for planning.

The examples presented in Exhibit 3 illustrate the basic fact that there are no hard and fast rules for weighing the relative importance of different types of hazards in the context of the planning process. Compare example hazards B and C in the exhibit. Hazard C involves a substance, methyl isocyanate (MIC), whose lethal and severe chronic effects were evident at Bhopal. As described in the example, an MIC release could affect 200 plant workers and 1000 children in a nearby school. By contrast, the ammonia in example hazard B is less lethal than MIC and threatens fewer people. With just this information in mind, a planner might be expected to assign the MIC a higher planning priority than he would the ammonia. Consider now the "probability of occurrence." In hazard C, plant safety and prevention measures are excellent, and an MIC incident is correspondingly unlikely to occur. On the other hand, poor highway construction and weather conditions that affect visibility make an ammonia incident (example hazard B) far more probable.

Planners must balance all factors when deciding whether to give planning priority to B or C. Both situations are dangerous and require emergency planning. Some would argue that the lethality of MIC outweighs the presence of good safety and prevention procedures; others would argue that the frequency of highway interchange accidents is reason enough to place greater emphasis on planning to deal with an ammonia incident. Each planning team must make such judgments on priorities in light of local circumstances.

Before initiating plan development, the planning team should complete an assessment of available response resources, including capabilities provided through mutual aid agreements. Guidance for conducting such an assessment is presented in the following section.

3.4 Capability Assessment

This section contains sample questions to help the planning team evaluate preparedness, prevention, and response resources and capabilities. The section is divided into three parts. The first part covers questions that the planning team can ask a technical representative from a facility that may need an emergency plan. The second part includes questions related to transportation.

The third part addresses questions to a variety of response and government agencies, and is designed to help identify all resources within a community. This information will provide direct input into the development of the hazardous materials emergency plan and will assist the planning team in evaluating what additional emergency response resources may be needed by the community.

3.4.1 Facility Resources

What is the status of the safety plan (also referred to as an emergency or contingency plan) for the facility? Is the safety plan consistent with any community emergency plan?

• Is there a list of potentially toxic chemicals available? What are their physical and chemical characteristics, potential for causing adverse health effects, controls, interactions with other chemicals? Has the facility complied with the community right-to-know provisions of Title III of SARA?

Exhibit 3   Example Hazard Analysis for a Hypothetical Community

	Hazard A	Hazard B	Hazard C
1. Hazards Identification (Major Hazards)			
a. Chemical	Chlorine	Ammonia	Liquid methyl isocyanate (MIC)
b. Location	Water treatment plant	Tank truck on local interstate highway	Pesticide manufacturing plant in nearby semi-rural area
c. Quantity	2000 lb	5000 lb	5000 lb
d. Properties	Poisonous; may be fatal if inhaled. Respiratory conditions aggravated by exposure. Contact may burn skin and eyes. Corrosive. Effects may be delayed.	Poisonous; may be fatal if inhaled. Vapors irritate eyes, respiratory tract. Liquid burns skin, eyes. Liquid may cause frostbite. Effects may be delayed. Burns within certain vapor concentration limits; increased fire hazard in presence of oil, other combustibles.	Caused death by respiratory distress after inhalation. Other health effects include permanent eye damage, respiratory distress, disorientation. Explosive. Extremely flammable.
2. Vulnerability Analysis			
a. Vulnerable zone	Spill of 2000 lb chlorine from storage tank could raise concentration of chlorine gas above level of concern within 1650-ft radius.	Spill of 5000 lb ammonia from tank truck collison could raise concentration of ammonia above level of concern within 1320-ft radius.	Spill of 5000 lb MIC could raise concentration of MIC vapors above level of concern within 3300-ft radius, assuming liquid is hot when spilled, tank is not diked, MIC is at 100% concentration.
b. Population within vulnerable zone	Approximately 500 residents of nursing home; workers at small factory.	Up to 700 persons in nearby residences, commercial establishments, vehicles. Influx of visitors to forest preserve in the fall.	Up to 200 workers at plant, 1000 children in school.

Exhibit 3, continued   Example Hazard Analysis for a Hypothetical Community

	Hazard A	Hazard B	Hazard C
c. Private and public property that may be damaged	Facility equipment, vehicles, structures susceptible to damage from corrosive fumes. Community's water supply may be temporarily affected since facility is primary supplier. Mixture with fuels may cause explosion.	25 residences, 2 fast food restaurants, 30-room motel, a truck stop, a gas station, a mini-market. Highway, nearby vehicles may be damaged by fire or explosion resulting from collision.	Runoff to a sewer may cause explosion hazard as MIC reacts violently with water.
d. Environment that may be affected	Terrestrial life.	Adjacent forest highly susceptible to fires, especially during drought.	Nearby farm animals.
3. Risk Analysis			
a. Probability of hazard occurrence	Low—because chlorine is stored in area with leak-detection equipment in 24-hour service with alarms. Protective equipment kept outside storage room.	High. Highway interchange has history of accidents due to poor visibility of exits, entrances.	Low. Facility has up-to-date containment facilities with leak detection equipment, and emergency plan for employees. Good security arrangements to deter tampering, accidents.
b. Consequences if people are exposed	High levels of chlorine gas in nursing home, factory could cause death, respiratory distress. Bedridden patients especially susceptible.	Release of vapors, subsequent fire may cause traffic accidents. Injured, trapped motorists subject to lethal vapors, possible incineration. Windblown vapors can cause respiratory distress.	If accident occurs while school is in session, children could be killed, blinded, and/or suffer chronic debilitating respiratory problems. Plant workers would be subject to similar effects.
c. Consequences for property	Possible superficial damage to equipment structures from corrosive fumes.	Repairable damage to highway. Potential destructon of nearby vehicles by fire, explosions.	Vapors may explode in confined space causing repairable property damage. Fires could cause repairable damage.
d. Consequences of environmental exposure	Possible destruction of surrounding fauna, flora.	Potential for fire damage to adjacent forest preserve due to combustible material (recoverable in the long term).	Farm animals, other fauna could die or suffer health effects necessitating their destruction or indirectly causing death.
e. Probability of simultaneous emergencies	Low	High	Low
f. Unusual environmental conditions	None	Hilly terrain prone to mists, creating adverse driving conditions.	Located in a 500-year river floor plain.

- Has a hazards analysis been prepared for the facility? If so, has it been updated? Has a copy been provided to the local emergency planning committee?
- What steps have been taken to reduce identified risks?
- How does the company reward good safety records?
- Have operation or storage procedures been modified to reduce the probability of a release and minimize potential effects?
- What release prevention or mitigation systems, equipment, or procedures are in place?
- What possibilities are there for safer substitutes for any acutely toxic chemicals used or stored at the facility?
- What possibilities exist for reducing the volume of the hazardous materials in use or stored at the facility?
- What additional safeguards are available to prevent accidental releases?
- What studies have been conducted by the facility to determine the feasibility of each of the following approaches for each relevant production process or operation: (a) input change, (b) product reformulation, (c) production process change, and (d) operational improvements?
- Are on-site emergency response equipment (e.g., fire fighting equipment, personal protective equipment, communications equipment) and trained personnel available to provide on-site initial response efforts?
- What equipment (e.g., self-contained breathing apparatus, chemical suits, unmanned fire monitors, foam deployment systems, radios, beepers) is available? Is equipment available for loan or use by the community on a reimbursable basis? (Note: Respirators should not be lent to any person not properly trained in their use.)
- Is there emergency medical care on site?
- Are the local hospitals prepared to accept and provide care to patients who have been exposed to chemicals?
- Who is the emergency contact for the site (person's name, position, and 24-hour telephone number) and what is the chain of command during an emergency?
- Are employee evacuation plans in effect and are the employees trained to use them in the event of an emergency?
- What kinds of notification systems connect the facility and the local community emergency services (e.g., direct alarm, direct telephone hook-up, computer hook-up) to address emergencies on site?
- What is the mechanism to alert employees and the surrounding community in the event of a release at the facility?

- Is there a standard operating procedure for the personal protection of community members at the time of an emergency?
- Does the community know about the meaning of various alarms or warning systems? Are tests conducted?
- How do facility personnel coordinate with the community government and local emergency and medical services during emergencies? Is overlap avoided?
- What mutual aid agreements are in place for obtaining emergency response assistance from other industry members? With whom?
- Are there any contacts or other prearrangements in place with specialists for cleanup and removal of releases, or is this handled in-house? How much time is required for the cleanup specialists to respond?
- What will determine concentrations of released chemicals existing at the site? (Are there toxic gas detectors, explosimeters, or other detection devices positioned around the facility? Where are they located?)
- Are wind direction indicators positioned within the facility perimeter to determine in what direction a released chemical will travel? Where are they located?
- Is there capability for modeling vapor cloud dispersion?
- Are auxiliary power systems available to perform emergency system functions in case of power outages at the facility?
- How often is the safety plan tested and updated? When was it last tested and updated?
- Does the company participate in CHEMNET or the CAER program?
- Does the company have the capability and plans for responding to off-site emergencies? Is this limited to the company's products?

What is the safety training plan for management and employees?

- Are employees trained in the use of emergency response equipment, personal protective equipment, and emergency procedures detailed in the plant safety plan? How often is training updated?
- Are simulated emergencies conducted for training purposes? How often? How are these simulations evaluated and by whom? When was this last done? Are the local community emergency response and medical service organizations invited to participate?
- Are employees given training in methods for coordinating with local community emergency response and medical services during emergencies? How often?
- Is management given appropriate training? How frequently?

HAZARDOUS MATERIALS RESPONSE HANDBOOK

- Is there an emergency response equipment and systems inspection plan?
- Is there a method for identifying emergency response equipment problems? Describe it.
- Is there testing of on-site alarms, warning signals, and emergency response equipment? How often is this equipment tested and replaced?

3.4.2 Transporter Resources

What cargo information and response organization do ship, train, and truck operators provide at a release?

- Do transport shipping papers identify hazardous materials, their physical and chemical characteristics, control techniques, and interactions with other chemicals?
- Do transports have proper placards?
- Are there standard operating procedures (SOPs) established for release situations? Have these procedures been updated to reflect current cargo characteristics?
- Who is the emergency contact for transport operators? Is there a 24-hour emergency contact system in place? What is the transport operation's chain of command in responding to a release?

What equipment and cleanup capabilities can transport operations make available?

- What emergency response equipment is carried by each transporter (e.g., protective clothing, breathing apparatus, chemical extinguishers)?
- Do transports have first-aid equipment (e.g., dressings for chemical burns and water to rinse off toxic chemicals)?
- By what means do operators communicate with emergency response authorities?
- Do transport operations have their own emergency response units?
- What arrangements have been established with cleanup specialists for removal of a release?

What is the safety training plan for operators?

- Are operators trained in release SOPs and to use emergency response equipment? How often is training updated?
- How often are release drills conducted? Who evaluates these drills and do the evaluations become a part of an employee's file?
- Are safe driving practices addressed in operator training? What monetary or promotional incentives encourage safety in transport operation?

Is there a transport and emergency response equipment inspection plan?

- What inspections are conducted? What leak detection and equipment readiness tests are done? What is the schedule for inspections and tests?
- Are problems identified in inspections corrected? How are maintenance schedules established?

3.4.3 Community Resources

What local agencies make up the community's existing response preparedness network? Some examples include:

- Fire department;
- Police/sheriff/highway patrol;
- Emergency medical/paramedic service associated with local hospitals or fire and police departments;
- Emergency management or civil defense agency;
- Public health agency;
- Environmental agency;
- Public works and/or transportation departments;
- Red Cross; and
- Other local community resources such as public housing, schools, public utilities, communications.

What is the capacity and level of expertise of the community's emergency medical facilities, equipment, and personnel?

Does the community have arrangements or mutual aid agreements for assistance with other jurisdictions or organizations (e.g., other communities, counties, or states; industry; military installations; federal facilities; response organizations)? In the absence of mutual aid agreements, has the community taken liability into consideration?

What is the current status of community planning and coordination for hazardous materials emergency preparedness? Have potential overlaps in planning been avoided?

- Is there a community planning and coordination body (e.g., task force, advisory board, interagency committee)? If so, what is the defined structure and authority of the body?
- Has the community performed any assessments of existing prevention and response capabilities within its own emergency response network?
- Does the community maintain an up-to-date technical reference library of response procedures for hazardous materials?

- Have there been any training sessions, simulations, or mock incidents performed by the community in conjunction with local industry or other organizations? If so, how frequently are they conducted? When was this last done? Do they typically have simulated casualties?

Who are the specific community points of contact and what are their responsibilities in an emergency?

- List the agencies involved, the area of responsibility (e.g., emergency response, evacuation, emergency shelter, medical/health care, food distribution, control access to accident site, public/media liaison, liaison with federal and state responders, locating and manning the command center and/or emergency operating center), the name of the contact, position, 24-hour telephone number, and the chain of command.
- Is there any specific chemical or toxicological expertise available in the community, either in industry, colleges and universities, poison control centers, or on a consultant basis?

What kinds of equipment and materials are available at the local level to respond to emergencies? How can the equipment, materials, and personnel be made available to trained users at the scene of an incident?

Does the community have specialized emergency response teams to respond to hazardous materials releases?

- Have the local emergency services (fire, police, medical) had any hazardous materials training, and if so, do they have and use any specialized equipment?
- Are local hospitals able to decontaminate and treat numerous exposure victims quickly and effectively?
- Are there specialized industry response teams (e.g., CHLOREP, AAR/BOE), state/federal response teams, or contractor response teams available within or close to the community? What is the average time for them to arrive on the scene?
- Has the community sought any resources from industry to help respond to emergencies?

Is the community emergency transportation network defined?

- Does the community have specific evacuation routes designated? What are these evacuation routes? Is the general public aware of these routes?
- Are there specific access routes designated for emergency response and services personnel to reach facilities or incident sites? (In a real incident, wind direction might make certain routes unsafe.)

Does the community have other procedures for protecting citizens during emergencies (e.g., asking them to remain indoors, close windows, turn off air-conditioners, tune into local emergency radio broadcasts)?

Is there a mechanism that enables responders to exchange information or ideas during an emergency with other entities, either internal or external to the existing organizational structure?

Does the community have a communications link with an Emergency Broadcast System (EBS) station? Is there a designated emergency communications network in the community to alert the public, update the public, and provide communications between the command center and/or emergency operating center, the incident site, and off-scene support? Is there a back-up system?

- What does the communications network involve (e.g., special radio frequency, network channel, siren, dedicated phone lines, computer hook-up)?
- Is there an up-to-date list, with telephone numbers, of radio and television stations (including cable companies) that broadcast in the area?
- Is there an up-to-date source list with a contact, position, and telephone number for technical information assistance? This can be federal (e.g., NRC, USCG CHRIS/HACS, ATSDR, OHMTADS), state, industry associations (e.g., CHEMTREC, CHLOREP, AAR/BOE, PSTN), and local industry groups (e.g., local AlChE, ASME, ASSE chapters).

Is there a source list with a contact, position, and telephone number for community resources available?

- Does the list of resources include: wreck clearing, transport, cleanup, disposal, health, analytical sampling laboratories, and detoxifying agents?

Have there been any fixed facility or transportation incidents involving hazardous materials in the community? What response efforts were taken? What were the results? Have these results been evaluated?

3.5 Writing an Emergency Plan

When the team has reviewed existing plans, completed a hazards identification and analysis, and assessed its preparedness, prevention, and response capabilities, it can take steps to make serious incidents less likely. Improved warning systems, increased hazardous materials training of industry and local response personnel, and other efforts at the local level, can all make a community better prepared to live safely with hazardous materials. The team should also begin to write an emergency plan if one does not already exist, or revise existing plans to include hazardous materials. Chapter 4 describes two approaches to developing or revising an emergency plan. Chapter 5 describes elements related to hazardous materials incidents that should be included in whichever type of plan the community chooses to write.

4. Developing the Plan

4.1 Introduction

Most communities have some type of written plan for emergencies. These plans range from a comprehensive multi-hazard approach as described in FEMA's CPG 1-8 (Guide for Development of State and Local Emergency Operations Plans) to a single telephone roster for call-up purposes, or an action checklist. Obviously the more complete and thorough a plan is, the better prepared the community should be to deal with any emergency that occurs.

As noted in Chapter 1, the "Emergency Planning and Community Right-to-Know Act of 1986" requires local emergency planning committees to develop local plans for emergency responses in the event of a release of an extremely hazardous substance. Those communities receiving FEMA funds are required to incorporate hazardous materials planning into their multi-hazard emergency operations plan (EOP). Other communities are encouraged to prepare a multi-hazard EOP in accord with CPG 1-8 since it is the most comprehensive approach to emergency planning. Not every community, however, may be ready for or capable of such a comprehensive approach. Because each community must plan in light of its own situation and resources, a less exhaustive approach may be the only practical, realistic way of having some type of near-term plan. Each community must choose the level of planning that is appropriate for it, based upon the types of hazard found in the community.

This chapter discusses two basic approaches to writing a plan: (1) development or revision of a hazardous materials appendix (or appendices to functional annexes) to a multi-hazard EOP following the approach described in FEMA's CPG 1-8, and (2) development or revision of a plan covering only hazardous materials. Each approach is discussed in more detail below.

4.2 Hazardous Materials Appendix to Multi-Hazard EOP

The first responders (e.g., police, fire, emergency medical team) at the scene of an incident are generally the same whatever the hazard. Moreover, many emergency functions (e.g., direction and control, communications, and evacuation) vary only slightly from hazard to hazard. Procedures to be followed for warning the public of a hazardous materials incident, for example, are not that different from procedures followed in warning the public about other incidents such as a flash flood. It is possible, therefore, to avoid a great deal of unnecessary redundancy and confusion by planning for all hazards at the same time. A multi-hazard EOP avoids developing separate structures, resources, and plans to deal with each type of hazard. Addressing the general aspects of all hazards first and then looking at each potential hazard individually to see if any unique aspects are involved result in efficiencies and economies in the long run. Multi-hazard EOPs also help ensure that plans and systems are reasonably compatible if a large-scale hazardous materials incident requires a simultaneous, coordinated response by more than one community or more than one level of government.

A community that does not have a multi-hazard plan is urged to consider seriously the advantages of this integrated approach to planning. In doing so, the community may want to seek state government advice and support.

CPG 1-8 describes a sample format, content, and process for state and local EOPs. It recommends that a multi-hazard EOP include three components—a basic plan, functional annexes, and hazard-specific appendices. It encourages development of a basic plan that includes generic functional annexes applicable to any emergency situation, with unique aspects of a particular hazard being addressed in hazard-specific appendices. It stresses improving the capabilities for simultaneous, coordinated response by a number of emergency organizations at various levels of government. Local communities that receive FEMA funds must incorporate hazardous materials planning into their multi-hazard EOP. In most of these communities, there are paid staff to do emergency operations planning as well as related emergency management tasks.

CPG 1-8 provides flexible guidance, recognizing that substantial variation in planning may exist from community to community. A community may develop a separate hazardous material appendix to each functional annex where there is a need to reflect considerations unique to hazardous materials not adequately covered in the functional annex. On the other hand, a community may develop a single hazardous materials appendix to the EOP, incorporating all functional annex considerations related to hazardous materials in one document. The sample plan format used in CPG 1-8 is a good one, but it is not the only satisfactory one. It is likely that no one format is the best for all communities of all sizes in all parts of the country. Planners should, therefore, use good judgment and common sense in applying CPG 1-8 principles to meet their needs. The community has latitude in formatting the plan but should closely follow the basic content described in CPG 1-8.

CPG 1-8 should be used in preparing the basic plan and functional annexes. This guide should be used as a supplement to CPG 1-8 to incorporate hazardous materials considerations into a multi-hazard EOP. Communities that want to develop standard operating procedures (SOP) manuals could begin with information included in the functional annexes of a multi-hazard EOP.

A community that is incorporating hazardous materials into a multi-hazard EOP should turn to Chapter 5 of this guide for a discussion of those elements that need to be taken into account in hazardous materials planning.

4.3 Single-Hazard Emergency Plan

If a community does not have the resources, time, or capability readily available to undertake multi-hazard planning, it may wish to produce a single-hazard plan addressing hazardous materials.

Exhibit 4 identifies sections of an emergency plan for hazardous materials incidents. The sample outline is not a model. It is not meant to constrain any community. Indeed, each community should seek to develop a plan that is best suited to its own circumstances, taking advantage of the sample outline where appropriate.

The type of plan envisioned in the sample outline would affect all governmental and private organizations involved in emergency response operations in a particular community. Its basic purpose would be to provide the necessary data and documentation to anticipate and coordinate the many persons and organizations that would be involved in emergency response actions. As such, the plan envisioned in this sample outline is intended neither to be a "hip-pocket" emergency response manual, nor to serve as a detailed standard operating procedures (SOP) manual for each of the many agencies and organizations involved in emergency response actions, although it could certainly be used as a starting point for such manuals. Agencies that want to develop an SOP manual could begin with the information contained under the appropriate function in Plan

Section C of this sample outline. If it is highly probable that an organization will be involved in a hazardous materials incident response, then a more highly detailed SOP should be developed.

Exhibit 4. Sample Outline of a Hazardous Materials Emergency Plan.

(NOTE: Depending upon local circumstances, communities will develop some sections of the plan more extensively than other sections. See page 5.1 for how the sample outline relates to SARA Title III requirements.)

A. Introduction

1. Incident Information Summary
2. Promulgation Document
3. Legal Authority and Responsibility for Responding
4. Table of Contents
5. Abbreviations and Definitions
6. Assumptions/Planning Factors
7. Concept of Operations
   a. Governing Principles
   b. Organizational Roles and Responsibilities
   c. Relationship to Other Plans
8. Instructions on Plan Use
   a. Purpose
   b. Plan Distribution
9. Record of Amendments

B. Emergency Assistance Telephone Roster

C. Response Functions*

1. Initial Notification of Response Agencies
2. Direction and Control
3. Communications (among Responders)
4. Warning Systems and Emergency Public Notification
5. Public Information/Community Relations

---

*These "Response Functions" are equivalent to the "functional annexes" of a multi-hazard emergency operations plan described in CPG 1-8.

6. Resource Management
7. Health and Medical Services
8. Response Personnel Safety
9. Personal Protection of Citizens
    a. Indoor Protection
    b. Evacuation Procedures
    c. Other Public Protection Strategies
10. Fire and Rescue
11. Law Enforcement
12. Ongoing Incident Assessment
13. Human Services
14. Public Works
15. Others

D. Containment and Cleanup

1. Techniques for Spill Containment and Cleanup
2. Resources for Cleanup and Disposal

E. Documentation and Investigative Follow-up

F. Procedures for Testing and Updating Plan

1. Testing the Plan
2. Updating the Plan

G. Hazards Analysis (Summary)

H. References

1. Laboratory, Consultant, and Other Technical Support Resources
2. Technical Library

5. Hazardous Materials Planning Elements

5.1 Introduction

This chapter presents and discusses a comprehensive list of planning elements related to hazardous materials incidents. Communities that are developing a hazardous materials appendix/plan need to review these elements thoroughly. Communities that are revising an existing appendix/plan need to evaluate their present appendix/plan and identify what elements need to be added, deleted, or amended in order to deal with the special problems associated with the accidental spill or release of hazardous materials.

Title III of SARA requires each emergency plan to include at least each of the following. The appropriate section of the plan as indicated in Exhibit 4 is shown in parentheses after each required Title III plan element.

(1) Identification of facilities subject to the Title III requirements that are within the emergency planning district; identification of routes likely to be used for the transportation of substances on the list of extremely hazardous substances; and identification of additional facilities contributing or subjected to additional risk due to their proximity to facilities, such as hospitals or natural gas facilities. (Exhibit 4, Sections A.6 and G)

(2) Methods and procedures to be followed by facility owners and operators and local emergency and medical personnel to respond to any releases of such substances. (Exhibit 4, Section C)

(3) Designation of a community emergency coordinator and facility emergency coordinators, who shall make determinations necessary to implement the plan. (Exhibit 4, Section A.7b)

(4) Procedures providing reliable, effective, and timely notification by the facility emergency coordinators and the community emergency coordinator to persons designated in the emergency plan, and to the public, that a release has occurred. (Exhibit 4, Sections C.1 and C.4)

(5) Methods for determining the occurrence of a release, and the area or population likely to be affected by such release. (Exhibit 4, Sections A.6 and G)

(6) A description of emergency equipment and facilities in the community and at each facility in the community subject to Title III requirements, and an identification of the persons responsible for such equipment and facilities. (Exhibit 4, Section C.6)

(7) Evacuation plans, including provisions for a precautionary evacuation and alternative traffic routes. (Exhibit 4, Section C.9b)

(8) Training programs, including schedules for training of local emergency response and medical personnel. (Exhibit 4, Sections C.6 and F.1)

(9) Methods and schedules for exercising the emergency plan. (Exhibit 4, Section F.1)

The various planning elements are discussed here in the same order as they appear in the sample outline for a hazardous materials emergency plan in Chapter 4. Community planners might choose, however, to order these planning elements differently in a multi-hazard plan following the model of CPG 1-8.

5.2 Discussion of Planning Elements

The remainder of this chapter describes in detail what sorts of information could be included in each element of the emergency plan. These issues need to be addressed in the planning process. In some cases, they will be adequately covered in SOPs and will not need to be included in the emergency plan.

Planning Element A: Introduction

A.1: Incident Information Summary

Develop a format for recording essential information about the incident:

- Date and time
- Name of person receiving call
- Name and telephone number of on-scene contact
- Location
- Nearby populations
- Nature (e.g., leak, explosion, spill, fire, derailment)
- Time of release
- Possible health effects/medical emergency information
- Number of dead or injured; where dead/injured are taken
- Name of material(s) released, if known

    Manifest/shipping invoice/billing label
    Shipper/manufacturer identification
    Container type (e.g., truck, rail car, pipeline, drum)
    Railcar/truck 4-digit identification numbers
    Placard/label information

- Characteristics of material (e.g., color, smell, physical effects), only if readily detectable
- Present physical state of the material (i.e., gas, liquid, solid)
- Total amount of material that may be released
- Other hazardous materials in area
- Amount of material released so far/duration of release
- Whether significant amounts of the material appear to be entering the atmosphere, nearby water, storm drains, or soil
- Direction, height, color, odor of any vapor clouds or plumes
- Weather conditions (wind direction and speed)
- Local terrain conditions
- Personnel at the scene

Comment: Initial information is critical. Answers to some of these questions may be unknown by the caller, but it is important to gather as much information as possible very quickly in order to facilitate decisions on public notification and evacuation. Some questions will apply to fixed facility incidents and others will apply only to transportation incidents. Some questions will apply specifically to air releases, while other questions will gather information about spills onto the ground or into water. Identification numbers, shipping manifests, and placard information are essential to identify any hazardous materials involved in transportation incidents, and to take initial precautionary and containment steps. First responders should use DOT's Emergency Response Guidebook to help identify hazardous materials. Additional information about the identity and characteristics of chemicals is available by calling CHEMTREC (800-424-9300). CHEMTREC and the Hazard Information Transmission (HIT) program are described in Appendix C.

This emergency response notification section should be:

Brief—never more than one page in length.

Easily accessible—located on the cover or first page of the plan. It should also be repeated at least once inside the plan, in case the cover is torn off.

Simple—reporting information and emergency telephone numbers should be kept to a minimum.

Copies of the emergency response notification form could be provided to potential dischargers to familiarize them with information needed at the time of an incident.

A.2:   Promulgation Document

Statement of plan authority

Comment: A letter, signed by the community's chief executive, should indicate legal authority and responsibility for putting the plan into action. To the extent that the execution of this plan involves various private and public-sector organizations, it may be appropriate to include here letters of agreement signed by officials of these organizations.

A.3:   Legal Authority and Responsibility for Responding

* Authorizing legislation and regulations
* Federal (e.g., CERCLA, SARA, Clean Water Act, National Contingency Plan, and Disaster Relief Act)
* State

- Regional
- Local
- Mandated agency responsibilities
- Letters of agreement

Comment: If there are applicable laws regarding planning for response to hazardous materials releases, list them here. Analyze the basic authority of participating agencies and summarize the results here. The community may choose to enact legislation in support of its plan. Be sure to identify any agencies required to respond to particular emergencies.

A.4:   Table of Contents

Comment: All sections of the plan should be listed here and clearly labeled with a tab for easy access.

A.5:   Abbreviations and Definitions

Comment: Frequently used abbreviations, acronyms, and definitions should be gathered here for easy reference.

A.6:   Assumptions/Planning Factors

Geography

- Sensitive environmental areas
- Land use (actual and potential, in accordance with local development codes)
- Water supplies

Public transportation network (roads, trains, buses)
Population density
Particularly sensitive institutions (e.g., schools, hospitals, homes for the aged)
Climate/weather statistics
Time variables (e.g., rush hour, vacation season)
Particular characteristics of each facility and the transportation routes for which the plan is intended

- On-site details
- Neighboring population
- Surrounding terrain
- Known impediments (tunnels, bridges)
- Other areas at risk

Assumptions

Comment: This section is a summary of precisely what local conditions make an emergency plan necessary. Information for this section will be derived from the hazards identification and analysis. Appropriate maps should be included in this section. Maps should show: water intake, environmentally sensitive areas, major chemical manufacturing or storage facilities, population centers, and the location of response resources.

Assumptions are the advance judgments concerning what would happen in the case of an accidental spill or release. For example, planners might assume that a certain percentage of local residents on their own will evacuate the area along routes other than specified evacuation routes.

A.7:   Concept of Operations

Governing Principles

Comment: The plan should include brief statements of precisely what is expected to be accomplished if an incident should occur.

Organizational Roles and Responsibilities

Municipal government

- Chief elected official
- Emergency management director
- Community emergency coordinator (Title III of SARA)
- Communications personnel
- Fire service
- Law enforcement
- Public health agency
- Environmental agency
- Public works

County government

- Officials of fixed facilities and/or transportation companies
- Facility emergency coordinators (Title III of SARA)

Nearby municipal and county governments

- Indian tribes within or nearby the affected jurisdiction
- State government
- Environmental protection agency

- Emergency management agency
- Public health agency
- Transportation organization
- Public safety organization

Federal government

- EPA
- FEMA
- DOT
- HHS/ATSDR
- USCG
- DOL/OSHA
- DOD
- DOE
- RRT

Predetermined arrangements
How to use outside resources

- Response capabilities
- Procedures for using outside resources

Comment: This section lists all those organizations and officials who are responsible for planning and/or executing the pre-response (planning and prevention), response (implementing the plan during an incident), and post-response (cleanup and restoration) activities to a hazardous materials incident. One organization should be given command and control responsibility for each of these three phases of the emergency response. The role of each organization/official should be clearly described. The plan should clearly designate who is in charge and should anticipate the potential involvement of state and federal agencies and other response organizations. (Note: The above list of organizations and officials is not meant to be complete. Each community will need to identify all the organizations/officials who are involved in the local planning and response process.)

This section of the plan should contain descriptions and information on the RRTs and the predesignated federal OSC for the area covered by the plan. (See Section 1.4.1 of this guidance.) Because of their distant location, it is often difficult for such organizations to reach a scene quickly; planners should determine in advance approximately how much time would elapse before the federal OSC could arrive at the scene.

This section should also indicate where other disaster assistance can be obtained from federal, state, or regional sources. Prearrangements can be made with higher-level government agencies, bordering political regions, and chemical plants.

Major hazardous materials releases may overwhelm even the best prepared community, and an incident may even cross jurisdictional boundaries. Cooperative arrangements are an efficient means of obtaining the additional personnel, equipment, and materials that are needed in an emergency by reducing expenditures for maintaining extra or duplicative resources. Any coordination with outside agencies should be formalized through mutual aid and Good Samaritan agreements or memoranda of understanding specifying delegations of authority, responsibility, and duties. These formal agreements can be included in the plan if desired.

Relationship to Other Plans

Comment: A major task of the planning group is to integrate planning for hazardous materials incidents into already existing plans. In larger communities, it is probable that several emergency plans have been prepared. It is essential to coordinate these plans. When more than one plan is put into action simultaneously, there is a real potential for confusion among response personnel unless the plans are carefully coordinated. All emergency plans (including facility plans and hospital plans) that might be employed in the event of an accidental spill or release should be listed in this section. The community plan should include the methods and procedures to be followed by facility owners and operators and local emergency response personnel to respond to any releases of such substances. The NCP, the federal regional contingency plan, any OSC plan for the area, and any state plan should be referenced. Of special importance are all local emergency plans.

Even where formal plans do not exist, various jurisdictions often have preparedness capabilities. Planners should seek information about informal agreements involving cities, counties, states, and countries.

A.8: Instructions on Plan Use

Purpose

Comment: This should be a clear and succinct statement of when and how the plan is meant to be used. It is appropriate to list those facilities and transportation routes explicitly considered in the plan.

Plan Distribution

List of organizations/persons receiving plan

Comment: The entire plan should be available to the public; it can be stored at a library, the local emergency management agency, or some other public place. The plan should be distributed to all persons responsible for response operations. The plan distribution list should account for all organizations receiving such copies of the plan. This information is essential when determining who should be sent revisions and updates to the plan.

A.9: Record of Amendments

Change record sheet

* Date of change
* Recording signature
* Page numbers of changes made

Comment: Maintaining an up-to-date version of a plan is of prime importance. When corrections, additions, or changes are made, they should be recorded in a simple bookkeeping style so that all plan users will be aware that they are using a current plan.

All that is necessary for this page is a set of columns indicating date of change, the signature of the person making the change, and the page number for identifying each change made.

Planning Element B: Emergency Assistance Telephone Roster

List of telephone numbers for:

* Participating agencies
* Technical and response personnel
* CHEMTREC
* Public and private sector support groups
* National Response Center

Comment: An accurate and up-to-date emergency telephone roster is an essential item. The name of a contact person (and alternate) and the telephone number should be listed. Briefly indicate the types of expertise, services, or equipment that each agency or group can provide. Indicate the times of day when the number will be answered; note all 24-hour telephone numbers. All phone numbers and names of personnel should be verified at least every six months. When alternate numbers are available, these should be listed. This section of the plan should stand alone so that copies can be carried by emergency response people and others. Examples of organizations for possible inclusion in a telephone roster are as follows:

Community Assistance _____

Police	Ambulance
Fire	Hospitals
Emergency Management Agency	Utilities:
Public Health Department	Gas
Environmental Protection Agency	Phone
Department of Transportation	Electricity
Public Works	Community Officials:
Water Supply	Mayor
Sanitation	City Manager
Port Authority	County Executive
Transit Authority	Councils of Government
Rescue Squad	

Volunteer Groups _____

Red Cross	Ham Radio Operators
Salvation Army	Off-Road Vehicle Clubs
Church Groups	

State Assistance _____

State Emergency Response Commission (Title III of SARA)	Department of Transportation
	Police
State Environmental Protection Agency	Public Health Department
	Department of Agriculture
Emergency Management Agency	

Response Personnel _____

Incident Commander	Response Team Members
Agency Coordinators	

Bordering Political Regions _____

Municipalities	Countries
Counties	River Basin Authorities
States	Irrigation Districts

Interstate Compacts	Sanitation Authorities/
Regional Authorities	Commissions
Bordering International Authorities	

Industry _____

Transporters	Spill Cooperatives
Chemical Producers/Consumers	Spill Response Teams

Media _____

Television	Radio
Newspaper	

Federal Assistance

(Consult regional offices listed in Appendix F for appropriate telephone numbers.)

Federal On-Scene Coordinator
U.S. Department of Transportation
U.S. Coast Guard
U.S. Environmental Protection Agency
Federal Emergency Management Agency
    (24 hours, 202-646-2400)
U.S. Department of Agriculture
Occupational Safety and Health Administration
Agency for Toxic Substances and Disease Registry
National Response Center
    (24 hours, 800-424-8802;
    in Washington, DC area,
    202-426-2675
    or 202-267-2675)
U.S. Army, Navy, Air Force
Bomb Disposal and/or Explosive Ordnance Team, U.S. Army
Nuclear Regulatory Commission
    (24 hours, 301-951-0550)
U.S. Department of Energy Radiological Assistance
    (24 hours, 202-586-8100)
U.S. Department of the Treasury Bureau of Alcohol, Tobacco, and Firearms

Other Emergency Assistance

CHEMTREC
    (24 hours, 800-424-9300)
CHEMNET
    (24 hours, 800-424-9300)
CHLOREP
    (24 hours, 800-424-9300)
NACA Pesticide Safety Team
Association of American Railroads/ Bureau of Explosives
    (24 hours, 202-639-2222)
Poison Control Center
Cleanup Contractor

Planning Element C: Response Functions

Comment: Each function should be clearly marked with a tab so that it can be located quickly. When revising and updating a plan, communities might decide to add, delete, or combine individual functions.

Each response "function" usually includes several response activities. Some communities prepare a matrix that lists all response agencies down the left side of the page and all response activities across the top of the page. Planners can then easily determine which response activities need interagency coordination and which, if any, activities are not adequately provided for in the plan.

Function 1:   Initial Notification of Response Agencies

24-hour emergency response hotline telephone numbers

- Local number to notify area public officials and response personnel
- Number to notify state authorities
- National Response Center (800-424-8802; 202-426-2675 or 202-267-2675 in Washington, DC area)

Other agencies (with telephone numbers) to notify immediately (e.g., hospitals, health department, Red Cross)

Comment: The local 24-hour emergency response hotline should be called first and therefore should have a prominent place in the plan. Provision should be made for notifying nearby municipalities and counties that could be affected by a vapor cloud or liquid plumes in a water supply.

Normally, the organization that operates the emergency response hotline will inform other emergency service organizations (e.g., health department, hospitals, Red Cross) once the initial notification is made. The plan should provide a method for notifying all appropriate local, state, and federal officials and agencies, depending upon the severity of the incident. To ensure that the appropriate federal on-scene coordinator (OSC) is notified of a spill or release, the NRC operated by the U.S. Coast Guard should be included in the notification listing. CERCLA requires that the NRC be notified by the responsible party of releases of many hazardous materials in compliance with the reportable quantity (RQ) provisions. The NRC telephone number is 800-424-8802 (202-426-2675 or 202-267-2675 in the Washington, DC area). If there is an emergency notification number at the state or regional level, it should be called before the NRC, and then a follow-up call made to the NRC as soon as practicable.

The plan should indicate how volunteer and off-duty personnel will be summoned. Similarly, there should be a method to notify special facilities (e.g., school districts, private schools, nursing homes, day care centers, industries, detention centers), according to the severity of the incident.

Function 2: Direction and Control

Name of on-scene authority
Chain of command (illustrated in a block diagram)
Criteria for activating emergency operating center
Method for establishing on-scene command post and communications network for response team(s)
Method for activating emergency response teams
List of priorities for response actions
Levels of response based on incident severity

Comment: Response to a hazardous materials spill or release will involve many participants: police, fire fighters, facility personnel, health personnel, and others. It is also possible to have more than one organization perform the same service; for example, local police, the county sheriff and deputies, as well as the highway patrol may respond to perform police functions. Because speed of response is so important, coordination is needed among the various agencies providing the same service. It is essential to identify (by title or position) the one individual responsible for each participating organization, and the one individual responsible for each major function and service. The plan might require that the responsible person establish an incident command system (ICS).

Work out, in advance, the following:

(1) Who will be in charge (lead organization)
(2) What will be the chain of command
(3) Who will activate the emergency operating center, if required
(4) Who will maintain the on-scene command post and keep it secure
(5) Who will have advisory roles (and what their precise roles are)
(6) Who will make the technical recommendations on response actions to the lead agency
(7) Who (if anyone) will have veto power
(8) Who is responsible for requesting assistance from outside the community

This chain of command should be clearly illustrated in a block diagram.

Response action checklists are a way of condensing much useful information. They are helpful for a quick assessment of the response operation. If checklists are used, they should be prepared in sufficient detail to ensure that all crucial activities are included.

Planners should consider whether to have categories of response actions based on severity. The severity of an incident influences decisions on the level (or degree) of response to be made. This will determine how much equipment and how many personnel will be called, the extent of evacuation, and other factors.

Function 3: Communications (among Responders)

Any form(s) of exchanging information or ideas for emergency response with other entities, either internal or external to the existing organizational structure.

Comment: This aspect of coordination merits special consideration. Different response organizations typically use different radio frequencies. Therefore, specific provision must be made for accurate and efficient communication among all the various organizations during the response itself. Several states have applied for one "on-scene" command radio frequency that all communities can use. At a minimum, it may be beneficial to establish radio networks that will allow for communication among those performing similar functions. The plan might specify who should be given a radio unit and who is allowed to speak on the radio. In order to avoid possible explosion/fire hazards, all communications equipment (including walkie-talkies) should be intrinsically safe.

The following chart summarizes who and what are involved in three typical emergency conditions. Information about the three response levels should be provided to special facilities (e.g., school districts, private schools, day care centers, hospitals, nursing homes, industries, detention centers).

Response Level	Description	Contact:
I. Potential Emergency Condition	An incident or threat of a release that can be controlled by the first response agencies and does not require evacuation of other than the involved structure or the immediate outdoor area. The incident is confined to a small area and does not pose an immediate threat to life or property.	Fire Department Emergency Medical Services Police Department Partial EOC Staff Public Information Office CHEMTREC National Response Center
II. Limited Emergency Condition	An incident involving a greater hazard or larger area that poses a potential threat to life or property and that may require a limited evacuation of the surrounding area.	All Agencies in Level 1 HAZMAT Teams EOC Staff Public Works Department Health Department Red Cross County Emergency Management Agency State Police Public Utilities
III. Full Emergency Condition	An incident involving a severe hazard or a large area that poses an extreme threat to life and property and will probably require a large-scale evacuation; or an incident requiring the expertise or resources of county, state, federal, or private agencies/ organizations.	All Level I and Level II Agencies plus the following, as needed: Mutual Aid, Fire, Police, Emergency Medical State Emergency Management Agency State Department of Environmental Resources State Department of Health EPA USCG ATSDR FEMA OSC/RRT

Function 4:   Warning Systems and Emergency Public Notification

Method for alerting the public

- Title and telephone number of person responsible for alerting the public as soon as word of the incident is received
- List of essential data to be passed on (e.g., health hazards, precautions for personal protection, evacuation routes and shelters, hospitals to be used)

Comment: This section should contain precise information on how sirens or other signals will be used to alert the public in case of an emergency. This should include information on what the different signals mean, how to coordinate the use of sirens, and the geographic area covered by each siren. (If possible, a backup procedure should be identified.) While a siren alerts those who hear it, an emergency broadcast is necessary to provide detailed information about the emergency and what people should do.

Sample Emergency Broadcast System messages should be prepared with blank spaces that can be filled in with precise information about the accident. One sample message should provide fundamental information about the incident and urge citizens to remain calm and await further information and instructions. Another sample message should be for an evacuation. Another sample message should describe any necessary school evacuations so that parents will know where their children are. Another sample message should be prepared to tell citizens to take shelter and inform them of other precautions they may take to protect themselves. The message should clearly identify those areas in which protective actions are recommended, using familiar boundaries. Messages might be developed in languages other than English, if customarily spoken in the area.

This section could be of urgent significance. When life-threatening materials are released, speed of response is crucial. It is not enough to have planned for alerting the community; one organization must be assigned the responsibility of alerting the public as soon as word of the accidental release is received. Delay in alerting the public can lead to the loss of life. In addition to sirens and the Emergency Broadcast System, it may be necessary to use mobile public address systems and/or house-by-house contacts. In this case, adequate protection must be provided for persons entering the area to provide such help.

Function 5:   Public Information/Community Relations

Method to educate the public for possible emergencies
Method for keeping the public informed

- Provision for one person to serve as liaison to the public
- List of radio and T.V. contacts

Comment: Many communities develop a public information program to educate citizens about safety procedures during an incident. This program could include pamphlets; newspaper stories; periodic radio and television announcements; and programs for schools, hospitals, and homes for the aged.

It is important to provide accurate information to the public in order to prevent panic. Some citizens simply want to know what is happening. Other citizens may need to be prepared for possible evacuation or they may need to know what they can do immediately to protect themselves. Because information will be needed quickly, radio and television are much more important than newspapers in most hazardous materials releases. In less urgent cases, newspaper articles can provide detailed information to enhance public understanding of accidental spills and procedures for containment and cleanup. One person should be identified to serve as spokesperson. It is strongly recommended that the individual identified have training and experience in public information, community relations, and/or media relations. The spokesperson can identify for the media individuals who have specialized knowledge about the event. The chain of command should include this spokesperson. Other members of the response team should be trained to direct all communications and public relations issues to this one person.

Function 6:   Resource Management

List of personnel needed for emergency response
Training programs, including schedules for training of local emergency
 response and medical personnel
List of vehicles needed for emergency response
List of equipment (both heavy equipment and personal protective equipment)
 needed for emergency response

Comment: This section should list the resources that will be needed, and where the equipment and vehicles are located or can be obtained. A major task in the planning process is to identify what resources are already available and what must still be provided. For information on the selection of protective equipment, consult the *Occupational Safety and Health Guidance Manual for Hazardous Waste Site Activities* prepared by NIOSH, OSHA, USCG, and EPA; and the EPA/Los Alamos "Guidelines for the Selection of Chemical Protective Clothing" distributed by the American Conference of Governmental Industrial Hygienists (Building B-7, 6500 Glynway Ave., Cincinnati, OH 45211).

This section should also address funding for response equipment and personnel. Many localities are initially overwhelmed by the prospect of providing ample funding for hazardous materials response activities. In large localities, each response agency is usually responsible for providing and maintaining certain equipment and personnel; in such cases, these individual agencies must devise funding methods, sources, and accounting procedures. In smaller localities with limited resources, officials frequently develop cooperative agreements with other jurisdictions and/or private industries. Some communities stipulate in law that the party responsible for an incident should ultimately pay the cost of handling it.

For a more detailed discussion of response training, consult Chapter 6 of this guide.

Function 7:   Health and Medical

Provisions for ambulance service
Provisions for medical treatment

Comment: This section should indicate how medical personnel and emergency medical services can be summoned. It may be appropriate to establish mutual aid agreements with nearby communities to provide backup emergency medical personnel and equipment. The community should determine a policy (e.g., triage) for establishing priorities for the use of medical resources during an emergency. Medical personnel must be made aware of significant chemical hazards in the community in order to train properly and prepare for possible incidents. Emergency medical teams and hospital personnel must be trained in proper methods for decontaminating and treating persons exposed to hazardous chemicals. Planners should include mental health specialists as part of the team assisting victims of serious incidents. Protective action recommendations for sanitation, water supplies, recovery, and reentry should be addressed in this section.

Function 8:   Response Personnel Safety

Standard operating procedure for entering and leaving sites
Accountability for personnel entering and leaving the sites
Decontamination procedures
Recommended safety and health equipment
Personal safety precautions

Comment: Care must be taken to choose equipment that protects the worker from the hazard present at the site without unnecessarily restricting the capacities of the worker. Although the emphasis in equipment choices is commonly focused on protecting the worker from the risks presented by the hazardous material, impaired vision, restricted movements, or excessive heat can put the worker at equal risk. After taking these factors into account, the planner should list the equipment appropriate to various degrees of hazard using the EPA Levels of Protection (A, B, C, and D). The list should include: the type of respirator (e.g., self-contained breathing apparatus, supplied air respirator, or air purifying respirator) if needed; the type of clothing that must be worn; and the equipment needed to protect the head, eyes, face, ears, hands, arms, and feet. This list can then be used as a base reference for emergency response. The specific equipment used at a given site will vary according to the hazard. In addition, the equipment list should be reevaluated and updated as more information about the site is gathered to ensure that the appropriate equipment is being used. Responders should receive ongoing training in the use of safety equipment.

This section can also address liability related to immediate and long-term health hazards to emergency responders. State and local governments may want to consider insurance coverage and/or the development of waivers for employees and contractors who may be on site during a hazmat incident.

Function 9:   Personal Protection of Citizens

Indoor Protection

Hazard-specific personal protection

Comment: The plan should clearly indicate what protective action should be taken in especially hazardous situations. Evacuation is sometimes, but not always, necessary. (See Evacuation Procedures.) For some hazardous materials it is safer to keep citizens inside with doors and windows closed rather than to evacuate them. It is perhaps appropriate to go upstairs (or downstairs). Household items (e.g., wet towels) can provide personal protection for some chemical hazards. Frequently a plume will move quickly past homes. Modern housing has adequate air supply to allow residents to remain safely inside for an extended period of time. Because air circulation systems can easily transport airborne toxic substances, a warning should be given to shut off all air circulation systems (including heating, air conditioning, clothes dryers, vent fans, and fire places) both in private and institutional settings.

In order for an indoor protective strategy to be effective, planning and preparedness activities should provide:

* An emergency management system and decision-making criteria for determining when an indoor protection strategy should be used;
* A system for warning and advising the public;
* A system for determining when a cloud has cleared a particular area;
* A system for advising people to leave a building at an appropriate time; and
* Public education on the value of indoor protection and on expedient means to reduce ventilation.

Evacuation Procedures

Title of person and alternate(s) who can order/recommend an evacuation
Vulnerable zones where evacuation could be necessary and a method for
  notifying these places
Provisions for a precautionary evacuation
Methods for controlling traffic flow and providing alternate traffic routes
Shelter locations and other provisions for evacuations (e.g., special assistance
  for hospitals)
Agreements with nearby jurisdictions to receive evacuees
Agreements with hospitals outside the local jurisdictions
Protective shelter for relocated populations
Reception and care of evacuees
Reentry procedures

Comment: Evacuation is the most sweeping response to an accidental release. The plan should clearly identify under what circumstances evacuation would be appropriate and necessary. DOT's *Emergency Response Guidebook* provides suggested distances for evacuating unprotected people from the scene of an incident during the initial phase. It is important to distinguish between general evacuation of the entire area and selective evacuation of a part of the risk zone. In either case, the plan should identify how people will be moved (i.e., by city buses, police cars, private vehicles). Provision must be made for quickly moving traffic out of the risk zone and also for preventing outside traffic from entering the risk zone. If schools are located in the risk zone, the plan must identify the location to which students will be moved in an evacuation and how parents will be notified of this location. Special attention must also be paid to evacuating hospitals, nursing homes, and homes for the physically and mentally disabled.

Maps (drawn to the same scale) with evacuation routes and alternatives clearly identified should be prepared for each risk zone in the area. Maps should indicate precise routes to another location where special populations (e.g., from schools, hospitals, nursing homes, homes for the physically or mentally disabled) can be taken during an emergency evacuation, and the methods of transportation during the evacuation.

Consideration of when and how evacuees will return to their homes should be part of this section.

This section on evacuation should include a description of how other agencies will coordinate with the medical community.

Copies of evacuation procedures should be provided to all appropriate agencies and organizations (e.g., Salvation Army, churches, schools, hospitals) and could periodically be published in the local newspaper(s).

Other Public Protection Strategies

Relocation
Water supply protection
Sewage system protection

Comment: Some hazardous materials incidents may contaminate the soil or water of an area and pose a chronic threat to people living there. It may be necessary for people to move out of the area for a substantial period of time until the area is decontaminated or until natural weathering or decay reduce the hazard. Planning must provide for the quick identification of a threat to the drinking water supply, notification of the public and private system operators, and warning of the users. Planners should also provide sewage system protection. A hazardous chemical entering the sewage system can cause serious and long-term damage. It may be necessary to divert sewage, creating another public health threat and environmental problems.

Function 10:   Fire and Rescue

Chain of command among fire fighters
List of available support systems
List of all tasks for fire fighters

Comment: This section lists all fire fighting tasks, as well as the chain of command for fire fighters. This chain of command is especially important if fire fighters from more than one jurisdiction will be involved. Planners should check to see if fire fighting tasks and the chain of command are mandated by their state law. Fire fighters should be trained in proper safety procedures when approaching a hazardous materials incident. They should have copies of DOT's *Emergency Response Guidebook* and know how to find shipping manifests in trucks, trains, and vessels. Specific information about protective equipment for fire fighters should be included here. (See Function 6, "Resource Management," and the *Occupational Safety and Health Guidance Manual for Hazardous Waste Site Activities*.)

This section should also identify any mutual aid or Good Samaritan agreements with neighboring fire departments, hazmat teams, and other support systems.

Function 11:   Law Enforcement

Chain of command for law enforcement officials
List of all tasks for law enforcement personnel

Comment: This section lists all the tasks for law enforcement personnel during an emergency response. Planners should check to see if specific law enforcement tasks are mandated by their state law. Because major emergencies will usually involve state, county, and local law enforcement personnel, and possibly the military, a clear chain of command must be determined in advance. Because they are frequently first on scene, law enforcement officials should be trained in proper procedures for approaching a hazardous materials incident. They should have copies of DOT's *Emergency Response Guidebook* and know how to find shipping manifests in trucks, trains, and vessels. Specific information about protective equipment for law enforcement officials should be included here. (See Function 6, "Resource Management," and the *Occupational Safety and Health Guidance Manual for Hazardous Waste Site Activities*.)

This section should include maps that indicate control points where police officers should be stationed in order to expedite the movement of responders toward the scene and of evacuees away from the scene, to restrict unnecessary traffic from entering the scene, and to control the possible spread of contamination.

Function 12:   Ongoing Incident Assessment

Field monitoring teams
Provision for environmental assessment, biological monitoring, and contamination surveys
Food/water controls

Comment: After the notification that a release has occurred, it is crucial to monitor the release and assess its impact, both on and off site. A detailed log of all sampling results should be maintained. Health officials should be kept informed of the situation. Often the facility at which the release has occurred will have the best equipment for this purpose.

This section should describe who is responsible to monitor the size, concentration, and movement of leaks, spills, and releases, and how they will do their work. Decisions about response personnel safety, citizen protection (whether indoor or through evacuation), and the use of food and water in the area will depend upon an accurate assessment of spill or plume movement and concentration. Similarly, decisions about containment and cleanup depend upon monitoring data.

Function 13:   Human Services

List of agencies providing human services
List of human services tasks

Comment: This section should coordinate the activities of organizations such as the Red Cross, Salvation Army, local church groups, and others that will help people during a hazardous materials emergency. These services are frequently performed by volunteers. Advance coordination is essential to ensure the most efficient use of limited resources.

Function 14:   Public Works

List of all tasks for public works personnel

Comment: This section lists all public works tasks during an emergency response. Public works officials should also be familiar with Plan Section D ("Containment and Cleanup").

Function 15:   Others

Comment: If the preceding list of functions does not adequately cover the various tasks to be performed during emergency responses, additional response functions can be developed.

Planning Element D: Containment and Cleanup

D.1:   Techniques for Spill Containment and Cleanup

Containment and mitigation actions

Cleanup methods
Restoration of the surrounding environment

Comment: Local responders will typically emphasize the containment and stabilization of an incident; state regulatory agencies can focus on cleanup details. Federal RRT agencies can provide assistance during the cleanup process. It is the releaser's legal and financial responsibility to clean up and minimize the risk to the health of the general public and workers that are involved. The federal OSC or other government officials should monitor the responsible party cleanup activities.

A clear and succinct list of appropriate containment and cleanup countermeasures should be prepared for each hazardous material present in the community in significant quantities. This section should be coordinated with the section on "Response Personnel Safety" so that response teams are subjected to minimal danger.

Planners should concentrate on the techniques that are applicable to the hazardous materials and terrain of their area. It may be helpful to include sketches and details on how cleanup should occur for certain areas where spills are more likely.

It is important to determine whether a fire should be extinguished or allowed to burn. Water used in fire fighting could become contaminated and then would need to be contained or possibly treated. In addition, some materials may be water-reactive and pose a greater hazard when in contact with water. Some vapors may condense into pools of liquid that must be contained and removed. Accumulated pools may be recovered with appropriate pumps, hoses, and storage containers. Various foams may be used to reduce vapor generation rates. Water sprays or fog may be applied at downwind points away from "cold" pools to absorb vapors and/or accelerate their dispersal in the atmosphere. (Sprays and fog might not reduce an explosive atmosphere.) Volatile liquids might be diluted or neutralized.

If a toxic vapor comes to the ground on crops, on playgrounds, in drinking water, or other places where humans are likely to be affected by it, the area should be tested for contamination. Appropriate steps must be taken if animals (including fish and birds) that may become part of the human food chain are in contact with a hazardous material. It is important to identify in advance what instruments and methods can be used to detect the material in question.

Restoration of the area is a long-range project, but general restoration steps should appear in the plan. Specific consideration should be given to the mitigation of damages to the environment.

D.2:   Resources for Cleanup and Disposal

Cleanup/disposal contractors and services provided
Cleanup material and equipment
Communications equipment
Provision for long-term site control during extended cleanups
Emergency transportation (e.g., aircraft, four-wheel-drive vehicles, boats)
Cleanup personnel
Personal protective equipment
Approved disposal sites

Comment: This section is similar to the yellow pages of the telephone book. It provides plan users with the following important information:

• What types of resources are available (public and private);
• How much is stockpiled;
• Where it is located (address and telephone number); and
• What steps are necessary to obtain the resources.

Organizations that may have resources for use during a hazardous materials incident include:

• Public agencies (e.g., fire, police, public works, public health, agriculture, fish and game);
• Industry (e.g., chemical producers, transporters, storers, associations; spill cleanup contractors; construction companies);
• Spill/equipment cooperatives; and
• Volunteer groups (ham radio operators, four-wheel-drive vehicle clubs).

Resource availability will change with time, so keep this section of the plan up-to-date.

Hazardous materials disposal may exceed the capabilities of smaller cities and towns; in such cases, the plan should indicate the appropriate state and/or federal agency that is responsible for making decisions regarding disposal.

Disposal of hazardous materials or wastes is controlled by a number of federal and state laws and regulations. Both CERCLA and RCRA regulate waste disposal and it is important that this section reflect the requirements of these regulations for on-site disposal, transportation, and off-site disposal. The plan should include an updated list of RCRA disposal facilities for possible use during an incident.

Many states have their own regulations regarding transport and ultimate disposal of hazardous waste. Usually such regulations are similar and substantially equal to federal regulations. Contact appropriate state agency offices for information on state requirements for hazardous waste disposal.

Planning Element E: Documentation and Investigative Follow-Up

List of required reports
Reasons for requiring the reports
Format for reports
Methods for determining whether the response mechanism worked properly
Provision for cost recovery

Comment: This section indicates what information should be gathered about the release and the response operation. Key response personnel could be instructed to maintain an accurate log of their activities. Actual response costs should be documented in order to facilitate cost recovery.

It is also important to identify who is responsible for the post-incident investigation to discover quickly the exact circumstances and cause of the release. Critiques of real incidents, if handled tactfully, allow improvements to be made based on actual experience. The documentation described above should help this investigation determine if response operations were effective, whether the emergency plan should be amended, and what follow-up responder and public training programs are needed.

Planning Element F: Procedures for Testing and Updating Plan

F.1:   Testing the Plan

Provision for regular tabletop, functional, and full-scale exercises

Comment: Exercises or drills are important tools in keeping a plan functionally up-to-date. These are simulated accidental releases where emergency response personnel act out their duties. The exercises can be tabletop and/or they can be realistic enough so that equipment is deployed, communication gear is tested, and "victims" are sent to hospitals with simulated injuries. Planners should work with local industry and the private medical community when conducting simulation exercises, and they should provide for drills that comply with state and local legal requirements concerning the content and frequency of drills. After the plan is tested, it should be revised and retested until the planning team is confident that the plan is ready. The public should be involved in or at least informed of these exercises. FEMA, EPA, and CMA provide guidance on simulation exercises through their training programs complementing this guide.

This section should specify:

(1) The organization in charge of the exercise;

(2) The types of exercises;

(3) The frequency of exercises; and

(4) A procedure for evaluating performance, making changes to plans, and correcting identified deficiencies in response capabilities as necessary. (See Chapter 6 of this guide.)

F.2:   Updating the Plan

Title and organization of responsible person(s)

Change notification procedures

How often the plan should be audited and what mechanisms will be used to change the plan

Comment: Responsibility should be delegated to someone to make sure that the plan is updated frequently and that all plan holders are informed of the changes. Notification of changes should be by written memorandum or letter; the changes should be recorded in the RECORD OF AMENDMENTS page at the front of the completed plan. Changes should be consecutively numbered for ease of tracking and accounting.

Following are examples of information that must regularly be checked for accuracy:

(1) Identity and phone numbers of response personnel

(2) Name, quantity, properties, and location of hazardous materials in the community. (If new hazardous materials are made, used, stored, or transported in the community, revise the plan as needed.)

(3) Facility maps

(4) Transportation routes

(5) Emergency services available

(6) Resource availability

This topic is considered in greater detail in Chapter 6 of this guidance.

Planning Element G: Hazards Analysis (Summary)

Identification of hazards
Analysis of vulnerability
Analysis of risk

Comment: This analysis is a crucial aspect of the planning process. It consists of determining where hazards are likely to exist, what places would most likely be adversely affected, what hazardous materials could be involved, and what conditions might exist during a spill or release. To prepare a hazards analysis, consult Chapter 3 of this guide, EPA's CEPP technical guidance, and DOT's *Community Teamwork* and *Lessons Learned*. Ask federal offices (listed in Appendix F) for information about available computer programs to assist in a hazards analysis.

Individual data sheets and maps for each facility and transportation routes of interest could be included in this section. Similar data could be included for recurrent shipments of hazardous materials through the area. This section will also assess the probability of damage and/or injury. In communities with a great deal of hazardous materials activity, the hazards analysis will be too massive to include in the emergency plan. In that case, all significant details should be summarized here.

Planning Element H: References

H.1:   Laboratory, Consultant, and Other Technical Support Resources

Telephone director of technical support services
Laboratories (environmental and public health)
Private consultants
Colleges or universities (chemistry departments and special courses)
Local chemical plants

Comment: This section should identify the various groups capable of providing technical support and the specific person to be contacted. Medical and environmental laboratory resources to assess the impact of the most probable materials that could be released should be identified. Note should be made about the ability of these laboratories to provide rapid analysis. These technical experts can provide advice during a disaster and also be of great service during the development of this plan. For this reason, one of the first planning steps should be gathering information for this section.

H.2:   Technical Library

List of references, their location, and their availability

* General planning references
* Specific references for hazardous materials
* Technical references and methods for using national data bases
* Maps

Comment: Industry sources can provide many specific publications dealing with hazardous materials. This section of the plan will list those published resources that are actually available in the community. Also list any maps (e.g., of facilities, transportation routes) that will aid in the response to an accidental spill or release.

The list of technical references in Appendix E could be helpful. Regional federal offices can also be contacted (see Appendix F).

It is important for planners to acquire, understand, and be able to use available hazardous materials data bases, including electronic data bases available from commercial and government sources. Planning guides such as DOT's *Community Teamwork*, CMA's CAER program, EPA's CEPP technical guidance, and this guide should also be available locally.

6. Plan Appraisal and Continuing Planning

6.1 Introduction

Any emergency plan must be evaluated and kept up-to-date through the review of actual responses, simulation exercises, and regular collection of new data. Effective emergency preparedness requires periodic review and evaluation, and the necessary effort must be sustained at the community level. Plans should reflect any recent changes in: the economy, land use, permit waivers, available technology, response capabilities, hazardous materials present, federal and state laws, local laws and ordinances, road configurations, population change, emergency telephone numbers, and facility location. This chapter describes key aspects of appraisal and provides specific guidance for maintaining an updated hazardous materials emergency plan.

6.2 Plan Review and Approval

Plan review and approval are critically important responsibilities of the planning team. This section discusses the various means by which a plan can be reviewed thoroughly and systematically.

6.2.1 Internal Review

The planning team, after drafting the plan, should conduct an internal review of the plan. It is not sufficient merely to read over the plan for clarity or to search for errors. The plan should also be assessed for adequacy and completeness. Appendix D is an adaptation of criteria developed by the National Response Team that includes questions useful in appraising emergency plans. Individual planning team members can use these questions to conduct self review of their own work and the team can assign a committee to review the total plan. In the case of a hazardous materials appendix (or appendices) to a multi-hazard EOP, the team will have to review the basic EOP as well as the functional annexes to obtain an overall assessment of content. Once the team accomplishes this internal review the plan should be revised in preparation for external review.

6.2.2 External Review

External review legitimizes the authority and fosters community acceptance of the plan. The review process should involve elements of peer review, upper level review, and community input. The planning team must devise a process to receive, review, and respond to comments from external reviewers.

A. Peer Review

Peer review entails finding qualified individuals who can provide objective reviews of the plan. Individuals with qualifications similar to those considered for inclusion on the planning team should be selected as peer reviewers. Examples of appropriate individuals include:

- The safety or environmental engineer in a local industry;
- Responsible authorities from other political jurisdictions (e.g., fire chief, police, environmental and/or health officers);
- A local college professor familiar with hazardous materials response operations; and

- A concerned citizen's group, such as the League of Women Voters, that provides a high level of objectivity along with the appropriate environmental awareness.

Exhibit 2 (Chapter 2) presents a comprehensive list of potential peer reviewers. Those selected as peer reviewers should use the criteria contained in Appendix D to develop their assessments of the plan.

B. Upper Level Review

Upper level review involves submitting the plan to an individual or group with oversight authority or responsibility for the plan. Upper level review should take place after peer review and modification of the plan.

C. Community Input

Community involvement is vital to success throughout the planning process. At the plan appraisal stage, such involvement greatly facilitates formal acceptance of the plan by the community. Approaches that can be used include:

- Community workshops with short presentations by planning team members followed by a question-and-answer period;
- Publication of notice "for comment" in local newspapers, offering interested individuals and groups an opportunity to express their views in writing;
- Public meetings at which citizens can submit oral and written comments;
- Invited reviews by key interest groups that provide an opportunity for direct participation for such groups that are not represented on the planning team; and
- Advisory councils composed of a relatively large number of interested parties that can independently review and comment on the planning team's efforts.

These activities do more than encourage community consensus building. Community outreach at this stage in the process also improves the soundness of the plan by increased public input and expands public understanding of the plan and thus the effectiveness of the emergency response to a hazardous materials incident.

D. State/Federal Review

After local review and testing through exercises, a community may want to request review of the plan by state and/or federal officials. Such a review will depend upon the availability of staff resources. Planning committees set up in accordance with Title III of SARA are to submit a copy of the emergency plan to the state emergency response commission for review to ensure coordination of the plan with emergency plans of other planning districts. Federal Regional Response Teams may review and comment upon an emergency plan, at the request of a local emergency planning committee. FEMA regional offices review FEMA-funded multi-hazard EOPs using criteria in CPG 1-8A.

6.2.3 Plan Approval

The planning team should identify and comply with any local or state requirements for formal plan approval. It may be necessary for local officials to enact legislation that gives legal recognition to the emergency plan.

6.3 Keeping the Plan Up-to-Date

All emergency plans become outdated because of social, economic, and environmental changes. Keeping the plan current is a difficult task, but can be achieved by scheduling reviews regularly. As noted in Chapter 5, the plan itself should indicate who is responsible for keeping it up-to-date. Outdated information should be replaced, and the results of appraisal exercises should be incorporated into the plan. The following techniques will aid in keeping abreast of relevant changes:

• Establish a regular review period, preferably every six months, but at least annually. (Title III of SARA requires an annual review.)

• Test the plan through regularly scheduled exercises (at least annually). This testing should include debriefing after the exercises whenever gaps in preparedness and response capabilities are identified.

• Publish a notice and announce a comment period for plan review and revisions.

• Maintain a list of individuals, agencies, and organizations that will be interested in participating in the review process.

• Make one reliable organization responsible for coordination of the review and overall stewardship of the plan. Use of the planning team in this role is recommended, but may not be a viable option due to time availability constraints of team members.

- Require immediate reporting by any facility of an increase in quantities of hazardous materials dealt with in the emergency plan, and require review and revision of plan if needed in response to such new information.

- Include a "Record of Amendments and Changes" sheet in the front section of the plan to help users of the plan stay abreast of all plan modifications.

- Include a "When and Where to Report Changes" notice in the plan and a request for holders of the plan to report any changes or suggested revisions to the responsible organization at the appropriate time.

- Make any sections of the plan that are subject to frequent changes either easily replaceable (e.g., looseleaf, separate appendix), or provide blank space (double- or triple-spaced typing) so that old material may be crossed out and new data easily written in. This applies particularly to telephone rosters and resource and equipment listings.

The organization responsible for review should do the following:

- Maintain a list of plan holders, based on the original distribution list, plus any new copies made or distributed. It is advisable to send out a periodic request to departments/branches showing who is on the distribution list and asking for any additions or corrections.

- Check all telephone numbers, persons named with particular responsibilities, and equipment locations and availability. In addition, ask departments and agencies to review sections of the plan defining their responsibilities and actions.

- Distribute changes. Changes should be consecutively numbered for ease of tracking. Be specific, e.g., "Replace page _____ with the attached new page _____ ." or "Cross out _____ on page _____ and write in the following" (new phone number, name, location, etc.). Any key change (new emergency phone number, change in equipment availability, etc.) should be distributed as soon as it occurs. Do not wait for the regular review period to notify plan holders.

- If possible, the use of electronic word processing is recommended because it facilitates changing the plan. After a significant number of individual changes, the entire plan should be redistributed to ensure completeness.

- If practical, request an acknowledgement of changes from those who have received changes. The best way to do this is to include a self-addressed postcard to be returned with acknowledgement (e.g., "I have received and entered changes _____ . Signed _____ ").

- Attend any plan critique meetings and issue changes as may be required.
- Integrate changes with other related plans.

6.4 Continuing Planning

In addition to the periodic updates described above, exercises, incident reviews, and training are necessary to ensure current and effective planning.

6.4.1 Exercises

The plan should also be evaluated through exercises to see if its required activities are effective in practice and if the evaluation would reveal more efficient ways of responding to a real emergency. As noted in Chapter 5, the plan itself should indicate who is responsible for conducting exercises. Simulations can be full-scale, functional, or tabletop exercises.

A full-scale exercise is a mock emergency in which the response organizations that would be involved in an actual emergency perform the actions they would take in the emergency. These simulations may focus on limited objectives (e.g., testing the capability of local hospitals to handle relocation problems). The responsible environmental, public safety, and health agencies simulate, as realistically as possible, notification, hazards identification and analysis, command structure, command post staging, communications, health care, containment, evacuation of affected areas, cleanup, and documentation. Responders use the protective gear, radios, and response equipment and act as they would in a real incident. These multi-agency exercises provide a clearer understanding of the roles and resources of each responder.

A functional exercise involves testing or evaluating the capability of individual or multiple functions, or activities within a function.

A low-cost, valuable version of an exercise is the staging of a tabletop exercise. In this exercise, each agency representative describes and acts out what he or she would do at each step of the response under the circumstances given.

Exercises are most beneficial when followed by a meeting of all participants to critique the performance of those involved and the strengths and weaknesses of the plan's operation. The use of an outside reviewer, free of local biases, is desirable. The emergency plan should be amended according to the lessons learned. Provisions should be made to follow up exercises to see that identified deficiencies are corrected.

Communities that want to help in preparing and conducting exercises should consult FEMA's four-volume "Exercise Design Course," which includes sample hazardous materials exercises. CMA's *Community Emergency Response Exercise Handbook* is also helpful. CMA describes four types of exercises: tabletop, emergency operations simulation, drill, and field exercise.

6.4.2 Incident Review

When a hazardous materials incident does occur, a review or critique of the incident is a means of evaluating the plan's effectiveness. Recommendations for conducting an incident review are:

- Assign responsibility for incident review to the same organization that is responsible for plan update, for example, the planning team.
- Conduct the review only after the emergency is under control and sufficient time has passed to allow emergency respondents to be objective about the incident.
- Use questionnaires, telephone interviews, or personal interviews to obtain comments and suggestions from emergency respondents. Follow-up on non-respondents.
- Identify plan and response deficiencies: items that were overlooked, improperly identified, or were not effective.
- Convene the planning team to review comments and make appropriate plan changes.
- Revise the plan as necessary. Communicate personal or departmental deficiencies informally to the appropriate person or department. Follow up to see that deficiencies are corrected.

6.4.3 Training

Training courses can help with continuing planning by sharpening response personnel skills, presenting up-to-date ideas/techniques, and promoting contact with other people involved in emergency response. Everyone who occupies a position that is identified in the plan must have appropriate training. This applies to persons at all levels who serve to coordinate or have responsibilities under the plan, both those directly and indirectly involved at the scene of an incident. One should not assume that a physician in the emergency room or a professional environmentalist is specifically trained to perform his/her assigned mission during an emergency.

The training could be a short briefing on specific roles and responsibilities, or a seminar on the plan or on emergency planning and response in general. However the training is conducted, it should convey a full appreciation of the importance of each role and the effect that each person has on implementing an effective emergency response.

Training is available from a variety of sources in the public and private sectors. At the federal level, EPA, FEMA, OSHA, DOT/RSPA and the USCG offer hazardous materials training. (In some cases, there are limits on attendance in these courses.) FEMA, EPA, and other NRT agencies cooperatively offer the inter-agency "train-the-trainer" course, Hazardous Materials Contingency Planning, at Emmitsburg, MD and in the field.

Title III of SARA authorizes federal funding for training. Communities seeking training assistance should consult appropriate state agencies. States may consult with the RRT and the various federal regional and district offices. (See Appendix F.)

In addition to government agencies, consult universities or community colleges (especially any fire science curriculum courses), industry associations, special interest groups, and the private sector (fixed facilities, shippers, and carriers). Many training films and slide presentations can be borrowed or rented at little cost. Many chemical companies and carriers provide some level of training free.

The Chemical Manufacturers Association has a lending library of audio-visual training aids for use by personnel who respond to emergencies involving chemicals. The training aids are available on a loan basis at no charge to emergency response personnel and the public sector.

Training aids can also be purchased from:

National Chemical Response and Information Center
Chemical Manufacturers Association
2501 M Street, N.W.
Washington, DC 20037

In addition to classroom training, response personnel will need hands-on experience with equipment to be used during an emergency.

Communities should provide for refresher training of response personnel. It is not sufficient to attend training only once. Training must be carried out on a continuing basis to ensure currency and capability. Some communities have found it effective to hold this refresher training in conjunction with an exercise.

The NRT, through its member agencies, is developing a strategy to address issues related to emergency preparedness and response for hazardous materials incidents. The training strategy includes: (1) improved coordination of available federal training programs and courses; (2) sharing information about available training, and lessons learned from responses to recent hazardous materials incidents; (3) the increased use of exercises as a training method; (4) the revision of existing core courses, and the development of any needed new core courses that prepare responders to do the actual tasks expected in their own communities; and (5) decentralizing the delivery of training so that it is more easily available to responders. Further information about this training strategy can be obtained from EPA or FEMA offices in Washington, DC.

Appendix A

Implementing Title III: Emergency Planning and Community Right-to-know

Superfund Amendments and Reauthorization Act of 1986

On October 17, 1986, the President signed the "Superfund Amendments and Reauthorization Act of 1986" (SARA) into law. One part of the new SARA provisions is Title III: the "Emergency Planning and Community Right-to-Know Act of 1986." Title III establishes requirements for Federal, State, and local governments and industry regarding emergency planning and community right-to-know reporting on hazardous chemicals. This legislation builds upon the Environmental Protection Agency's (EPA's) Chemical Emergency Preparedness Program (CEPP) and numerous State and local programs aimed at helping communities to meet their responsibilities in regard to potential chemical emergencies.

Title III has four major sections: emergency planning (§301-303), emergency notification (§304), community right-to-know reporting requirements (§311, 312), and toxic chemical release reporting–emissions inventory (§313). The sections are interrelated in a way that unifies the emergency planning and community right-to-know provisions of Title III. (See Exhibit 6.)

In addition to increasing the public's knowledge and access to information on the presence of hazardous chemicals in their communities and releases of these chemicals into the environment, the community right-to-know pro-visions of Title III will be important in preparing emergency plans.

This appendix includes a summary of these four major sections, followed by a discussion of other Title III topics of interest to emergency planners.

Sections 301-303: Emergency Planning

The emergency planning sections are designed to develop state and local government emergency preparedness and response capabilities through bet-ter coordination and planning, especially at the local level.

Title III required that the governor of each state designate a state emer-gency response commission (SERC) by April 17, 1987. While existing state organizations can be designated as the SERC, the commission should have broad-based representation. Public agencies and departments concerned with issues relating to the environment, natural resources, emergency management, public health, occupational safety, and transportation all have important roles in Title III activities.

Various public and private sector groups and associations with interest and experience in Title III issues can also be included on the SERC.

The SERC must designate local emergency planning districts, and appoint local emergency planning committees (LEPCs) within one month after a dis-trict is designated. The SERC is responsible for supervising and coordinat-ing the activities of the LEPCs, for establishing procedures for receiving and processing public requests for information collected under other sections of Title III, and for reviewing local emergency plans.

The LEPC must include elected state and local officials, police, fire, civil defense, public health professionals, environmental, hospital, and transpor-tation officials as well as representatives of facilities, community groups, and the media. Interested persons may petition the SERC to modify the mem-bership of an LEPC.

No later than September 17, 1987, facilities subject to the emergency plan-ning requirements were to notify the LEPC of a representative who will par-ticipate in the planning process as a facility emergency coordinator.

Facility emergency coordinators will be of great service to LEPCs. For example, they can provide technical assistance, an understanding of facil-ity response procedures, information about chemicals and their potential effects on nearby persons and the environment, and response training opportuni-ties. CEPP experience revealed that, as a result of CMA's CAER initiative, there already exist a large number of plant managers and other facility per-sonnel who want to cooperate with local community planners.

The LEPC must establish rules, give public notice of its activities, and establish procedures for handling public requests for information.

The LEPC's primary responsibility is to develop an emergency response plan. In developing this plan, the local committee will evaluate available resources for preparing for and responding to a potential chemical accident. The plan must include:

- Identification of facilities and extremely hazardous substances transportation routes;
- Emergency response procedures, on site and off site;
- Designation of a community coordinator and facility coordinator(s) to implement the plan;
- Emergency notification procedures;
- Methods for determining the occurrence of a release and the probable affected area and population;
- Description of community and industry emergency equipment and facilities, and the identity of persons responsible for them;
- Evacuation plans;
- Description and schedules of a training program for emergency response to chemical emergencies; and
- Methods and schedules for exercising emergency response plans.

To assist the LEPC in preparing and reviewing plans, Congress required the National Response Team (NRT), composed of 14 federal agencies with emergency preparedness and response responsibilities, to publish guidance on emergency planning. This *Hazardous Materials Emergency Planning Guide* is being published by the NRT to fulfill this requirement.

The emergency plan must be reviewed by the SERC upon completion and reviewed annually by the LEPC. The Regional Response Teams (RRTs), composed of federal regional officials and state representatives, may review the plans and provide assistance if the LEPC so requests.

The emergency planning activities of the LEPC and facilities should initially be focused on, but not limited to, the extremely hazardous substances published as an interim final rule in the November 17, 1986, *Federal Register*. The list included the threshold planning quantity (TPQ) for each substance. EPA can revise the list and TPQs but must take into account the toxicity, reactivity, volatility, dispersability, combustibility, or flammability of a substance. Consult EPA regional offices for a copy of the Title III (Section 302) list of extremely hazardous substances.

Any facility that produces, uses, or stores any of the listed chemicals in a quantity greater than the TPQ must meet all emergency planning requirements. In addition, the SERC or the governor can designate additional facilities, after public comment, to be subject to these requirements. Facilities must notify the SERC that they are subject to these requirements. If, after such notification, a facility first begins to produce, use, or store an extremely hazardous substance in an amount exceeding the threshold planning quantity, it must notify the SERC and LEPC within 60 days.

Each SERC must notify EPA regional offices of all facilities subject to Title III planning requirements.

In order to complete information on many sections of the emergency plan, the LEPC will require data from the facilities covered under the plan. Title III provides authority for the LEPC to secure from a facility information that it needs for emergency planning and response. This is provided by Section 303(d)(3), which states that:

"Upon request from the emergency planning committee, the owner or operator of the facility shall promptly provide information to such committee necessary for developing and implementing the emergency plan."

Within the trade secret restrictions contained in Section 322, LEPCs should be able to use this authority to secure from any facility subject to the planning provisions of the law information needed for such mandatory plan contents as: facility equipment and emergency response capabilities, facility emergency response personnel, and facility evacuation plans.

Some of the facilities subject to Section 302 planning requirements may not be subject to Sections 311-12 reporting requirements, which are currently limited to manufacturers and importers in SIC codes 20-39. LEPCs may use Section 303(d)(3) authority to gain information such as name(s), MSDSs, and quantity and location of chemicals present at facilities subject to Section 302.

Section 304: Emergency Notification

If a facility produces, uses, or stores one or more hazardous chemical, it must immediately notify the LEPC and the SERC if there is a release of a listed hazardous substance that exceeds the reportable quantity for that substance. Substances subject to this notification requirement include substances on the list of extremely hazardous substances published in the *Federal Register* on November 17, 1986, and substances subject to the emergency notification requirements of CERCLA Section 103(a).

Information included in this initial notification (as well as the additional information in the follow-up written notice described below) can be used by the LEPC to prepare and/or revise the emergency plan. This information should be especially helpful in meeting the requirement to list methods for determining if a release has occurred and identifying the area and population most likely to be affected.

The initial notification of a release can be by telephone, radio, or in person. Emergency notification requirements involving transportation incidents may be satisfied by dialing 911 or, in the absence of a 911 emergency number, calling the operator.

This emergency notification needs to include: the chemical name; an indication of whether the substance is an extremely hazardous substance; an estimate of the quantity released into the environment; the time and duration of the release; the medium into which the release occurred; any known or anticipated acute or chronic health risks associated with the emergency and, where appropriate, advice regarding medical attention necessary for exposed individuals; proper precautions, such as evacuation; and the name and telephone number of a contact person.

Section 304 also requires a follow-up written emergency notice after the release. The follow-up notice or notices shall update information included in the initial notice and provide additional information on actual response actions taken, any known or anticipated data on chronic health risks associated with the release, and advice regarding medical attention necessary for exposed individuals.

The requirement for emergency notification comes into effect with the establishment of the SERC and LEPC. If no SERC is established by April 17, 1987, the governor becomes the SERC and notification should be made to him/her. If no LEPC is established by August 17, 1987, local notification must be made to the appropriate local emergency response personnel, such as the fire department.

Sections 311-312: Community Right-to-Know Reporting Requirements

As noted above, Section 303(d)(3) gives LEPCs access to information from facilities subject to Title III planning requirements. Sections 311-12 provide information about the nature, quantity, and location of chemicals at many facilities not subject to the Section 303(d)(3) requirement. For this reason, LEPCs will find information in Sections 311-12 especially helpful when preparing a comprehensive plan for the entire planning district.

There are two community right-to-know reporting requirements. Section 311 requires a facility which must prepare or have available material safety data sheets (MSDSs) under the Occupational Safety and Health Administration (OSHA) hazard communications regulations to submit either copies of its MSDSs or a list of MSDS chemicals to the LEPC, the SERC, and the local fire department. Currently, only facilities in Standard Industrial Classification (SIC) Codes 20-39 (manufacturers and importers) are subject to these OSHA regulations.

The initial submission of the MSDSs or list was required no later than October 17, 1987, or 3 months after the facility is required to prepare or have available an MSDS under OSHA regulations. A revised MSDS must be provided to update an MSDS which was originally submitted if significant new information regarding a chemical is discovered.

EPA encourages LEPCs and fire departments seriously to consider contacting facilities prior to the deadline of October 17, 1987 to request the submission of lists rather than MSDS forms. In communities with a large number of facilities, handling large numbers of chemicals, and in communities with limited capabilities to store and manage the MSDSs, the list of MSDS chemicals from the facility would be more useful than the forms themselves, and likely to be more easily produced.

LEPCs also have the option of using the chemical names provided to develop additional data on each of the chemicals, using a variety of data sources, including several on-line data bases maintained by agencies of the federal government.

Specific MSDSs could be requested on chemicals that are of particular concern. In general every MSDS will provide the LEPC and the fire departments in each community with the following information on each of the chemicals covered:

- The chemical name;
- Its basic characteristics, for example:
  - toxicity, corrosivity, reactivity,
  - known health effects, including chronic effects from exposure,
  - basic precautions in handling, storage, and use,
  - basic countermeasures to take in the event of a fire, explosion, leak, and
  - basic protective equipment to minimize exposure.

In any case, these data should be useful for the planning to be accompanied by the LEPC and first responders, especially fire departments and hazmat teams. Both hazards analysis and the development of emergency countermeasures should be facilitated by the availability of MSDS information.

If the facility owner or operator chooses to submit a list of MSDS chemicals, the list must include the chemical name or common name of each substance and any hazardous component as provided on the MSDS. This list must be organized in categories of health and physical hazards as set forth in OSHA regulations or as modified by EPA.

If a list is submitted, the facility must provide the MSDS for any chemical on the list upon the request of the LEPC. Under Section 311, EPA may establish threshold quantities for hazardous chemicals below which no facility must report.

The reporting requirement of Section 312 requires facilities to submit an emergency and hazardous chemical inventory form to the LEPC, the SERC, and the local fire department. The hazardous chemicals covered by Section 312 are the same chemicals for which facilities are required to submit MSDS forms or the list for Section 311.

Under Sections 311-12, EPA may establish threshold quantities for hazardous chemicals below which no facility is subject to this requirement. See the proposed rule in the January 27, 1987 *Federal Register*. The Final Rule will be published before October 1987.

The inventory form incorporates a two-tier approach. Under Tier I, facilities must submit the following aggregate information for each applicable OSHA category of health and physical hazard:

- An estimate (in ranges) of the maximum amount of chemicals for each category present at the facility at any time during the preceding calendar year;
- An estimate (in ranges) of the average daily amount of chemicals in each category; and
- The general location of hazardous chemicals in each category.

Tier I information shall be submitted on or before March 1, 1988 and annually thereafter on March 1.

The public may also request additional information for specific facilities from the SERC and LEPC. Upon the request of the LEPC, the SERC, or the local fire department, the facility must provide the following Tier II information for each covered substance to the organization making the request:

- The chemical name or the common name as indicated on the MSDS;
- An estimate (in ranges) of the maximum amount of the chemical present at any time during the preceding calendar year;
- A brief description of the manner of storage of the chemical;
- The location of the chemical at the facility; and
- An indication of whether the owner elects to withhold information from disclosure to the public.

The information submitted by facilities under Sections 311 and 312 must generally be made available to the public by local and state governments during normal working hours.

As in the case of the MSDS data, this Section 312 information may be useful for LEPCs interested in extending the scope of their planning beyond the facilities covered by Section 302, and for reviewing and updating existing plans. Section 312 information about the quantity and location of chemicals can be of use to fire departments in the development of pre-fire plans. Section 312 data may be of limited use in the initial planning process, given the fact that initial emergency plans are to be completed by October 17, 1988, but they will be useful for the subsequent review and update of plans. Facility owners or operators, at the request of the fire department, must allow the fire department to conduct an on-site inspection and provide specific information about the location of hazardous chemicals.

Section 313: Toxic Chemical Release Reporting

Section 313 of Title III requires EPA to establish an inventory of toxic chemical emissions from certain facilities. Facilities subject to this reporting requirement must complete a toxic chemical release form (to be prepared by EPA by June 1987) for specified chemicals. The form must be submitted to EPA and those state officials designated by the governor on or before July 1, 1988, and annually thereafter on July 1, reflecting releases during each preceding calendar year.

The purpose of this reporting requirement is to inform government officials and the public about releases of toxic chemicals into the environment. It will also assist in research and the development of regulations, guidelines, and standards.

The reporting requirement applies to owners and operators of facilities that have 10 or more full-time employees, that are in Standard Industrial Classification (SIC) Codes 20 through 39, and that manufactured, processed, or otherwise used a listed toxic chemical in excess of specified threshold quantities. The SIC Codes mentioned cover basically all manufacturing industries.

Facilities using listed toxic chemicals in quantities over 10,000 pounds in a calendar year are required to submit toxic chemical release forms by July 1 of the following year. Facilities manufacturing or processing any of these chemicals in excess of 75,000 pounds in 1987 must report by July 1, 1988. Facilities manufacturing or processing in excess of 50,000 pounds in 1988 must report by July 1, 1989. Thereafter, facilities manufacturing or processing more than 25,000 pounds in a year are required to submit the form. EPA can revise these threshold quantities and the SIC categories involved.

The list of toxic chemicals subject to reporting consists initially of chemicals listed for similar reporting purposes by the states of New Jersey and Maryland. There are over 300 chemicals and categories on these lists. EPA can modify this combined list. In adding a chemical to the combined Maryland and New Jersey lists, EPA must consider the following factors:

(1) Is the substance known to cause cancer or serious reproductive or neurological disorders, genetic mutations, or other chronic health effects?
(2) Can the substance cause significant adverse acute health effects as a result of continuous or frequently recurring releases?
(3) Can the substance cause an adverse effect on the environment because of its toxicity, persistence, or tendency to bioaccumulate?

Chemicals can be deleted if there is not sufficient evidence to establish any of these factors. State governors or any other person may petition the EPA administrator to add or delete a chemical from the list for any of the above reasons. EPA must either publish its reasons for denying the petition, or initiate action to implement the petition within 180 days.

Through early consultation with states or EPA regions, petitioners can avoid duplicating previous petitions and be assisted in locating sources of data already collected on the problem of concern and data sources to support their petitions. EPA will conduct information searches on chemicals contained in a petition, focusing on the effects the petitioners believe warrant addition or deletion.

The toxic chemical release form includes the following information for released chemicals:

• The name, location, and type of business;
• Whether the chemical is manufactured, processed, or otherwise used and the general categories of use of the chemical;
• An estimate (in ranges) of the maximum amounts of the toxic chemical present at the facility at any time during the preceding year;
• Waste treatment and disposal methods and the efficiency of methods for each wastestream;
• The quantity of the chemical entering each environmental medium annually; and
• A certification by a senior official that the report is complete and accurate.

EPA must establish and maintain a national toxic chemical inventory based on the data submitted. This information must be computer accessible on a national database.

In general these Section 313 reports appear to be of limited value in emergency planning. Over time, however, they may contain information that can be used by local planners in developing a more complete understanding of the total spectrum of hazards that a given facility may pose to a community. These reports will not be available to states until July 1, 1988. These reports do not go to the LEPCs directly but they are likely to become available if the LEPCs request them from the states.

Other Title III Provisions

In addition to these four major sections of Title III, there are other provisions of interest to local communities.

Preemption

Section 321 stipulates that (with the exception of the MSDS format and content required by Section 311) Title III does not preempt any state and local laws. In effect, Title III imposes minimum planning and reporting standards where no such standards (or less stringent standards) exist, while permitting states and localities to pursue more stringent requirements as they deem appropriate.

Trade Secrets

Section 322 of Title III addresses trade secrets and applies to Section 303 emergency planning and Sections 311, 312, 313 regarding planning information, community right-to-know reporting requirements, and toxic chemical release reporting. Any person may withhold the specific chemical identity of an extremely hazardous substance or toxic chemical for specific reasons. Even if the chemical identity is withheld, the generic class or category of the chemical must be provided. Such information may be withheld if the facility submits the withheld information to EPA along with an explanation of why the information is a trade secret. The information may not be withheld as a trade secret unless the facility shows each of the following:

- The information has not been disclosed to any other person other than a member of the LEPC, a government official, an employee of such person, or someone bound by a confidentiality agreement, and that measures have been taken to protect the confidentiality;

- The information is not required to be disclosed to the public under any other federal or state law;
- The information is likely to cause substantial harm to the competitive position of the person; and
- The chemical identity could not reasonably be discovered by anyone in the absence of disclosure.

Even if information can be legally withheld from the public, Section 323 requires it not to be withheld from health professionals who require the information for diagnostic purposes or from local health officials who require the information for assessment activities. In these cases, the person receiving the information must be willing to sign a confidentiality agreement with the facility.

Information claimed as trade secret and substantiation for that claim must be submitted to EPA. People may challenge trade secret claims by petitioning EPA, which must then review the claim and rule on its validity.

EPA will publish regulations governing trade secret claims. The regulations will cover the process for submission of claims, petitions for disclosure, and a review process for these petitions.

Enforcement

Section 325 identifies the following enforcement procedures:

- Civil penalties for facility owners or operators who fail to comply with emergency planning requirements;
- Civil, administrative, and criminal penalties for owners or operators who fail to comply with the emergency notification requirements of Section 304;
- Civil and administrative penalties for owners or operators who fail to comply with the reporting requirements in Sections 311-313;
- Civil and administrative penalties for frivolous trade secret claims; and
- Criminal penalties for the disclosure of trade secret information.

In addition to the federal government, state and local governments and individual citizens may enforce the provisions of Title III through the citizen suit authority provided in Section 326.

Training

Section 305 mandates that federal emergency training programs must emphasize hazardous chemicals. It also authorizes the Federal Emergency Management Agency (FEMA) to provide $5 million for each of fiscal years 1987, 1988, 1989, and 1990 for training grants to support state and local governments. These training grants are designed to improve emergency planning, preparedness, mitigation, response, and recovery capabilities. Such programs must give special emphasis to hazardous chemical emergencies. The training grants may not exceed 80 percent of the cost of any such programs. The remaining 20 percent must come from non-federal sources. Consult FEMA and/or EPA Regional offices for a list of training courses.

Review of Emergency Systems

Under Section 305, EPA has initiated a review of emergency systems for monitoring, detecting, and preventing releases of extremely hazardous substances at representative facilities that produce, use, or store these substances. It also is examining public alert systems. EPA will report interim findings to the Congress no later than May 17, 1987 and issue a final report of findings and recommendations to the Congress by April 17, 1988.

The report must include EPA's findings regarding each of the following:

- Status of current technological capabilities to 1) monitor, detect, and prevent significant releases of extremely hazardous substances; 2) determine the magnitude and direction of the hazard posed by each release; 3) identify specific substances; 4) provide data on the specific chemical composition of such releases; and 5) determine relative concentrations of the constituent substances;
- Status of public emergency alert devices or systems for effective public warning of accidental releases of extremely hazardous substances into any media; and
- The technical and economic feasibility of establishing, maintaining, and operating alert systems for detecting releases.

The report must also include EPA's recommendations for the following:

- Initiatives to support development of new or improved technologies or systems that would assist the timely monitoring, detection, and prevention of releases of extremely hazardous substances; and
- Improving devices or systems for effectively alerting the public in the event of an accidental release.

Exhibit 5. Key Title III Dates

The following is a list of some key dates relative to the implementation of the "Emergency Planning and Community Right-to-Know Act of 1986."

Date	Event
November 17, 1986	EPA publishes interim final List of Extremely Hazardous Substances and their Threshold Planning Quantities in *Federal Register* [§302(a)(2–3)]
November 17, 1986	EPA initiates comprehensive review of emergency systems [§ 305(b)]
January 27, 1987	EPA publishes proposed formats for emergency inventory forms and reporting requirements in *Federal Register* (§311–12)
March 17, 1987	National Response Team publishes guidance for preparation and implementation of emergency plans [§ 303 (f)]
April 17, 1987	State Governors appoint SERCs [§ 301(a)]
May 17, 1987	Facilities subject to Section 302 planning requirements notify SERC [§ 302(c)]
June 1, 1987	EPA publishes toxic chemicals release (i.e., emissions inventory) form [§ 302(c)]
July 17, 1987	SERC designates emergency planning districts [§ 301(b)]
August 17, 1987 (or 30 days after designation of districts, whichever is sooner)	SERC appoints members of LEPCs [§ 301(c)]
September 17, 1987 (or 30 days after local committee is formed, whichever is first)	Facility notifies LEPC of selection of a facility representative to serve as facility emergency coordinator [§303(d)(1)]
October 17, 1987	MSDSs or list of MSDS chemicals submitted to SERC, LEPC, and local fire department [§ 311(d)]
March 1, 1988	Facilities submit their initial emergency inventory forms to SERC, LEPC, and local fire department [§ 312(a) (2)]
April 17, 1988	Final report on emergency systems study due to Congress [§ 305(b)]
July 1, 1988 (and annually hereafter)	Facilities to submit initial toxic chemical release forms to EPA and designated State officials [§ 313(a)]
October 17, 1988	LEPCs complete preparation of an emergency plan [§ 303(a)]

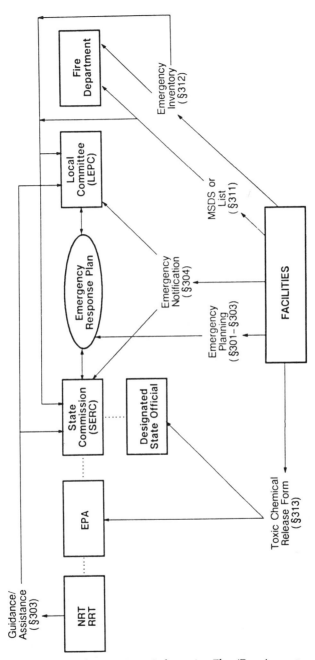

Exhibit 6. Title III—Major Information Flow/Requirements.

Exhibit 7. Information from Facilities Provided by Title III in Support of LEPC Plan Development

Information Generated by Title III Compliance	Authority	How LEPC Can Use the Information
Facilities subject to Title III planning requirements (including those designated by the Governor or SERC)	Section 302; Notice from Governor/SERC	Hazards analysis—Hazards identification
Additional facilities near subject facilities (such as hospitals, natural gas facilities, etc.)	Sections 302(b)(2); 303(c)(1)	Hazards analysis—Vulnerability analysis
Transportation routes	Sections 303(c)(1); 303(d)(3)	Hazards analysis—Hazards identification
Major chemical hazards (chemical name, properties, location, and quantity	Section 303(d)(e) for extremely hazardous substances used, produced, stored	Hazards analysis—Hazards identification
	Section 311 MSDSs for chemicals manufactured or imported	
	Section 312 inventories for chemicals manufactured or imported	
Facility and community response methods, procedures, and personnel	Sections 303(c)(2); 303(d)(3)	Response functions
Facility and community emergency coordinators	Sections 303(c)(3); 303(d)(1)	Assistance in preparing and implementing the plan
Release detection and notification procedures	Sections 303(c)(4); 303(d)(3)	Initial notification Warning system
Methods for determining release occurrence and population affected	Sections 303(c)(5); 303(d)(3)	Hazards analysis—Vulnerability analysis and risk analysis
Facility equipment and emergency facilities; persons responsible for such equipment and facilities	Sections 303(c)(6); 303(d)(3)	Resource management
Evacuation plans	Sections 303(c)(7); 303(d)(3)	Evacuation planning
Training programs	Sections 303(c)(8); 303(d)(3)	Resource management
Exercise methods and schedules	Sections 303(c)(9); 303(d)(3)	Testing and updating

Exhibit 8. Title III Chemical Lists and Their Purposes

List	Required in Section	Purpose
Extremely Hazardous Substances (*Federal Register* 11/7/86—initially 402 chemicals listed in CEPP Interim Guidance)	Section 302: Emergency Planning	Facilities with more than established planning quantities of these substances must notify the SERC. Initial focus for preparation of emergency plans by LEPCs.
	Section 304: Emergency Notification	Certain releases of these chemicals trigger Section 304 notification to SERC and LEPC.
Substance requiring notification under Section 103(a) of CERCLA (717 chemicals)	Section 304: Emergency Notification	Certain releases of these chemicals trigger Section 304 notification to SERC and LEPC as well as CERCLA Section 103(a) requirement to notify National Response Center.
	Section 304: Emergency Notification	Identifies facilities subject to emergency notification requirements
Hazardous Chemicals considered physical or health hazards under OSHA's *Hazard Communication Standard* (This is a performance standard, there is no specific list of chemicals.)	Section 311: Material Safety Data Sheets	MSDS or list of MSDS chemicals provided by facilities to SERC, LEPC, and local fire department
	Section 312: Emergency and Hazardous Chemical Inventory	Covered facilities provide site-specific information on the quality and location of chemicals to SERC, LEPC, and local fire departments to inform the community and assist in plan preparation.
Toxic Chemicals identified as chemicals of concern by States of New Jersey and Maryland (329 chemicals/chemical categories)	Section 313: Toxic Chemical Release Reporting	These chemicals are reported on an emissions inventory to inform government officials and the public about releases of toxic chemicals in the environment.

Appendix B

List of Acronyms and Recognized Abbreviations

AAR/BOE	Association of American Railroads/Bureau of Explosives
AIChE	American Institute of Chemical Engineers
ASCS	Agricultural Stabilization and Conservation Service
ASME	American Society of Mechanical Engineers
ASSE	American Society of Safety Engineers
ATSDR	Agency for Toxic Substances and Disease Registry (HHS)
CAER	Community Awareness and Emergency Response (CMA)
CDC	Centers for Disease Control (HHS)
CEPP	Chemical Emergency Preparedness Program
CERCLA	Comprehensive Environmental Response, Compensation, and Liability Act of 1980 (PL 96-510)
CFR	Code of Federal Regulations
CHEMNET	A mutual aid network of chemical shippers and contractors
CHEMTREC	Chemical Transportation Emergency Center
CHLOREP	A mutual aid group comprised of shippers and carriers of chlorine.
CHRIS/HACS	Chemical Hazards Response Information System/Hazard Assessment Computer System
CMA	Chemical Manufacturers Association
CPG 1-3	Federal Assistance Handbook: Emergency Management, Direction and Control Programs
CPG 1-8	Guide for Development of State and Local Emergency Operations Plans
CPG 1-8A	Guide for the Review of State and Local Emergency Operations Plans
CWA	Clean Water Act
DOC	U.S. Department of Commerce
DOD	U.S. Department of Defense
DOE	U.S. Department of Energy
DOI	U.S. Department of the Interior
DOJ	U.S. Department of Justice
DOL	U.S. Department of Labor
DOS	U.S. Department of State

DOT	U.S. Department of Transportation
EENET	Emergency Education Network (FEMA)
EMA	Emergency Management Agency
EMI	Emergency Management Institute
EOC	Emergency Operating Center
EOP	Emergency Operations Plan
EPA	U.S. Environmental Protection Agency
ERD	Emergency Response Division (EPA)
FEMA	Federal Emergency Management Agency
FEMA-REP-5	Guidance for Developing State and Local Radiological Emergency Response Plans and Preparedness for Transportation Accidents
FWPCA	Federal Water Pollution Control Act
HAZMAT	Hazardous Materials
HAZOP	Hazard and Operability Study
HHS	U.S. Department of Health and Human Services
ICS	Incident Command System
IEMS	Integrated Emergency Management System
LEPC	Local Emergency Planning Committee
MSDS	Material Safety Data Sheet
NACA	National Agricultural Chemicals Association
NCP	National Contingency Plan
NCRIC	National Chemical Response and Information Center (CMA)
NETC	National Emergency Training Center
NFA	National Fire Academy
NFPA	National Fire Protection Association
NIOSH	National Institute of Occupational Safety and Health
NOAA	National Oceanic and Atmospheric Administration
NRC	U.S. Nuclear Regulatory Commission; National Response Center
NRT	National Response Team
NUREG 0654/ FEMA-REP-1	Criteria for Preparation and Evaluation of Radiological Emergency Response Plans and Preparedness in Support of Nuclear Power Plants
OHMTADS	Oil and Hazardous Materials Technical Assistance Data System
OSC	On-Scene Coordinator
OSHA	Occupational Safety and Health Administration (DOL)
PSTN	Pesticide Safety Team Network

RCRA	Resource Conservation and Recovery Act
RQs	Reportable Quantities
RRT	Regional Response Team
RSPA	Research and Special Programs Administration (DOT)
SARA	Superfund Amendments and Reauthorization Act of 1986 (PL 99-499)
SCBA	Self-Contained Breathing Apparatus
SERC	State Emergency Response Commission
SPCC	Spill Prevention Control and Countermeasures
TSD	Treatment, Storage, and Disposal Facilities
USCG	U.S. Coast Guard (DOT)
USDA	U.S. Department of Agriculture
USGS	U.S. Geological Survey
USNRC	U.S. Nuclear Regulatory Commission

Appendix C

Glossary

CAER   Community Awareness and Emergency Response program developed by the Chemical Manufacturers Association. Guidance for chemical plant managers to assist them in taking the initiative in cooperating with local communities to develop integrated (community/industry) hazardous materials response plans.

CEPP   Chemical Emergency Preparedness Program developed by EPA to address accidental releases of acutely toxic chemicals.

CERCLA   Comprehensive Environmental Response, Compensation, and Liability Act regarding hazardous substance releases into the environment and the cleanup of inactive hazardous waste disposal sites.

CHEMNET   A mutual aid network of chemical shippers and contractors. CHEMNET has more than fifty participating companies with emergency teams, twenty-three subscribers (who receive services in an incident from a participant and then reimburse response and cleanup costs), and several emergency response contractors. CHEMNET is activated when a member shipper cannot respond promptly to an incident involving that company's product(s) and requiring the presence of a chemical expert. If a member company cannot go to the scene of the incident, the shipper will authorize a CHEMNET-contracted emergency response company to go. Communications for the network are provided by CHEMTREC, with the shipper receiving notification and details about the incident from the CHEMTREC communicator.

CHEMTREC   Chemical Transportation Emergency Center operated by the Chemical Manufacturers Association. Provides information and/or assistance to emergency responders. CHEMTREC contacts the shipper or producer of the material for more detailed information, including on-scene assistance when feasible. Can be reached 24 hours a day by calling 800-424-9300. (Also see *HIT*.)

CHLOREP   Chlorine Emergency Plan operated by the Chlorine Institute. 24-hour mutual aid program. Response is activated by a CHEMTREC call to the designated CHLOREP contact, who notifies the appropriate team leader, based upon CHLOREP's geographical sector assignments for teams. The team leader in turn calls the emergency caller at the incident scene and determines what advice and assistance are needed. The team leader then decides whether or not to dispatch his team to the scene.

CHRIS/HACS   Chemical Hazards Response Information System/Hazard Assessment Computer System developed by the U.S. Coast Guard. HACS is a computerized model of the four CHRIS manuals that contain chemical-specific data. Federal OSCs use HACS to find answers to specific questions during a chemical spill/response. State and local officials and industry representatives may ask an OSC to request a HACS run for contingency planning purposes.

CPG 1-3   Federal Assistance Handbook: Emergency Management, Direction and Control Programs, prepared by FEMA. Provides states with guidance on administrative and programmatic requirements associated with FEMA funds.

CPG 1-5   Objectives for Local Emergency Management, prepared by FEMA. Describes and explains functional objectives that represent a comprehensive and integrated emergency management program. Includes recommended activities for each objective.

CPG 1-8   Guide for Development of State and Local Emergency Operations Plans, prepared by FEMA (see *EOP* below).

CPG 1-8A   Guide for the Review of State and Local Emergency Operations Plans, prepared by FEMA. Provides FEMA staff with a standard instrument for assessing EOPs that are developed to satisfy the eligibility requirement to receive Emergency Management Assistance funding.

CPG 1-35   Hazard Identification, Capability Assessment, and Multi-Year Development Plan for Local Governments, prepared by FEMA. As a planning tool, it can guide local jurisdictions through a logical sequence for identifying hazards, assessing capabilities, setting priorities, and scheduling activities to improve capability over time.

EBS   Emergency Broadcasting System to be used to inform the public about the nature of a hazardous materials incident and what safety steps they should take.

EMI   The Emergency Management Institute is a component of FEMA's National Emergency Training Center located in Emmitsburg, Maryland. It conducts resident and nonresident training activities for federal, state, and local government officials, managers in the private economic sector, and members of professional and volunteer organizations on subjects that range from civil nuclear preparedness systems to domestic emergencies caused by natural and technological hazards. Nonresident training activities are also conducted by State Emergency Management Training Offices under cooperative agreements that offer financial and technical assistance to establish annual training programs that fulfill emergency management training requirements in communities throughout the nation.

ERT   Environmental Response Team, a group of highly specialized experts available through EPA 24 hours a day.

EOP   Emergency Operations Plan developed in accord with the guidance in CPG 1-8. EOPs are multi-hazard, functional plans that treat emergency management activities generically. EOPs provide for as much generally applicable capability as possible without reference to any particular hazard; then they address the unique aspects of individual disasters in hazard-specific appendices.

Fault-Tree Analysis   A means of analyzing hazards. Hazardous events are first identified by other techniques such as HAZOP. Then all combinations of individual failures that can lead to that hazardous event are shown in the logical format of the fault tree. By estimating the individual failure probabilities, and then using the appropriate arithmetical expressions, the top-event frequency can be calculated.

FEMA-REP-5   Guidance for Developing State and Local Radiological Emergency Response Plans and Preparedness for Transportation Accidents, prepared by FEMA. Provides a basis for state and local governments to develop emergency plans and improve emergency preparedness for transportation accidents involving radioactive materials.

Hazardous Materials   Refers generally to hazardous substances, petroleum, natural gas, synthetic gas, acutely toxic chemicals, and other toxic chemicals.

HAZOP   Hazard and operability study, a systematic technique for identifying hazards or operability problems throughout an entire facility. One examines each segment of a process and lists all possible deviations from normal operating conditions and how they might occur. The consequences on the process are assessed, and the means available to detect and correct the deviations are examined.

HIT   Hazard Information Transmission program provides a digital transmission of the CHEMTREC emergency chemical report to first responders at the scene of a hazardous materials incident. The report advises the responder on the hazards of the materials, the level of protective clothing required, mitigating action to take in the event of a spill, leak, or fire, and first aid for victims. HIT is a free public service provided by the Chemical Manufacturers Association. Reports are sent in emergency situations only to organizations that have preregistered with HIT. Brochures and registration forms may be obtained by writing: Manager, CHEMTREC/CHEMNET, 2501 M Street, N.W., Washington, DC 20037.

ICS   Incident Command System, the combination of facilities, equipment, personnel, procedures, and communications operating within a common organizational structure with responsibility for management of assigned resources to effectively accomplish stated objectives at the scene of an incident.

IEMS   Integrated Emergency Management System, developed by FEMA in recognition of the economies realized in planning for all hazards on a generic functional basis as opposed to developing independent structures and resources to deal with each type of hazard.

NCP   National Oil and Hazardous Substances Pollution Contingency Plan (40 CFR Part 300), prepared by EPA to put into effect the response powers and responsibilities created by CERCLA and the authorities established by Section 311 of the Clean Water Act.

NFA   The National Fire Academy is a component of FEMA's National Emergency Training Center located in Emmitsburg, Maryland. It provides fire prevention and control training for the fire service and allied services. Courses on campus are offered in technical, management, and prevention subject areas. A growing off-campus course delivery system is operated in conjunction with state fire training program offices.

NHMIE   National Hazardous Materials Information Exchange, provides information on hazmat training courses, planning techniques, events and conferences, and emergency response experiences and lessons learned. Call toll-free 1-800-752-6367 (in Illinois, 1-800-367-9592). Planners with personal computer capabilities can access NHMIE by calling FTS 972-3275 or (312) 972-3275.

NRC   National Response Center, a communications center for activities related to response actions, is located at Coast Guard headquarters in Washington, DC. The NRC receives and relays notices of discharges or releases to the appropriate OSC, disseminates OSC and RRT reports to the NRT when appropriate, and provides facilities for the NRT to use in coordinating a national response action when required. The toll-free number (800-424-8802, or 202-426-2675 or 202-267-2675 in the Washington, DC area) can be reached 24 hours a day for reporting actual or potential pollution incidents.

NRT   National Response Team, consisting of representatives of 14 government agencies (DOD, DOI, DOT/RSPA, DOT/USCG, EPA, DOC, FEMA, DOS, USDA, DOJ, HHS, DOL, Nuclear Regulatory Commission, and DOE), is the principal organization for implementing the NCP. When the NRT is not activated for a response action, it serves as a standing committee to develop and maintain preparedness, to evaluate methods of responding to discharges or releases, to recommend needed changes in the response organization, and to recommend revisions to the NCP. The NRT may consider and make recommendations to appropriate agencies on the training, equipping, and protection of response teams; and necessary research, development, demonstration, and evaluation to improve response capabilities.

NSF   National Strike Force, made up of three Strike Teams. The USCG counterpart to the EPA ERTs.

NUREG 0654/FEMA-REP-1   Criteria for Preparation and Evaluation of Radiological Emergency Response Plans and Preparedness in Support of Nuclear Power Plants, prepared by NRC and FEMA. Provides a basis for state and local government and nuclear facility operators to develop radiological emergency plans and improve emergency preparedness. The criteria also will be used by federal agency reviewers in determining the adequacy of State, local, and nuclear facility emergency plans and preparedness.

OHMTADS Oil and Hazardous Materials Technical Assistance Data System, a computerized data base containing chemical, biological, and toxicological information about hazardous substances. OSCs use OHMTADS to identify unknown chemicals and to learn how to best handle known chemicals.

OSC On-Scene Coordinator, the federal official predesignated by EPA or USCG to coordinate and direct federal responses and removals under the NCP; or the DOD official designated to coordinate and direct the removal actions from releases of hazardous substances, pollutants, or contaminants from DOD vessels and facilities. When the NRC receives notification of a pollution incident, the NRC Duty Officer notifies the appropriate OSC, depending on the location of an incident. Based on this initial report and any other information that can be obtained, the OSC makes a preliminary assessment of the need for a federal response. If an on-scene response is required, the OSC will go to the scene and monitor the response of the responsible party or state or local government. If the responsible party is unknown or not taking appropriate action, and the response is beyond the capability of state and local governments, the OSC may initiate federal actions, using funding from the FWPCA Pollution Fund for oil discharges and the CERCLA Trust Fund (Superfund) for hazardous substance releases.

PSTN Pesticide Safety Team Network operated by the National Agricultural Chemicals Association to minimize environmental damage and injury arising from accidental pesticide spills or leaks. PSTN area coordinators in ten regions nationwide are available 24 hours a day to receive pesticide incident notifications from CHEMTREC.

RCRA Resource Conservation and Recovery Act (of 1976) established a framework for the proper management and disposal of all wastes. RCRA directed EPA to identify hazardous wastes, both generically and by listing specific wastes and industrial process waste streams. Generators and transporters are required to use good management practices and to track the movement of wastes with a manifest system. Owners and operators of treatment, storage, and disposal facilities also must comply with standards, which are generally implemented through permits issued by EPA or authorized states.

RRT   Regional Response Teams composed of representatives of federal agencies and a representative from each state in the federal region. During a response to a major hazardous materials incident involving transportation or a fixed facility, the OSC may request that the RRT be convened to provide advice or recommendations in specific issues requiring resolution. Under the NCP, RRTs may be convened by the chairman when a hazardous materials discharge or release exceeds the response capability available to the OSC in the place where it occurs; crosses regional boundaries; or may pose a substantial threat to the public health, welfare, or environment, or to regionally significant amounts of property. Regional contingency plans specify detailed criteria for activation of RRTs. RRTs may review plans developed in compliance with Title III, if the local emergency planning committee so requests.

SARA   The "Superfund Amendments and Reauthorization Act of 1986." Title III of SARA includes detailed provisions for community planning.

Superfund   The trust fund established under CERCLA to provide money the OSC can use during a cleanup.

Title III   The "Emergency Planning and Community Right-to-Know Act of 1986." Specifies requirements for organizing the planning process at the state and local levels for specified extremely hazardous substances; minimum plan content; requirements for fixed facility owners and operators to inform officials about extremely hazardous substances present at the facilities; and mechanisms for making information about extremely hazardous substances available to citizens. (See Appendix A.)

Appendix D

Criteria for Assessing State and Local Preparedness

C.1 Introduction

The criteria in this appendix, an adaptation of criteria developed by the Preparedness Committee of the NRT in August 1985, represent a basis for assessing a state or local hazardous materials emergency response preparedness program. These criteria reflect the basic elements judged to be important for a successful emergency preparedness program.

The criteria are separated into six categories, all of which are closely interrelated. These categories are hazards analysis, authority, organizational structure, communications, resources, and emergency planning.

These criteria may be used for assessing the emergency plan as well as the emergency preparedness program in general. It must be recognized, however, that few state or local governments will have the need and/or capability to address all these issues and meet all these criteria to the fullest extent. Resource limitations and the results of the hazards analysis will strongly influence the necessary degree of planning and preparedness. Those governmental units that do not have adequate resources are encouraged to seek assistance and take advantage of all resources that are available.

Other criteria exist that could be used for assessing a community's preparedness and emergency planning. These include FEMA's CPG 1-35 (Hazard Identification, Capability Assessment and Multi-Year Development Plan for Local Governments) and CPG 1-8A. Additionally, states may have issued criteria for assessing capability.

C.2 The Criteria

C.2.1 Hazards Analysis

"Hazards Analysis" includes the procedures for determining the susceptibility or vulnerability of a geographical area to a hazardous materials release, for identifying potential sources of a hazardous materials release from fixed facilities that manufacture, process, or otherwise use, store, or dispose of materials that are generally considered hazardous in an unprotected environment. This also includes an analysis of the potential or probable hazard of transporting hazardous materials through a particular area.

A hazards analysis is generally considered to consist of identification of potential hazards, determination of the vulnerability of an area as a result of the existing hazards, and an assessment of the risk of a hazardous materials release or spill.

The following criteria may assist in assessing a hazards analysis:

- Has a hazards analysis been completed for the area? If one exists, when was it last updated?
- Does the hazards analysis include the location, quantity, and types of hazardous materials that are manufactured, processed, used, disposed, or stored within the appropriate area?

- Was it done in accordance with community right-to-know laws and prefire plans?
- Does it include the routes by which the hazardous materials are transported?
- Have areas of public health concern been identified?
- Have sensitive environmental areas been identified?
- Have historical data on spill incidents been collected and evaluated?
- Have the levels of vulnerability and probable locations of hazardous materials incidents been identified?
- Are environmentally sensitive areas and population centers considered in analyzing the hazards of the transportation routes and fixed facilities?

C.2.2   Authority

"Authority" refers to those statutory authorities or other legal authorities vested in any personnel, organizations, agencies, or other entities in responding to or being prepared for responding to hazardous materials emergencies resulting from releases or spills.

The following criteria may be used to assess the existing legal authorities for response actions:

- Do clear legal authorities exist to establish a comprehensive hazardous materials response mechanism (federal, state, county, and local laws, ordinances, and policies)?
- Do these authorities delegate command and control responsibilities between the different organizations within the same level of government (horizontal), and/or provide coordination procedures to be followed?
- Do they specify what agency(ies) has (have) overall responsibility for directing or coordinating a hazardous materials response?
- Do they specify what agency(ies) has (have) responsibility for providing assistance or support for hazardous materials response and what comprises that assistance or support?
- Have the agency(ies) with authority to order evacuation of the community been identified?
- Have any limitations in the legal authorities been identified?

C.2.3   Organizational Structure

"Organizational" refers to the organizational structure in place for responding to emergencies. This structure will, of course, vary considerably from state to state and from locality to locality.

There are two basic types of organizations involved in emergency response operations. The first is involved in the planning and policy decision process similar to the NRT and RRT. The second is the operational response group that functions within the precepts set forth in the state or local plan. Realizing that situations vary from state to state and locality to locality and that emergency planning for the state and local level may involve the preparation of multiple situation plans or development of a single comprehensive plan, the criteria should be broadly based and designed to detect a potential flaw that would then precipitate a more detailed review.

Are the following organizations included in the overall hazardous materials emergency preparedness activities?

- Health organizations (including mental health organizations)
- Public safety
    fire
    police
    health and safety (including occupational safety and health)
    other responders
- Transportation
- Emergency management/response planning
- Environmental organizations
- Natural resources agencies (including trustee agencies)
- Environmental agencies with responsibilities for:
    fire
    health
    water quality
    air quality
    consumer safety
- Education system (in general)
    public education
    public information
- Private sector interface
    trade organizations
    industry officials
- Labor organizations

Have each organization's authorities, responsibilities, and capabilities been determined for pre-response (planning and prevention), response (implementing the plan during an incident), and post-response (cleanup and restoration) activities?

Has one organization been given the command and control responsibility for these three phases of emergency response?

Has a "chain of command" been established for response control through all levels of operation?

Are the roles, relationships, and coordination procedures between government and non-government (private entities) delineated? Are they understood by all affected parties? How are they instituted (written, verbal)?

Are clear interrelationships, and coordination procedures between government and non-government (private entities) delineated? Are they understood by all affected parties? How are they instituted (written, verbal)?

Are the agencies or departments that provide technical guidance during a response the same agencies or departments that provide technical guidance in non-emergency situations? In other words, does the organizational structure vary with the type of situation to be addressed?

Does the organizational structure provide a mechanism to meet regularly for planning and coordination?

Does the organizational structure provide a mechanism to regularly exercise the response organization?

Has a simulation exercise been conducted within the last year to test the organizational structure?

Does the organizational structure provide a mechanism to review the activities conducted during a response or exercise to correct shortfalls?

Have any limitations within the organizational structure been identified?

Is the organizational structure compatible with the federal response organization in the NCP?

Have trained and equipped incident commanders been identified?

Has the authority for site decisions been vested in the incident commanders?

Have the funding sources for a response been identified?

How quickly can the response system be activated?

C.2.4 Communication

"Communication" means any form or forms of exchanging information or ideas for emergency response with other entities, either internal or external to the existing organizational structure.

Coordination

Have procedures been established for coordination of information during a response?

Has one organization been designated to coordinate communications activities?

Have radio frequencies been established to facilitate coordination between different organizations?

Information Exchange

Does a formal system exist for information sharing among agencies, organizations, and the private sector?

Has a system been established to ensure that "lessons learned" are passed to the applicable organizations?

Information Dissemination

Has a system been identified to carry out public information/community relations activities?

Has one organization or individual been designated to coordinate with or speak to the media concerning the release?

Is there a communication link with an Emergency Broadcast System (EBS) point of entry (CPCS-1) station?

Does a communications system/method exist to disseminate information to responders, affected public, etc.?

Is this system available 24 hours a day?

Have alternate systems/methods of communications been identified for use if the primary method fails?

Does a mechanism exist to keep telephone rosters up-to-date?

Are communications networks tested on a regular basis?

Information Sources and Data Base Sharing

Is a system available to provide responders with rapid information on the hazards of chemicals involved in an incident?

Is this information available on a 24-hour basis? Is it available in computer software?

Is a system in place to update the available information sources?

Notification Procedures

Have specific procedures for notification of a hazardous materials incident been developed?

Are multiple notifications required by overlapping requirements (e.g., state, county, local each have specific notification requirements)?

Does the initial notification system have a standardized list of information that is collected for each incident?

Does a network exist for notifying and activating necessary response personnel?

Does a network exist for notifying or warning the public of potential hazards resulting from a release? Does this network have provisions for informing the public of what hazards to expect, what precautions to take, whether evacuation is required, etc.?

Has a central location or phone number been established for initial notification of an incident?

Is the central location or phone number accessible on a 24-hour basis?

Does the central location phone system have the ability to expand to a multiple line system during an emergency?

Clearinghouse Functions

Has a central clearinghouse for hazardous materials information been established with access by the public and private sector?

C.2.5   Resources

"Resource" means the personnel, training, equipment, facilities, and other sources available for use in responding to hazardous materials emergencies. To the extent that the hazards analysis has identified the appropriate level of preparedness for the area, these criteria may be used in evaluating available resources of the jurisdiction undergoing review.

Personnel

Have the numbers of trained personnel available for hazardous materials been determined?

Has the location of trained personnel available for hazardous materials been determined? Are these personnel located in areas identified in the hazards analysis as:

• heavily populated;
• high hazard areas—i.e., numbers of chemical (or other hazardous materials) production facilities in well-defined areas;
• hazardous materials storage, disposal, and/or treatment facilities; and
• transit routes?

Are sufficient personnel available to maintain a given level of response capability identified as being required for the area?

Has the availability of special technical expertise (chemists, industrial hygienists, toxicologists, occupational health physicians, etc.) necessary for response been identified?

Have limitations on the use of above personnel resources been identified?

Do mutual aid agreements exist to facilitate interagency support between organizations?

Training

Have the training needs for the state/local area been identified?

Are centralized response training facilities available?

Are specialized courses available covering topics such as:

* organizational structures for response actions (i.e., authorities and coordination);
* response actions;
* equipment selection, use, and maintenance; and
* safety and first aid?

Does the organizational structure provide training and cross training for or between organizations in the response mechanism?

Does an organized training program for all involved response personnel exist? Has one agency been designated to coordinate this training?

Have training standards or criteria been established for a given level of response capability? Is any certification provided upon completion of the training?

Has the level of training available been matched to the responsibilities or capabilities of the personnel being trained?

Does a system exist for evaluating the effectiveness of training?

Does the training program provide for "refresher courses" or some other method to ensure that personnel remain up-to-date in their level of expertise?

Have resources and organizations available to provide training been identified?

Have standardized curricula been established to facilitate consistent State-wide training?

Equipment

Have response equipment requirements been identified for a given level of response capability?

Are the following types of equipment available?

- personal protective equipment
- first aid and other medical emergency equipment
- emergency vehicles available for hazardous materials response
- sampling equipment (air, water, soil, etc.) and other monitoring devices (e.g., explosivity meters, oxygen meters)
- analytical equipment or facilities available for sample analyses
- fire fighting equipment/other equipment and material (bulldozers, boats, helicopters, vacuum trucks, tank trucks, chemical retardants, foam)

Are sufficient quantities of each type of equipment available on a sustained basis?

Is all available equipment capable of operating in the local environmental conditions?

Are up-to-date equipment lists maintained? Are they computerized?

Are equipment lists available to all responders?

Are these lists broken down into the various types of equipment (e.g., protective clothing, monitoring instruments, medical supplies, transportation equipment)?

Is there a mechanism to ensure that the lists are kept up to date?

Have procedures necessary to obtain equipment on a 24-hour basis been identified?

Does a program exist to carry out required maintenance of equipment?

Are there maintenance and repair records for each piece of equipment?

Have mutual aid agreements been established for the use of specialized response equipment?

Is sufficient communications equipment available for notifying personnel or to transmit information? Is the equipment of various participating agencies compatible?

Is transportation equipment available for moving equipment rapidly to the scene of an incident, and its state of readiness assured?

Facilities

Have facilities capable of performing rapid chemical analyses been identified?

Do adequate facilities exist for storage and cleaning/reconditioning of response equipment?

Have locations or facilities been identified for the storage, treatment, recycling, and disposal of wastes resulting from a release?

Do adequate facilities exist for carrying out training programs?

Do facilities exist that are capable of providing medical treatment to persons injured by chemical exposure?

Have facilities and procedures been identified for housing persons requiring evacuation or temporary relocation as a result of an incident?

Have facilities been identified that are suitable for command centers?

C.2.6 Emergency Plan

The emergency plan, while it relates to many of the above criteria, also stands alone as a means to assess preparedness at the state and local level of government, and in the private sector. The following questions are directed more toward evaluating the plan rather than determining the preparedness level of the entity that has developed the plan. It is not sufficient to ask if there is a plan, but rather to determine if the plan that does exist adequately addresses the needs of the community or entity for which the plan was developed.

Have the levels of vulnerability and probable locations of hazardous materials incidents been identified in the plan?

Have areas of public health concern been identified in the plan?

Have sensitive environmental areas been identified in the plan?

For the hazardous materials identified in the area, does the plan include information on the chemical and physical properties of the materials, safety and emergency response information, and hazard mitigation techniques? (NOTE: It is not necessary that all this information be included in the emergency plan; the plan should, however, at least explain where such information is available.)

Have all appropriate agencies, departments, or organizations been involved in the process of developing or reviewing the plan?

Have all the appropriate agencies, departments, or organizations approved the plan?

Has the organizational structure and notification list defined in the plan been reviewed in the last six months?

Is the organizational structure identified in the plan compatible with the federal response organization in the NCP?

Has one organization been identified in the plan as having command and control responsibility for the pre-response, response, and post-response phases?

Does the plan define the organizational responsibilities and relationships among city, county, district, state, and federal response agencies?

Are all organizations that have a role in hazardous materials response identified in the plan (public safety and health, occupational safety and health, transportation, natural resources, environmental, enforcement, educational, planning, and private sector)?

Are the procedures and contacts necessary to activate or deactivate the organization clearly given in the plan for the pre-response, response, and post-response phases?

Does the organizational structure outlined in the plan provide a mechanism to review the activities conducted during a response or exercise to correct shortfalls?

Does the plan include a communications system/method to disseminate information to responders, affected public, etc.?

Has a system been identified in the plan to carry out public information/community relations activities?

Has a central location or phone number been included in the plan for initial notification of an incident?

Have trained and equipped incident commanders been identified in the plan?

Does the plan include the authority for vesting site decisions in the incident commander?

Have government agency personnel that may be involved in response activities been involved in the planning process?

Have local private response organizations (e.g., chemical manufacturers, commercial cleanup contractors) that are available to assist during a response been identified in the plan?

Does the plan provide for frequent training exercises to train personnel or to test the local contingency plans?

Are lists/systems that identify emergency equipment available to response personnel included in the plan?

Have locations of materials most likely to be used in mitigating the effects of a release (e.g., foam, sand, lime) been identified in the plan?

Does the plan address the potential needs for evacuation, what agency is authorized to order or recommend an evacuation, how it will be carried out, and where people will be moved?

Has an emergency operating center, command center, or other central location with the necessary communications capabilities been identified in the plan for coordination of emergency response activities?

Are there follow-up response activities scheduled in the plan?

Are there procedures for updating the plan?

Are there addenda provided with the plan, such as: laws and ordinances, statutory responsibilities, evacuation plans, community relations plan, health plan, and resource inventories (personnel, equipment, maps [not restricted to road maps], and mutual aid agreements)?

Does the plan address the probable simultaneous occurrence of different types of emergencies (e.g., power outage and hazardous materials releases) and the presence of multiple hazards (e.g., flammable and corrosive) during hazardous materials emergencies?

Appendix E

Bibliography

General Emergency Planning for Hazardous Materials

American Institute of Chemical Engineers, Center for Chemical Plant Safety. *Guidelines for Hazard Evaluation Procedures*. Washington, DC: A.I.Ch.E., 1985.

American Society of Testing & Materials. *Toxic and Hazardous Industrial Chemicals Safety Manual*. 1983.

Association of Bay Area Governments. *San Francisco Bay Area: Hazardous Spill Prevention and Response Plan*. Volumes I & II. Berkeley, CA: 1983.

Avoiding and Managing Environmental Damage from Major Industrial Accidents. Proc. of Conference of the Air Pollution Control Association. 1985.

Bretherick, L. *Handbook of Reactive Chemical Hazards*. 2nd ed. Butterworth, 1979.

Brinsko, George A. et al. *Hazardous Material Spills and Responses for Municipalities*. (EPA-600/2-80-108, NTIS PB80-214141). 1980.

Cashman, John R. *Hazardous Materials Emergencies: Response and Control*. 1983.

Chemical Manufacturers Association. *Community Awareness and Emergency Response Program Handbook*. Washington, DC: CMA, 1985.

Chemical Manufacturers Association. *Community Emergency Response Exercise Program*. Washington, DC: CMA, 1986.

Chemical Manufacturers Association. *Risk Analysis in the Chemical Industry— Proceedings of a Symposium*. Rockville, MD: Government Institutes, Inc., 1985.

Chemical Manufacturers Association. *Site Emergency Response Planning*. Washington, DC: CMA, 1986.

Copies of the CMA guides can be obtained by writing to:

    Publications Fulfillment
    Chemical Manufacturers Association
    2501 M Street, N.W.
    Washington, DC 20037

Emergency Management and Civil Defense Division, Consolidated City of Indianapolis. *Final Report: Demonstration Project to Develop a Hazardous Materials Accident Prevention and Emergency Response Program, Phases I, II, III, IV.* Indianapolis: 1983.

Energy Resources Co., Inc.; Cambridge Systematics, Inc.; Massachusetts Department of Environmental Quality Engineering. *Demonstration Project to Develop a Hazardous Materials Accident Prevention and Emergency Response Program for the Commonwealth of Massachusetts.* Volumes I & II. Cambridge and Boston, MA: 1983.

Environmental and Safety Design, Inc. *Development of a Hazardous Materials Accident Prevention and an Emergency Response Program.* Memphis, TN: 1983.

Federal Emergency Management Agency. *Disaster Operations: A Handbook for Local Governments.* Washington, DC: 1981.

Federal Emergency Management Agency. *Hazard Identification, Capability Assessment, and Multi-Year Development Plan for Local Governments.* CPG 1-35, Washington, DC: 1985.

Federal Emergency Management Agency. *Objectives for Local Emergency Management.* CPG 1-5, Washington, DC: 1984.

Federal Emergency Management Agency. *Professional Development Series: Emergency Planning — Student Manual.* Washington, DC.

Federal Emergency Management Agency. *Professional Development Series: Introduction to Emergency Management — Student Manual.* Washington, DC.

Gabor, T. and Griffith, T.K. *The Assessment of Community Vulnerability to Acute Hazardous Materials Incidents.* Newark, DE: University of Delaware, 1985.

Government Institutes, Inc. *R.C.R.A. Hazardous Waste Handbook.* Volumes 1 & 2. 1981.

Green, Don W., ed. *Perry's Chemical Engineers' Handbook.* 6th ed. McGraw-Hill, 1984.

Hawley, Gessner G., ed. *Condensed Chemical Dictionary.* 10th ed. New York: Van Nostrand Reinhold, 1981.

Hildebrand, Michael S. *Disaster Planning Guidelines for Fire Chiefs.* Washington, DC: International Association of Fire Chiefs, 1980.

Multnomah County Office of Emergency Management. *Hazardous Materials Management System: A Guide for Local Emergency Managers.* Portland, OR: 1983.

National Fire Protection Association. *Fire Protection Guide on Hazardous Materials.* Boston: NFPA, 1986.

National Institute of Occupational Safety and Health. *Pocket Guide to Chemical Hazards*. Washington, DC: DHEW (NIOSH) 78-210, 1985. (GPO Stock No. 017-033-00342-4)

New Orleans, City of. *Demonstration Project to Develop a Hazardous Materials Accident Prevention and Emergency Response Program for the City of New Orleans, Phases I, II, III, IV*. New Orleans: 1983.

Portland Office of Emergency Management. *Hazardous Materials Hazard Analysis*. Portland, OR: 1981.

Puget Sound Council of Governments. *Hazardous Materials Demonstration Project Report: Puget Sound Region*. Seattle, WA: 1981.

Sax, N. Irving. *Dangerous Properties of Industrial Materials*. 6th ed. New York: Van Nostrand Reinhold, 1984.

Sittig, Marshall. *Handbook of Toxic and Hazardous Chemicals and Carcinogens*. Noyes, 1985.

Smith, A. J. *Managing Hazardous Substances Accidents*. 1981.

U.S. Department of Transportation. *CHRIS: Manual I, A Condensed Guide to Chemical Hazards*. U.S. Coast Guard, 1984.

U.S. Department of Transportation. *CHRIS: Manual II, Hazardous Chemical Data*. U.S. Coast Guard, 1984.

U.S. Department of Transportation. *Emergency Response Guidebook*. Washington, DC: 1984.

U.S. Environmental Protection Agency. *Community Relations in Superfund: A Handbook*. Washington, DC.

U.S. Environmental Protection Agency. *The National Oil and Hazardous Substances Pollution Contingency Plan*. 40 CFR 300.

Verschuaren, Karel. *Handbook of Environmental Data on Organic Chemicals*. 2nd ed. New York: Van Nostrand Reinhold, 1983.

Waste Resource Associates, Inc. *Hazmat—Phases I, II, III, IV: Demonstration Project to Develop a Hazardous Materials Accident Prevention and Emergency Response Program*. Niagara Falls, NY: 1983.

Zajic, J.E. and Himmelman, W.A. *Highly Hazardous Material Spills and Emergency Planning*. Dekker, 1978.

Transportation Emergency Planning

American Trucking Association. *Handling Hazardous Materials*. Washington, DC: 1980.

Association of American Railroads. *Emergency Action Guides*. Washington, DC: 1984.

Association of American Railroads. *Emergency Handling of Hazardous Materials in Surface Transportation.* Washington, DC: 1981.

Battelle Pacific Northwest Laboratories. *Hazardous Material Transportation Risks in the Puget Sound Region.* Seattle, WA: 1981.

Portland Office of Emergency Management. *Establishing Routes for Trucks Hauling Hazardous Materials: The Experience in Portland, Oregon.* Portland, Oregon: 1984.

Portland Office of Emergency Management. *Hazardous Materials Highway Routing Study: Final Report.* Portland, OR: 1984.

Russell, E.R., Smaltz, J.J., et al. *A Community Model for Handling Hazardous Materials Transportation Emergencies: Executive Summaries.* Washington, DC: U.S. Department of Transportation, January 1986.

Russell, E.R., Smaltz, J.J., et al. *Risk Assessment/Vulnerability Users Manual for Small Communities and Rural Areas.* Washington, DC: U.S. Department of Transportation, March 1986.

Russell, E.R., Brumgardt, W., et al. *Risk Assessment/Vulnerability Validation Study Volume 2: 11 Individual Studies.* Washington, DC: U.S. Department of Transportation, June 1983.

Transportation Research Board. *Transportation of Hazardous Materials: Toward a National Strategy.* Volumes 1 & 2. Washington, DC: 1983.

Urban Consortium Transportation Task Force. *Transportation of Hazardous Materials.* Washington, DC: U.S. Department of Transportation, September 1980.

Urban Systems Associates, Inc., St. Bernard Parish Planning Commission. *St. Bernard Parish:Hazardous Materials Transportation and Storage Study.* New Orleans, LA: 1981.

Urganek, G. and Barber, E. *Development of Criteria to Designate Routes for Transporting Hazardous Materials.* Springfield, VA: National Technical Information Service, 1980.

U.S. Department of Transportation. *Community Teamwork: Working Together to Promote Hazardous Materials Transportation Safety.* Washington, DC: 1983.

U.S. Department of Transportation. *A Guide for Emergency Highway Traffic Regulation.* Washington, DC: 1985.

U.S. Department of Transportation. *A Guide to the Federal Hazardous Transportation Regulatory Program.* Washington, DC: 1983.

U.S. Department of Transportation. *Guidelines for Selecting Preferred Highway Routes for Highway Route Controlled Quantity Shipments of Radioactive Materials.* Washington, DC: 1984.

U.S. Department of Transportation. Three-Phase/Four-Volume Report: Volume I, *A Community Model for Handling Hazardous Materials Transportation Emergencies*; Volume II, *Risk Assessment Users Manual for Small Communities and Rural Areas*; Volume III, *Risk Assessment/Vulnerability Model Validation*; and Volume IV, *Manual for Small Towns and Rural Areas to Develop a Hazardous Materials Emergency Plan*. 7/81–12/85. Document is available to the U.S. public through the National Technical Information Service, Springfield, VA 22161.

U.S. Department of Transportation and U.S. Environmental Protection Agency. *Lessons Learned from State and Local Experiences in Accident Prevention and Response Planning for Hazardous Materials Transportation*. Washington, DC: December 1985.

Spill Containment and Cleanup

Guswa, J.H. *Groundwater Contamination and Emergency Response Guide*. Noyes, 1984.

U.S. Environmental Protection Agency. *State Participation in the Superfund Remedial Program*. Washington, DC: 1984.

Personal Protection

International Association of Fire Chiefs. *Fire Service Emergency Management Handbook*. Washington, DC: 1985.

National Institute of Occupational Safety and Health. *Occupational Safety and Health Guidance Manual for Hazardous Waste Site Activities*. Washington, DC: DHHS Publication No. 85-115, 1985.

U.S. Environmental Protection Agency. *Standard Operating Safety Guides*. Washington, DC: 1984.

Videotapes

The following videotapes are available from the Chemical Manufacturers Association:

- CAER: "Reaching Out"
- CAER: "How a Coordinating Group Works"
- CAER: "Working with the Media"
- CAER: "Planning and Conducting Emergency Exercises"
- NCRIC: "First on the Scene"

The following videotapes are available from FEMA's National Emergency Training Center/Emergency Management Information Center:

- "Livingston, LA, Hazardous Materials Spills" (September 28, 1982)

- "Waverly, TN, Hazardous Materials Blast" (February 22, 1978)

   Also available for purchase from FEMA's National Emergency Training Center (see Appendix F for address and telephone number) are videotapes of teleconferences produced by FEMA's Emergency Education Network (EENET). One available teleconference is:

- "Emergency Exercises — Getting Involved in Community Preparedness," originally seen on December 11, 1986, and co-sponsored by FEMA, EPA, DOT/RSPA, USCG, and CMA.

   The following documentary videotape (produced by the League of Women Voters of California and available from Bullfrog Films, Oley, PA 19547) provides public education on the nature and need for local emergency planning and hazardous materials data bases from a citizen's perspective.

- "Toxic Chemicals: Information Is the Best Defense"

Appendix F

Federal Agency Addresses

1. National Offices

Federal Emergency Management Agency
Technological Hazards Division
Federal Center Plaza
500 C Street, S.W.
Washington, DC 20472
(202) 646-2500

FEMA National Emergency Training Center
16825 S. Seaton Ave.
Emmitsburg, MD 21727
(301) 447-6771

U.S. Environmental Protection Agency
OSWER Preparedness Staff
401 M Street, S.W.
Washington, DC 20460

(202) 475-8600
CEPP Hotline: 1-800-535-0202
(479-2449 in Washington, DC area)

U.S. Environmental Protection Agency
OERR Emergency Response Division
401 M Street, S.W.
Washington, DC 20460
(202) 382-2090

Agency for Toxic Substances
and Disease Registry
Department of Health & Human Services
Bldg. 1, 1600 Clifton Rd., N.E.
Atlanta, GA 30303
(404) 639-3291

U.S. Department of Energy
1000 Independence Avenue, S.W.
Washington, DC 20585
(202) 586-5000

Department of Agriculture
Forest Service
Administration Building, 12th Street & Jefferson Drive, S.W.
Washington, DC 20250
(202) 447-6661

Department of Labor
Occupational Safety & Health Admin.
Frances Perkins Building
200 Constitution Avenue, N.W.
Washington, DC 20210
(202) 523-6091

U.S. Coast Guard (G-MER)
Marine Environmental Response Division
2100 2nd Street, S.W.
Washington, DC 20593-0001
(202) 267-0518

National Response Center
1-800-424-8802
(202-426-2675 or 202-267-2675 in Washington, DC area)

U.S. Department of Transportation
Research and Special Programs Admin.
Office of Hazardous Materials
Transportation (Attention: DHM-50)
Nassif Building
400 7th Street, S.W.
Washington, DC 20590
(202) 366-4433

Department of Justice
Environmental Enforcement Section
10th and Constitution Avenue, N.W.
Washington, DC 20530
(202) 514-5217

Department of the Interior
18th and C St., N.W.
Washington, DC 20240
(202) 208-3100

Department of Commerce
National Oceanic and Atmospheric Administration
11400 Rockville Pike
Rockville, MD 20852
(301) 377-3436

Department of Defense
The Pentagon
Washington, DC 20301-8000
(703) 545-6700

Department of State
Office of Oceans Affairs
2201 C St., N.W.
Washington, DC 20520
(202) 647-3262

Nuclear Regulatory Commission
One White Flint North Building
11555 Rockville Pike
Rockville, MD 20555
(301) 492-7000

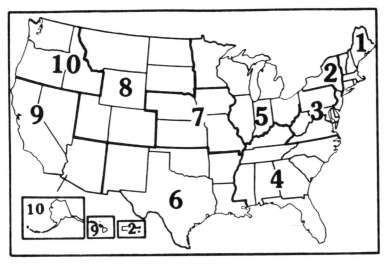

EPA, FEMA, HHS, ATSDR, OSHA

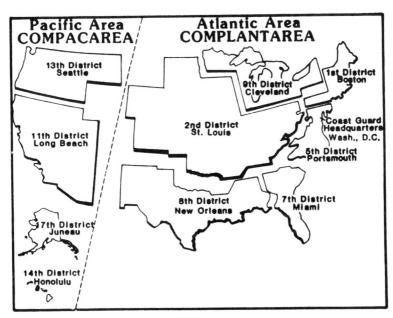

U.S. COAST GUARD DISTRICTS

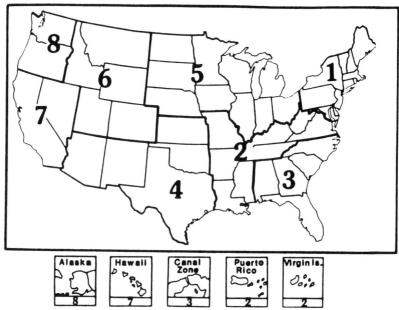

Department of Energy Regional Coordinating Offices for Radiological Assistance and Geographical Areas of Responsibility

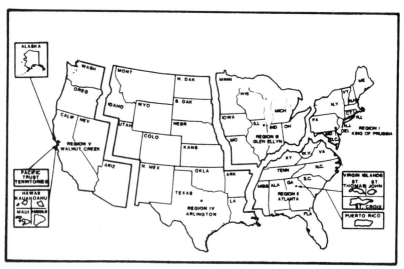

United States Nuclear Regulatory Commission

2. Regional Offices

A. EPA Regional Offices

(Note: Direct all requests to the "EPA Regional Preparedness Coordinator" (RPC) of the appropriate EPA Regional Office.)

Region I (Connecticut, Maine, Massachusetts, New Hampshire, Rhode Island, Vermont)

John F. Kennedy Building
1 Congress Street
Boston, MA 02203
(617) 565-3420

Region II (New Jersey, New York, Puerto Rico, Virgin Islands)

26 Federal Plaza
New York, NY 10278
(212) 264-2525

Region III (Delaware, Washington DC, Maryland, Pennsylvania, Virginia, West Virginia)

841 Chestnut Street
Philadelphia, PA 19107
(215) 597-9800

Region IV (Alabama, Florida, Georgia, Kentucky, Mississippi, North Carolina, South Carolina, Tennessee)

345 Courtland Street, N.E.
Atlanta, GA 30365
(404) 257-4727

Region V (Illinois, Indiana, Michigan, Minnesota, Ohio, Wisconsin)

230 S. Dearborn Street
Chicago, IL 60604
(312) 353-2000

Region VI (Arkansas, Louisiana, New Mexico, Oklahoma, Texas)

Suite 1200
1445 Ross Avenue
Dallas, TX 75202-2733
(214) 255-2100

Region VII (Iowa, Kansas, Missouri, Nebraska)

726 Minnesota Avenue
Kansas City, KS 66101
(913) 276-7006

Region VIII (Colorado, Montana, North Dakota, South Dakota, Utah, Wyoming)

Suite 500
999 18th Street
Denver, CO 80202-2405
(303) 330-1603

Region IX (Arizona, California, Hawaii, Nevada, American Samoa, Guam)

215 Fremont Street
San Francisco, CA 94105
(415) 556-6478

Region X (Alaska, Idaho, Oregon, Washington)

1200 6th Avenue
Seattle, WA 98101
(206) 399-5810

B. FEMA Regional Offices

(Note: Direct all requests to the "Hazmat Program Staff" of the appropriate FEMA Regional office.)

Region I (Connecticut, Maine, Massachusetts, New Hampshire, Rhode Island, Vermont)

442 J.W. McCormack POCH
Boston, MA 02109-4595
(617) 223-9540

Region II (New Jersey, New York, Puerto Rico, Virgin Islands)

Room 1337
26 Federal Plaza
New York, NY 10278-0002
(212) 225-7209

Region III (Delaware, Washington DC, Maryland, Pennsylvania, Virginia, West Virginia)

Liberty Square Building
105 S. 7th Street
Philadelphia, PA 19106-3316
(215) 931-5500

Region IV (Alabama, Florida, Georgia, Kentucky, Mississippi, North Carolina, South Carolina, Tennessee)

1371 Peachtree Street, N.E.
Atlanta, GA 30309-3108
(404) 853-4200

Region V (Illinois, Indiana, Michigan, Minnesota, Ohio, Wisconsin)

4th Floor
175 W. Jackson Blvd.
Chicago, IL 60604-2698
(312) 408-5500

Region VI (Arkansas, Louisiana, New Mexico, Oklahoma, Texas)

Federal Regional Center
800 N. Loop 288
Denton, TX 76201-3698
(817) 898-9399

Region VII (Iowa, Kansas, Missouri, Nebraska)

911 Walnut Street, Room 200
Kansas City, MO 64106-2085
(816) 283-7061

Region VIII (Colorado, Montana, North Dakota, South Dakota, Utah, Wyoming)

Denver Federal Center, Building 710
Box 25267
Denver, CO 80225-0267
(303) 235-4811

Region IX (Arizona, California, Hawaii, Nevada, American Samoa, Guam)

Building 105
San Francisco Presidio, CA 94129-1250
(415) 923-7100

Region X (Alaska, Idaho, Oregon, Washington)

Federal Regional Center
130 228th St., S.W.
Bothell, WA 98021-9796
(206) 487-4600

C. HHS Regional Offices

(Note: Consult the map on Page 645 to determine which states are assigned to each Region.)

Region I

Division of Preventive Health Services, Room 1856
J. F. Kennedy Building
1 Congress Street
Boston, MA 02203
(617) 565-1440

Region II

Division of Preventive Health Services
Jacob K. Javits Federal Building
26 Federal Plaza
New York, NY 10278
(212) 264-2485

Region III

Division of Preventive Health Services
3535 Market Street
Philadelphia, PA 19104
(215) 596-6650

Region IV

Division of Preventive Health Services
101 Marietta Tower
Atlanta, GA 30323
(404) 331-2313

Region V

Division of Preventive Health Services
105 W. Adams
Chicago, IL 60603
(312) 353-3652

Region VI

Division of Preventive Health Services
1200 Main Tower
Dallas, TX 75202
(214) 767-3916

Region VII

Division of Preventive Health Services
601 East 12th Street
Kansas City, MO 64106
(816) 374-3491

Region VIII

Division of Preventive Health Services
1961 Stout Street
Denver, CO 80294
(303) 844-6166, ext. 28

Region IX

Division of Preventive Health Services
Federal Office Building
50 United Nations Plaza
San Francisco, CA 94102
(415) 556-2219

Region X

Division of Preventive Health Services
2201 Sixth Avenue
Seattle, WA 98121
(206) 442-0502

D. ATSDR Public Health Advisors Assigned to EPA Regional Offices

(Note: Consult the map on Page 645 to determine which States are assigned to each Region.)

Region I

ATSDR Public Health Advisor
EPA Building
60 Westview
Lexington, MA 02173
(617) 860-4314

Region II

ATSDR Public Health Advisor
26 Federal Plaza
New York, New York 10278
(212) 264-7662

Region III

ATSDR Public Health Advisor
841 Chestnut Street, 6th Floor
Philadelphia, PA 19106
(215) 597-7291

Region IV

ATSDR Public Health Advisor
345 Courtland Street, N.E.
Atlanta, GA 30365
(404) 347-3043

Region V

ATSDR Public Health Advisor
77 West Jackson Blvd.
 6th Floor — MHS-6J
Chicago, IL 60604
(312) 886-0327

Region VI

ATSDR Public Health Advisor
Office of Health Response
First Interstate Tower
EPA Region VI (6H-E)
1445 Ross Ave.
Dallas, TX 75270
(214) 655-2245

Region VII

ATSDR Public Health Advisor
Waste Management Branch
726 Minnesota Avenue
Kansas City, KS 66101
(913) 551-7692

Region VIII

ATSDR Public Health Advisor
6th Floor, North Tower
999 18th Street
Denver, CO 80202
(303) 294-1063

Region IX

ATSDR Public Health Advisor
75 Hawthorne Street
Rm 09261 Mail Code H-1-2
San Francisco, CA 94105
(415) 744-2194

Region X

ATSDR Public Health Advisor
1200 6th Avenue
Seattle, WA 98101
(206) 553-2113

E.   OSHA Regional Offices

(Note: Consult the map on Page 645 to determine which States are assigned to each Region.)

Region I

133 Portland Street
First Floor
Boston, MA 02114
(617) 565-7164

Region II

201 Varick Street
Room 607
New York, NY 10014
(212) 337-2325

Region III

3535 Market Street
Philadelphia, PA 19104
(215) 596-1201

Region IV

1375 Peachtree Street, N.E.
Suite 587
Atlanta, GA 30367
(404) 347-3573

Region V

230 Dearborn Street
Chicago, IL 60604
(312) 353-2220

Region VI

555 Griffin Square Building
Griffin and Young Streets
Dallas, TX 75202
(214) 767-4731

Region VII

911 Walnut Street
Kansas City, MO 64106
(816) 374-5861

Region VIII

Federal Building
1961 Stout Street
Denver, CO 80294
(303) 844-3061

Region IX

Room 415
Federal Office Building
71 Stevenson Street
San Francisco, CA 94102
(415) 744-6670

Region X

Federal Office Building
909 First Avenue
Seattle, WA 98174
(206) 442-5930

F. U.S. Coast Guard District Offices

Atlantic Area

Governors Island
New York, NY 10004
(212) 668-7196

Pacific Area

Coast Guard Island
Alameda, CA 94501-5100
(415) 437-3196

District 1 (Maine, Massachusetts, New York, New Hampshire, Connecticut, Rhode Island, Vermont, Northern Pennsylvania, Northern New Jersey)

Commander
408 Atlantic Avenue
Boston, MA 02110-3350
(617) 223-8480

District 2 (Alabama, Arkansas, Colorado, Illinois, Indiana, Iowa, Kansas, Kentucky, Minnesota, Mississippi, Missouri, Nebraska, North Dakota, Ohio, Western Pennsylvania, South Dakota, Tennessee, West Virginia, Wyoming)

Commander
1430 Olive Street
St. Louis, MO 63103
(314) 539-7601

District 5 (Maryland, Delaware, North Carolina, Southern Pennsylvania, Southern New Jersey, Virginia)

Commander
431 Crawford Street
Portsmouth, VA 23705
(804) 398-6287

District 7 (Georgia, Florida, South Carolina, Puerto Rico, Virgin Islands)

Commander
909 S.E. 1st Avenue
Miami, FL 33131-3050
(305) 536-5654

District 8 (Alabama, Florida, Georgia, Louisiana, Mississippi, New Mexico, Texas)

Commander
501 Magazine Street
New Orleans, LA 70130
(504) 589-6298

District 9 (Indiana, Illinois, Michigan, Minnesota, Ohio, Pennsylvania, New York, Wisconsin)

Commander
1240 East 9th Street
Cleveland, OH 44199
(216) 522-3910

District 11 (Arizona, California, Nevada, Utah)

Commander
400 Oceangate Blvd.
Long Beach, CA 90822-5399
(213) 499-5201

District 13 (Idaho, Montana, Oregon, Washington)

Commander
915 Second Avenue
Seattle, WA 98174
(206) 442-5078

District 14 (Hawaii, Guam, American Samoa, Trust Territory of the Pacific Island, Commonwealth of Northern Mariana Islands)

Commander
300 Ala Moana Boulevard, 9th Floor
Honolulu, HI 96850
(808) 541-2051

District 17 (Alaska)

Commander
P.O. Box 3-5000
Juneau, AK 99802
(907) 586-7346

G. Department of Energy (DOE) Regional Coordinating Offices for Radiological Emergency Assistance Only

Region 1 (Connecticut, Delaware, District of Columbia, Maine, Maryland, Massachusetts, New Hampshire, New Jersey, New York, Pennsylvania, Rhode Island, Vermont)

Brookhaven Area Office:
Upton, NY 11973
(516) 282-2200

Region 2 (Arkansas, Kentucky, Louisiana, Mississippi, Missouri, Puerto Rico, Tennessee, Virgin Islands, Virginia, West Virginia)

Oak Ridge Operations Office:
P.O. Box E
Oak Ridge, TN 37831
(615) 576-4444

Region 3 (Alabama, Canal Zone, Florida, Georgia, North Carolina, South Carolina)

Savannah River Operations Office:
P.O. Box A
Aiken, SC 29808
(803) 725-2277

Region 4 (Arizona, Kansas, New Mexico, Oklahoma, Texas)

Albuquerque Operations Office:
P.O. Box 5400
Albuquerque, NM 87115
(505) 846-7231

Region 5 (Illinois, Indiana, Iowa, Michigan, Minnesota, Nebraska, North Dakota, Ohio, South Dakota, Wisconsin)

Chicago Operations Office:
9800 South Cass Avenue
Argonne, IL 60439
(312) 972-2100 (duty hours)

Region 6 (Colorado, Idaho, Montana, Utah, Wyoming)

Idaho Operations Office:
785 D.O.E. Place
Idaho Falls, ID 83415
(208) 526-1322

Region 7 (California, Hawaii, Nevada)

San Francisco Operations Office:
1333 Broadway
Oakland, CA 94612
(415) 273-7111

Region 8 (Alaska, Oregon, Washington)

Richland Operations Office:
825 Jadwin Avenue, P.O. Box 550
Richland, WA 99352
(509) 376-7395

H. Department of Transportation, Regional Pipeline Offices

Eastern Region (Connecticut, Delaware, District of Columbia, Maine, Maryland, Vermont, Massachusetts, New Hampshire, New Jersey, New York, Pennsylvania, Rhode Island, Virginia, West Virginia, Puerto Rico)

400 7th Street, S.W.
Washington, DC 20590
(202) 366-4585

Southern Region (Alabama, Florida, Georgia, Kentucky, North Carolina, South Carolina, Tennessee)

1720 Peachtree Road, N.W.
Atlanta, GA 30309
(404) 347-2632

Central Region (Iowa, Illinois, Indiana, Kansas, Michigan, Minnesota, Ohio, Missouri, Nebraska, Wisconsin)

911 Walnut Street, Room 1802
Kansas City, MO 64106
(816) 374-2654

Southwest Region (Arkansas, Louisiana, New Mexico, Oklahoma, Texas)

2320 La Branch
Houston, TX 77704
(713) 750-1746

Western Region (Arizona, California, Colorado, Idaho, Montana, Nevada, North Dakota, Oregon, South Dakota, Utah, Washington, Wyoming, Alaska, Hawaii)

555 Zang Street
Lakewood, CO 80228
(303) 236-3424

I. U.S. Nuclear Regulatory Commission Regional Offices

Region 1 (Connecticut, Delaware, District of Columbia, Maine, Maryland, Massachusetts, New Hampshire, New Jersey, New York, Pennsylvania, Rhode Island, Vermont)

USNRC
475 Allendale Road
King of Prussia, PA 19406
(215) 337-5299

Region 2 (Alabama, Florida, Georgia, Kentucky, Mississippi, North Carolina, Puerto Rico, South Carolina, Tennessee, Virginia, Virgin Islands, West Virginia)

USNRC
Suite 2900
101 Marietta Street, NW
Atlanta, GA 30323
(404) 331-5500

Region 3 (Illinois, Indiana, Iowa, Michigan, Minnesota, Missouri, Ohio, Wisconsin)

USNRC
799 Roosevelt Road
Glen Ellyn, IL 60137
(708) 790-5681

Region 4 (Arkansas, Colorado, Idaho, Kansas, Louisiana, Montana, Nebraska, New Mexico, North Dakota, Oklahoma, South Dakota, Texas, Utah, Wyoming)

USNRC
Suite 1000, Parkway Central Plaza Building
611 Ryan Plaza Drive
Arlington, TX 76011
(817) 860-8225

Region 5 (Alaska, Arizona, California, Hawaii, Nevada, Oregon, Pacific Trust Territories, Washington)

USNRC
Suite 210
1450 Maria Lane
Walnut Creek, CA 94596
(415) 975-0335

Supplement 5: Criteria for Review of Hazardous Materials Emergency Plans

This supplement is reprinted from a U.S. government publication detailing the organization and planning of a community hazardous materials incident response plan. This plan was developed by the National Response Team of the National Oil and Hazardous Substances Contingency Plan, 2100 2nd Street S.W., Washington, DC 20593.

Introduction

This document contains a set of criteria which may be used by the Regional Response Teams (RRTs) in the review of local plans under the provisions of Section 303(g) of the Superfund Amendments and Reauthorization Act of 1986 (SARA). These criteria also may be used by local emergency planning committees (LEPCs) for preparing plans as required under Section 303(a) and by state emergency response commissions (SERCs) for reviewing plans as required under Section 303(e) of the Act. This review guide is intended as a companion document to the *Hazardous Materials Emergency Planning Guide* (NRT-1), and can be viewed as a supplement to the planning process as implemented by local emergency planning committees.

Background

Section 303(a) of the Superfund Amendments and Reauthorization Act of 1986 requires each local emergency planning committee to prepare comprehensive hazardous substances emergency response plans by October 1988. The local emergency planning committee is required to review the plan once a year, or more frequently as changed circumstances in the community or at any facility may require.

Section 303(b) requires each local emergency planning committee to evaluate the need for resources necessary to develop, implement, and exercise the emergency plan, and to make recommendations with respect to additional resources that may be required and the means for providing these additional resources.

Section 303(c) specifically states that "Each emergency management plan shall include (but is not limited to) each of the following:

(1) Identification of facilities subject to the requirements of this subtitle that are within the emergency planning district, identification of routes likely to be used for the transportation of substances on the list of extremely hazardous substances referred to in Section 303(a), and identification of additional facilities contributing or subjected to additional risk due to their proximity to facilities subject to the requirements of this subtitle, such as hospitals or natural gas facilities.

(2) Methods and procedures to be followed by facility owners and operators and local emergency and medical personnel to respond to any release of such substances.

(3) Designation of a community emergency coordinator and facility emergency coordinators, who shall make determinations necessary to implement the plan.

(4) Procedures providing reliable, effective, and timely notification by the facility emergency coordinators and the community emergency coordinator to persons designated in the emergency plan, and to the public, that a release has occurred (consistent with the emergency notification requirements of Section 304).

(5) Methods for determining the occurrence of a release, and the area or population likely to be affected by such release.

(6) A description of emergency equipment and facilities in the community and at each facility in the community subject to the requirements of this subtitle, and an identification of the persons responsible for such equipment and facilities.

(7) Evacuation plans, including provisions for a precautionary evacuation and alternative traffic routes.

(8) Training programs, including schedules for training of local emergency response and medical personnel.

(9) Methods and schedules for exercising the emergency plan."

Under Section 303(e) of the Act, state emergency response commissions are required to review and make recommendations on each plan to ensure "coordination" with the plans of other local emergency planning districts.

Under Section 303(g) of the Act, the Regional Response Teams "may review and comment upon an emergency plan or other issues related to preparation, implementation, or exercise of such a plan upon request of a local emergency planning committee." This review is viewed by the National Response Team to be a form of technical assistance to the local emergency planning committees and the state emergency response commissions, and is not to be considered as an approval of these plans.

Finally, under Section 303(f), the National Response Team is required to issue guidance documents for the preparation and implementation of emergency plans. In March 1987 the National Response Team published and distributed the first such guidance document by issuing NRT-1, the *Hazardous Materials Emergency Planning Guide*. NRT-1 contains extensive discussion of both the planning process and the elements or contents required for an effective hazardous materials emergency response plan. The following plan review criteria are issued as supplemental technical guidance to NRT-1.

Basis for the Criteria

The review criteria are based on the guidelines for plans as contained in Section 303(c) of the Act, NRT-1, and CPG 1-8, *Guide for Development of State and Local Emergency Operations Plans*, published by the Federal Emergency Management Agency. Section 303(c) outlines the minimum requirements for local emergency response plans. The criteria which address these minimum requirements are introduced throughout the document by the phrase "the plan shall." The criteria based upon NRT-1, which expand upon the above minimum requirements, include all the elements that the NRT considers essential for an effective hazardous materials emergency response plan. While local emergency planning committee plans are not required to contain all these elements, the NRT believes that they should. Accordingly, the criteria based on the planning elements in NRT-1 (and many of the elements in CPG 1-8) are introduced by the phrase "the plan should." In those cases where a plan may be improved by including other considerations, the criteria are introduced by the phrase "the plan might." In these cases, the criteria are not recommended either by Title III, NRT-1, or CPG 1-8.

There also are criteria included in this document that are considered to be of such merit that they are placed under the category "the plan should," but cannot be specifically cited from Title III, NRT-1, or CPG 1-8. These criteria will be included in subsequent revisions to NRT-1 and are highlighted in this document with an asterisk (*).

CPG 1-8 (used as one of the sources for these criteria) is used by local governments to develop emergency operations plans which are required by the Comprehensive Cooperative Agreements between the Federal Emergency Management Agency and the states. CPG 1-8a, *Guide for the Review of State and Local Emergency Operations Plans*, also was used as a resource for developing these review criteria. The planning elements in CPG 1-8 have been incorporated into NRT-1, and most of the review criteria in CPG 1-8a are included in the attached RRT Plan Review Document. The relevant sections of Title III, NRT-1, and CPG 1-8 are indexed in the review document for informational purposes.

The plan criteria outlined below are structured to correspond to the sequence of plan elements suggested by Chapters 4 through 6 of NRT-1 and Chapters 2 and 3 of CPG 1-8.

Use of Criteria

The NRT expects that the primary use of these criteria will be in plan review by Regional Response Teams. Through the use of these criteria and the development of comments related thereto, the RRTs can both conduct organized and systematic reviews of local plans and ensure that plan elements of particular interest to RRTs are covered. The RRTs should also use the criteria as a basis for ensuring coordination between federal plans developed under the National Contingency Plan (e.g., Regional Contingency Plans and OSC Plans) and plans developed at the local level.

As mentioned above, the local emergency planning committees may find the criteria useful in the development of plans required under Section 303(c) of the Act. These criteria are concise statements of the contents of plan elements covered in NRT-1 and CPG 1-8, and all of the plan elements required in Section 303(c). It is essential, however, that the criteria be used by local emergency planning committees only in concert with the full range of available guidance.

State emergency response commissions may find the plan criteria useful in the coordination of local emergency planning committees and in the review of each local plan. The criteria offer a useful guide for all of the planning elements which may require coordinated and consistent treatment among the local emergency planning committees within a state. They also provide the basis for a more general review of plans.

RRT Consideration of the LEPC Planning Process

One of the major themes of NRT-1 is that the way in which a local hazardous materials emergency plan is developed is as important as the actual contents of such a plan. Thus, the Regional Response Teams may find it useful to secure the following information pertaining to the local emergency planning committee under review:

1. A list of the names and affiliations of the members of the LEPC;
2. A description of the activities and accomplishments (with completion dates) of the committee in compliance with Section 301, including:

Appointment of a chairperson;

Establishment of rules for committee operations;

Development of methods for public notification of committee activities;

Conduct of public meetings on the emergency plan;

Receiving and responding to public comments;

Public notice of availability of emergency response plan, Material Safety Data Sheets, and inventory forms under Section 324;

Dealing with public requests for information under Sections 311, 312 and 313; and

Securing information from facilities covered by the plan.

3. A description of the major activities of the committee in completing the tasks for the hazard analysis and capability assessment.
4. A summary of the data produced by these tasks, if not already described in the plan.
5. A summary of the resources expended in developing the plan, including local funds, staff effort and technical expertise, plus a summary of resources required for maintaining and revising the plan.
6. A description of any findings on ways to fund hazardous materials emergency planning within the district.

	Criteria for Plans	Documentation		
		Title III	NRT-1	CPG 1-8
1.0	INCIDENT INFORMATION SUMMARY			
	The Plan should[1] contain:			
1.1	Detailed description of the essential information that is to be developed and recorded by the local response system in an actual incident, e.g., date, time, location, type of release, and material released;		A.1	
2.0	PROMULGATION DOCUMENT			
	The Plan should contain:			
2.1	A document signed by the chairperson of the LEPC, promulgating the plan for the district;		A.2	2.3(a)(1)
2.2	Documents signed by the chief executives of all local jurisdictions within the district[2]; and		*	
2.3	Letters from affected facilities endorsing the plan.		*	
	The Plan might contain:			
2.4	Letters of agreement between the affected facilities and local jurisdictions for emergency response and notification responsibilities.		A.2	
3.0	LEGAL AUTHORITY AND RESPONSIBILITY FOR RESPONSE			
	The Plan should:			
3.1	Describe, reference, or include legal authorities of the jurisdictions whose emergency response roles are described in the plan, including authorities of the emergency planning district and the local jurisdictions within the district; and		A.3	2.3(h)
3.2	List all other authorities the LEPC regards as essential for response within the district, including state and federal authorities.		"	
4.	TABLE OF CONTENTS			
	The plan should:			
4.1	List all elements of the plan, provide tabs for each, and provide a cross-reference for all of the nine required elements in Sectin 303 of the Act. Plans that are prepared in the context of requirements of CPG 1-8 should contain an index to the location of both NRT-1 and Section 303 elements.		A.4	2.3(a)(3)

[1](a) All criteria with a Title III reference are required by Section 303(c) of the Act and are introduced by the phrase "The Plan shall."

(b) All criteria with a NRT-1 reference are not required by Title III, but are regarded as essential by the NRT for an effective hazardous materials emergency response plan. They can be found in NRT-1, Chapter 5, "Planning Elements," and are introduced by the phrase "The Plan should."

(c) All CPG 1-8 references include those criteria that address requirements for emergency operations plans prepared under the provisions of the Comprehensive Cooperative Agreements with the Federal Emergency Management agency.

[2] These criteria are considered to be of such merit that they are included under the heading "the plan should," but cannot specifically be cited from Title III, NRT-1, or CPG 1-8. They are designated in the documentation section by an asterisk (*). They will be included in subsequent revisions to NRT-1.

	Criteria for Plans	Documentation		
		Title III	NRT-1	CPG 1-8
5.	ABBREVIATIONS AND DEFINITIONS			
	The Plan should:			
5.1	Explain all abbreviations and define all essential terms included in the plan text.		A.5	2.3(i)
6.0	PLANNING FACTORS			
	Assumptions: Assumptions are the advance judgments concerning what might happen in the case of an accidental spill or release.			
	The Plan should:			
6.1	List all of the assumptions about conditions that might develop in the district in the event of accidents from any of the affected facilities or along any of the transportation routes.		A.6	2.3(c)
	Planning Factors: The planning factors consist of all the local conditions that make an emergency plan necessary.			
	The Plan shall:			
6.2	Identify and describe the facilities in the district that possess extremely hazardous substances and the transportation routes along which such substances may move within the district;	303(c)(1)	A.6	2.3(c)
6.3	Identify and describe other facilities that may contribute to additional risk by virtue of their proximity to the above mentioned facilities;	"	"	"
6.4	Identify and describe additional facilities included in the plan that are subject to additional risks due to their proximity to facilities with extremely hazardous substances; and	303(c)(1)	A.6	2.3(c)
6.5	Include methods for determining that a release of extremely hazardous substances has occurred, and the area of population likely to be affected by such release.	303(c)(5)	A.6	2.3(c)
	The Plan should:			
6.6	Include the major findings from the hazard analysis (date of analysis should be provided), which should consist of:		A.6	2.3(a)(4)
6.6.1	Major characteristics of affected facilities/ transportation routes impacting on the types and levels of hazards posed, including the types, identities, characteristics, and quantities of hazardous materials related to facilities and transportation routes;		"	"
6.6.2	Potential release situations with possible consequences beyond the boundaries of facilities, or adjacent to transportation routes. Use may be made of historical data on spills and any data secured from facilities under Section 303(d)(3) of the Act;		"	"
6.6.3	Maps showing locations of facilities, transportation routes, and special features of district, including vulnerable areas;		"	"

	Criteria for Plans	Documentation		
		Title III	NRT-1	CPG 1-8
6.7	Geographical features of the district, including sensitive environmental areas, land use patterns, water supplies, and public transportation;		"	"
6.8	Major demographic features of the district, including those features that impact most on emergency response, e.g., population density, special populations, and particularly sensitive institutions;		"	"
6.9	The district's climate and weather as they affect airborne distribution of chemicals; and		"	"
6.10	Critical time variables impacting on emergencies, e.g., time of day and month of year in which they would be most likely to occur.		"	"
7.0	CONCEPT OF OPERATIONS			
	The Plan shall:			
7.1	Designate a community emergency coordinator and facility emergency coordinators, who shall make determinations necessary to implement the plan.	303(c)(3)		
	The Plan should:			
7.2	Identify, by title, the individual designated as the community emergency coordinator and each of the facility emergency coordinators;		A.7b	2.3(d)
7.3	Explain the relationships between these coordinators, their organizations, and the other local governmental response authorities within the district, e.g., the county emergency management authority;		A.7b	2.3(d)
7.4	Describe the relationship between this plan and other response plans within the district which deal in whole or in part with hazardous materials emergency response, e.g., the county Emergency Operations Plan and plans developed by fire departments under OSHA Regulation CFR 29 Part 1910.120;		A.7c	
7.5	List all the facility emergency plans within the district that apply to hazardous materials emergency response, including all plans developed under OSHA Regulation on Hazardous Waste Operations and Emergency Response (CFR 29 Part 1910.120);		"	
7.6	Describe the way in which the above plans are integrated with local response plans;		"	
7.7	Describe the functions and responsibilities of all the local response organizations within the district, including public and private sector as well as volunteer and charitable organizations;		A.7b	2.3(e)
7.8	List mutual aid agreements or other arrangements for sharing data and response resources;		A.3	
7.9	Describe conditions under which the local government will coordinate its response with other districts and the means or sequence of activities to be followed by districts in interacting with other districts;		A.7b	
7.10	Describe the relationship between plans of the district and related state plans;		A.7c	
7.11	Describe the relationship between local and state emergency response authorities; and		A.7b	

Criteria for Plans		Documentation		
		Title III	NRT-1	CPG 1-8
7.12	Describe the relationships between emergency response plans and activities in the district and response plans and activities by federal agencies, including all plans and responses outlined in the National Contingency Plan.		A.7b; A.7c	
	[Emphasis should be given to the allocation of responsibilities among federal emergency response agencies, including listing the names and the responsibilities of federal response agencies and entities such as the RRT, describing the means for their notification, the types of resources to be sought from them, the means for obtaining them, the conditions under which assistance is to be provided, and the methods of coordination during a response.]			
8.0	INSTRUCTIONS FOR PLAN USE			
	The Plan should:			
8.1	Contain a discussion of the purpose of the plan; and		A.8a	2.3(a)(5)(b)
8.2	Contain a list of organizations and persons receiving the plan or plan amendments and the date that the plan was transmitted, as well as a specific identification number for each plan.		A.8b	2.3(a)(5)(c)
9.0	RECORD OF AMENDMENTS			
	The Plan should:			
9.1	Contain a section that describes methods for maintaining and revising the plan and recording all changes in the plan, including a method for controlling distribution.		A.9	2.3(a)(6)
10.0	EMERGENCY NOTIFICATION PROCEDURES			
	The Plan shall:			
10.1	Include procedures for providing reliable, effective, and timely notification by the facility emergency coordinators and the community emergency coordinator to persons designated in the emergency plan, and to the public, that a release has occured.	303(c)(4)	C.4	3.3(c)
	The Plan should:			
10.2	Include procedures for immediately notifying the appropriate 24-hour hotline first, and should locate these procedures in a prominent place in the plan;		C.1	
10.3	List the 24-hour emergency hotline number(s) for the local emergency response organization(s) within the district;		"	
10.4	Contain an accurate and up-to-date Emergency Assistance Telephone Roster that includes numbers for the:		B	
10.4.1	Technical and response personnel;		"	
10.4.2	Community emergency coordinator, and all facility emergency coordinators;		*	

	Criteria for Plans	Documentation		
		Title III	NRT-1	CPG 1-8
10.4.3	CHEMTREC		B	
10.4.4	National Response Center[3]		"	
10.4.5	Other participating agencies;		"	
10.4.6	Community emergency coordinators in neighboring emergency planning districts;		*	
10.4.7	Public and private sector support groups; and the		B	
10.4.8	Points of contact for all major carriers on transportation routes within the district;		*	
10.5	List all local organizations to be notified of a release, and the order of their notification, and list names and telephone numbers of primary and alternate points of contact;		*	
10.6	List all local institutions to be notified of the occurrence of a release and the order of their notification, and the names and telephone numbers of contacts;		*	
10.7	List all state organizations to be notified, and list the names and telephone numbers of contacts; and		*	
10.8	List all federal response organizations to be notified, and the names and telephone numbers of the contacts.		*	
11.0	INITIAL NOTIFICATION OF RESPONSE AGENCIES			
	The Plan should:			
11.1	Describe methods or means to be used by facility emergency coordinators (FECs) within the district to notify community emergency coordinators (CEDs) of any potentially affected districts, and SERCs of any potentially affected states, and any other persons to whom the facility is to give notification of any release, in compliance with Section 304 of Title III;		C.1	
11.2	Describe methods by which the CECs and local response organizations will be notified of releases from transportation accidents, following notification through 911 systems or specified alternative means;		"	
11.3	Describe methods by which the CEC, or his/her designated agent, will ensure that contents of notification match the requirements of Section 304, including the regulations contained in 40 CFR Part 355 (Notification Requirements, Final Rule);		"	
11.4	List procedures by which the CEC will assure that both the immediate and follow-on notifications from facility operators are made within the time frames specified by Notification of Final Rule in 40 CFR Part 355; and		"	

[3] Spills exceeding the CERCLA reportable quantities are required to be reported to the National Response Center.

	Criteria for Plans	Documentation		
		Title III	NRT-1	CPG 1-8
11.5	Identify the person or office responsible for receiving the notification for the community emergency coordinator or his/her designated agent and list the telephone number;		"	
12.0	DIRECTION AND CONTROL			
	The Plan shall:			
12.1	Include methods and procedures to be followed by facility owners and operators and local emergency and medical personnel to respond to a release of extremely hazardous substances.	303(c)(2)		
	The Plan should:			
12.2	Identify the organization within the district responsible for providing direction and control to the overall emergency response system described in the Concept of Operations;		C.2	3.3(a)
12.3	Identify persons or offices within each response organization who provide direction and control to each of the organizations;		"	"
12.4	Identify persons or offices providing direction and control within each of the emergency response functions;		"	"
12.5	Describe persons or offices responsible for the performance of incident command functions and the way in which the incident command system is used in hazardous substances incidents;		"	"
12.6	Describe the chain of command for the total response system, for each of the major response functions, and for the organization controlled by the incident commander; and		"	"
12.7	Identify persons responsible for the activation and operations of the emergency operations center, the on-scene command post, and the methods by which they will coordinate their activities;		"	"
12.8	List three levels of incident severity and associated response levels;		"	"
12.9	Identify the conditions for each level; and		"	"
12.10	Indicate the responsible organizations at each level.		"	"
13.0	COMMUNICATION AMONG RESPONDERS			
	The Plan should:			
13.1	Describe all the methods by which identified responders will exchange information and communicate with each other during a response, including the communications networks and common frequencies to be used; [At a minimum, these methods should be described for each function. Both communications among local response units and between these units and facilities where incidents occur should be described.]		C.3	3.3(b)

	Criteria for Plans	Documentation		
		Title III	NRT-1	CPG 1-8
13.2	Describe the methods by which emergency respnoders can receive information on chemical and related response measures; and		C.3	
	[May include a description of computer systems with on-line data bases.]			
13.3	Describe primary and back-up systems for all communicaton channels and systems.		*	3.3(b)
	The Plan might:			
13.4	Contain a diagram or matrix showing the flows of information within the response system.		*	
14.0	WARNING SYSTEMS AND EMERGENCY PUBLIC NOTIFICATION			
	The Plan should:			
14.1	Identify responsible officials within the district and describe the methods by which they will notify the public of a release from any facility or along any transportation route, including sirens or other signals, and use of the broadcast media and the Emergency Broadcast System. This should include a description of:		C.4	3.3(c)
14.1.1	The sirens and other signals to be employed, their meaning, their methods of coordination, and their geographical coverage;		"	"
14.1.2	Other methods, such as door-to-door alerting, that may be employed to reach segments of the population that may not be reached by sirens or other signals; and		"	"
14.1.3	Time frames within which notification to the public can be accomplished;		"	"
14.2	Describe methods for the coordination of emergency public notification during a response; and		"	"
14.3	Describe any responsibilities or activities of facilities covered by the Act for emergency public notification during a response.	"	"	
15.	PUBLIC INFORMATION AND COMMUNITY RELATIONS			
	The Plan should:			
15.1	Describe the methods used by local governments, prior to emergencies, for educating the public about possible emergencies and planned protective measures;		C.5	3.3(d)
15.2	Describe the role and organizational position of the public information officer during emergencies;		"	
15.3	Designate a spokesperson and describe the methods for keeping the public informed during an emergency situation, including a list of all radio, TV, and press contacts, and;		"	
15.4	Describe any related public information activities of affected facilities, both prior to an emergency and during an emergency.		*	

Criteria for Plans		Documentation		
		Title III	NRT-1	CPG 1-8
16.0	RESOURCE MANAGEMENT			
	The Plan shall:			
16.1	Include a description of emergency equipment and facilities in the community and at each facility in the community subject to the requirements of this subtitle and an identification of the persons responsible for such equipment and facilities.	303(c)(6)		
	The Plan should:			
16.2	List personnel resources available for emergency response by major categories, including governmental, volunteer, and the private sector;		C.6	3.3(n)
16.3	Describe the types, quantities, capabilities and locations of emergency response equipment available to the local emergency response units, including fire, police and emergency medical response units.		"	"
	[Categories of equipment should include transportation, communications, monitoring and detection, containment, decontamination, removal, and cleanup.]			
16.4	List the emergency response equipment available to each of the affected facilities and describe them in the same way as community equipment is described;		"	"
16.5	Describe the emergency operating centers or other facilities available to the local community and the facility emergency coordinators and other response coordinators, such as incident commanders;		*	
16.6	Describe emergency response equipment and facilities available to each affected facility and the conditions under which they are to be used in support of local responders;		"	"
16.7	Describe significant resource shortfalls and mutual support agreements with other jurisdictions whereby the district might increase its capabilities in an emergency;		"	"
	[This may be discussed under the Concept of Operations.]		"	"
16.8	Describe procedures for securing assistance from federal and state agencies and their emergency support contractors;		"	"
	[This may be discussed under the Concept of Operations.]			
16.9	Describe emergency response capabilities and the expertise in the private sector that might be available to assist local responders, facility managers, and transportation companies during emergencies.		C.6	3.3(n)

	Criteria for Plans	Documentation		
		Title III	NRT-1	CPG 1-8
17.0	HEALTH AND MEDICAL			
	The Plan shall:			
17.1	Include methods and procedures to be followed by facility owners and operators and local emergency and medical personnel to respond to a release of extremely hazardous substances.	303(c)(2)		
	The Plan should:			
17.2	Describe the procedures for summoning emergency medical and health department personnel;		C.7	3.3(h)
17.3	Describe the procedures for the major types of emergency medical services, including first aid, triage, ambulance service, and emergency medical care, using both the resources available within the district and those that can be secured in neighboring districts;		"	"
17.4	Describe the procedures to be followed for decontamination of exposed people;		"	"
17.5	Describe the procedures for providing sanitation, food, water supplies, and safe re-entry of persons to the accident area;		"	"
17.6	Describe procedures for conducting health assessments upon which to base protective action decisions;		*	
17.7	Describe the level and types of emergency medical capabilities in the district to deal with exposure of people to extremely hazardous substances;		C.7	
17.8	Describe the provisions for emergency mental health care; and		"	
17.9	Indicate mutual aid agreements with other communities to provide back-up emergency medical and health department personnel, and equipment.		*	
18.0	RESPONSE PERSONNEL SAFETY			
	The Plan should:			
18.1	Describe initial and follow-up procedures for entering and leaving incident sites, including personnel accountability, personnel safety precautions, and medical monitoring.		C.8	
18.2	Describe personnel and equipment decontamination procedures; and		"	
18.3	List sampling, monitoring and personnel protective equipment appropriate to various degrees of hazards based on EPA levels of protection (A, B, C, & D)		"	
	[Just prior to publication of NRT-1, the Occupational Safety and Health Administration (OSHA) published proposed rules (29 CFR Part 1910.120) to provide more definitive requirements to plan for emergency response personnel safety. If the LEPC plans include a section on this function, the plan elements listed in the OSHA regulation should be used.]			

	Criteria for Plans	Documentation		
		Title III	NRT-1	CPG 1-8
19.0	PERSONAL PROTECTION OF CITIZENS/ INDOOR PROTECTION			
	The Plan shall:			
19.1	Describe methods in place in the community and in each of the affected facilites for determining the areas likely to be affected by a release.	303(c)(5)		
	The Plan should:			
19.2	Include methods to predict the speed, direction, and concentration of plumes resulting from airborne releases, and methods for modeling vapor cloud dispersion as well as methods to monitor the release and concentration in real time;		C.9a	3.3(g)
	[19.1 and 19.2 may be considered in the hazard analysis, included in Section 6, Planning Factors.]			
19.3	Identify the decision-making processing, including the decision-making authority for indoor protection;		"	"
19.4	Describe the roles and activities of affected facilities in the decision-making for indoor protection decisions, including the determination that indoor sheltering is no longer required;		*	
19.5	Indicate the conditions under which indoor protection would be recommended, including the decision-making criteria;		C.9a	3.3(g)
19.6	Describe the methods for indoor protection that would be recommended for citizens, including provisions for shutting off ventilation systems; and		*	
19.7	Describe the methods for educating the public on indoor protective measures;		C.9a	
	[May be discussed in the section on public information.]			
20.0	PERSONAL PROTECTIVE MEASURES/ EVACUATION PROCEDURES			
	The Plan shall:			
20.1	Describe evacuation plans, including those for precautionary evacuations and alternative traffic routes;	303(c)(7)	C.9b	3.3(e)
	The Plan should:			
20.2	Describe the authority for ordering or recommending evacuation, including the personnel authorized to recommend evacuation.		C.9b	
20.3	Describe the authority and responsibility of various governmental agencies and supporting private sector organizations, such as the Red Cross, and the chain of command among them;		"	
20.4	Describe the role of the affected facilities in the evacuation decision-making		*	

	Criteria for Plans	Documentation		
		Title III	NRT-1	CPG 1-8
20.5	Describe methods to be used in evacuation, including methods for assisting the movement of mobility impaired persons and in the evacuation of schools, hospitals, prisons, and other facilities;		C.9b	
20.6	Describe the relationship of evacuation procedures to other protective measures.		*	
20.7	Describe potential conditions requiring evacuation, i.e, the types of accidental release and spills that may require evacuation;		C.9b	
20.8	Describe evacuation routes, including primary and alternative routes; [These may be either established routes for the community or special routes appropriate to the location of facilities.]		"	3.3(e)
20.9	Describe evacuation zones and distances and the basis for their determination; [These should be related to the location of facilities and transportation routes and the potential pathways to exposure.]		"	"
20.10	Describe procedures for precautionary evacuations of special populations;		"	
20.11	List the mass care facilities for providing food, shelter, and medical care to relocated populations; [This may be discussed under the human services section.]		"	3.3(f)
20.12	Describe procedures for providing security for the evacuation, for evacuees, and of the evacuated areas; [May be covered under the law enforcement discussions.]		C.11	3.3(i)
20.13	Describe methods for managing the flow of traffic along evacuation routes and for keeping the general public from entering threatened areas, including maps with traffic and other control points; and [May be covered in the law enforcement section.]		C.9b	3.3(e)
20.14	Describe the procedures for managing an orderly return of people to the evacuated area;		C.9b	3.3(e)
21.0	FIRE AND RESCUE The Plan should:			
21.1	List the major tasks to be performed by fire fighters in coping with releases of extremely hazardous substances;		C.10	3.3(k)
21.2	Identify the public and private sector fire protection organizations with a response capability and responsibility for hazardous materials incidents;		"	"

Criteria for Plans		Documentation		
		Title III	NRT-1	CPG 1-8
21.3	Describe the command structure of multi-agency, multi-jurisdictional incident management systems in place, and identify applicable mutual aid agreements and good samaritan provisions in place;		"	"
21.4	List available support systems, e.g., protective equipment and emergency response guides, DOT Emergency Response Guidebook, mutual aid agreements, and good samaritan provisions; and		"	"
	[May be covered under resource management.]			
21.5	List and describe any HAZMAT teams in the district.		*	
	[May be covered in Section 21.2 above.]			
22.0	LAW ENFORCEMENT			
	The Plan should:			
22.1	Describe the command structure of multi-agency, multi-jurisdictional incident management systems in place, and identify applicable mutual aid agreements and good samaritan provisions in place;		C.11	3.3(i)
22.2	List the major law enforcement tasks related to responding to releases of extremely hazardous materials, including those related to security for the accident site and for evacuation activities; and		"	"
22.3	List the locations of control points for the performance of tasks, with appropriate maps,		*	
23.0	ONGOING INCIDENT ASSESSMENT			
	The Plan should:			
23.1	Describe methods in place in the community and/or each of the affected facilities for determining the areas likely to be affected by an ongoing release.		C.12	3.3(p)
23.2	Describe methods for determining the private and public property that may be in the affected areas and the nature of the impact of the release on this property;		"	"
23.3	Describe methods and capabilities of both local response organizations and facilities for monitoring the size, concentration, and migration of leaks, spills, and releases, including sampling around the site; and		C.12	
23.4	Describe provisions for environmental assessments, biological monitoring, and contamination surveys;		"	
24.0	HUMAN SERVICES			
	The Plan should:			
24.1	List the agencies responsible for providing emergency human services, e.g, food, shelter, clothing, continuity of medical care, and crisis counseling; and		C.13	3.3(m)

	Criteria for Plans	Documentation		
		Title III	NRT-1	CPG 1-8
24.2	Describe the major human services activities and the means for their accomplishment.		"	"
25.	PUBLIC WORKS			
	The Plan should:			
25.1	Describe the chain of command for the performance of public works actions in an emergency; and		C.14	3.3(j)
25.2	List all major tasks to be performed by the public works department in a hazardous materials incident.		"	"
26.0	TECHNIQUES FOR SPILL CONTAINMENT AND CLEANUP			
	The Plan should:			
26.1	Explain the allocation of responsibilities among local authorities and affected facilities and responsible parties for these activities:		D.1	
26.2	Describe the major containment and mitigation activities for all major types of HAZMAT incidents;		"	
26.3	Describe cleanup and disposal services to be provided by the responsible parties and/or local community;		"	
26.4	Describe major methods for cleanup;		"	
26.5	Describe methods to restore the surrounding environment, including natural resource areas, to pre-emergency conditions;		"	
26.6	Describe the provisions for long-term site control;		D.2	
26.7	List the location of approved disposal sites;		"	
26.8	List cleanup material and equipment available within the district;		"	
	[May be covered in the resource management section.]			
26.9	Describe the capabilities of cleanup personnel; and		D.2	
26.10	List the applicable regulations governing disposal of hazardous materials in the district.		"	
27.	DOCUMENTATION AND INVESTIGATIVE FOLLOW-UP			
	The Plan should:			
27.1	List all reports required in the district and all offices and agencies that are responsible for preparing them following a release;		E	
27.1	Describe the methods of evaluating responses and identify persons responsible for evaluations; and		"	
27.3	Describe provisions for cost recovery.		"	

	Criteria for Plans	Documentation		
		Title III	NRT-1	CPG 1-8
28.0	PROCEDURES FOR TESTING AND UPDATING THE PLAN			
	The Plan shall:			
28.1	Include methods and schedules for exercising the emergency plan.	303(c)(9)	F.1	
	The Plan should:			
28.2	Describe the nature of the exercises for testing the adequacy of the plan;		"	
28.3	List the frequency of such exercises, by type;		"	
28.4	Include an exercise schedule for the current year and for future years;		"	
28.5	Describe the role of affected facilities or transportation companies in these exercises; and		"	
28.6	Describe the procedures by which performance will be evaluated in the exercise, revisions will be made to plans, and deficiencies in response capabilities will be corrected.		F.6	
29.0	TRAINING			
	The Plan shall:			
29.1	Include the training programs, including schedules, for training of local emergency response and medical personnel.	303(c)(8)	6.4.3	
	The Plan should:			
29.2	Describe training requirements for LEPC members and all emergency planners within the district;		*	
29.3	Describe training requirements for all major categories of hazardous materials emergency response personnel, including the types of courses and the number of hours;		*	
29.4	List and describe the training programs to support these requirements, including all training to be provided by the community, state and federal agencies, and the private sector; and		*	
29.5	Contain a schedule of training activities for the current year and for the following three years.		*	

Supplement 6: Hazardous Materials Emergency Response Plans

The following supplement contains two excellent examples of working response plans for hazardous materials emergencies, one developed by the City of Sacramento Fire Department and the other by the Fire Department of New York. Our thanks to both fire departments, especially to Sacramento Battalion Chief Jan Dunbar and to New York Deputy Chief Joseph Gallagher, for generously allowing us to use these documents in the Hazardous Materials Response Handbook. For more information on these response plans, contact either the Sacramento Fire Department, 1231 I Street, Suite 401, Sacramento, CA 95814 or the Fire Department of New York, Special Operations Command, 750 Main Street, New York, NY 10044.

City of Sacramento Fire Department

1. Planning Basis.

1.1 Purpose.

To establish an organization to mitigate hazardous materials incidents within the City Limits of Sacramento and on mutual aid hazardous materials incident outside the City Limits of Sacramento.

1.2 Objectives.

1.2.1 To describe operational concepts, organization, and support systems required to implement the plan.

1.2.2 To identify authority, responsibilities, and actions of Federal, State, Local, Private Industry, and Utility agencies necessary to minimize damage to human health, natural systems, property, and to aid in the mitigation of the hazard.

1.2.3 To establish an operational structure that has the ability to function not only within the City of Sacramento, but also on any mutual aid call where Sacramento City Fire Department equipment responds to a hazardous materials incident outside of the City Limits of Sacramento.

1.2.4 To utilize Fire Department officers and members who have been trained to handle hazardous materials incidents.

1.2.5 To establish lines of authority and management for a hazardous materials incident.

1.2.6 To establish and provide an overall response plan that adheres to and addresses the provisions of SARA TITLE III (OSHA), Title 29 CFR Section 1910 that affect the operations and functions of the fire service and hazardous materials response teams who will or might be engaged in various mitigation activities at the scene of a hazardous materials incident.

2. Administration

2.1 Scope.

2.1.1 Geographical Factors. This plan is directed to those hazardous materials incidents which occur within the Limits of the City of Sacramento, and on any mutual aid call to any location or agency outside the City of Sacramento.

2.1.2 The Hazards Factor. The *hazards* shall include actual or the threat of fires, spills, leaks, ruptures, container failure, contamination, any threat to life safety, property, or the environment involving hazardous materials.

2.1.3 The Hazardous Materials Factor. The hazardous material itself may include but not be limited to explosives, flammables, combustibles, compressed gases, cryogenics, poisons, toxics, reactive and oxidizing agents, radioactive materials, corrosives, carcinogenics, etiological agents, ORMs, hazardous substance, hazardous waste, or any combination thereof, or any material that may pose a hazard to health or the environment in the opinion of the officer of the Hazardous Materials Response Team.

2.1.4 This plan is for any hazardous materials incident associated with any mode of transportation, any industrial process, storage or storage sites, waste disposal procedures, manufacturing, usage, abandonent, and illegal usage and disposal.

2.2 Authority.

2.2.1 Charter of the City of Sacramento.

2.2.2 Sacramento City Code, Chapter 15, Article 2.

2.2.3 Uniform Fire Code: Article 10, Division 1.

2.2.4 California Government Code: Chapter 7, Division 1, Title 2.

2.2.5 California Health and Safety Code: Sections 25115 and 2517, and Sections 2560 through 25610.

2.2.6 California Vehicle Code.

2.2.7 California State Office of Emergency Services Fire and Rescue Mutual Aid Plan.

2.3 References.

2.3.1 California State Hazardous Materials Incident Contingency Plan, 1982.

2.3.2 California State Oil Spill Contingency Plan.

2.3.3 California State Radiological Emergency Assistance Plan.

2.3.4 Federal Response Plan.

2.3.5 Sacramento County Hazardous Materials Emergency Response Plan.

3. Hazardous Materials Incident Classification (Levels)

3.1 There are three (3) levels of hazardous materials incident classification. The bases used for the establishment of the concept of classifying hazardous materials incidents into LEVELS are:

3.1.1 Level of technical expertise required to abate the incident.

3.1.2 Extent of Local, State, and Federal Government, and Private Industry involvement required to assist in abating the hazard.

3.1.3 Extent of evacuation of civilians.

3.1.4 Extent of injuries and/or deaths related to the hazardous materials incident.

3.1.5 Extent and involvement of decontamination procedures.

3.2 Level I Incident (Known as a Level I HMI).

3.2.1 Spills, leaks, ruptures, and/or fires involving hazardous materials which can be contained, extinguished, and/or abated utilizing equipment, supplies, and resources immediately available to the first responders of the fire department having jurisdiction; and

3.2.2 The incident can properly be handled by fire department personnel whose qualifications are limited to and do not exceed the scope of training explained in SARA TITLE III (OSHA), Title 29 CFR Section 1910 with reference to *first responder*.

3.2.3 Hazardous materials incidents which do not require evacuation of civilians beyond the perimeter of incident scene isolation.

3.3 Level II Incident (Known as a Level II HMI).

Any Sacramento Fire Department Officer can upgrade a Level I HMI to a Level II HMI.

3.3.1 A hazardous materials incident which can only be identified, tested, sampled, contained, extinguished, and/or abated utilizing the expertise and resources of the Sacramento Fire Department Hazardous Materials Response Team; a hazardous materials incident which requires the use of any kind of specialized protective gear, tools, equipment or knowledge beyond the normal scope of a *first responder*; and/or

3.3.2 Hazardous materials incident which requires the evacuation of civilians within the area of the fire department having jurisdiction; and/or

3.3.3 Fires involving hazardous materials that are permitted to burn for a controlled period of time, or are allowed to consume themselves; and/or

3.3.4 The incident can only be properly handled by fire department personnel whose qualifications meet or exceed the scope of training explained in SARA TITLE III (OSHA), Title 29 CFR Section 1910 with reference to *hazardous materials specialist.*

3.4 Level III Incident (Known as a Level III HMI).

Any Sacramento Fire Department Officer can upgrade a Level I HMI or a Level II HMI to a Level III HMI.

3.4.1 Actual or the threat of spills, leaks, or ruptures which can or must be contained and/or abated only by utilizing the highly specialized equipment and supplies available to environmental and industrial response personnel. Such equipment, techniques, and qualified personnel are in excess of or are in addition to those available from the on-scene hazardous materials response team; and/or

3.4.2 Fires involving hazardous materials that are allowed to burn due to the ineffectiveness or dangers of the use of any kind of extinguishing agent, or the unavailability of the proper extinguishing agent; and/or there is a real threat of large container failure; and/or an explosion, detonation, BLEVE or container failure has already occurred; and/or

3.4.3 Hazardous materials incidents which require evacuation of civilians from a large geographical area, or evacuation has extended across jurisdictional boundaries; and/or there are serious civilian injuries and/or deaths as a result of the hazardous materials incident; and/or

3.4.4 Hazardous materials incidents which require at least two Sacramento Fire Department Hazardous Materials Response Teams on scene; and/or decontamination of equipment, civilians, or personnel is required; and/or the Sacramento Fire Department De-Con Team is required on scene; and/or

3.4.5 The hazardous materials incident has become one of a multi-agency involvement.

3.4.6 The incident can only be properly handled by fire department personnel whose qualifications meet or exceed the scope of training explained in SARA TITLE III (OSHA), Title 29 CFR Section 1910 with reference to *hazardous materials specialist.*

4. Dispatch Procedures for HAZ-MAT Incidents

4.1 Within Jurisdiction of Sacramento Fire Department.

Dispatch of equipment to a reported hazardous materials incident shall be as follows:

4.1.1 Level I HMI.
Initial dispatch — One engine co. (as a first responder)

4.1.2 Level II HMI.
Initial dispatch — One engine co. (if not already on scene)
One HMRT
One BC
One investigator

Upgrade of a Level I HMI to a Level II HMI (engine is already on scene)
— One HMRT
One BC
One investigator

4.1.3 Level III HMI.
Upgrade of a Level II HMI to a Level III HMI (Level II response is already on scene)
— One additional HMRT
One De-Con Team
One additional BC

4.2 Mutual Aid to Outside Sacramento Fire Department.

4.2.1 Level I HMI. (Handled by fire department having jurisdiction on a first responder level)
Initial dispatch — None

4.2.2 Level II HMI. (Upon request of outside agency already on scene)
Initial dispatch — One HMRT

> BC is apprised of incident, and may elect to respond; otherwise, BC is *not* dispatched.

4.2.3 Level III HMI. (From HMRT or IC already on scene)
Upgrade of a Level II HMI to a Level III HMI (Level II response is already on scene)

> — One additional HMRT
> One De-Con Team
> One BC

5. Incident Command and Scene Management

5.1 Incident Commander.

5.1.1 The *incident commander* (IC) shall be the designated fire department officer responsible for all operations directed toward the containment and mitigation of the hazards at the scene of a hazardous materials incident. Upon arrival, the IC shall secure and maintain immediate control until the situation has been corrected or abated.

5.1.1.1 The Sacramento Fire Department shall accept and provide the position of incident commander for the scene of all hazardous materials incidents within the *City of Sacramento*. The Sacramento Fire Department shall coordinate and direct within its jurisdiction and responsibility to include, but not be limited to: rescue and first aid, product identification, scene stabilization and management, agency notification, scene isolation, personnel protection, safety, decontamination, and enforcement of all applicable regulations, laws, and fire department procedures.

5.1.1.2 The captain of the Sacramento Fire Department Hazardous Materials Response Team shall report to and function through the on-scene fire department incident commander.

5.1.1.3 The captain of the Hazardous Materials Response Team shall not assume nor be given the responsibility of incident command of any hazardous materials incident. Only when first on scene shall the captain of the Hazardous Materials Response Team function as the incident commander. Immediately upon arrival of the next fire department company, the Captain of the HMRT shall pass the command of the incident to that fire officer.

5.1.2 The incident commander shall report to and function through the scene manager when appropriate.

5.2 Scene Manager.

5.2.1 The *scene manager* (SM) shall employ overall management and coordination of the hazardous materials incident. The scene manager shall be responsible for the identification of incident resources and needs, the procurement of these resources, and the coordination of the resources so as to abate the incident and protect life, property, and the environment.

5.2.2 The scene manager shall not be responsible for the detailed direction of technical or specialized procedures, but shall oversee that these procedures are followed and implemented when suggested. Scene management decisions are to be made with the assistance of expert and technical advisors and specialists, never to exclude the officer of the HMRT.

5.2.2.1 Freeways and State Roads. For all hazardous materials incidents that occur on any freeway or any state road, including those within the City of Sacramento, the scene manager shall be the *California Highway Patrol*, in accordance with the provisions of Section 2454 of the California Vehicle Code.

5.2.2.2 Within Sacramento City Limits.

a. City Streets and Roads. For all hazardous materials incidents that occur within the City Limits of Sacramento, the *Sacramento Fire Department* shall function as the scene manager in accordance with the "Memorandum of Understanding for Hazardous Materials Incidents" as signed by the Police Department and the Fire Department.
b. Public and Private Property (Off Road). For all other hazardous materials incidents that occur within the City Limits of Sacramento, the *Sacramento Fire Department* shall function as the scene manager. The scene manager's position shall be provided for by the ranking chief officer of the fire department on scene.

NOTE: Under all of the above conditions, the incident commander of the fire department having jurisdiction shall provide direct control and authority of all the fire department related activities at the scene of any hazardous materials incident.

5.2.2.3 Outside the City Limits of Sacramento.

a. Freeways and State Roads. For hazardous materials incidents that occur on any state road and all freeways in any country, and all surface streets and public roads in all unincorporated portions of any country, the *California Highway Patrol* shall function as the scene manager in accordance with Section 2454 of the California Vehicle Code.

b. Streets and Public Roads, Incorporated Areas. For hazardous materials incidents that occur on streets and public roads in an incorporated area of any county, the local *law enforcement agency* of that jurisdiction shall function as the scene manager in accordance with Section 2454 of the California Vehicle Code.

c. Public and Private Property (Off-Road), Sacramento County. For hazardous materials incidents that occur on public and private property in the County of Sacramento, the *Sacramento County Sheriff* shall function as the scene manager in accordance with the Sacramento County Hazardous Materials Response Plan.

6. Incident Command Operational Flow Charts

6.1 Level 1 Hazardous Materials Incident.

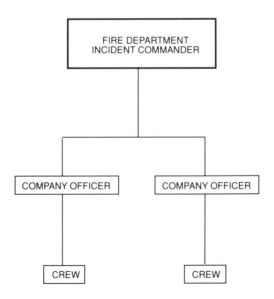

6.2 Level 2 Hazardous Materials Incident.

6.2.1 Reflects compliance to FIRESCOPE ICS HM-120/1990

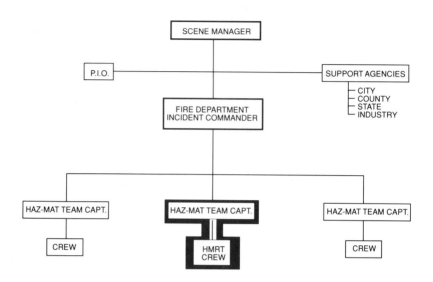

6.3 Level 3 Hazardous Materials Incident.

6.3.1 Reflects compliance to FIRESCOPE ICS HM-120/1990 (*see opposite page*)

7. Hazardous Materials Response Team (HMRT)

7.1 Emergency Response Program.

7.1.1 The Sacramento Fire Department shall maintain three (3) specially trained and equipped "four member" haz-mat teams, referred to in this document as HMRT.

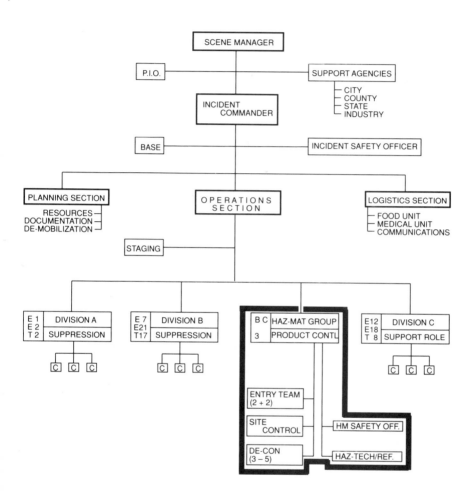

NOTE: THE MODULE HIGHLIGHTED BY A BORDER IS THAT PORTION OF THE INCIDENT COMMAND SYSTEM THAT THE HAZARDOUS MATERIALS RESPONSE TEAM SHALL ALWAYS IMPLEMENT. THIS PORTION OF THE INCIDENT COMMAND STRUCTURE SHALL BE IMPLEMENTED BY AND SHALL FUNCTION WITH SACRAMENTO FIRE DEPARTMENT HAZARDOUS MATERIALS MEMBERS ONLY.

7.2 Response Discipline.

7.2.1 The Sacramento Fire Department HMRT is available to respond anywhere within the City and County of Sacramento, a geographical area which encompasses 1,000 square miles, as dispatched by the City of Sacramento Comm Center, and in accordance with the Sacramento County Hazardous Materials Response Plan. Response into the County of Sacramento is at no cost to the respective county fire departments.

7.2.2 The Sacramento Fire Department HMRT is available to respond to any other county to a pre-approved distance by the City Council not to exceed 50 miles radius on "Mutual Aid." This distance is measured from the interchange of I-80 / HWY 99 / HWY 50. This response into any county other than the County of Sacramento will be based on a current fee structure chargeable to the agency requesting assistance.

7.2.3 The Sacramento Fire Department HMRT is available to respond to any distance exceeding 50 miles when requested by an outside agency (for example, by OES or CHP). This response must be approved *before dispatch* by the Sacramento Fire Department duty chief. This response will be based on a current fee structure chargeable to the agency requesting assistance.

7.3 Equipment Discipline.

7.3.1 All equipment, instruments, protective clothing, tools, and kits assigned to a HMRT *are not* to be loaned to or used by any other fire fighter, fire department, civilian, or agency without the express consent of the HMRT captain on scene.

7.3.2 Any equipment, instruments, protective clothing, tools, and kits used and damaged, including that which might have to be discarded or destroyed at a hazardous materials incident will be inventoried and documented for the purpose of cost recovery. This will be a concern for every hazardous materials incident.

7.4 Basic Incident Discipline.

7.4.1 When appropriate it shall be the responsibility of the HMRT captain to initiate and place into position at all hazardous materials incidents the following actions:

a. Upon arrival, identify themselves as the Officer of the HMRT to all other agencies on scene, the incident commander and the comm center.

b. Immediately arrange for a briefing with the key lead agency representatives on scene.

c. Function within the FIRESCOPE HM-120 Incident Command System.

d. Modify as necessary the boundaries for the *warm zone* (limited access zone), or identify and mark if not yet established.

e. Identify and mark as necessary the boundaries for the *hot zone* (restricted zone).

f. Determine if a "Level III HMI" is required to provide appropriate staffing to function within the guidelines of FIRESCOPE HM-120, and SARA Title III, and initiate a "Level III" dispatch.

g. Assign a "Recorder" to maintain a written log of all pertinent on-scene activities and decisions. It is preferable that this person be obtained from an on-scene engine company so as not to deplete the staffing of the on-scene HMRT.

h. Establish and assign a *site access control leader* as necessary in accordance with FIRESCOPE HM-120 ICS.

i. Establish and assign an *Entry Team* and leader as necessary in accordance with FIRESCOPE HM-120 ICS.

j. Establish and assign a *De-Con Team* and leader as necessary in accordance with FIRESCOPE HM-120 ICS.

7.4.2 The HMRT captain shall work with, and be subordinate to, the incident commander. If it becomes necessary to establish the position within the incident command system known as the Hazardous Materials Group, the HMRT captain shall work with, and be subordinate to, the Hazardous Materials Group supervisor.

7.5 Agency Notification Protocols of HMRT.

7.5.1 The HMRT shall determine what governmental, industrial, utility, or other outside agency(s) needs to be *notified* of the incident. The exact number and type of agencies that require *notification only* shall be dependent upon the nature and magnitude of the incident, and the jurisdictional location of the incident.

NOTE: An "outside agency" is any agency other than the fire department having jurisdiction.

7.5.1.1 A listing of all agencies that require *notified only* shall be compiled by the HMRT captain for each incident. This list shall be made available to the incident commander and the Hazardous Materials Group supervisor so that contact of these agencies is coordinated. It should be suggested by the captain of the HMRT that permission be given by the IC or the SM on scene to allow the HMRT to have the authority to initiate these contacts.

NOTE: The HMRT may be in need of specific and technical information from various sources and contacts. It is prudent to allow the HMRT to establish the required contacts, as they will be in the best position to explain the circumstances of the incident, and to ask the necessary technical questions.

7.5.1.2 The HMRT *must contact* the Office of Emergency Services (OES) for every Level II and Level III HMI. The OES must be briefed on the incident, and shall issue a "CHIMERS" code number to the HMRT.

7.5.1.2.1 Notification of all State and Federal agencies can be accomplished by making one call to the California Office of Emergency Services. Requests to have representatives of various State or Federal agencies return contacts to the HMRT shall be done at this time.

7.5.1.2.2 Insure that the Office of Emergency Services is always notified of the following regarding each hazardous materials incident:

a. *Any* incident involving spillage of a suspected or known hazardous material that has *contacted the ground.*

b. *Any* incident involving leakage or venting of a flammable, toxic, corrosive, or reactive substance into the atmosphere.

c. *Any* incident involving radioactive or etiological materials with or without package or container failure.

d. *Any* incident that involves civilians that have, or are suspected to have, been *injured or contaminated* due to the nature of the incident.

e. *Any* incident that involves a *wildlife kill* or the imminent threat of a wildlife kill.

7.5.1.2.3 When appropriate, ask that the Office of Emergency Services contact the County Health Department in compliance with the requirements of "Proposition #65."

7.5.2 Determine what governmental, industrial, utility or outside agencies need to be *summoned to the scene* of the incident. The specific agencies that need to be summoned to the scene shall be dependent upon the nature and magnitude of the incident, the jurisdictional responsibilities, and the legal ramifications that exist or that may become likely.

7.5.2.1 A listing of all agencies that should be *summoned to the scene* shall be compiled by the HMRT captain for each hazardous materials incident. This list shall be made available to the incident commander and the Hazardous Materials Group supervisor so that contact of these agencies be coordinated. It should be suggested by the captain of the HMRT that permission be given by the IC or the SM on scene to the HMRT to have the authority to make these calls.

NOTE: Review Appendix 2 for a detailed explanation of services and responsibilities of governmental agencies.

8. Decontamination Team (De-Con)

8.1 Response Discipline.

8.1.1 The Sacramento Fire Department shall maintain one (1) specially trained and equipped "four member" Decontamination Team referred to as De-Con.

8.1.2 Response shall be the same as explained in Section 6.2 of this plan.

8.1.3 The De-Con Team shall respond to all "Level III HMI" calls.

8.2 Equipment Discipline.

8.2.1 The De-Con Team can provide expertise and special equipment specially prepared for the purpose of decontaminating equipment, tools, personnel, and civilians. This includes a portable "De-Con Trailer" that will respond as part of the De-Con Team at all times.

8.3 Basic Incident Discipline.

8.3.1 When appropriate, it shall be the responsibility of the De-Con captain to initiate and place into position at all hazardous materials incidents the following actions:

a. Upon arrival, immediately arrange for a briefing with the HMRT captain.

b. Locate the De-Con Trailer in a safe location and place it into operation as necessary.

c. Be responsible for identifying the location of the De-Con Corridor, then getting it set up.

d. Function as the De-Con Team for the duration of the incident in accordance with FIRESCOPE HM-120 incident command.

e. Inform the HMRT captain as to the best recommended method(s) of decontamination suitable for the incident, and the reasons for these recommendations.

f. Inform the HMRT captain, and the IC, of those items that cannot or should not be decontaminated, and the reasons.

g. Be subordinate to, and function through, the incident commander.

8.3.2 Work with, and be subordinate to, the incident commander and the Hazardous Materials Group supervisor when established within the incident command structure.

9. Control Zones

9.1 Warm Zone (Limited Access Zone).

9.1.1 The *warm zone* shall be a designated area to define where some potential or real danger exists with respect to safety and health to the public and harm to the environment.

9.1.2 Identification of the warm zone shall be done by the first arriving fire department company officer. Yellow "banner" tape shall be used to identify the boundaries of the warm zone.

9.1.3 Access into the warm zone shall be controlled as necessary by the site access control leader. Access shall be limited to only those members of agencies on scene who are protected and are directly engaged in incident ground activities at the direction of the incident commander, the hazardous materials group supervisor, or the HMRT captain.

9.2 Hot Zone (Restricted Zone).

9.2.1 The *hot zone* shall be designated as necessary by the HMRT captain, and shall be utilized to identify and define an area of exceptional danger or potential danger, including extreme danger to life safety.

9.2.2 Identification of the boundaries of the hot zone shall be with the use of red "banner" tape.

9.2.3 Access into the hot zone shall be controlled by the site access control leader. Access shall be allowed only by permission of the HMRT captain or entry team leader. Only properly protected and equipped members of the HMRT and other designated workers will be allowed to enter the hot zone.

9.2.3.1 An "Entry Point" into the hot zone shall be identified by the Site Access Control Team. All entry into the hot zone shall be made at this designated location.

9.2.4 Exiting from the hot zone shall be controlled and supervised by the De-Con Team. The De-Con Corridor may become part of the exiting from the hot zone.

9.2.5 A log shall be maintained by the Site Access Control Team of all personnel who are allowed to enter the hot zone, including times of entry and inspection of protective clothing.

9.3 Decontamination Corridor.

9.3.1 The De-Con Corridor shall be designated as necessary to establish a controlled environment to decontaminate personnel, civilians, equipment, and tools in an effort to reduce or stop the spread of suspected contamination.

9.3.2 Workers entering into the De-Con Corridor to assist in procedures shall do so only as directed by the De-Con captain, and only when properly equipped and protected.

9.3.3 Decontamination procedures shall be implemented and executed only by trained members of a Hazardous Materials Response Team or by the De-Con Team, in keeping with the requirements of NFPA Standard 472.

The Operational Annex

10. Activation

This plan will become operational when the Sacramento Fire Department receives notification of any Level II or Level III HMI.

11. Notification and Dispatch

11.1 For a Level II HMI or a Level III HMI dispatch, the Communications Center (Comm Center) shall dispatch the appropriate response according to Chapter 4 of this document.

11.2 The Comm Center shall dispatch the closest available HMRT and battalion chief to all Level II HMI.

11.3 Response shall be "Code 3" unless specified otherwise during dispatch, or as revised after dispatch.

11.4 Location of the teams are as follows:

HMRT 5 consists of TRUCK 5 and LP 5, in Battalion One
HMRT 6 consists of TRUCK 6 and LP 6, in Battalion Two
HMRT 20 consists of TRUCK 20 and LP 20, in Battalion Three
De-CON 19 consists of ENGINE 19 and De-CON TRAILER 19, in Battalion Three

11.5 Notification of Administrative Officers.

11.5.1 For Level II HMI the Comm Center no longer is required to notify the duty chief for incidents within the City or County of Sacramento. However, if there is a request for *mutual aid* of the HMRT to any other county, the following contacts shall be made:

a. The duty chief.
b. Haz-Mat Division captain.

11.5.2 For a LEVEL III HMI, the Comm Center shall additionally notify the following:

a. The duty chief.
b. The Haz-Mat Division chief.
c. The Haz-Mat Division captain.

11.5.3 The Comm Center shall advise on the nature of the incident, the type, the magnitude, the location, and the units responding.

11.6 Agency Call Lists.

11.6.1 The Comm Center shall maintain current telephone contact numbers for all appropriate local and state government agencies.

11.6.2 The Comm Center shall maintain a file of private industry contacts and telephone numbers. These contacts should represent a resource of expertise and abilities beyond the normal scope of public agencies, and can be notified and summoned to the scene to assist in the containment, control, transfer, and removal of the hazardous material.

11.7 Upgrade of a Hazardous Materials Incident.

11.7.1 In keeping with this plan, the Sacramento County Hazardous Materials Plan, and the FIRESCOPE HM-120 Incident Command System adopted by the Sacramento Fire Department Hazardous Materials Response Team, the following must be administered in order to insure proper hazmat team dispatch:

a. The lead officer functioning in the capacity of incident commander of the fire department having jurisdiction can upgrade an incident to a Level II HMI.

b. The lead officer functioning in the capacity of scene manager of the law enforcement agency having jurisdiction can upgrade an incident to a Level II HMI.

12. Immediate On-Scene Actions

12.1 Fire Department First on Scene.

12.1.1 Identification.

12.1.1.1 The first fire department unit on scene will usually be a "first responder" (engine or truck comany) with respect to their training and capabilities in handling hazardous materials incidents. They shall determine if hazardous materials are involved in the incident. If confirmed, the company captain must, in a safe manner, attempt to identify:

a. The type of material involved.

b. The quantity of material involved.

c. The possibility of contamination.

d. The immediate exposure problem.

e. The threat to life safety.

12.1.1.2 This information is to be broadcast via radio to the Comm Center immediately.

12.1.1.3 If the Captain of the first arriving unit determines that hazardous materials are indeed involved, but the incident is a Level I HMI which is within the capabilities of a "first responder" to handle correctly, they shall:

a. Inform the Comm Center that the unit on scene can and will handle.

b. Proceed to handle the incident and mitigate the problem.

12.1.1.4 If the captain of the first arriving unit determines that the incident is beyond the capabilities of a "first responder," the incident shall be upgraded to a Level II HMI. This will initiate the dispatch of a HMRT.

12.1.1.5 The captain shall also do the following:

a. Isolate the scene and protect it using yellow "banner" tape to identify the boundaries of the warm zone (limited access zone).

b. Assure that crew assignments are within the expertise and limitations of their equipment, protective gear, and training.

c. Begin gathering information regarding the incident and the product.

d. Initiate containment techniques within the scope of their capabilities.

e. Initiate procedures to protect or remove civilians from the immediate area.

12.1.1.6 In all cases, the captain of the first arriving unit shall inform all incoming companies of the situation and the actions being initiated. The captain shall advise the responding HMRT and battalion chief of which access routes they should use when approaching the incident.

12.1.1.7 The captain shall determine what other assistance would be necessary to call to the scene, such as:

a. Additional fire units for manpower and support.

b. Law enforcement units for traffic control, street blockage, civilian control.

c. Sand from the Corp Yard (or from Cal/Trans if incident is on state highway).

d. EMS assistance (ambulance) for injured.

12.1.1.7.1 It is *not advisable* to call for "Life Flight" or other helicopters at a hazardous materials incident because:

a. High possibility of chemicals or fire being spread over a wide area by the strong backwash of the rotor blades.

b. Unavailability of suitable landing location.

c. Possibility of contamination to helicopter interior, especially when considering transportation of injured.

12.1.2 Command Post and Battalion Chief.

12.1.2.1 The battalion chief, upon arrival, shall establish a *command post* location in the most strategically desirable and safe location.

12.1.2.2 The location of the command post shall be indicated via radio broadcast to the Comm Center.

12.1.2.3 The battalion chief shall assume the position of incident commander upon arrival at the scene. As necessary, the battalion chief shall also institute other portions of the Incident Command System for the incident which may include some or all of the following:

a. Divisions

b. Groups, including Hazardous Materials Group and Medical Group

c. Public Information Officer

d. Incident Safety Officer

e. Hazardous Materials Safety Officer

f. Entry Team and Leader

g. Site Access Control Team and Leader

h. De-Con Team and Leader

12.1.3 Base.

12.1.3.1 The location of a base shall be established as necessary outside of the anticipated hazard area and outside of the warm zone. This area and its resources shall be coordinated by assigned personnel.

12.1.3.2 The base location is to be transmitted to the Comm Center as soon as possible, and all responding units dispatched to this incident shall report to base unless otherwise directed during response.

12.2 Fire Department Not First on Scene.

12.2.1 When another agency is on the scene of a hazardous materials incident prior to the arrival of the fire department, it generally will be a law enforcement agency. The arrival of the first fire department officer shall establish contact and communications with the agency first on scene.

12.2.2 The first fire department officer on scene shall ascertain, if possible, what level the incident is. If it is a Level II HMI, the fire officer shall call for the proper dispatch. The fire department officer shall then proceed to gather as much information as possible about the incident to relay and pass on to other responding units.

12.2.3 Until a higher ranking chief officer of the fire department having jurisdiction arrives, the fire department officer on the scene shall be the incident commander and will establish the necessary incident command structure to deal with the hazardous materials incident.

12.2.4 A Level III shall follow the same operational procedures as outlined in this plan for a Level II HMI. It should be understood that a Level III HMI will by its nature involve a large number of outside agencies.

12.3 Mutual Aid and Automatic Aid.

12.3.1 The Sacramento Fire Department has a special written agreement with the County of Sacramento to provide hazardous materials emergency response to any area within the boundaries of the County of Sacramento. This special automatic aid agreement between the City of Sacramento and the County of Sacramento is negotiated yearly, or as necessary.

12.3.1.1 Activation of emergency response of the Sacramento Fire Department Hazardous Materials Response Team is initiated by a call from the county dispatch center requesting a Level II HMI response. The Sacramento City dispatch center shall dispatch a standard Level II HMI dispatch according to Section 4.2 of the Administrative chapter of this plan.

12.3.2 Other features and guidelines explaining the policy regarding manual aid and automatic aid dispatch of the HMRT are explained in Section 6.2 of the administrative chapter of this plan.

Special Circumstances Annex

13. Financial Responsibility

13.1 Primary Responsible Party

13.1.1 The "primary responsible" for the assumption of all costs for the cleanup and disposal of any hazardous materials incident shall be in the following order:

13.1.1.1 The person, persons, or agency whose neglect, action, inaction, or willful or deliberate act caused or allowed to perpetuate said incident;

13.1.1.2 The person, persons, or agency who owned or had custody or control of the transportation vehicle which was used to move the hazardous material and its containers at the time the incident occurred or was discovered and reported;

13.1.1.3 The person, persons, or agency who owned or had custody or control of the hazardous material and its containers at the time the incident occurred or was discovered and reported;

13.1.1.4 The person, persons, or agency who owns the property or who has immediate financial and tax liability to the property on which the incident occurred.

13.2 The scene manager, incident commander, and the captain of the HMRT shall work together to identify the "primary responsible" party for every hazardous materials incident. The person, persons, or agency identified as the "primary responsible" shall be liable for all costs associated with control, identification, and cleanup.

13.2.1 In the event that the "primary responsible" is known but cannot be located or is incapable of assuming total financial responsibility, or the "primary responsible" is unknown, then procedures must be initiated to ascertain what City, County, or State governmental agency is to assume financial responsibility.

13.3 City Property — Spiller Is Not City of Sacramento.

13.3.1 If the incident is on City property (street, curb, city parking lot, sidewalk, parks, etc.) and the spiller is unknown, or is known but is unable to assume financial liability, the responsibility for assuming all costs associated with control, identification and cleanup may be the City of Sacramento. The specific City Division or Department shall be determined based on exact location of the incident, and that Division or Department shall be liable for all costs associated with the incident.

13.3.1.1 For incidents occurring on city streets, parking lots, sidewalks, city owned property, city landfill, water systems, waste water systems, sewer facilities, flood control facilities, storm drainage systems, the financial liability for all costs will belong to the *Sacramento City Public Works Department*.

13.3.1.2 For incidents occurring in city parks, museums, cemeteries, community centers, city swimming pool facilities, ball parks, city zoo, city marinas, the financial liability for all costs will belong to the *Sacramento City Public Works Department*.

13.3.1.3 For incidents occurring on Convention Center property, the financial liability for all costs will belong to the *convention center*.

13.3.1.4 For incidents occurring on Fire Department property, the financial liability for all costs will belong to the *Fire Department*.

13.3.1.5 For incidents occurring on Police Department property, the financial liability for all costs will belong to the *Police Department*.

13.3.2 In the event a spiller is known for the circumstances described above in Section 12.1.1, they will be the "primary responsible" and the City of Sacramento may elect to pursue total cost recovery from the "primary responsible" by civil action, court action, property lien, pursuant to the provisions of the City Charter.

13.4 City Property — Spiller Is City of Sacramento.

13.4.1 If the incident is on city property, then the responsible party is the City of Sacramento, and will be the "primary responsible." The specific division or department shall be determined based on exact location of the incident and what division or department caused the incident.

13.4.1.1 For incidents caused by City equipment (i.e., hydraulic failure), regardless as to where it occurred, the financial liability for all costs will be the division or department who is responsible for the piece of equipment which caused the incident.

13.4.1.2 For incidents caused by City employees (i.e., accidents, traffic or industrial), regardless as to where it occurred, the financial liability for all costs will be the division or department who employs the employee.

13.5 Private Property.

13.5.1 If the incident occurred on private property within the City Limits of Sacramento, then the "primary responsible" for all financial liability shall be that as outlined in Section 12.1.1 of this plan.

13.5.2 In the event that the "primary responsible" is known but is unable to assume financial liability, or is unknown, then the following shall apply with regard to ascertaining who will initiate procedures for cleanup and cost recovery:

13.5.2.1 If the "primary responsible" has been identified and is known, but is unavailable or unable to assume immediate financial liability for the costs incurred for the hazardous materials incident, including cleanup, the *Sacramento Fire Department* shall assume these costs and initiate appropriate procedures to abate the incident in as timely a manner as is deemed necessary. The Sacramento Fire Department may elect to pursue total cost recovery from the "primary responsible" in accordance with Section 12.3.2 of this plan.

13.5.2.2 If the "primary responsible" cannot be identified at the time of the incident, the Sacramento Fire Department shall assume all costs associated to clean up and abating the incident. If and when it becomes known to the Sacramento Fire Department at a later date that a "primary responsible" has been located, the Sacramento Fire Department may elect to pursue total cost recovery from the "primary responsible" in accordance with Section 12.3.2 of this plan.

13.5.3 Special Case — Drug Lab/Crime-Related Dumping.

13.5.3.1 Clandestine hazardous materials incidents including those associated with drug labs, drug lab chemicals and precursors, drug lab waste products, glassware, hardware, paraphernalia, and abandoned chemicals, and all associated containers, it is imperative that the HMRT work closely with the law enforcement agency having jurisdiction on scene to determine if there is sufficient evidence to connect or declare the hazardous materials incident scene a "crime-related" incident. If it is the opinion of the law enforcement agencies on scene that it is "crime-related," then the total cost associated to the incident regarding control, identification, and cleanup shall be assumed by the lead law enforcement agency having jurisdiction on scene.

a. All sampling will only be done as necessary by law enforcement task force, most likely DEA.

b. Authority and control of all evidence, including chemicals and hazardous materials, is now the responsibility of the law enforcement task force. The HMRT may assist as necessary only to the extent of removal of chemicals to a safe location for sample taking, and stabilization of certain situations to allow law enforcement personnel to proceed with their investigation.

c. Notification of a reputable cleanup agency shall be done by the lead law enforcement agency having jurisdiction on scene, and that agency shall bear all associated costs of cleanup and abatement.

13.5.3.2 If it is the opinion of the law enforcement agency on scene that the hazardous materials incident is NOT "crime-related," or that there is insufficient evidence to allow the law enforcement agency to follow through with a "crime-related" scene, thereby absolving law enforcement of any relationship to the scene, then the Sacramento Fire Department shall assume the financial liability of all costs of abatement.

13.5.4 State Property.

13.5.4.1 For hazardous materials incidents that occur on state roads or state freeways, the agency that shall be responsible for the abatement of the incident is *Cal/Trans* (California Department of Transportation). Cal/Trans shall make the necessary arrangements for contact with a licensed hazardous waste hauler.

13.5.4.2 For hazardous materials incidents that occur on all other state property and in state buildings, the agency that shall be responsible for the abatement of these incidents is *California State Police*. State Police shall make all of the necessary arrangements for contact with a licensed hazardous waste hauler.

13.5.5 Sacramento County.

13.5.5.1 If the incident occurs on private property in the *County of Sacramento*, then the "primary responsible" for all financial liability shall be that as outlined in Section 12.1 of this plan.

13.5.5.2 In the event that the "primary responsible" is known but is unable to assume financial liability, or is unknown, then arrangements will be made by the Sacramento County Sheriff (functioning as scene manager) to contact the Sacramento County Public Works Department and request for financial assistance to abate the incident.

13.6 Authority to Abate a Hazard.

13.6.1 Within the City Limits of Sacramento, the Sacramento Fire Department has the authority to order the immediate cleanup, removal, and total abatement of any hazardous materials incident.

13.6.2 When the Sacramento Fire Department is to assume the financial liability for abatement costs, the captain of the HMRT shall make the appropriate arrangements by contacting the appropriate duty chief. Only duty chief can grant final approval to release city treasury monies for this purpose.

14. Cost Recovery

14.1 In the event that the Sacramento Fire Department ultimately must pay for cleanup and abatement costs of a hazardous materials incident, and when the HMRT and the De-Con Teams suffer loss and damage of equipment, the Sacramento Fire Department may pursue all appropriate legal avenues to initiate action, including civil action, against all parties stated in Sections 12.1 and 12.3 of this plan in order to recover all costs.

14.2 The agency acting in the capacity of scene manager, when it is not the Sacramento Fire Department, shall assist the Sacramento Fire Department in initiating all actions necessary to recover all costs incurred in the handling and abatement of the hazardous materials incident.

15. Cleanup, Re-Packing, Disposal — General Restrictions

15.1 It shall not be the responsibility nor the policy of the Sacramento Fire Department HMRT to engage fire department personnel in the cleanup and/or removal of hazardous materials incident when such actions circumvent the responsibilities of a licensed hazardous waste hauler and circumvent existing state and federal law regarding cleanup and abatement requirements.

15.1.1 The HMRT may elect to abate certain small spills when within the capabilities of the team, and when within the constraints of law and outside agency policy guidelines.

15.1.2 The HMRT may elect to assist as a matter of safety a hazardous waste hauler as is deemed prudent and necessary by the HMRT captain.

Appendix 1

A-1 Definitions.

A-1.1 Assisting Agencies: Any outside agency that assists at the scene of a hazardous materials incident that provides supporting services and within the responsibility or scope of the Sacramento Fire Department. Such services would include, but not be limited to, road closures and detours, technical advice, sampling and monitoring capabilities, cleanup operations, off-loading, tank righting, disposal, and other supportive tasks as required.

A-1.2 B.L.E.V.E.: An acronym for "Boiling Liquid Expanding Vapor Explosion."

A-1.3 Cleanup: Incident scene activities directed to removing the hazardous material, and all contaminated debris including dirt, water, road surface material, containers, vehicles, contaminated articles, tools, and clothing, and returning the scene to as near as normal as it existed prior to the incident. Cleanup is not the function of the Sacramento Fire Department, but overseeing and observing cleanup operations would be the responsibility of the incident commander. Technical advice and guidance for cleanup can be provided by the HMRT captain.

A-1.4 Command: To direct and delegate authoritatively through application of the incident command system so as to insure effective implementation of departmental control procedures.

A-1.5 Command Post — Location: When positioned in a safe strategic location, the Command Post provides a base for the incident commander and support staff in order to manage all aspects of the incident. Representatives of all other agencies involved at the incident should provide liaison officers to the Command Post.

A-1.6 Command Post — Vehicle: When positioned in a safe strategic location, the vehicle provides a facility for tactical planning, includes resources such as multiple radio frequency access, technical references and resources, maps, reports, cellular phones, access to the CAD computer.

A-1.7 Containment: Includes all activities necessary to bring the scene of the hazardous materials incident to a point of stabilization, and to the greatest degree of safety as possible.

A-1.8 Coordination: The administering and management of several tasks so as to act together in a smooth concerted manner. To bring together in a uniform manner the function of several agencies.

A-1.9 Cost Recovery: A process that enables an agency, such as the Sacramento Fire Department, to be reimbursed for the costs incurred at a hazardous materials incident.

A-1.10 Explosion: A sudden release of a large amount of energy in a destructive manner. It can be the result of the ignition of powders, mists, or gases, or it can be the sudden decomposition of liquids and solids, or it can be a container or vessel undergoing container failure as the result of overpressure. All explosions generally produce tremendous heat, cause severe structural damage, occasionally produce a shock wave, and propel shrapnel.

A-1.11 Hazardous Material: A material or substance in a quantity or form that, when not properly controlled or contained, may pose an unreasonable risk to health, safety, and the environment. The standard Federal Department of Transportation definitions for the individual hazardous materials "Hazard Classes" are adopted as the standard for this document (CFR 49 Section 173).

A-1.12 HAZ-MAT: An abbreviation for the term "hazardous material."

A-1.13 HMRT: An abbreviation for "Hazardous Materials Response Team."

A-1.14 Hazardous Materials Incident: Any spill, leak, rupture, fire, or accident that results or has the potential to result in the intentional or unintentional loss or escape of a hazardous material from its container or confinement.

A-1.15 Incident Command: A system of unified command and control designed to assure the smooth implementation of immediate and contained operational procedures by all agencies on scene in a coordinated manner until the incident has been contained or abated.

A-1.16 Incident Commander: A representative of the agency having jurisdiction that is responsible for implementing the incident command system and provides guidance and management of the incident in a coordinated manner.

A-1.17 Leak: A leak will be considered to be the release or generation of a toxic, poisonous, or noxious liquid or gas in a manner that poses a threat to air and ground quality and to health safety.

A-1.18 Rupture: A rupture will be considered to be the physical failure of a container, releasing or threatening to release a hazardous material. Physical failure may be due to forces acting upon the container in such a manner as to cause punctures, creases, tears, corrosion, breakage, or collapse.

A-1.19 Spill: A spill will be considered to be the release of a liquid, powder or solid from its original or intended container or confinement in a manner that poses an unreasonable threat to air and ground quality and to health safety.

A-1.20 Stabilization: Incident scene activities directed to channel, restrict, and/or halt the spread of a hazardous material; to control the flow of a hazardous material to an area of lesser hazard; to implement procedures to insure against ignition; to control a fire in such a manner as to be safe, such as a controlled burn, flaring off, or extinguishment by consumption of the fuel.

A-1.21 Transportation: Methods of transporting or moving commodities and materials by any mode which include highway, railroad, pipeline, water, and air.

Appendix 2

A-2 Responsibilities of Agencies

A-2.1 City and County Municipal Agencies

A-2.1.1 County Agricultural Commissioner.

Responsibilities: Enforces all state and federal regulations relating to the use of pesticides.

Incident Support: This county office can provide technical advice by responding to the scene when requested. They can provide information on the toxicity of pesticides, their affect to the environment, and can suggest mitigation methods and cleanup suggestions.

Notification: The County Ag. Department *must* be notified in the event of any incident involving agricultural chemicals.

Reports: Other than in-house, none.

Emergency Funding: None.

A-2.1.2 Fire Department.

Responsibilities: Providing routine fire and rescue support services at all incidents. The fire department having jurisdiction shall in most cases assume the position of incident commander (IC) at the scene of a hazardous materials incident.

Incident Support: The fire department having jurisdiction shall coordinate and effect appropriate rescue efforts, first aid, containment, and immediate hazard reduction activities within the scope of *awareness* or *operational* hazardous materials training, as well as the implementation of all other normal fire department related activities and responsibilities.

The Sacramento Fire Department manages a Hazardous Materials Response Team program, which provides a capability to assist at a hazardous materials incident at a level of *specialist* exceeding the minimum training requirements of SARA Title III, Section 1910.120, and NFPA Standard #472.

Notification: Reporting is required by the spiller for any hazardous materials incident occurring within their jurisdiction.

Reports: Normal in-house report system. Some reports must be forwarded to various state agencies depending on circumstances of incident.

Emergency Funding: Varies. None for county fire departments. Sacramento City has as a last resort an emergency procedure in place to access funds after all other avenues of financial responsibilities have been exhausted. Available only for incidents within the City Limits of Sacramento.

A-2.1.3 Law Enforcement Agency.

Police Department

Responsibilities: Civil, criminal, and traffic enforcement. In some cities the police department will be charged with the responsibility of "scene management" of a hazardous materials incident, unless a contract has been entered and agreed delegating the management of the scene to the fire department having jurisdiction.

Incident Support: Typical activities include the closure of roads, establishing detours, assisting in establishing security around specified boundaries, assisting in evacuation, and the access to a bomb squad task force, an explosives disposal team, and the Clandestine Drug Lab Task Force.

Sheriff Department

Responsibilities: In some counties, the local sheriff department having jurisdiction has voluntarily assumed the responsibility of "scene management" for all hazardous materials incidents that occur "off-road" and on private property.

Incident Support: Activities include the establishing of security around specified boundaries, assisting in evacuation, and giving assistance to the California Highway Patrol in road closures, establishing detours and controlling traffic. Additional responsibilities might include functioning as the county liaison or representative to access county money as necessary through the appropriate County Public Works Division or Department.

Reports: None required, unless functioning as the incident commander/scene manager, in which appropriate reports are required to be forwarded to OES.

Emergency Funding: For hazardous materials, none. For verified clandestine drug lab operations where criminal violations and arrests are made, cost of cleanup is borne by the drug lab task force. For explosives and bombs, the bomb task force handles all removal and costs associated with the incident.

A-2.1.4 Traffic and Engineering (Public Works).

Responsibilities: Normal responsibilities include repair and maintenance of city (or county) roads, provision of road barricades and directional devices.

Incident Support: Can usually assist on small road spills in cleanup if within their capability. Some counties have a contract with a hazardous waste hauler for larger spills on county roads. Can provide to the scene of an emergency other services such as sand, water, special heavy equipment (i.e., tractors, backhoe, loaders, sweepers).

Notification: As necessary, and when incident has disrupted normal traffic flow, and when there is damage to Public Works property, including streets, overpasses, etc.

Reports: None required.

Emergency Funding: Sacramento City — except for being responsible for financial liability of their own spills or spills on city streets, none. Sacramento County — maintains an emergency fund and has a procedure in place to access funds after all other avenues of financial responsibilities have been exhausted. Available only for incidents within the County of Sacramento.

A-2.1.5 Water and Sewer Departments.

Responsibilities: In some city or county governments these are two separate departments. Normally shall be responsible for the design and maintenance of palatable water, storm drain, and waste water systems including filtration plants and distribution systems.

Incident Support: Most Water and/or Sewer Departments operate a testing laboratory which has the capability of analyzing water contaminants. Other resources include technical expertise regarding hazards of many water borne contaminants, mapping of underground systems, control valves, and emergency analysis of water base unknown samples. These departments have the final word with regard to the possibility of flushing contaminants into storm drains, and clearance must be obtained from them *before* this action is taken.

Notification: If any contaminant or hazardous material has come into contact with any water channel, storm drain, ditch, canal, creek, etc., which is normally maintained by the appropriate water or sewer department, notification is *mandatory* and *immediate*. Agency *must* respond to inspect.

Reports: None.

Emergency Funding: None.

A-2.1.6 Health Department (Including Environmental Health Section).

Responsibilities: Normal activities include daily ambient air monitoring, collection and tabulation of "right-to-know" ordinance data from hazardous materials business inventories, and monitoring normal "waste-stream" procedures of industry generated wastes.

Incident Support: Include response for the purpose of providing quick on-scene technical advice, considerable knowledge on abatement and removal suggestions, monitoring of air and water ways, sample taking and analysis — particularly in support of on site activities in unincorporated areas of a county, and the access to (county) funds to help mitigate the incident. If there has been an evacuation made as a result of a hazardous materials incident, it is important to note that *only the county health department representative* can announce the area clear and allow civilians or employees to re-enter.

Notification: County Health *must* be notified of any hazardous materials incident that has, might, or could cause contamination or make contact employees or civilians. If there are chemically injured people, County Health will respond a toxicologist. Environmental Health Section *must* be notified for any hazardous materials incident in the County of Sacramento. In all likelihood they will respond.

Emergency Funding: Available through County Department of Public Works. See A-2.1.4, "Traffic and Engineering."

A-2.1.7 Air Pollution Control Board.

Responsibilities: Routine activities include daily monitoring of local ambient air for pollutants, pollen, and molds.

Incident Support: Have portable monitors that can be used for placement down-wind of large airborne chemical releases for remote monitoring. They work in concert with and report to State Air Resources and have access to their incident support activities. Also may have technical expertise useful to the incident commander, such as predicting or modeling dispersion patterns for airborne pollutants. Normally has no remedial hands-on expertise or equipment.

Notification: They *must* be notified regarding any hazardous materials incident that is releasing to the atmosphere a toxic substance or pollutant.

Reports: Must file report with State Resources.

Emergency Funding: None.

A-2.1.8 City (County) Office of Emergency Planning.

Responsibilities: This office is normally responsible for maintaining the various emergency action plans and area plans for natural and manmade disasters. This office has close and direct ties with many response oriented agencies, and with the State Office of Emergency Services.

Incident Support: This office can be used to coordinate procurement of needed resources of other agencies. They can also obtain the services of the OES. For very large incidents, or incidents of a complex nature, OEP can activate the emergency operations center, and institute various plans and contingencies in support of the incident.

Notification: Only as necessary depending on nature and complexity of incident.

Reports: Other than in-house, none required.

Emergency Funding: The OEP does have access to emergency funding and special budget lines in support of local agency expenses outside the realm of the hazardous materials incident.

A-2.1.9 Poison Control Center.

Responsibilities: Provides human poison exposure information, and medical health related hazardous material information. Maintains 24 hour emergency numbers. Is staffed by specially trained Poison Information Specialists.

Incident Support: Provides a toxicologist available 24 hours, and has an extensive toxilogical library. Can FAX information to the scene. Has access to and is well prepared to pull together numerous toxilogical resource recommendations for evaluating, assessing and medically managing health exposures associated with hazardous materials exposure.

Notification: Highly advisable if contact to a valuable source of information is desired. Highly advisable if there are chemically injured or contaminated victims being prepared for transit to medical facilities.

Reports: Medical reports are confidential. Statistics generated may be issued and shared.

Emergency Funding: None.

A-2.1.10 District Attorney's Office.

Responsibilities: Investigates crimes handed over to them from law enforcement. Pursues possible arrests and prosecution of various crimes.

Incident Support: Gathers all reports from allied and assisting agencies in the effort to coordinate additional environmental and criminal crimes investigation. Special "Consumer and Environmental Crimes Division" will assist during an incident which demonstrates obvious or highly suspected criminal intent. Will respond an attorney or criminal investigator to the scene upon request of the incident commander. Encourages all agencies to call for any assistance in investigating hazardous materials incidents.

Notification: Maintains a task force call list of available investigators for this purpose 24 hours. Contact is made through county operator 366-2913.

Reports: None to State OES. However, generates a "file" of forms and investigations for each haz-mat incident investigated. All agencies should forward all appropriate reports to the DA Office.

Emergency Funding: None.

A-2.2 State Government Agencies.

WARNING: State agencies may be contacted in an emergency by calling State OES Center: local 427-4341; outside 916 — 1-800-852-7550.

The Warning Center will contact the appropriate state agencies, and some specified federal and local agencies, upon notification.

A-2.2.1 Air Resources Board (ARB).

Responsibilities: The Air Resources Board is responsible for managing air quality in California, and providing to the public daily monitoring readings. They fulfill this responsibility through local air pollution control boards.

Incident Support: They can provide technical advice regarding the potential harm or toxicity of airborne contaminants. They also have portable field monitoring devices that can be set up for off-site remote monitoring. They can also provide *air modeling* services. They can be a source for weather predictions, current weather status, and provide information for future ambient air movement and wind drift.

Notification: They *must* be notified in the event a toxic or hazardous material has the potential to adversely affect ambient air quality.

Reports: None required.

A-2.2.2 Attorney General's Office.

Responsibilities: Represents state and local agencies in civil litigation and has supervisory and enforcement powers under criminal laws.

Incident Support: May provide legal advice to state and local agencies as necessary during hazardous materials incidents. Pursues litigation arising from hazardous materials incidents. Has joined forces in some communities with local law enforcement agencies and federal DEA to form the Clandestine Drug Lab task force. Task force will respond when requested to ascertain condition of drug lab. If arrests can be made, will assume total responsibility of identifying all chemicals, taking of samples, breaking down the lab, contracting with hazardous waste haulers, and cleanup.

Notification: None required, unless services are requested.

Reports: None.

Emergency Funding: Clandestine Lab Cleanup Fund is available to state and local law enforcement agencies in counties with population less than 1.2 million.

A-2.2.3 California Highway Patrol (CHP).

Responsibilities: Routine traffic supervision and control on all state roads, state owned bridges, and on all public roads within unincorporated areas. They shall provide traffic control, traffic re-routing, road closure, prevention of unauthorized entry into restricted areas, and shall when requested assist local law enforcement agencies.

Incident Support: Function as the *incident commander* (formerly the scene manager) for traffic and hazardous materials incidents occurring within their response jurisdiction. CHP officers have the authority to enforce all California criminal statutes and specified sections of the Health and Safety Code, and Title 49 of the federal DOT Code of Regulations. Operate a special "Environmental Crimes Investigator" who are subject to call-out by all allied agencies for the following activities:

a. Investigate specific haz-mat incidents for traffic violations and illegal dumping (reference sections 25180, 25189.5 of Health and Safety Code).

b. Provide assistance to CHP area commanders and all other agencies during extensive investigations.

c. Conduct inspections of hazmat vehicles and terminals.

d. Provide liaison with agencies at local, State, and Federal levels for information exchange and case development.

e. Prepare search warrants for the collection of evidence in environmental crime cases.

f. Perform intelligence gathering concerning illegal transportation/dumping activities that occur within the CHP Valley Division.

g. Provide liaison and assistance to local District Attorney for the arrest and prosecution of suspects.

The key in requesting assistance from the CHP Environmental Crimes Investigators is to determine that the hazardous materials incident, regardless of where it occurs, is transportation related with respect to violations and criminal intent thereof.

Notification: Must be notified of any incident occurring within their jurisdiction. They in turn notify OES and Cal/Trans.

Reports: CHP hazardous materials incident report submitted to OES.

Emergency Funds: None.

A-2.2.4 Department of Transportation (Cal/Trans).

Responsibilities: Cal/Trans has the responsibility for maintaining the State Highway System. They maintain a file of licensed hazardous waste haulers who are on contract to Cal/Trans for emergency response to any state road.

Incident Support: They can assist (as requested through the CHP) in the identification, containment, and cleanup of some less toxic hazardous materials. They can provide highway closure and traffic management equipment. On occasion they can assist local police to close exiting from freeways to surface streets. Cal/Trans is responsible for the opening of any freeway. They are responsible for the repair or replacement of any freeway due to contamination or damage. They coordinate their on-scene activities through the IC.

Notification: Cal/Trans *must* be notified of any hazardous materials spill or incident affecting a state highway. This can be done through OES or the CHP.

Reports: None required, other than in-house reporting.

Emergency Funding: None. Internal funding for state highway cleanup only. Not legally or financially responsible for contamination or cleanup outside the state right of way even though the incident commences from within the right of way.

A-2.2.5 Department of Fish and Game (DFG).

Responsibilities: The DFG has the responsibility for protecting the State's natural living and wildlife resources and their habitat.

Incident Support: They function as the law enforcement voice for these lands and can be very instrumental in investigation, sample analysis, issuance of citations, and arrest. They can assist local authorities in administering immediate corrective directives to facilities including complete shutdown until compliance is demonstrated. They can provide recommendations and guidelines when a hazardous material has or may contaminate streams or waterways. The DFG can, at their selection, function as the IC for incidents occurring within their scope of authority, thus fulfilling the role of lead agency in determining the completion of cleanup when natural resources are affected. They are specified as the "state agency coordinator" for off-highway incidents. They have law enforcement powers within their jurisdiction. They can provide damage assessment, criminal and civil investigation, and they can advise and counsel the IC regarding corrective actions to be taken or contemplated to mitigate the causes of the hazard or pollution, including cleanup operations.

Notification: When an incident occurs that threatens or has caused harm to plant, animal, and aquatic life, or has caused contamination to their habitat, the DFG *must* be notified. They elect to respond depending on the circumstances, or will respond when requested.

Reports: None required, only in-house reports.

Emergency Funding: They do have access to some cleanup funds from the Fish and Wildlife Pollution Cleanup Account. The account may be accessed by their employees for expenditures related to control and recovery actions related to a hazardous materials incident in which they are involved and not fundable by the Department of Health Services' own agency.

A-2.2.6 Department of Water Resources (DWR).

Responsibilities: DWR is responsible for maintaining the California State Water Project, and protecting it from pollutants. The project includes aqueducts, reservoirs, pumps, dams, and natural channels.

Incident Support: When requested, they can provide advice and assistance concerning corrective actions to mitigate any incident affecting the state water system. When notified, they may elect to respond, depending on the circumstances at the scene. They can assist by providing documentation regarding violations of California Codes.

Notification: The DWR *must* be notified of any contaminant that encroaches upon their property or threatens to disrupt the operation of the state water project including the delivery of water via the project.

Reports: None required, other than in-house reporting.

Emergency Funding: Funding and resources exist only for minor self-generated hazardous materials incidents.

A-2.2.7 State Water Resources Control Board (SWRCB).

Responsibilities: The water resources control board is responsible for protecting surface and ground water throughout the state, including all rivers and their tributaries, and lakes.

Incident Support: They can provide advice concerning the potential impact of a hazardous material incident entering their waterways. They can conduct sampling and analysis services, and can advise critical downstream water users of condition of palatable water. They can assist in scene investigation and documentation regarding violations of California Codes, assess fines, and pursue recovery of costs for abatement, mitigation, and cleanup.

Notification: The SWRCB *must* be notified of any contaminant that threatens or has entered their waterways. They may elect to respond.

Reports: Damage Assessment Report and Remedial Action Plan may be required of the responsible party. Normal in-house reporting.

Emergency Funding: Administers the Water Pollution Cleanup and Abatement Account. This account is available to public agencies to cleanup oil and hazardous material releases which could pose a substantial threat to surface and ground water and to abate actual damage to surface and ground water.

A-2.2.8 Department of Food and Agriculture (Dept. of Ag.).

Responsibilities: The Department of Ag. is responsible to provide for proper and safe pesticide control in the state, and for protecting the public and the environment from potential adverse effects due to agricultural chemicals. They have responsibility for regulating the registration, sale, and use of agricultural chemicals prior to entering the waste stream, but do not have any on-site regulatory authority during an emergency.

Incident Support: Upon request, they can provide helpful technical assistance on pesticide related incidents, and can advise other state and local authorities of the potential of contamination to farm lands, feed, animals, and to the environment. The Dept. of Ag. Chemistry Laboratory Services, accessed through the Pesticide Enforcement Branch, may be utilized for emergency substance identification purposes if pesticides or fertilizers are suspected. They can provide expertise on medical and toxicological risk assessment regarding active pesticide ingredients, and provide information regarding the environmental fate of pesticides in water, air, and the soil.

Notification: This agency *must* be notified (sometimes through the County Department of Ag.) for any incident involving the accidental, unintentional, or intentional release of agricultural chemicals. The State Dept. of Ag. is not a response agency and normally will not respond. However, the local County Department of Ag. in all likelihood will respond when requested. The spiller or licensed pesticide handler is also *required* to notify the Dept. of Ag.

Reports: None required, other than in-house.

Emergency Funding: None.

A-2.2.9 Department of Forestry (CDF).

Responsibilities: The CDF performs fire prevention and suppression duties in the California state forested lands, and to some local jurisdictions on a contract basis. They function as the incident commander for all emergencies that occur within their jurisdictions.

Incident Support: Emergency support activities that they can provide other agencies include feeding operations of state and local workers, communication support as requested, environmental monitoring as requested, and support of local fire fighting activities in accordance with the State Mutual Aid Plan.

Notification: None required if outside of their jurisdiction.

Reports: None.

Emergency Funding: None.

A-2.2.10 Department of Parks and Recreation (DPR).

Responsibilities: Responsible for the maintenance and upkeep of all State Parks and other recreation lands throughout the state.

Incident Support: They can provide assistance to local authorities in evacuation of their lands, assist the CDF in setting up emergency feeding stations, and provide emergency living facilities for evacuees and emergency workers at large campaign incidents.

Notification: *Must* be notified of any incident occurring on their jurisdictional lands.

Reports: Other than internal reports, none required.

Emergency Funding: None.

A-2.2.11 Department of Industrial Relations (DIR).

Responsibilities: To prevent and regulate occupational exposures to hazardous materials by enacting standards which designate maximum allowable toxic threshold levels. The DIR has the responsibility for investigating and compiling information regarding all industrial accidents where workers are seriously injured or killed.

Incident Support: They can assist in providing technical advice regarding how to safely handle toxic materials at the scene of an incident. They can provide some evaluation of the health hazards associated to some toxic chemicals at the site of an industrial accident. They may elect to respond when requested, depending on the circumstances of the incident. Has the capability of evaluating the adequacy of health and safety measures designed to protect employees from exposure to hazardous materials during the normal course of employment, during hazardous materials incidents, and during cleanup and recovery operations.

Notification: When there is a report of any industrial injury the DIR *must* be notified. When there is an exposure to a regulated carcinogen, or there is serious injury, illness, or death of an employee during any work activity associated to hazardous materials, notification is *required* by the emergency response agency and by the employer.

Reports: None required.

Emergency Funding: None.

A-2.2.12 Department of Health Services (DOHS).

Responsibilities: DOHS is normally responsible for the protection of public health from chemicals and low level radioactivity. They are responsible for regulating the treatment, storage, transportation, and disposal of all hazardous waste. Their further responsibilities include protecting packaged or prepared food and water supplies from the effects of hazardous materials, and designating a location for the disposal of hazardous waste. They license hazardous waste sites and hazardous waste haulers.

Incident Support:

a. Hazardous Materials Management Section (HMMS) can provide technical advice regarding what protective measures should be exercised by emergency personnel. It is this section that provides advice to hazardous waste haulers as to where to take and dispose of hazardous wastes. Can provide advice on suitable disposal of toxic chemicals. Can respond to incidents involving facilities where the Section has enforcement responsibilities.

b. Radiological Health Section (RHS) has the responsibility to provide emergency response (when requested) to all accidents involving radioactive materials. Can assist local authorities in monitoring contamination of environment, personnel, and equipment. Establish activities to mitigate the radiological impact, and can recommend measures to limit the spread of the radioactive contaminant. Assist in defining and marking areas of contamination. Can provide or identify laboratory capable of radiological analysis, of food, and of feed when necessary (in concert with the DOHS). Can establish and direct measures to mitigate the radiological impact on public health. They can request federal (DOE) radiological assistance. Assist the local health officer (County Health) in assessing the impact on the public's health.

Notification: All incidents involving possible contamination of packaged or prepared foodstuffs *must* be reported to DOHS, and it is very likely they will respond a representative. DOHS is totally responsible for making the decisions regarding disposal of contaminated foodstuffs.

All incidents regarding radiological sources *must* be reported to DOHS.

Reports: Required as outlined in Title 17, Calif. Code of Regulations, copy forwarded to OES.

Emergency Funding: Maintains the Emergency Reserve Account (State Superfund) for hazardous materials incidents to assist local agencies. Specified instructions must be followed in order to qualify.

A-2.2.13 Department of General Services (DGS).

Responsibilities: The DGS is responsible for providing security to all state buildings and property, not including state parks. California State Police are managed by this department.

Incident Support: In all likelihood, they *will not function* as incident commander for incidents within their jurisdiction due to lack of training and experience. Local authorities, most notably the fire department having jurisdiction, should insure that the incident commander position is provided for. State police can ensure security of evacuated buildings, and they assist local law enforcement agencies very readily. They should have a good capability of access into state buildings after hours, and should have the ability in most cases to contact responsible parties for state property.

Notification: State police *must* be notified of any incident occurring on their property.

Reports: None.

Emergency Funding: None.

A-2.2.14 Division of Oil and Gas (DOG).

Responsibilities: The Division of Oil and Gas (of the Department of Conservation) is responsible for monitoring all regular oil and gas drilling and well operations within the territorial boundaries of the state. They require and approve oil spill contingency plans that should provide prevention, containment, and cleanup procedures.

Incident Support: They can, during an emergency, provide assistance to determine the seriousness of a drilling operation or pipeline emergency, including serious water drilling incidents. They can suggest appropriate actions to take, particularly in the event of a hazardous materials incident emanating from a well or drilling operation. They can authorize emergency drilling of a relief well.

Notification: Required if drilling operation, abandonment of wells, or underground pipeline is involved. When notified, they may elect to respond.

Emergency Funding: Small internal fund that can be accessed for mitigating the impact of an environmental release.

A-2.2.15 Emergency Medical Services Authority (EMSA).

Responsibilities: Develops general guidelines for the triage and handling of contaminated/exposed patients. Provides funding and management of the State Regional Poison Control Centers. (See *Poison Control Center*, Section 2.4.4).

Incident Support: Identifies medical facilities outside the affected county capable of handling injured and contaminated persons. Can arrange for the emergency procurement, distribution, and handling of supplementary medical supplies and equipment in support of local government response. Identify and coordinate procurement of medical assistance from other state departments, hospitals, and ambulance providers. Helps mobilize medical mutual aid. Can assist in the coordination of the evacuation of casualties from affected area to definitive care facilities (hospitals). Most likely will respond a Regional Disaster Medical Coordinator to the scene when notified.

Notification: Is *required* if activation of the regional disaster medical health coordinator is necessary. It is suggested if services of the authority are needed, and if a significant number of human exposures is expected.

Reports: Internal only.

Emergency Funding: None.

A-2.2.16 Fire Marshal's Office (SFM).

Responsibilities: Develops and enforces fire and life safety codes. Provides fire training and prevention information. Investigates all fires in state owned/ occupied buildings. Manages the interstate hazardous liquid pipeline system through the Pipeline Safety Division. Maintains Explosive Ordinance Disposal technicians who are available to assist with suspected explosive devices.

Incident Support: Can respond an engineer to pipeline emergencies, assists in incident investigation, and maintains a response unit for such events.

Notification: Is *required* for reports of underground liquid pipeline emergencies, through OES.

Reports: Must submit report to Federal Office of Pipeline Safety.

Emergency Funding: None.

A-2.2.17 (Governor's) Office of Emergency Services (OES).

Responsibilities: OES is responsible for the notification and coordinated response of all state agencies that may become involved in a hazardous materials incident through its State Warning Center. They maintain statistics of haz-mat incidents in the state. Prepares situation reports for the Governor's Office. Publishes the State Hazardous Materials Incident Contingency Plan, and assists local agencies in preparing their own emergency response plans. Maintains and distributes radiological monitoring devices. Maintains the State Mutual Aid Plan and the Incident Command System (ICS).

Incident Support: They shall contact state agencies as appropriate for the incident, or as requested. They can have key state agency representatives make contact with on-scene units. The OES has access to state owned equipment and materials. Mobile incident command post and communications unit is available to local agencies.

Notification: They shall have reported to them all hazardous materials incidents of a Level II magnitude or greater. It is state law that the "spiller" is mandated to notify OES for any significant release. Upon notification, they issue to the emergency response agency a "CHIMERS" report number.

Reports: The State OES "CHIMERS" report is fowarded to OES. Spiller is also required to submit a report.

Emergency Funding: In the event of a State or Federally declared disaster some federal and state disaster relief funding is released via the Natural Disaster Assistance Act. No funding for routine haz-mat incidents.

A-2.2.18 Public Utilities Commission (PUC).

Responsibilities: Has responsibility and authority for instigating all railroad accidents. Conducts daily inspection of railroad facilities for compliance to Federal Title 49.

Incident Support: Can provide field investigator to conduct on-site investigation for major hazardous materials related transportation accidents and has responsibility for investigating all railroad accidents.

Notification: The PUC *must* be notified of any and all railroad accidents.

Reports: Investigation report is required, which might result in the further formal Commission investigation.

Emergency Funding: None.

A-2.3   Federal Government.

A-2.3.1 Environmental Protection Agency (EPA).

Responsibilities: The EPA has responsibility to assure protection of the environment from all types of contamination. They must test and then grant approval for sale of all pesticides.

Incident Support: They may elect to respond depending on the seriousness of the threat of contamination. When they do respond, they can be a source of additional on-scene technical advice. When appropriate, the EPA representatives should be capable of assisting local authorities in identifying violations of federal law. On very large incidents or operations involving long drawn-out cleanup within their jurisdictions, they may elect to become the on-scene coordinator (OSC) in accordance to the National Contingency Plan.

Notification: They *must* be notified of any hazardous material incident that is confirmed to have caused ground contamination. In some locations in California, the EPA and not the USCG is notified of water contamination.

Reports: Same responsibilities as USCG.

Emergency Funding: Federal OSC may access the Hazardous Substance Response Trust Fund (Superfund).

A-2.3.2 Department of Defense (DOD).

Responsibilities: To defend our country.

Incident Support: Can assist in investigations to evaluate the magnitude and severity of discharges or releases on or adjacent to resources under the jurisdiction of the DOD. Responds to incidents involving nuclear weapons and institutes procedures in accordance with the Nuclear Weapons Accident Response Procedure manual. In accordance with the National Contingency Plan, they can and most likely would become the on-scene coordinator on very serious incidents involving military equipment and property.

Notification: The DOD *must* be notified in the event of any accident involving nuclear weapons or nuclear fissile material, and of any accident involving military transportation vehicles. Depending on the circumstances, they may elect to respond. Local military authorities would be immediately dispatched to secure the area until DOD representatives arrive.

Reports: Unknown.

Emergency Funding: Only with respect to the mitigation and removal of their property and equipment, and the damage incurred at the scene as the result of the accident.

A-2.3.3 Department of Energy (DOE).

Responsibilities: The DOE has the responsibility to monitor the movement of all major civilian radiological sources, such as reactor fuel.

Incident Support: They can assist with technical information and prudent on-scene handling advice. In all likelihood, the DOE will assume overall responsibility of major accidents. They are responsible for providing information on the disposal of these civilian nuclear materials.

Notification: They *must* be notified in the event of an accident involving these sources and in most cases will respond, however response may be severely delayed.

Reports: Unknown.

Emergency Funding: None.

A-2.3.4 Department of Transportation (DOT).

Responsibilities: DOT has the responsibility to regulate the construction of all containers intended to package and move hazardous materials, including vehicles; to regulate the labeling and placarding system; to regulate the shipping documents that are required; and to keep the Title 49 of the Code of Regulations up to date. Publishes the DOT *Emergency Response Guidebook*.

Incident Report: Through the National Transportation Safety Board, they may as requested or as needed investigate and report on very serious transportation related hazardous materials incidents.

A-2.3.5 U.S. Coast Guard (USCG).

Responsibilities: The USCG has a responsibility encompassing the nation's coastline and for 100 miles inland, although this will depend on local capabilities and policies. Enforces the federal DOT Title 49 statutes for water transportation.

Incident Support: Their support of local authority is similar to the EPA where the EPA has elected not to have jurisdiction. They do have the ability to provide for some containment, decontamination, and cleanup of major water-borne spills. The National Contingency Plan specifies that for major incidents the U.S. Coast Guard will function as the on-site on-scene coordinator (OSC) within their jurisdiction if they elect to incorporate this position.

A-2.3.6 National Response Center (NRC)

They operate the National Response Center which is available for dissemination of technical information regarding chemicals, mitigation technics, and cleanup suggestions.

Notification: For major contaminants into major waterways, rivers, lakes, the USCG *must* be notified. Depending on the circumstances at the scene, they may elect to respond when requested, or the EPA may cover. Spiller is obligated to report release or spillage of any "reportable quantity" (RQ) of a hazardous material to the National Response Center of the USCG.

Reports: Detailed reports for large incidents, forwarded to the National Response Team. Some incidents require follow-up reports one year after incident date.

Funding: None, except that if they assume the position of on-scene coordinator they also assume all mitigation and cleanup costs from that point on.

A-2.4 Non-Governmental Agencies.

A-2.4.1 American Red Cross.

Incident Support: Can provide relief for persons affected by disaster. For those temporarily in need, can provide food and clothing allowance and temporary lodging for a few days. Will try to supplement medical needs and nursing assistance and some family services. Operates independently from but coordinates with local emergency response agencies. Can be very instrumental in helping set up temporary shelter ares.

Notification: Only as their assistance is requested.

Reporting: None.

Emergency Funding: Can fund victims with a few days of financial support for food, clothing, lodging, medical only.

A-2.4.2 CHEMTREC.

Incident Support: Immediate emergency action information for spill, leak, exposure, or fire control. Can assist with identification of hazardous materials, especially if the manufacturer is known or the shipping papers are present. Can immediately notify manufacturers or shippers through their emergency contacts or notification of industry mutual aid networks. Can notify other federal agencies as is necessary or as is required depending on circumstances at scene.

Notification: Not mandatory, a service organization only.

Reporting: For those incidents that require notification of the USCG National Response Center regarding spills of an "RQ," CHEMTREC will pass on this notification.

Funding: None.

A-2.4.3 Community Awareness and Emergency Response (CAER).

Incident Support: Because chemical industry representatives can be especially knowledgeable during the planning process, they may be willing and able to share equipment and personnel during response operations.

Notification: Only as necessary based on incident needs.

Reporting: None.

Emergency Funding: May provide some specific supplies without cost to the responding agencies on scene.

A-2.4.4 Poison Control Center.

Incident Support: Located regionally throughout the State and managed by the Emergency Medical Authority, they can provide toxilogical information concerning hazardous materials incidents. They can suggest immediate on-scene protocols to initiate with respect to handling civilians contaminated with dangerous chemicals. Local jurisdictions should have arrangements for direct contact with their closest center.

Notification: Only as necessary.

Reporting: None.

Emergency Funding: None.

A-2.4.5 National Poison Antidote Center (NPAC).

Incident Support: This center is now a working part of the CHEMTREC system. It provides immediate information for treatment of most known poisons. It maintains communications with most major hospitals, particularly those identified as *poison control centers*. Contact with the NPAC can be made through CHEMTREC, or through the local poison control center.

Notification: Only as necessary.

Reporting: None.

Emergency Funding: None.

A-2.4.6 Underground Service Alert (USA - 800-642-2444).

Incident Support: A service subscribed to by most major public utilities and private contractors which has the capability of providing the location of any underground structure and piping which could impact the response to hazardous materials incidents.

Notification: They *must* be notified before any digging or excavating is done by any contractor to verify that no underground lines will be disturbed. They *should* be notified in the event of an accident so that they may assist in contacting the correct responsible party.

Reporting: None.

Emergency Funding: None.

Appendix 3: Agency Responsibilities Matrix

(*Note: See key on pages 734–735.*)

A-3.1 Local Agencies

Function	1	2	3	4	5	6	7	8	9	10	11	12	13	14	15	16	17	18	19	20	21	22
Agri Commissioner	S	T	y	N	n	N	n	T	n	2	N	t	n	n	t	T	C	r	n	N	t	n
Air Pollution Board	C	T	y	N	n	N	n	T	n	Y	N	t	n	n	t	T	C	r	n	N	t	n
County Health Dept.	S	T	y	N	n	N	T	T	y	2	N	1	n	y	2	T	T	R	n	y	t	T
District Attorney	1	n	n	N	n	N	n	t	y	1	N	n	n	n	1	n	n	R	n	y	n	n
Emergency Operations	2	n	y	N	y	1	y	n	n	2	y	2	y	2	y	y	y	r	y	2	y	Y
Environmental Health Div.	2	y	y	N	Y	2	T	T	n	2	N	1	y	n	y	t	C	R	t	y	t	y
Fire Dept. of Jurisdiction	2	y	2	N	y	1	y	n	y	2	y	1	y	1	2	1	y	r	y	1	y	Y
Haz-Mat Team (Specialist)	2	Y	1	N	Y	1	Y	y	y	2	Y	2	n	y	y	y	1	r	Y	1	1	Y
Hospitals	N	N	y	y	n	N	T	n	n	n	n	t	n	n	n	n	n	R	n	n	n	C
Police	1	N	n	N	n	N	n	n	n	1	n	1	n	2	1	2	n	r	n	1	t	n
Public Works/Street	2	N	y	y	Y	y	t	y	n	2	t	t	n	n	t	y	n	R	n	t	n	n
Sheriff	1	N	n	N	n	N	n	n	n	1	n	1	n	1	1	2	n	r	n	1	n	n
Schools	N	N	n	n	n	N	n	n	n	n	n	n	n	n	n	n	n	R	n	n	n	n
Water/Sewer	T	y	y	Y	Y	y	t	y	n	2	t	t	n	n	t	y	y	R	n	t	t	N

A-3.2 Federal Agencies

Function	1	2	3	4	5	6	7	8	9	10	11	12	13	14	15	16	17	18	19	20	21	22
Environmental Prot. Agency	y	n	n	n	n	n	n	y	y	1	n	n	n	y	Y	t	y	R	n	y	n	n
Dept. of Defense (DOD)	n	n	t	n	y	n	n	y	n	Y	n	n	n	n	y	t	N	r	n	n	T	n
Dept of Energy (DOE)	n	n	t	n	y	n	n	y	n	y	n	n	n	n	y	t	N	r	n	n	T	n
Dept of Trans. (DOT)	t	n	n	n	n	n	n	n	n	Y	n	n	n	n	y	N	N	R	n	n	n	n
Drug Enforcement Agency (DEA)	1	T	t	y	Y	y	n	y	y	1	Y	t	n	n	1	t	y	r	n	1	Y	n
National Response Center	n	n	C	n	C	C	C	n	n	n	n	C	n	n	n	C	n	R	C	n	C	T
U.S. Coast Guard (USCG)	C	t	t	n	y	y	y	y	n	n	y	y	t	n	y	y	t	y	r	n	y	y	n

A-3.3 State Agencies

Function	1	2	3	4	5	6	7	8	9	10	11	12	13	14	15	16	17	18	19	20	21	22
Air Resources Control Board	n	n	n	n	n	n	n	n	n	t	n	t	n	N	t	t	Y	r	n	n	n	n
Attorney General	y	n	n	n	n	n	n	n	n	T	n	n	n	N	t	n	n	n	n	n	n	n
California Hwy Patrol (CHP)	1	n	n	n	n	n	n	n	n	1	n	2	n	1	1	N	n	r	n	1	n	s
Cal/TRANS	y	n	t	n	t	y	n	y	n	2	n	t	n	y	2	y	y	R	n	y	y	s
Dept. of Conservation	y	n	t	n	n	n	n	n	n	t	n	n	n	N	t	N	n	E	N	n	n	n
Dept. Fish and Game (DFG)	Y	n	y	y	y	y	t	T	y	1	y	T	n	2	1	T	y	R	n	2	t	n
Dept. Food and Ag. (DFA)	y	n	y	y	t	t	t	T	n	2	y	T	n	n	t	N	t	R	n	y	t	n
Dept. of Forestry (CDF)	Y	n	n	n	y	y	y	y	N	n	1	y	y	n	1	2	y	R	n	1	y	Y
Dept. General Services (GSA)	n	n	n	n	y	n	n	n	n	T	n	n	n	n	n	n	N	R	n	s	n	n
Dept. Health Services (DOHS)	2	n	T	n	C	t	t	C	Y	2	n	C	n	n	2	t	Y	R	n	n	T	n
Dept. Indust. Rel. (DIR)	y	n		n	t	n	N	t	n	2	n	n	n	n	t	n	n	R	n	n	n	n
Dept. of Parks and Rec.	y	n	n	n	n	n	n	N	n	t	n	y	n	n	t	t	n	R	n	n	n	n
Dept. Water Resources (DWR)	Y	n	t	y	y	t	n	y	n	T	n	t	n	n	t	t	y	R	n	y	t	n
Division of Oil and Gas	y	n	t	n	y	t	n	y	y	t	n	t	n	n	t	t	y	R	n	y	t	n
Emergency Medical Authority	n	n	n	n	n	n	t	n	n	n	n	n	n	n	n	n	n	n	n	n	n	T
Office of Emergency Services	n	n	n	n	n	n	n	n	n	n	n	n	n	y	n	n	n	R	n	y	n	n
Office of State Fire Marshal	Y	n	n	n	n	n	n	n	n	T	n	n	n	n	n	n	n	R	n	y	n	n
Public Utilities Commission	t	n	n	n	n	n	n	n	n	t	n	n	n	n	n	n	n	R	n	y	n	n
State Water Quality Control	t	n	t	y	y	t	n	y	y	2	n	n	n	n	n	t	y	R	n	y	n	n

A-3.4 Non-Governmental Agencies

Function	1	2	3	4	5	6	7	8	9	10	11	12	13	14	15	16	17	18	19	20	21	22
American Red Cross	N	N	N	N	N	N	N	N	N	N	N	S	Y	N	N	N	N	N	N	N	N	N
CHEMTREC	N	N	C	N	N	C	C	C	N	N	N	N	N	N	N	C	N	N	N	N	C	C
Chemical Waste Removers	N	C	C	S	S	C	T	S	N	N	y	N	N	N	N	s	S	N	N	N	C	N
Community Awareness (CAER)	N	E	E	E	E	E	E	E	E	N	N	E	E	E	E	N	N	N	N	N	E	E
Electrical Utilities	N	C	C	s	s	T	T	S	y	N	y	N	y	N	t	s	s	r	N	N	T	N
Hospitals, Main Trauma Cntr	N	N	T	s	N	N	C	N	N	N	N	N	N	N	N	N	N	R	N	N	N	T
National Poison Antidote	N	N	C	N	N	N	C	N	N	N	N	N	N	N	N	N	N	R	N	N	N	T
Poison Control Center (SMC)	N	N	C	N	N	N	C	N	N	N	N	t	N	N	N	N	N	R	N	N	N	T
Private Indust. Resp. Team	N	C	C	s	s	C	C	s	N	N	y	t	y	N	t	s	s	N	N	N	T	N
Underground Service Alert	N	N	N	N	N	N	N	N	N	y	N	N	N	N	N	N	N	r	N	Y	N	N

Function Areas of Agency Involvement

Locate agency in left column, then locate type of function relative to incident operations across top (numbered). Follow down the column to note the code entered for types and levels of functions for that agency.

1. Arrest/prosecute
2. Chemical Haz-cat
3. Chemical ID
4. Chemical lab analysis
5. Cleanup
6. Containment
7. Decontamination
8. Disposal/removal
9. Emergency funding
10. Enforcement/citation
11. Entry to hot zone

12. Evacuation/relocation
13. Funding (outside of emergency)
14. Incident command
15. Investigation—criminal
16. Isolation/security
17. Monitoring capacilities
18. Notification required
19. Rescue
20. Scene authority
21. Stabilization of product
22. Trauma care/EMS

Types and Levels of Functions

1 = Primary Responsibility: Initiator, control of incident or segment of it, legal authority vested

2 = Secondary Responsibility: legally can and should support others who have this vested authority

S = Support: can provide work, services, activities in support of on-scene authorities

s = Minor Support: limited support services, not primary choice, usually limited training

N = No Authority: provides assistance as requested and allowed

R = Function Required by Responder: per legal mandates, protocols, departmental requirements

r = Function Required by Spiller: per legal mandates, laws, legislation

C = Technical Assistance Only: can assist technically or verbally, highly suggested

T = Technical Assistance Only: good source but should not get involved directly

t = Technical Assistance Only: limited source, should never get involved directly

Y = Yes: this service provided, or this function is mandatory

y = Yes: but limited; there are restrictions

N = Has no on-scene responsibility, should not be given any

E = Education: before- or after-the-fact education capabilities

Appendix 4: Haz-Mat Incident Command Worksheet

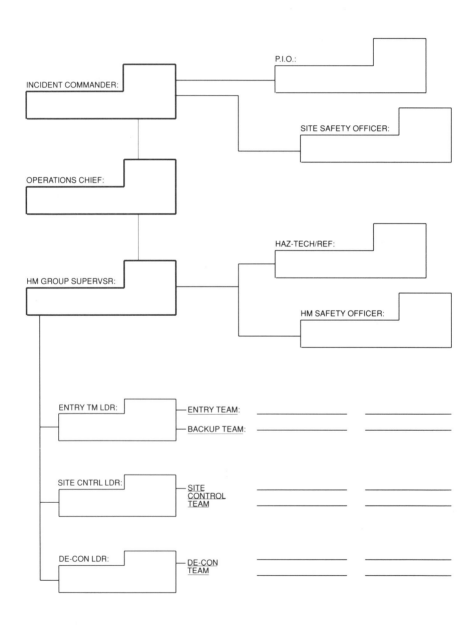

Appendix 5: Notification of Agencies

A-5 Typical Agency Contacts, by Type of Incident.

	EXTENT OF PROBLEM	AGENCIES TO BE NOTIFIED
FOR: ALL LEVEL II HMI INCIDENTS*	**INVOLVING:** ALL CHEMICALS, TRANSPORTATION, HANDLING	**CONTACT:** OES, COUNTY ENVIRONMENTAL HEALTH *POSSIBLY DFG, EPA, RWQCB also hmrt *OES WILL MAKE THESE CONTACTS
MATERIAL INVOLVED*	PESTICIDE OR AG CHEMICAL, FERT AND PEST MIXTURE	DFAg, RWQCB, DOHS, County Ag
	RADIOACTIVE MATERIAL, COMMERCIAL	DOHS, County Health, +ARB
	RADIOACTIVE MATERIAL, MILITARY, FISSILE	DOD, DOE, +NRC, DOHS, County Health
	EXPLOSIVES, EXPLOSIVE-SENSITIVE CHEMICALS	LAW ENF, BOMB SQUAD
	OIL, LARGE FUEL SPILLS	DFG, +RWQCB, +SWRCB
	OIL WELL, BLOW-OUT	DFG, DOG, SLC, +RWQCB, +SWRCB
	ON HIGHWAY, CITY OR COUNTY ROAD	CHP, CAL/TRANS, APD, ASTREET DEPT
MODE OR LOCATION*	OFF ROAD ON PRIVATE OR PUBLIC PROPERTY	+DFG, DPW
	WATER ROUTES, UTILITY AND DRAINAGE	DPW, CWD, SD, +DFG
	WATER ROUTES, NATURAL OR FLOOD CONTROL	DFG, DWR, +RWQCB, +USCG
	INDUSTRIAL (WITH INJURY)	+DFG, +DFA, +DOHS, (DIR)
	PIPELINE	CSFM, DFG, +RWQCB, DWR
	RAILROAD	PUC, AAR, CSFM, +DOT
TYPE OF PROBLEM	HEALTH EFFECTS, LG NUMBER OF EXPOSURES	EMSA, County Health, PCC
	AIR POLLUTION, (DOWNWIND EVACUATION)	ARB, (County Health)
	FISH AND WILDLIFE AFFECTED	DFG, USFWS
	WATER POLLUTION	RWQCB, +DWR, +DOHS, DFG
	DRUG LAB, CHEMICALS, RESIDUE	PD, DEA/DOJ

(See key next page.)

HAZARDOUS MATERIALS RESPONSE HANDBOOK

A-5.1 Codes for Chart A-5.

ARB	= Air Resources Board
AAR	= Association of American Railroads
CAL/TRANS	= California Department of Transportation
CHP	= California Highway Patrol
CSFM	= California State Fire Marshal
CWD	= City Water Department
DEA	= U.S. Drug Enforcement Task Force
DFA	= Department of Food and Agriculture
DFG	= Department of Fish and Game
DIR	= Department of Industrial Relations
DOD	= U.S. Department of Defense
DOE	= U.S. Department of Energy
DOG	= Division of Oil and Gas
DOHS	= Department of Health Services
DOT	= U.S. Department of Transportation
DPW	= (Local) Department of Public Works
DWR	= Department of Water Resources
EMSA	= Emergency Medical Services Authority
EPA	= U.S. Environmental Protection Agency
NRC	= National Response Center
OES	= Office of Emergency Services
PD	= Police Department
PUC	= Public Utilities Commission
RWQCB	= Regional Water Quality Control Board
SD	= Sewer Treatment Facility, Department
SLC	= State Land Commission
SWRCB	= State Water Resources Control Board
USCG	= U.S. Coast Guard
USFWS	= U.S. Fish and Wildlife Service
+	= Must be called, depending on circumstances
*	= OES will make many of these calls, depending on circumstances and requests, except for local agency contacts
@	= Depending on location, some or many law agencies may become involved

Appendix 6: Funding Assistance and Sources

A-6.1 Clandestine Laboratory Enforcement Program.

Funding Authority: Health and Safety Code Section 11642(c).

Annual Total: Up to $300,000.

Administered By: State controller.

How Contacted: Local law enforcement must, within 24 hours, notify the local health officer who shall contact DOHS.

Types of Releases Funded: Must be a prosecutable case, must include removal and disposal of waste, chemicals, end product, lab ware of a proportion to support arrest and prosecution. Discovery of controlled substance improves case for funding.

Limitations: Counties with a population *less than 1,250,000*. Also does not address cleanup of soil, property, dwellings.

A-6.2 California Super-Fund.

Funding Authority: Health and Safety Code Section 25354.

Annual Total: $1,000,000.

Administered By: DOHS, Toxic Substances Control Program.

How Contacted: Through OES, or 916-324-2445. Contact OES and inform that we are seeking approval for assistance from the *State Superfund Emergency Reserve Account*.

Maximum Award: $20,000.

Types of Releases Funded: Only after all other avenues of funding (i.e., responsible party) have been exhausted. That which is a "threat to the public" including clandestine dumping or discharge, abandonment, highly toxic or corrosive materials. Usually flammables are not considered unless a part of the total manifest. On occasion includes drug lab waste not covered by Clandestine Laboratory Enforcement Program, or by the DEA Drug Lab Task Force federal funding. Corrosives should be a pH 11 or > or a pH 3 or <. Criteria also include the following:

a. Verify hazard and degree of all substances.
b. Estimate total quantities in containers, released to ground.
c. The hazard characterization, exposure potentials.
d. Location relative to waterways, public, thoroughfares.
e. Alternative funding (responsibles) is not available.

Limitations: Waste oils, fuel tanks, fuel spills from accidents, radioactive materials not covered unless highly unusual circumstances exist; drug lab waste related to law enforcement involvement; incidents on federal property.

Cost Recovery: Will be made by DOHS at every site where there is a culpable responsible party. Includes all incident costs plus cleanup, plus administrative costs, pursuant to Section 25360.

A-6.3 Fish and Wildlife Pollution Cleanup Account.

Funding Authority: Fish and Game Code Section 12017.
Annual Total: $500,000.
Administered By: DFG, Wildlife Protection Division.
How Contacted: OES.
Types of Releases Funded: Cleanup actions of chemicals and containers threatening to pollute, contaminate, or obstruct waters in such a way to the detriment of fish, game, plant and animal life and their habitat. Eligibility usually determined by DFG on scene.
Limitations: Impact must be to F and G and their habitat. When DFG account could cover costs, the State Superfund will not. DFG will work with responsible party to encourage responsible to remedy in accordance with guidelines provided by DFG.
Cost Recovery: All costs associated to control and cleanup, removal, and disposal, plus administrative costs, plus civil damages, pursuant to Sections 5655, 12015, and 12016.

A-6.4 Oil Spill Response Trust Fund.

Funding Authority: California Government Code Sections 8670.46-8670.53.95.
Annual Total: $100,000,000.
Administered By: DFG, Oil Spill Administrator.
How Contacted: OES.
Types of Releases Funded: Marine oil spills only.

Limitations: Marine related only. Responsible is unknown, unwilling, or unable to provide adequate and reasonable assistance. Federal oil spill funds will not cover all or part of spill cleanup.

A-6.5 Water Pollution Cleanup and Abatement Account

Funding Authority: California Water Code Sections 13440-13442.
Annual Total: ??
Administered By: State Water Resources Control Board, SWRCB.
How Contacted: OES.
Types of Releases Funded: Usually *non-emergency* situations, assistance given local agencies for cleanup of waste threatening to contaminate or impacting surface or groundwater. Site usually must be inspected first.
Limitations: Assistance not provided if other funds available, local agency is capable of funding. Oral approval of $50,000 can be given. Details available from SWRCB. If hazardous materials incident impacted a State Water Project, SWRCB should be on scene and they will handle fund release.

A-6.6 Cal/Trans.

Annual Total: ??
How Contacted: CHP must be on scene.
Types of Releases Funded: Cal/Trans administers a fund for hazardous incidents that impact state highways and right-of-ways. Several cleanup firms are on yearly contract.
Limitations: Cal/Trans *will not* finance cleanup of hazardous materials that migrate beyond the right-of-way and off of state highway property.

Fire Department of the City of New York Emergency Response Plan: Hazardous Materials

1.0 Introduction and Statement of Command Concepts

1.1 Planning Basis

In order to carry out the responsibilities delineated in the New York City Hazardous Materials Response Plan (Mayoral Directive 82-2 Revised) and to comply with standards established by Federal legislation the FDNY Emergency Response Plan for Hazardous Materials is established. This plan will guide and direct the action of all responding Fire Department personnel at hazardous materials operations.

In the event of a significant accident or other incident involving a toxic or hazardous material where, because of fire, explosion, radioactivity, toxic air release or chemical reaction there exists danger to the health or safety of emergency personnel or to the general public, the senior Fire Department representative, referred to as the Incident Commander (IC), at the scene will implement the FDNY Emergency Response Plan (ERP) for Hazardous Materials.

1.2 Agency Coordination

This plan is built on the fundamental organization and strategy of the FDNY Incident Command System (ICS) insuring maximum safety and efficiency of the operating forces while fulfilling our responsibility to the public.

It is assumed that all hazardous materials incidents will be managed under Unified Command principles because in virtually all cases fire, police, and public health agencies will have some statutory functional responsibility for incident mitigation.

Implementing the ERP will facilitate the coordination and control of all tasks and functions with the senior on-site representative of the Police Department who has been designated as the "On Scene Coordinator." Using the

ERP and working closely with the On-scene coordinator, the Incident Commander will manage the command and control of all mitigation tasks and functions. When the Interagency Command Center (ICC) is established the IC will represent the Fire Dept. at the Command Center.

All mitigation operations at the incident will be managed by the Fire Department. Tactical operations will be managed by the Fire Department "Operations Chief."

Specific tactical objectives will be carried out by Group/Sector Supervisors. Other needs will be met by staffing ICS positions.

2.0 Organizational Development

2.1 Response Objectives and Strategy

Any hazardous materials incident represents a potentially dangerous situation. Chemicals that are combustible, explosive, corrosive, toxic, or reactive, along with biological and radioactive materials can affect the general public or the environment as well as the emergency responder. Emergency responders may be subject to additional dangers operating in this abnormal environment. While the response activities needed at each incident are unique, there are similarities. One is that every response requires protecting the health and ensuring the safety of the responders.

This document identifies the functions needed to control and mitigate a hazardous materials incident. It also describes the lines of authority, responsibility, and communication between and among the various responders. It defines the interface of the Fire Dept. with other agencies. Finally, it specifies the authority of each responder in directing specific operations.

This plan will provide instructions on how to accomplish specific tasks in a safe manner. In concept and principle, standard operating safety procedures are independent of the type of incident. Their applicability at a particular incident must be determined and necessary modifications made to match prevailing conditions. However, in the case of hazardous materials operations, the specific requirements of training, equipment, and competence preclude the generalized statements that might apply to other types of emergency work.

The ERP will provide guidance for Fire Dept. responders in areas related to response, site control, entry, and mitigation of hazardous materials incidents.

The guidance is not meant to be a comprehensive treatment of each of the subjects discussed. Formal training in these areas will complement this document. Specific training will provide more information for the technical, administrative, and management oriented skills needed to fulfill our mission. This document will provide standard operating guides to develop more specific procedures.

The priority of the instructions in this plan is as follows:

A. Life Safety and health risks to the public and the emergency responders are the most important concern.
B. The Fire Department must stabilize the incident scene and prevent further escalation of the incident with minimum personal risk.
C. The Fire Department's response efforts should be directed toward protecting property and minimizing or lessening the impact of the event on the environment.

2.2 Incident Characterization

In the same way that units now call for additional assistance at a fire, hazardous materials incidents require that a method be established to determine the degree of severity for various types of releases. This avoids the need for calling out full resources for every incident. Personnel arriving at an incident can, of course, request additional resources and thereby raise the level of response based on the actual circumstances of the event. Response planning, procedures, and notifications to Federal, State, and City agencies will be determined by the following standard designations.

2.2.1 Level 1

An incident which can be controlled by the responding unit or units up to 3 Engines, 2 Ladders and 2 Battalion Chiefs and does not require evacuation of other than the involved structure or the immediate outdoor area. The incident is confined to a small area and does not pose an immediate threat to life or property.

A Level 1 incident includes the units on the scene up to and including the full first alarm assignment. It is *not* a request for additional units or specialized resources. It signifies the degree of hazard and the ability of the units on the scene to safely manage the incident. Additional resources or greater alarms may be transmitted at the discretion of the Incident Commander. This may indicate a need to upgrade the incident to Level 2.

2.2.2 Level 2

An incident involving a greater hazard or a larger area which poses a potential threat to life or property and which may require a limited evacuation of the surrounding area.

A Level 2 incident requires the response of a Deputy Chief, Hazardous Materials Unit, Safety Operating Battalion and SOC-Rescue Liaison and a total first alarm assignment of 3 Engines, 2 Ladders and 2 Battalion Chiefs.

2.2.3 Level 3

An incident involving a severe hazard or a large area which poses an extreme threat to life and property and will probably require a large-scale evacuation; or an incident requiring the expertise or resources of City, State, Federal, or private agencies and organizations.

A Level 3 incident requires activation of the full NYC Emergency Response Plan. Additional alarms or other Fire Dept. resources may be called by the Fire Dept. Incident Commander.

2.2.4 The following table is presented for *guidance* in determining these levels. The highest level for any single condition will determine the incident level. For example, poison gas (such as cyanide or phosgene) could initially require a Level 3 response due to the nature of the danger to the public. An incident level and the response can always be downgraded when additional information or resources become available. However, it is much more difficult to upgrade and obtain control when situations are going beyond the capabilities of the on scene resources.

2.3 Emergency Alerting and Response

Initial notification of a hazardous materials incident will be through transmission of the radio signal 10-80. Transmission of the signal 10-80 will serve to warn responders to proceed with caution to avoid entering a restricted area.

As soon as possible, the Officer assuming Command will briefly describe the nature of the incident, identify a preliminary hazard perimeter and transmit this information to the dispatcher for the benefit of other responding units and agencies that will be notified. If possible, a preliminary Staging area should

CONDITION	LEVEL ONE	LEVEL TWO	LEVEL THREE
PRODUCT	NO DOT PLACARD REQUIRED ORM A, B, C, D	DOT PLACARD PCBs/NO FIRE EPA REGULATED WASTE** ANY UNIDENTIFIED SUBSTANCE	POISON A, EXPLOSIVE A/B, ORG. PEROXIDE, FL. SOLIDS (WR), CHLORINE, FLUORINE, ANHYDROUS AMMONIA, RADIOACTIVES, PCBs ON FIRE •
NFPA #704	0 OR 1 ALL CATEGORIES	2 FOR ANY CATEGORY	3 OR 4 ANY CATEGORY INCL. SPECIAL HAZARDS
CONTAINER SIZE *	SMALL	MEDIUM	LARGE
CONTAINER INTEGRITY	NOT DAMAGED	DAMAGED BUT SERVICEABLE FOR HANDLING OR TRANSFER OF PRODUCT	DAMAGED, CATASTROPHIC RUPTURE POSSIBLE
LEAK SEVERITY	NO OR SMALL RELEASE CONTAINED OR CONFINED W/ AVAILABLE RESOURCES	NOT CONTROLLABLE W/O SPECIAL RESOURCES OR "REPORTABLE QUANTITIES"	MAY NOT BE CONTROLLABLE EVEN WITH SPECIAL RESOURCES
LIFE SAFETY	NO LIFE HAZARD	LOCAL AREA, LIMITED EVACUATION	LARGE AREA, MASS EVACUATION
IMPACT ON ENVIRONMENT	MINIMAL	MODERATE	SEVERE

*Small = Pail, drum cylinder, package bag.

Medium = One ton containers, portable containers, nurse tanks, multiple packages.

Large = Tank cars/trucks, stationary tanks, hopper cars/trucks, multiple medium containers.

**e.g., asbestos

be designated. The intent is to limit unnecessary exposure of personnel and equipment. More specific information should be relayed to the dispatcher in progress reports as it becomes available.

The following notifications of a hazardous materials incident will be made by the Dispatcher upon receipt of all 10-80 code 2 signals (Level 2 or 3 incidents) and when the Hazardous Materials Unit responds.

A. Police Dept. Operations Unit.
B. Dept. of Environmental Protection.
C. NYC Emergency Medical Services.
D. NYC Dept. of Health — for incidents involving radioactive materials.
E. NYS Dept. of Environmental Conservation — for fuel oil spills.

2.3.1 On transmission of a 10-80 code 2 signal (Level 2 or 3 incident), the dispatcher will order the response of a Deputy Chief, Hazardous Materials Unit, Safety Operating Battalion and Special Operation Command-Rescue Liaison and a total first alarm assignment of 3 Engines, 2 Ladders AND 2 Battalion Chiefs.

The following Fire Department units will also be notified:

1. Fire Command or the City Wide Command Chief during non-operating hours of the Fire Command.
2. Notifications Desk, Bureau of Operations.
3. Public Information Office.
4. Medical Officer on Duty.

As is the case in other fire and emergency operations, the Incident Commander may call for additional units and activate and staff any part of the ICS appropriate to the apparent circumstances.

2.4 Response Organization

Under the Incident Command System, the incident organization will develop in a modular progression depending on the exact nature and specific conditions prevailing at the scene.

2.4.1 The first response of the Fire Dept. will be managed by the senior ranking officer on the initial alarm who will be the Incident Commander. This could be the Lieutenant or Captain of the first company to arrive. Responsibility for command will be transferred to succeeding commanders using the established lines of authority within the Incident Command structure.

The Incident Commander on the first alarm will implement the ERP for hazardous materials and assume responsibility for all command and command staff functions necessary to manage the initial response. The IC will call additional resources as incident needs dictate.

First responders at this level have been trained to implement offensive and defensive control measures for common hazardous materials involving limited quantities of certain materials as specified in standard operating procedures. They will be capable of performing the following basic procedures:

A. Initiate the Emergency Response Plan including isolating the site, establishing an initial Command Post, and establishing a site safety plan.
B. Notify the Dispatcher of the need for additional resources. Communicate essential information abut the incident to other responders through the Dispatcher.
C. Initiate evacuation, if appropriate.
D. Initiate basic hazard and risk assessment activities including the use of personal protective equipment; preliminary identification of materials; containment and confinement of the materials within the limits of the resource and protective equipment capabilities available on site.
E. Understand and comply with "decon" procedures.

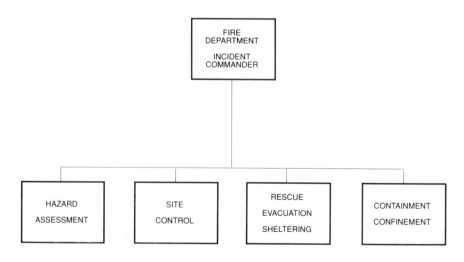

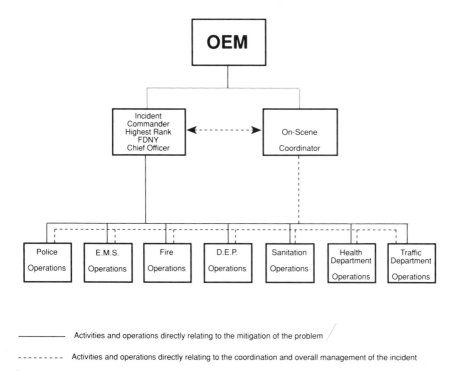

Activities and operations directly relating to the mitigation of the problem

- - - - - - - - Activities and operations directly relating to the coordination and overall management of the incident

2.4.2 If an incident is beyond the capability of the first responders, the ICS will be expanded to include the Hazardous Materials Group and other responders with a higher level of training and more specialized skills and equipment.

As the incident progresses it may be necessary to modify/expand the organization structure so tasks may be accomplished as efficiently and safely as possible. Changes will be recorded in the Incident Log and the Command/Control Chart with affected positions notified.

Responders at this next level will be operating in a more complex environment that requires the coordination of several different agencies.

2.4.3 A more complex operation requires a unified command structure composed of the senior officials of various agencies. They will determine strategies and objectives that will fulfill their individual responsibilities while coordinating the actions of their respective personnel through the On-scene Coordinator.

Tactical activities for the Fire Dept. will be managed by the Operations Chief.

The Fire Dept. Hazardous Materials Group will manage all activities in the contamination zones. Only specially trained and equipped personnel will enter the contaminated zones.

Fire Dept. personnel in the contaminated zones will be operating in a more aggressive manner than the first responders in that they will actually approach a hazardous material, its container and associated devices attempting to control, contain, confine, or prevent a release.

3.0 General Incident Procedures and Site Safety

3.1 Hazard Assessment

Early recognition of incident hazards and potential risk is essential. The initial responsibility for assessment of incident hazards lies with first responding units.

Responding units will gather and communicate to the Incident Commander pertinent information regarding the presence or release of hazardous materials or chemicals.

Each member should be alert to the signs, evidence, and indications of the presence of hazardous substances during fires and emergencies and report such information to the next higher level of command.

3.1.1 Persons reporting emergencies will usually describe the *kind* of incident creating a hazardous condition. They generally are unable to accurately describe a hazardous materials incident.

Since accurate information about the incident or site might not be available when responding, special attention should be focused on the possibility of exposure in the following circumstances:

A. Transportation accidents.

B. Industrial accidents.

C. Leaks, spills, or suspicious odors.

D. Medical emergencies involving chemical inhalation.

E. Explosions.

F. Structural collapses.

3.1.2 Information regarding hazardous occupancies or locations that has been obtained through inspection or preplanning activities should be available from "CIDS Information," described in the Communications Manual. This type of response information prior to arrival will prevent premature entry into dangerous environments and unnecessary exposure to responding personnel. It will also offer a significant measure of protection to responders unfamiliar with the location or occupancy.

3.1.3 On-site information gathering must be limited to that which can be obtained within the limits of each responder's level of training and protective equipment. It is not in the best interest of the public or the responders to become part of the emergency problem instead of the solution.

First responders should gather, evaluate, and report information prior to entering into or undertaking activities that would place them in a contaminated environment.

3.1.4 The following are environments that must be evaluated before any commitment of personnel for any reason:

A. Large containers or tanks that must be entered.
B. Confined spaces (manholes, trenches, etc.) that must be entered.
C. Potentially explosive or flammable situations indicated by gas generation or gas release or over pressurization of containers (BLEVE).
D. Presence of *extremely hazardous materials* such as cyanide, phosgene, or radioactive materials.
E. Visible vapor clouds.
F. Areas where biological indicators such as "Unconscious persons" (refer to Section 3.6), dead animals or vegetation are located.

3.2 Site Security and Control

An incident generally involves the escape of normally controlled substances and response activities involve actions to minimize and prevent these discharges. Site Control is preventing or reducing the exposure of any person and the transfer of hazardous substances (contaminants) from the site by civilians, department members and equipment. Site control involves two major activities:

A. Physical arrangements and control of the site work areas.
B. The removal of contaminants from people and equipment.

HAZARDOUS MATERIALS RESPONSE HANDBOOK

Control is needed to reduce the possibility of transport from the site of contaminants, which may be present on personnel and equipment. This can be accomplished in a number of ways including:

A. Establishing physical barriers to exclude the public and unnecessary personnel.

B. Establishing checkpoints with limited access to and from the site, or areas within the site.

C. Minimizing personnel and equipment on-site consistent with effective operations.

D. Establishing containment zones.

E. Undertaking decontamination procedures.

F. Conducting operations in a manner to reduce possibility of contamination.

3.3 Decontamination

Decontamination (DECON) is the process of making personnel, equipment and supplies safe by reducing present levels of poisonous or otherwise harmful substances. This process is one of the most important steps in ensuring personal safety at a hazardous materials emergency. The extent of its success depends on the ability of the IC to maintain control of personnel at the site.

The detail of decontamination operations required at an incident depends on the safety and health hazards of the contaminants. An uncontaminated light oil, for example, that presents a minimal hazard can be partially decontaminated by flushing it from protective clothing. In contrast, a poisonous material will require a careful and detailed course of action.

See "SAFE WORK PRACTICES — Decontamination Procedures," Section 5.3 of this document for more detailed information.

3.4 Emergency Medical Treatment

Teams from the Emergency Medical Service (EMS) are available to assist in medical treatment, and monitor the response personnel and others exposed to hazardous materials.

The Fire Department Medical Officer on duty will also be notified of all Level 2 or greater incidents and will respond in accordance with Dept. policy.

3.5 Personal Protective Equipment (PPE)

3.5.1 *Structural firefighting gear* is designed to protect firefighters from heat and flame.

Hazardous materials can contaminate protective clothing, respiratory equipment, tools, vehicles, and other equipment used at an emergency scene.

First responders should avoid leaks, spills, and obvious sources of hazards, as well as indirect contact with potentially contaminated areas. Full firefighting gear and SCBA will be used at all times as a minimum of protection against exposure. Safe work practices WILL MINIMIZE exposure and contamination.

3.5.2 The use of *chemical protective clothing and equipment* requires specific skills acquired through training. *It is only available to members of the Hazardous Materials Unit.*

This type of special clothing may only protect against one chemical, yet be readily penetrated by other chemicals for which it was not designed. It offers little or no thermal protection in case of fire. No one suit offers protection from all hazardous materials.

3.5.3 The level of *special* protection required in each zone at an incident will be determined by the Officer in charge of the Hazardous Materials Unit based on information available.

3.5.4 The levels of protection available include:

A. Level A — highest level of protection to the responder.
B. Level B — high level of protection to the respiratory tract but a lower level of skin protection than level A.

Level A and B protective equipment is only available for use by members of the Hazardous Materials Unit.

C. Level C — does not require maximum skin or respiratory protection. This level presupposes that the types of air contaminants have been identified, concentrations measured, and the atmosphere is not oxygen deficient.
D. Level D — provides minimal protection and is used to guard against nuisance contamination only.

Structural firefighter protective clothing, i.e. turnout or bunker gear, is not classified as chemical protective clothing. The highest level of chemical protection that this level will provide is Level D.

3.6 Rescue

In most situations emergency personnel can protect the public by isolating and denying entry to contaminated areas. Initial rescue actions should concentrate on removing able-bodied persons from immediate danger. Involvement in complicated rescue problems or situations should be evaluated before being attempted.

When the probability is high that the victim cannot be saved or is already dead, rescue should not be attempted if it will place the rescuer at unnecessary risk. The danger of exposure to unknown chemicals or a potential explosion may make the risk unacceptable.

The following should be considered in attempting a rescue during a hazardous materials emergency:

A. Has the presence of a victim been confirmed visually or by other credible sources?
B. Is the person conscious or responsive?
C. How long has the victim been trapped or exposed? Is he/she viable?
D. Is the leaking material pooling or vaporizing in the area of the victim?
E. What are the properties of the material involved? What is the concentration of the material around the victim?
F. What special equipment is available to assist in this effort?

Consideration of these questions will help in weighing the likelihood of a successful rescue against the overall risk to the rescuer.

3.7 Evacuation/Sheltering

There are essentially two ways to protect the public from the effects of hazardous material discharges into the environment.

A. Evacuation: involves moving threatened persons to shelter in another area.
B. Sheltering in place: involves giving instructions to people to remain where they are until the danger passes.

Evacuation is clearly safer with respect to the hazards, but has certain limitations and may pose new problems. Evacuation takes time and may not be possible if large numbers of persons or a large volume of vapor is present. Evacuation through a toxic atmosphere may actually cause more harm than good in some cases.

3.7.1 Evacuation is best considered when:

A. There is an immediate danger of fire or explosion.
B. The potential for discharge is great, it has not taken place, and there is time available to relocate people.
C. The discharge has taken place but people are sufficiently protected to permit time for evacuation.
D. People not yet in the path of a release will be threatened by changing conditions.

3.7.2 Large scale evacuation will be directed by the On Scene Coordinator and will require the coordinated efforts of several agencies.

3.7.3 The decision to shelter in place is appropriate when the hazardous material will not affect the structure or its occupants or the hazard will pass a structure with little infiltration.

3.7.4 Sheltering in place is the alternative when:

A. Pre-planning has identified options for problem areas such as hospitals, jails, nursing homes, public assemblies, etc.
B. Evacuation cannot be properly managed with the manpower, resources, and facilities presently available.
C. The hazardous material displays the following characteristics:
   1. Low to moderate toxicity;
   2. Totally released and dissipating;
   3. Small quantity solid or liquid leak;
   4. A migrating vapor of low toxicity and quantity and people are safer indoors than outside; and
   5. Release can be rapidly controlled at the source.

3.7.5 The success of either option will depend on the preplans and effectiveness of communication resources, notification and public information.

3.8 Emergency Equipment

Private corporations, outside agencies, and other city agencies maintain specialized equipment and services that may be placed at the disposal of the Fire Department for use at these incidents. A catalogue of these services and equipment is maintained by the Bureau of Fire Communications. Calls for the use of these services or cooperation of these agencies shall be transmitted to the Dispatcher by the IC. Approval will come from the Commissioner and Chief of Department or their designee.

Liability, costs to the provider, and the availability of certain types of equipment preclude irregular or other informal arrangements.

The IC will designate the exact location to which the special equipment will respond.

4.0 Roles and Responsibilities of Fire Department Responders

4.1 Incident Commander

Responsible for all Fire Department incident activities, planning and checklist functions, preparation of reports, incident termination, and post incident analysis.

The senior ranking Officer first on the scene will determine the need for the following actions and implement those that are necessary and within the capability of the resources and equipment available at the time.

Succeeding Commanders will further evaluate the needs of the incident and refine, reinforce, or expand the scope and depth of these response activities.

A. Command

1. Establish a Site Safety Plan
   a. Set up isolation, control, and support zones. Refer to site plans, if available.
   b. Limit and control site access.
   c. Set up and manage staging areas.
   d. Coordinate F.D. personnel, on site personnel and other support agency activities.
   e. Call for resources/support units.

2. Establish Lines of Communication
   a. Announce location of Command Post, Staging areas.
   b. Report incident and site information to Dispatcher.
   c. Communicate plan of action to on scene personnel.
   d. Establish contact/communications with other agencies responding.
   e. Consult on site personnel/specialists.

3. Establish a Command Post

   a. Establish F.D. Command Post.
   b. Set up/maintain Command & Control Chart.
   c. Establish and maintain incident log.

4. Staff ICS Functional Positions (as needed)

   a. Safety, Liaison/Information.
   b. Staging.
   c. Operations, Planning, Logistics, Finance.
   d. Establish appropriate Sectors/Groups.

B. Operations

1. Conduct a Site Survey/Size-up

   a. Determine the nature, source, extent of problem.
   b. Estimate incident level.
   c. Determine need for evacuation.

2. Formulate/Implement Plan of Action

   a. Implement initial plan of action as described in the Emergency Response Guidebook and Hazard Action Guides.
   b. Communicate plan of action to on scene personnel.
   c. If necessary, implement alternate plans as information, personnel, resources and equipment are secured.

3. Terminate Incident Activities

   a. Evaluate decontamination needs and support activities.
   b. Establish disengagement/demobilization plan including post incident information and transfer of responsibilities.
   c. Document remedial efforts during clean up operations.
   d. Evaluate compliance with Fire Dept. regulations and directives and take appropriate enforcement action.

4. Post Incident Analysis

   a. Conduct a post incident analysis and review.
   b. Evaluate response re: Response plan.
   c. Make recommendations.

C. Planning

1. Evaluate Hazards and Assess Needs

   a. Determine strategic goals and tactical objectives.
   b. Assess current status and intervention options.
   c. Formulate, review, revise goals, objectives, action plan.

D. Logistics

Assess logistical support requirements for incident activities.

E. Finance

Record all information re: costs involved in mitigating the incident incurred by the Fire Department.

4.2 First Responders (Operational Level)

Respond to the site for the purpose of protecting persons, property, or the environment from the effects of a hazardous materials release or potential release. *They are not trained or equipped to use specialized chemical protective clothing or special control equipment. Their actions are limited to those which may be accomplished using SCBA and structural firefighter protective clothing as described in Department SOP and the Hazard Action Guides (HAG).*

In the event that the public or any emergency responder has been exposed prior to recognition of a hazardous substance, those persons exposed should be isolated to avoid spreading any contamination and so that proper medical treatment and monitoring can be arranged. Those not exposed must be kept out of the contaminated area.

4.2.1 Initial Actions

A. Given the limited capability of units arriving first at the scene of a hazardous materials release or potential release, premature commitment of companies and personnel to unknown, potentially hazardous situations/locations must be avoided.

B. Cautious, methodical, and deliberate size-up combined with immediate site security limits needless exposure of responding personnel and reduces the vulnerability of the public and the environment.

C. As soon as the presence of a hazardous substance is detected or suspected, First Responders should take the following actions:

1. Use full structural firefighting clothing and SCBA as a minimum to protect themselves from unnecessary exposure to contaminants.

2. Isolate the hazard.
3. Control access to the hazard area.
4. Communicate available information to the IC or Dispatcher.
5. Alert other responders and call for assistance, if needed.
6. Establish an initial staging area for other responders.
7. Take steps to identify the hazard and evaluate the risk without endangering on scene personnel.
8. Consult Emergency Response Guidebook and Hazard Action Guides (HAG).

D. Whenever possible, first arriving units must alert other responding units of the hazardous nature of the incident by transmitting signal 10-80 with a brief description of known hazard information. This will prevent units from unknowingly responding into the hazard area.
E. Develop a plan of action consistent with the Department's Emergency Response Plan and Hazard Action Guides for hazardous materials (HAG).
F. Implement the plan within the capability of the available personnel, protective clothing and equipment.

NOTE: The following assignments (Sections 4.2.2 through 4.2.6) are prescribed as standard procedure for units responding to a known or suspected hazardous materials incident. When the presence of a hazardous material is discovered during fire or other emergency operations, the Incident Commander must isolate units, redeploy or withdraw units and adjust the strategic and tactical objectives to address the hazard affecting the public and the operating emergency personnel.

Individual units must be assigned or reassigned to Hazard assessment, site control for public and personnel protection, and limited containment/confinement or isolation strategies as are possible. Given the limited capabilities of responders trained and equipped to the operational level this could include:

A. Withdrawing from or isolate selected areas.
B. Adjusting tactical objectives. Redeploy or reassign units and personnel.
C. Continuing current strategy, with additional resources called for specific assignments, e.g., the hazardous material problem.

The priorities should continue to be focused on limiting the exposure of personnel and the public; stabilizing the incident until additional resources are available to gain control; limiting the impact of the event by protecting property and the environment from additional harm.

4.2.2 First Arriving Ladder Company — Hazard Assessment

The first arriving Ladder Co. is responsible for hazard assessment. Some members of this company will be divided into teams, each team equipped with at least one radio. The Officer of this unit will designate team members and a radio contact person.

The purpose of the team concept is to provide for the safety and accountability of all the members. No person involved in the operational activities of these teams is to work alone.

A. Officer: Supervise hazard assessment activity of assigned members.

1. Obtains briefing from IC.
2. Assigns and instructs teams regarding hazard assessment activities and limitations.
3. Coordinates apparatus placement with other responding units.
4. Reports information to IC.
5. Monitors Handie-talkie/Radio.

B. Ladder Chauffeur: Apparatus placement.

1. Positions apparatus out of the hazard area and as a barrier for site access control as ordered.
2. Coordinates apparatus placement with other responding units.
3. Remains with apparatus until further ordered.
4. Establishes physical barriers to area access using apparatus, barrier tape, traffic cones, traffic signs, etc., as ordered.
5. Monitors Handie-talkie/Radio.

C. Team 1: Hazard Identification.

1. Identify occupancy and location hazards, if possible.
2. Identify container shape and approximate size, if possible.
3. Identify any visible markings, colors, placards/labels, etc.
4. Report information available and stand-by.
5. Monitor Handie-talkie.

D. Team 2: Information resources.

1. Access HAG & DOT GUIDEBOOK locating appropriate Guide, isolation and evacuation recommendations.

2. Confirm Guide Number, isolation distance, and evacuation information with Second Arriving Ladder Co.
3. Report information to officer.
4. Monitor Handie-talkie.

4.2.3 Second Arriving Ladder Company — Site Access Control

The second arriving Ladder Co. is responsible for CONTROLLING ACCESS to the site and for establishing the initial exclusion area.

Some members of this company will be divided into teams, each team equipped with at least one radio. The Officer of this unit will designate team members and a radio contact person.

The purpose of the team concept is to provide for the safety and accountability of all the members. No person involved in the operational activities of these teams is to work alone.

A. Officer: Coordinate/limit site access.

1. Obtains briefing from IC.
2. Assigns and instructs teams regarding hazard assessment activities and limitations.
3. Advises IC re: need for Police assistance in area security and control.
4. Reports information to IC.
5. Monitors Handie-talkie/Radio.

B. Ladder Chauffeur: Apparatus placement.

1. Positions apparatus out of the hazard area and as a barrier for site access control as ordered.
2. Coordinates apparatus placement with other responding units.
3. Remains with apparatus until further ordered.
4. Establishes physical barriers to area access using apparatus, barrier tape, traffic cones, traffic signs, etc., as ordered.
5. Monitors Handie-talkie/radio.

C. Team 1: Information resources.

1. Using information from the first arriving ladder company or the IC, access HAG & DOT GUIDEBOOK locating appropriate Guide, isolation and evacuation recommendations.

2. Confirm Guide Number, isolation distance, and evacuation information with First Arriving Ladder Co.
3. Report information to Officer.
4. Monitor Handie-talkie.

D. Team 2: Initial Isolation Zone.

1. On orders of the Officer establish physical barriers as an initial isolation/exclusion zone using barrier tape, utility rope, traffic cones, etc., as ordered.
2. Remain outside isolation zone to deny access until relieved or as ordered.
3. Monitor Handie-talkie.

4.2.4 First Arriving Engine Company — Water Supply

The first arriving Engine will establish the initial water supply.

Based on the existing conditions, a line should be stretched but not placed into operation until ordered by the IC, pending the outcome of the initial hazard assessment. The premature, unnecessary, or incompatible use of water can compound the hazard. The unit will stand-by out of the hazard area.

Members of this unit will remain together as a unit unless otherwise ordered by the Incident Commander to perform duties appropriate to specific tactical objectives. This might include such things as assisting in site control or evacuation.

If and when necessary, some members of this company may be divided into teams, each team equipped with at least one radio. The Officer of this unit will designate team members and radio contact person.

The purpose of the team concept is to provide for the safety and accountability of all the members. No person involved in the operational activities of these teams is to work alone.

A. Officer: Supervise hazard assessment activity of assigned members.

1. Obtains briefing from IC.
2. Assigns and instructs teams regarding hazard assessment activities and limitations.
3. Supervises water supply and line placement activities.
4. Coordinates apparatus placement with other responding units.
5. Reports information to IC.
6. Monitors Handie-talkie/radio.

B. Engine Chauffeur: Apparatus placement.

1. Positions apparatus out of the hazard area.
2. Establishes a water supply.
3. Coordinates apparatus placement with other responding units.
4. Remains with apparatus until further ordered.
5. Establishes physical barriers to area access using apparatus, barrier tape, traffic cones, traffic signs, etc., as ordered.
6. Monitors Handie-talkie/radio.

4.2.5 Second Arriving Engine Company — Site Access Control

The second arriving Engine Co. is responsible for assisting and coordinating with the second ladder company to control access to the site.

Members of this unit will remain together as a unit unless otherwise ordered by the Incident Commander to perform duties appropriate to specific tactical objectives. This might include such things as assisting the first arriving Engine in water supply or line placement, site control or evacuation.

If and when necessary, some members of this company may be divided into teams, each team equipped with at least one radio. The Officer of this unit will designate team members and a radio contact person.

The purpose of the team concept is to provide for the safety and accountability of all the members. No person involved in the operational activities of these teams is to work alone.

A. Officer: Assist Second Ladder Co. in site control.

1. Obtains briefing from IC.
2. Assigns and instructs teams regarding hazard assessment activities and limitations.
3. Coordinates site activities with other units engaged in site control.
4. Reports information to IC.
5. Monitors Handie-talkie/radio.

B. Engine Chauffeur: Apparatus placement.

1. Positions apparatus out of the hazard area and as a barrier for site access control as ordered.

2. Coordinates apparatus placement with other responding units.
3. Remains with apparatus until further ordered.
4. Establishes physical barriers to area access using apparatus, barrier tape, traffic cones, traffic signs, etc., as ordered.
5. Monitors Handie-talkie/radio.

4.2.6 Other Units on Initial Alarm

Members of these units will remain together as a unit unless otherwise ordered by their Officer as directed by the Incident Commander to perform duties appropriate to specific tactical objectives. This might include assisting other units in site control or evacuation.

If and when necessary, members of these companies may be divided into teams, each team equipped with at least one radio. The Officer of this unit will designate team members and a radio contact person.

The purpose of the team concept is to provide for the safety and accountability of all the members. No person involved in the operational activities of these teams is to work alone.

A. Third Arriving Engine Co.

1. Reports to Incident Commander.
   a. Obtain briefing from IC.
   b. Assigns and instructs teams regarding hazard assessment activities and limitations.
   c. Coordinates apparatus placement with other responding units.
   d. Reports information to IC.
2. Stands-by to assist other responders as directed by the IC.
3. Monitors Handie-talkie/radio.

B. Rescue Co., Squad, Other Units, Etc.

1. Report to Incident Commander.
   a. Obtain briefing from IC.
   b. Assign and instruct teams regarding hazard assessment activities and limitations.
   c. Coordinates apparatus placement with other responding units.
   d. Report information to IC.
2. Stand-by to assist other responders as directed by the IC.
3. Monitor Handie-talkie/radio.

C. Additional units responding to the scene of a known hazardous materials incident should be given specific information regarding the location of staging areas and tactical objectives assigned to the unit. If, due to the nature of the event, the Officers of the responding units do not receive this information, they should:

1. Request this information from the dispatcher.
2. Keep their units intact and establish contact with the next higher level of supervision.
3. Members of the units will remain together as a unit unless otherwise ordered by their Officer who is directed by the Incident Commander to perform duties appropriate to specific tactical objectives. If and when necessary, members of these companies may be divided into teams, each team equipped with at least one radio. The Officers of these units will designate team members and a radio contact person.

4.3 Hazardous Materials Group

Individuals who respond to releases or potential releases for the purpose of stopping or controlling them. They assume a more aggressive role than a first responder at the operational level in that they will approach the point of the release. Their duties require a more directed or specific knowledge of the various substances they may be called upon to contain. Their training includes knowledge of procedures for the use of specialized chemical protective clothing, survey equipment, and special procedures for containing a chemical hazard and decontamination.

Some members of the Hazardous Materials Unit will be divided into teams, each team equipped with at least one radio. The Officer of this unit will designate team members and the radio contact person.

The purpose of the team concept is to provide for the safety and accountability of all the members. No person involved in the mitigation activities of these teams is to work alone.

The Haz-Mat Group consists of the Hazardous Materials Unit and members of the Special Operations Command. Each has a specific role and objective to accomplish at an incident. They are generally described as follows:

A. Analysis

1. Verify, identify, or classify hazardous materials and determine their concentration.

2. Collect and interpret hazard and response information.
3. Estimate damage to containment systems.

B. Planning

1. Select appropriate PPE for incident activities.
2. Develop decontamination procedures.
3. Develop plan of action consistent with the ERP and within the capability of the personnel, PPE, and equipment available.

C. Operations

1. Carry out planned activities under the direction of the IC as per the ERP.

4.3.1 Haz-Mat Group Supervisor

On duty Officer of Hazardous Materials Unit who will supervise the "hands-on" mitigation operation. He will coordinate the following activities as directed by and report to the IC or a level of supervision assigned by the IC as per FDNY-ICS.

A. Obtains briefing from the Incident Commander.
B. Confirms the development of Control Zones and Access Control Points and the placement of appropriate control lines.
C. Ensures that a Site Safety Plan is developed and implemented.
D. Participates in the development of the Incident Action Plan; develops the Hazardous Materials portion of the Incident Action Plan.
E. Evaluates and recommends evacuation, sheltering, and decontamination options to the Incident Commander.
F. Conducts safety meetings with the Hazardous Materials Unit members. The Hazardous Materials Group Supervisor is responsible for site safety and operations inside the area bounded by the Contamination Control Line including the Exclusion Area or "Hot Zone."
G. Ensures that the proper Personal Protective Equipment is selected and used.
H. Ensures that current weather data and future weather predictions are obtained.
I. Supervises the activities of and maintains communications with:

1. Haz-Mat Unit Team members.

2. Incident Commander or next higher level of supervision, if a position has been staffed.
J. Establishes environmental monitoring of the hazard site for contaminants.
K. Ensures that appropriate agencies are notified through the Incident Commander.
L. Terminates operations.

4.3.2 Entry Team

The Entry Team consists of a minimum of two members assigned to the Haz-Mat Unit. One member will be designated the Team Leader. The Team Leader will, on orders from the Haz-Mat Group Supervisor, perform the following functions:

A. Obtain briefing from the Hazardous Materials Group Supervisor.
B. Supervise entry operations.
C. Recommend actions to mitigate the situation within the Exclusion Zone.
D. Carry out actions, as directed by the Hazardous Materials Group Supervisor, to mitigate the hazardous materials release or threatened release.
E. Maintain communications and coordinate operations with:

1. Hazardous Materials Group Supervisor.
2. Decontamination Team Leader.
3. Resource Technician/Hazardous Materials Reference.
4. Backup Team.

F. Maintain control of the movement of people and equipment within the Exclusion Zone, including contaminated victims.
G. Direct rescue operations, as needed, in the Exclusion Zone.
H. Terminate operations.

4.3.3 Backup Team

The Backup Team consists of a minimum of two members assigned to the Haz-Mat Unit. One member will be designated the Team Leader. The Team Leader will, on orders from the Haz-Mat Group Supervisor, perform the following functions:

A. Obtain briefing from the Hazardous Materials Group Supervisor.
B. Assist Entry Team in site survey and product identification.

C. Carry out actions as directed by the Hazardous Materials Group Supervisor.

D. Ensure all equipment is prepared for use according to action plan.

E. Maintain sight contact with Entry Team whenever possible.

F. In proper PPE, act as a backup Entry Team from safe area. Be prepared to rescue the Entry Team.

G. Complete operations in Exclusion Zone should Entry Team fail to complete assignment.

H. Maintain communications and coordinate operations with:

1. Entry Team Leader.

2. Hazardous Materials Group Supervisor.

3. Resource Technician/Hazardous Materials Reference.

4. Decon Team Leader.

I. Furnish additional equipment or supplies to the Entry Team as needed. Replace equipment used.

J. Terminate operations.

4.3.4 Decontamination (Decon) Team

The Decon Team consists of a minimum of two members assigned to the Haz-Mat Unit. One member will be designated the Team Leader. The Team Leader will, on orders from the Haz-Mat Group Supervisor, perform the following functions:

A. Obtain briefing from the Hazardous Materials Group Supervisor.

B. Manage the Control Zones and Access Control Points and the placement of appropriate control lines.

C. Ensure appropriate action is taken to prevent spread of contaminants.

D. Set up Decon area. Implement decontamination process of Haz-Mat Team.

E. Decon, as possible, all equipment used.

F. Maintain communications and coordinate operations with:

1. Entry Team Leader.

2. Hazardous Materials Group Supervisor.

3. Resource Technician/Hazardous Materials Reference.

4. Backup Team.

G. Overpack exposed equipment for further decontamination or disposal.
H. Ensure that all equipment is placed back in service.
I. Terminate operations.

4.3.5 Resource Technician

A Member of the Haz-Mat Unit designated to provide technical information and assistance to the Hazardous Materials Group using various reference sources such as computer data bases, technical library, CHEMTREC, and phone contact with facility representatives.

The Resource Technician may provide product identification using tests kits and/or any other means of identifying unknown materials.

The Resource Technician will, on orders from the Haz-Mat Group Supervisor, perform the following functions:

A. Obtain briefing from the Hazardous Materials Group Supervisor.
B. Provide support to the Hazardous Materials Group Supervisor.

    1. Monitors SCBA use time for suit operations.
    2. Determines personal protective equipment compatibility to hazardous material.

C. Maintain communications and coordinate operations with:

    1. Entry Team Leader.
    2. Hazardous Materials Group Supervisor.
    3. Backup Team Leader.
    4. Decon Team Leader.

D. Provide technical information management with public and private agencies. Interpret environmental monitoring information.
E. Provide analysis of hazardous material samples.
F. Document operations and notifications; e.g., document serial numbers of chemical protective suits worn by members.
G. Provide technical information of the incident for documentation.
H. Determine proper decon requirements.
I. Complete all necessary reports and forms and terminate operations.

4.3.6 Special Operations Decon Officer

Member of Special Operations Command assigned to the Hazardous Materials Group designated responsible for on-site and off-site decontamination.

The SOC Decon Officer will respond upon the request of the Incident Commander. The SOC Decon Officer reports directly to the Rescue Liaison Officer. The SOC Decon Officer supervises the SOC Decon Team.

The SOC Decon Officer, as directed, will:

A. Obtain briefing from IC, advise IC of decon needs, and formulate plan of action. The Decon procedure to be followed will be determined by DEP, and the SOC Decon Officer.
B. Identify items of equipment and clothing to be decontaminated.
C. Establish a Decon area.
D. Supervise the set-up and operation of all tools and equipment required for the decon procedure.
E. Arrange for collection, transportation, storage, cleaning, and temporary replacement of exposed or contaminated equipment.
F. Insure proper disposition of all processed items.

Procedures and documentation for this phase of operations will be as per policy developed by the Special Operations Command. See Section 5.3 for information on the Decon process.

The SOC Decon Officer will respond with the Decontamination Support Vehicle and the SOC Decontamination Team. Decontamination will, if possible, be accomplished at the scene to place apparatus, clothing and equipment back in service. If this is not possible, decontamination will be conducted at a time and site designated by the SOC Decon Officer.

4.3.7 Special Operations Command Decon Team

Consists of members assigned to Special Operations Command. They report directly to the SOC Decon Officer.

The SOC Decon Team is trained and designated by the Fire Department to perform decontamination of apparatus, tools, and equipment. They are qualified to execute specific procedures and operate decontamination equipment wearing the protective clothing and respiratory protection appropriate to the hazardous substance being processed.

4.3.8 Rescue Liaison Officer

Captain assigned to the Special Operations Command who responds on notification of Level 2 and 3 incidents. The Rescue Liaison Officer reports to the IC or a level of supervision that has been assigned by the IC as per FDNY-ICS.

The Rescue Liaison Officer is responsible for the following activities at a hazardous materials incident:

A. Obtaining a briefing from IC.
B. Determining the location of the Decon Area.
C. Initiating procedures for collecting contaminated clothing, tools and equipment.
D. Providing for proper recording and documentation of clothing and equipment.
E. Coordinating the activities of members of the Special Operations Command.

5.0 Safe Work Practices

5.1 General Procedures

The following work practices must be enforced to ensure a safe work site for all personnel.

A. Always consider the possibility that hazardous materials may possess multiple hazards.
B. Use full structural firefighting clothing and SCBA as minimum protection from unnecessary exposure to contaminants.
C. Each member should be alert to the signs, evidence and indications of the presence of hazardous substances during fires and emergencies and report such information to the next higher level of command.
D. If and when necessary, members may be divided into teams, each team should be equipped with at least one radio. The Officer of the unit will designate team members and radio contact person.
E. Mark all work zones and access points with barricade tape, flagging, or traffic cones.
F. Access to the site should remain free of unnecessary equipment and apparatus to facilitate other types of emergency access to and egress from the site.
G. Persons entering or leaving a work zone must check in/out at the access control point.

H. Access and means of egress must be secured and protected for the safety of persons in work zones. Doors, stairways and ladders should be secured. Ramps, ditches and excavations should be made as secure as possible in the event that rapid escape is required from the site.

I. No eating, drinking, or smoking is allowed in any contaminated area.

J. Implement decontamination based on an analysis of the hazards and risks involved.

(See "Decontamination," Section 5.3)

5.2 Work Zones

The method of reducing the potential for transfer of contamination is to delineate work areas within the incident site based upon expected or known levels of contamination. Within the areas assigned, personnel will utilize appropriate personal protective equipment. Movement between areas is controlled at checkpoints. Three contiguous areas will be established.

A. Exclusion area (contaminated) — "HOT ZONE"
B. Contamination reduction area — "WARM ZONE"
C. Support area (non-contaminated) — "COLD ZONE"

5.2.1 Exclusion Area — Hot Zone

The Exclusion Area is the innermost area and is considered contaminated or "hot." Within the Exclusion Area, prescribed levels of protection must be worn by all entering personnel. A check point must be established at the periphery of the Exclusion Area to control the flow of personnel and equipment between contiguous areas and to insure that the procedures established to enter and exit the areas are followed.

The Exclusion Area boundary would be established initially based on the type of released/spilled materials, initial instrument readings, and a safe distance from any potential exposure.

Subsequently, the boundary may be readjusted based on additional observation and/or measurements. The area should be physically secured by barrier tape into well-defined boundaries.

In the event that the public or any emergency responder has been exposed prior to recognition of a hazardous substance, those persons exposed should be isolated to avoid spreading any contamination and so that proper medical treatment and monitoring can be arranged.

Those not exposed must be kept out of the contaminated area.

Unless otherwise demonstrated, everything leaving the Exclusion Area should be considered contaminated and appropriate methods established for decontamination should be implemented.

5.2.2 Contamination Reduction Area — Warm Zone

Between the Exclusion Area and the Support Area is the Contamination Reduction Area. The purpose of this zone is to provide an area to prevent or reduce the transfer of contaminants which may have been picked up by personnel or equipment returning from the Exclusion Area. All decontamination activities occur in this area.

The boundary between the Support Area and the Contamination Reduction Area is the "Contamination Control Line." This boundary separates the possibly contaminated area from the clean zone. Entry into the Contamination Reduction Area from the clean area will be through an access control point. Personnel entering at this location will be wearing the prescribed level of protection for working in the Contamination Reduction Area. Exiting the Contamination Reduction Area to the clean area requires the removal of any suspected or known contaminated protective clothing and/or equipment and that appropriate decontamination procedures be followed.

At the boundary between the Contamination Reduction Area and the Exclusion Area is the "Hot Line" and access control station. Entrance into the Exclusion Area requires the wearing of the prescribed chemical protective clothing which may be different than the equipment requirements for working in the Contamination Reduction Area.

At a point close to the "Hot Line," a personnel and/or equipment decontamination station is established for those exiting the Exclusion Area. *Unless otherwise demonstrated, everything leaving the Exclusion Area should be considered contaminated and appropriate methods established for decontamination should be implemented.*

5.2.3 Support Area — Cold Zone

The Support Area is the outermost area of the site and is considered a non-contaminated or "clean" area. It is designated as a controlled area for authorized support personnel and the location for support equipment (Command Post, Equipment, etc.). Since normal firefighting clothing is appropriate within this zone, potentially contaminated personnel clothing, equipment, etc., are not permitted.

5.2.4 Area Dimensions

Considerable judgment is needed to assure safe working distances for each area balanced against practical work considerations. During long-term operations zones may be adjusted.

The following criteria are to be considered in determining the area dimensions:

A. Physical and topographical barriers;
B. Weather conditions;
C. Monitoring measurements;
D. Explosion/exposure potential;
E. Physical, chemical, toxicological, etc., characteristics of the contaminant(s); and
F. Cleanup activities.

5.2.5 Other Considerations

The use of a three-zone system of area designation, access control points and exacting decontamination procedures provides a reasonable assurance against the translocation of contaminating substance. This control system is based on a "worst case" situation. Less stringent site control and decontamination procedures than described may be utilized based upon more accurate information on the types of contaminants involved and the contaminating hazards they present. This information can be obtained through air monitoring, instrument survey, etc., and technical data concerning the characteristics and behavior of material present. Site control requirements can be modified within the limits of safety for specific situations once more reliable data has been analyzed.

5.3 Decontamination Procedures

As part of the system to prevent or reduce the physical transfer of contaminants by people and/or equipment from the site, procedures will be instituted for decontaminating anything leaving the Exclusion Area and Contamination Reduction Area. These procedures include the decontamination of personnel, protective equipment, monitoring equipment, clean-up equipment, etc.

In the event that the public or any emergency responder has been exposed prior to recognition of a hazardous substance, those persons exposed should be isolated to avoid spreading the contamination and so that proper medical treatment and monitoring can be arranged.

Those not exposed must be kept out of the contaminated area.

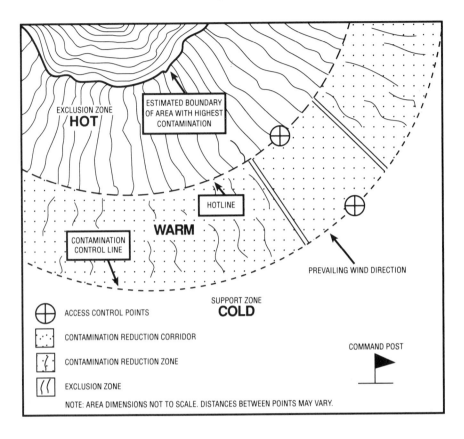

EXCLUSION ZONE
HOT

ESTIMATED BOUNDARY
OF AREA WITH HIGHEST
CONTAMINATION

HOTLINE

WARM

CONTAMINATION
CONTROL LINE

PREVAILING WIND DIRECTION

SUPPORT ZONE
COLD

⊕ ACCESS CONTROL POINTS

CONTAMINATION REDUCTION CORRIDOR

CONTAMINATION REDUCTION ZONE

EXCLUSION ZONE

NOTE: AREA DIMENSIONS NOT TO SCALE. DISTANCES BETWEEN POINTS MAY VARY.

COMMAND POST

5.3.1 The FDNY maintains a Hazardous Materials Decontamination Unit. The Unit consists of the Decon trailer and a support unit. The Police Department has a similar unit that can be utilized as a backup. (See Diagram on following page.)

5.3.2 The following steps have been designed to deal with personnel decontamination in worst case incidents. In some situations it will not be necessary to go through the entire procedure. The decision to implement all or part of the decontamination process should be based on a field analysis of the hazards and risks of the hazardous materials involved. The FDNY Hazardous Materials Unit and the N.Y. City Departments of Health and Environmental Protection supply the toxicological and reactivity information that determines which techniques apply.

OVERHEAD VIEW OF DECON UNIT

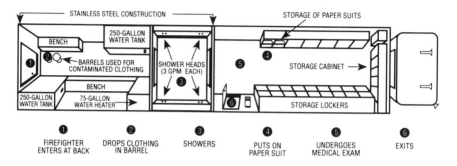

Step 1: An "entry point" will be established and marked in order to guide contaminated personnel into the Decon area.

Step 2: Protective clothing will be removed and isolated.
Contaminated coats, helmets, etc., should be placed in plastic bags to isolate contaminants. Bagged clothing should be sealed and placed in recovery drums outside the trailer for further analysis and for transportation to another location for laundering or disposal.

Step 3: Members enter the rear of the Decon Unit. EMS personnel will supervise the decon procedures. All personal effects and work clothing are removed and placed in bags and drums.

Step 4: In almost all cases, members will shower for a period determined by the type of substance involved. The exceptions occur when the substance involved is one made more active by water.

Step 5: After showering, members are issued towels and clothing to enable them to proceed to areas designated by EMS personnel. Members exit Decon Unit.

Step 6: EMS personnel examine members, taking a medical history, checking vital signs, and follow up on any physical complaints. This will usually take place in the EMS MERVAN on the clean side or Support Zone.

Step 7: Any members that require further evaluation or treatment will be transported by EMS.

Step 8: The Decon Unit will disconnect and store the unit's equipment and apparatus.

Step 9: Waste water from the showers will be analyzed by DEP.

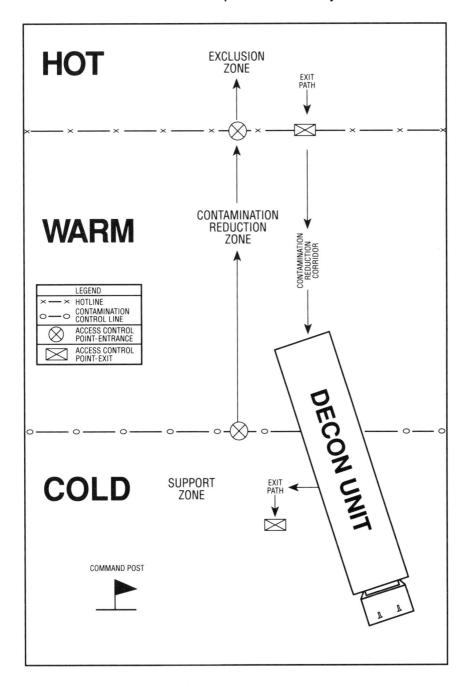

HAZARDOUS MATERIALS RESPONSE HANDBOOK

5.3.3 A member of the Special Operations Command will identify apparatus, equipment, and clothing to be decontaminated.

All clothing, personal items, apparatus, tools, and equipment will remain at the site pending this determination.

The Special Operations Command (SOC) will arrange for the collection, removal, transportation, storage, cleaning, and temporary replacement of exposed or contaminated clothing and equipment.

On receipt of members' original clothing, Company Commanders will contact SOC to arrange return of replacement clothing.

6.0 Incident Command System Functions and Responsibilities

NOTE: These activities are specific to hazardous materials incidents and are in addition to the usual requirements of each position. This does not restrict the IC from staffing other ICS positions or from specifying additional duties that might be required by the incident. ANY FUNCTIONS NOT DELEGATED REMAIN THE RESPONSIBILITY OF THE INCIDENT COMMANDER.

6.1 Safety — reports directly to the Incident Commander

The Safety Operating Battalion will respond to all 10-80 code 2 signals (Level 2 or 3 incidents) and as requested by the Incident Commander. At a hazardous materials incident, they will monitor conditions and activities in the *Support Zone (Cold Zone)*.

The Safety Operating Battalion Chief at the scene is also designated to confirm that all exposed members participate in the decontamination procedure conducted by the Decon Team and the Special Operations Decon Officer. If for any reason the Safety Operating Battalion does not respond to the scene, the Special Operations Decon Officer will perform this function.

The Safety Chief is responsible for recording the name, rank, Social Security number and unit identification of all members decontaminated with a description of the decontamination procedures employed.

The Safety Chief designated will, as soon as practicable after the incident, forward a report to the Bureau of Health Services and the Bureau of Operations.

6.1.1   The Hazardous Materials Group Supervisor is responsible for site safety and operations inside the area bounded by the Contamination Control Line including the Exclusion Area or "Hot Zone."

6.2   Information — reports directly to the Incident Commander

The Fire Dept. Public Information Office will be notified upon transmission of signal 10-80 code 2 (Level 2 or 3 incident).

Firefighters should not make any statements to anyone outside the F.D. chain of command about the character or nature of a hazardous materials incident. Inquiries should be directed to the Incident Commander or Information Officer, if the position has been staffed.

Information to the public will be released in conjunction with the Office of the Mayor as per the NYC Response Plan.

6.3   Liaison — reports directly to the Incident Commander

This is a senior Fire Dept. Officer who is the contact point for representatives of assisting and cooperating agencies. The Liaison has the authority given by the IC to commit Fire Dept. resources to specific incident activities and support functions.

The Liaison assists the IC in assigning Technical Specialists to appropriate functions within ICS.

6.4   Operations Chief — reports directly to the Incident Commander

The Operations Chief directs, coordinates, and controls all tactical activities and functions necessary to carry out the objectives of Command. Reports directly to the Fire Dept. IC.

A. Responsibilities of the Operations Chief.

1. Directs the management of all Fire Dept. on-scene tactical efforts needed to gain control of the incident.
2. Works directly with the Fire Dept. IC and the On-Scene Coordinator developing both primary and alternative strategies.
3. Determines operational necessities and requests additional support when required.
4. Establishes Sectors and Groups as determined by operational needs.

5. Assigns objectives to Sector and Group Supervisors.
6. Reports changing conditions to the IC.
7. Supervises the Staging Area Manager.

6.4.1 Technical Specialists (Assigned to Operations)

May be assigned by Command or Liaison Officer to advise or assist in the tactical activities under the direction of the Operations Chief.

6.5 Planning Chief — reports directly to the Incident Commander

A. Gathers, evaluates, disseminates information for the development of alternative strategies.
B. Maintains records of incident activities for analysis and review, tracks current resource and situation status.
C. Establishes and maintains communications network.

6.5.1 Resource and Situation Status

This function will include preparation and maintenance of a Command and Control Chart as part of the overall site safety plan.

A. The Chart should identify the topographic features of the site, prevailing wind direction, drainage, location of buildings, physical barriers, tanks, etc. This information is helpful in:
   1. Planning activities.
   2. Assigning personnel.
   3. Identifying access routes, evacuation routes, and any problem areas.
   4. Identifying areas at the site that require the use of personal protective equipment.
   5. As a visual aid during briefing of on-site emergency personnel.

B. The Chart should be updated throughout the course of the operation and should reflect:
   1. Changes in site activities;
   2. Hazards not previously identified;
   3. Weather conditions; and
   4. Activities of other agencies on site.

6.5.2 Technical Specialists (Assigned to Planning)

A. May be assigned by Command or Liaison Officer in an advisory capacity from various parties and agencies with interest or jurisdiction. These might include representatives of the material owner/shipper, chemists, field monitoring personnel from other agencies, code and regulation enforcement officials, members of City agencies with expertise or resource capability of a unique nature.

   They will supply technical information, data interpretation, sample analysis, forecasts, etc.

B. Persons assigned at this level generally do not have decision making authority for their particular agency unless specifically authorized. They act in an advisory capacity for incident planning.

C. Information gathered will be used by Command and Operations Chief for the development and execution of various incident activities.

6.5.3 Documentation

A. The IC will designate a member to keep a Log. Log sheets have been supplied to all Divisions, Battalions, and the Haz/Mat Unit. The Log should be retained by the Fire Dept. IC.

   Records of these incidents are always important but especially when the incident results in personal injury, property damage, or damage to the environment.

B. Certain events require more complex documentation. These incidents include but are not limited to:

1. Prolonged multi-agency operation.
2. Multiple changes at the Command level.
3. Major events involving several jurisdictions, private corporations, public utilities, etc.
4. Events involving high risk operations.
5. Multiple injuries or mass exposure.
6. Extensive isolation or evacuation of the public or public areas.
7. Isolation or contamination of environmentally sensitive areas or facilities.

C. Transcripts of this information can be taken from video tapes, photographs, and sound recordings made during the event or from notes kept in a bound log book (not looseleaf). Log books are carried in each Command Chief's car and in the Field Communications Unit.

D. All information should be recorded objectively. Each person making an entry should date and sign the document. The number of persons recording the information should be kept to a minimum.

E. Entries should be made in a timely way to insure accuracy and thoroughness. Neatness and legibility are essential.

F. The following information should be recorded:

1. Chronological history of the event.
2. Facts about the incident and when they became available including names, descriptions, source, quantity, and cause of release, if known.
3. Names and assignments of key personnel and resources.
4. Actions, decisions, orders and directions given: by whom, to whom, and when.
5. Actions taken: who did what, when, where, and how.
6. Possible exposure of personnel.
7. Records of all injuries and illnesses during, or as a result of, the emergency.

6.6 Logistics Chief — reports directly to the Incident Commander

A. Logistics is the procurement, distribution, maintenance of facilities, services, materials and personnel to support the incident.

B. The Service and Support functions would include the following responsibilities:

1. Service
   a. Fire Department Medical Officer:
      1. Determine a site for medical facilities.
      2. Assist in setting up and recording the transportation of the injured.
      3. Provide medical treatment for Fire Dept. personnel.
   b. Facilities
      1. Assist in the operation of the Command Post, rest and rehabilitation areas and secondary staging for apparatus. Procuring shelter facilities for personnel.

2. Support
   a. Mask Service
      Supply air cylinders or air lines.

NOTE: POTENTIALLY CONTAMINATED CYLINDERS MUST NOT BE EXCHANGED.

b. Special Units (Supplies)

1. Order and supply specialized extinguishing agents (CO_2, Foam, Dry Powder etc.).
2. Order and supply diking, absorbent, or neutralizing materials.
3. Facilitate specialized tools or heavy duty equipment (shoring, cranes, etc.), pumps, etc.

c. Fleet Maintenance

1. Repair and refueling of vehicles at extended operations.
2. Supply vehicles for transportation.

C. Liability, costs to the provider, and the availability of certain types of equipment preclude irregular or informal arrangements for the procurement of equipment and supplies. (See "Emergency Equipment," Section 3.8)

6.6.1 Technical Specialists (Assigned to Logistics)

Members from the Special Operations Command (SOC) will respond to the scene of each event that requires Decontamination to supervise and facilitate the procedure for collecting, documenting, and tacking clothing and equipment that is to be confiscated and replaced.

Procedures and documentation for this phase of operations will be as per Department publications.

(See Hazardous Materials Group, Sections 4.3.6, 4.3.7, 4.3.8)

6.7 Finance

Cost recovery/Restitution from the party responsible for the incident, or recovery from governmental funding, involves documenting all information pertinent to the costs involved in mitigating the incident incurred by the Department or the City. All costs must be documented and supported with written materials such as damage reports, time records, receipts, and invoices.

7.0 Interagency Support System

As part of the response system, various agencies will be notified and will respond based on the specific nature of the incident. It is essential that each responder be familiar with New York City's broader plan for response to a hazardous materials emergency since each agency has a specific role in the total response.

The following agency resources are available and will respond as needed and according to the NYC Hazardous Materials Response Plan. Their duties include but are not necessarily limited to the following:

A. Office of Emergency Management (OEM), Police Department

- Coordinate all City agency activities.
- Coordinate evacuation, if necessary.
- Consult with Dept. of Environmental Protection Response Team and Poison Control Center to evaluate the extent of the hazard.
- Staff and support Interagency Command Center (ICC).
- Contact and maintain liaison with NY State Office of Emergency Management.

B. On-Scene Coordinator (OSC), Police Department

- Establish Interagency Command Center (ICC).
- Establish Communications capability between the site and the ICC.
- Establish and maintain overall site security.
- Coordinate all on site personnel.
- Direct activities of Police Dept. Operations Unit.
- Coordinate and select Emergency Response Routes.
- Coordinate "off loading," if necessary.

C. Police Department (PD)

- Control traffic and emergency vehicle access in site vicinity.
- Crowd control and evacuation, if appropriate.
- Provide central information and notification center for all City agencies.

D. Dept. of Environmental Protection (DEP)

- Dispatch response Team to provide technical personnel and equipment.
- Sample and analyze substances and the environment.
- Gather information, provide technical guidance to OEM.
- Obtain financial approval for and arrange private sector containment, control, cleanup, and disposal of hazardous materials, where necessary.

E. Sanitation Department

- Maintain supply materials for containment and confinement operations.
- Cleanup and disposal of materials (except liquids) rendered harmless as determined by DEP and Health Dept.
- Investigate/enforce environmental/dumping regulations.

F. Department of Health

- Provide technical information on materials hazards.
- Assess effects to public health.
- Provide informational/precautionary press statements concerning public health.
- Endure proper mitigation and disposal of radioactive and etiological agents.

G. Emergency Medical Services (EMS)

- On-site medical services.
- Assist in Fire Department decontamination.

H. Department of Traffic

- Facilitate and control traffic flow in area.

I. Mayor's Press Office

- Exclusive authority for public information and warnings.

8.0 Incident Termination Procedures

8.1 Cleanup Operations

Incident scene activities include removing the hazardous material, all contaminated debris (including dirt, water, containers, vehicles, tools, and equipment) and returning the scene to as near normal as it existed prior to the incident.

Cleanup operations are not a Fire Department function. However, assuring that this cleanup takes place is a responsibility of the Fire Department.

The Fire Department will cooperate with the Department of Environ-mental Protection (DEP) in arranging cleanup activities. The Senior representative of DEP will arrange for authorization to obtain financial approval for and arrange private sector containment, control, cleanup, and disposal of hazardous materials, where necessary.

In some cases, the Hazardous Materials Unit will take samples of materials for testing. These samples will be picked up by DEP to be tested and analyzed. *Under no circumstances is any Fire Department unit, including the Hazardous Materials Unit, to transport these samples or any hazardous material, even if properly contained, to any Fire Department location or facility or other location or agency.*

If operations are concluded and DEP has not arrived at the scene, these samples will remain at the scene for pickup by DEP. The Police Department representative at the scene will safeguard these samples. If the Police Department is not represented at the scene, the Police Department Operations Unit should be contacted through the dispatcher. The Fire Dept. IC should request priority response and be prepared to give all particulars regarding the need for Police assistance.

The Hazardous Materials Unit shall not be detained at the scene for purposes of safeguarding materials after control and containment of the substance has been completed.

Arrangements for the collection, transportation, cleaning, and temporary replacement of exposed or contaminated Fire Department equipment shall be the responsibility of the Special Operations Decon Officer at the scene. (See 4.3.6)

8.2   Prior to leaving the scene, the IC shall review the particulars of the incident to make a preliminary determination regarding compliance with fire prevention regulations and directives. Non-compliance should be addressed through violation orders, summonses, vacate orders etc. If necessary, consult the Bureau of Fire Prevention for assistance.

Additionally, within ten (10) days from the completion of the post incident review, the IC shall notify the Chief of the Bureau of Fire Prevention of any violation or suspected violation of statutes or regulations that he or she became aware of during the course of the response to the hazardous material incident or during the post incident review process.

8.3   Specific incident termination instructions will be listed in the incident checklist.

9.0 Post Incident Procedures

9.1 Incident Analysis

A. The Incident Commander's responsibility following each hazardous materials operation includes a review of actions taken. Corrective actions should be taken at the lowest possible level. Recommendations regarding procedures or the elements of the response plan should be forwarded through the chain of command with endorsements to the Chief of Department.

B. Prior to leaving the scene, the IC shall review the particulars of the incident to make a preliminary determination regarding compliance with fire prevention regulations and directives. Non-compliance should be addressed through violation orders, summonses, vacate orders etc. If necessary, consult the Bureau of Fire Prevention for assistance.

Additionally, within ten (10) days from the completion of the post incident review, the IC shall notify the Chief of the Bureau of Fire Prevention of any violation or suspected violation of statutes or regulations that he or she became aware of during the course of the response to the hazardous material incident or during the post incident review process.

C. As standard part of this Emergency Response Plan the Fire Dept. will review all hazardous materials incidents in order to:

1. Identify and correct any deficiencies in the response plan.

2. Identify trends, patterns, and deficiencies in procedures that need to be addressed in revised operations, safety, or training programs.

3. Develop a database to establish profiles for decisions re: tools & equipment, training, budgeting, operations, hazard prevention legislation etc.

4. Identify incidents where the Fire Dept. may seek cost recovery or that may qualify for reimbursement under EPA regulations.

5. Document efforts of proactive intervention to reduce losses.

D. The process begins with a preliminary review of the following incidents:

1. Citywide 10-80 response record.

2. Exposure reports CD-73; CD-72 medical reports indicating hazmat exposure.

3. Any incident involving "Decon."

4. Level 2 and 3 incidents.

E. Consolidated reports and descriptions of these incidents will be reviewed on a monthly basis by:

1. Bureau of Operations
2. Division of Safety
3. Division of Training
4. OSHA Coordinator

F. Findings and recommendations, if any, from any individual review will be forwarded through the chain of command to the Chief of Department for consideration.
G. A formal critique may be ordered by the Chief of Department. The formal critique report is included in the appendix of this document.

9.2 Critique Outline

When a formal critique is ordered, the IC for the particular incident shall prepare a narrative report addressed to the Chief of Department, containing at least the following information. Include copies of the incident log to reduce preparation time.

1. Identifying information including the date, box number, location.
2. Description of conditions found and actions taken by the first responding units.
3. Description of conditions found and actions taken by the first Chief Officer to arrive.
4. Describe hazard identification and site control measures taken and results.
5. Describe the personal protective equipment and special chemical protective equipment used by each group of responders at the incident. Describe decision made re: changes in these levels of protection.
6. A timeline describing the notification, change of command, and arrival of special units, outside agencies, etc.
7. Describe the ICS organizational structure developed. (Include positions staffed, Command post, sectors/groups, control zones, etc.)
8. A description of the key phases of the incident including sufficient detail to explain changes in incident activities or other significant events during the life of the incident. Include description of alternative plans considered, if any.
9. Description of decon procedures.
10. Description of problems encountered and lessons learned.

9.3 Emergency Response Plan Review

A formal cyclical review process and a method of keeping this document current with existing laws and standards will be developed and implemented at the time this document is finalized and published.

The review process will be developed with the intent described in the Incident Analysis portion of this document.

10.0 Appendix

10.1 References

The following list of materials will provide greater insight on the individual subject areas discussed in this document:

A. Occupational Safety and Health Administration 29 CFR 1910.120 — Hazardous Waste Operations and Emergency Response
B. Environmental Protection Agency — 40 CFR 300
C. Superfund Amendment and Reauthorization Act
D. N.Y.C. Office of the Mayor
   Mayoral Directive 82-2 (revised 1986)
E. Office of Emergency Management
   Memorandum of Understanding — Asbestos (12/15/89)

10.2 Abbreviations

CFR	Code of Federal Regulations
CP	Command Post
DEP	Department of Environmental Protection
DOT	Department of Transportation
ERG	Emergency Response Guidebook (DOT)
ERP	Emergency Response Plan
EMS	Emergency Medical Services
EPA	Environmental Protection Agency (Federal)
HAG	Hazard Action Guide
IC	Incident Commander
ICS	Incident Command System
NFPA	National Fire Protection Association

OEM   Office of Emergency Management (NYC)
ORM   Other Regulated Materials
OSHA   Occupational Safety and Health Administration
PPE   Personal Protective Equipment
SCBA   Self Contained Breathing Apparatus
SOC   Special Operations Command
SOP   Standard Operating Procedure

Supplement 7: Hazardous Materials Incident Command System

This supplement is one example of an incident command system designed to organize and coordinate hazardous materials incidents. It was developed by the California Fire and Rescue Advisory Board of the Governor's Office of Emergency Service as part of the FIRESCOPE program and published as ICS-HM-120-1, "Hazardous Materials Operational System Description," in September 1990. This document contains information about the Incident Command System component of the National Interagency Incident Management System. Additional information and documentation can be obtained from the State Board of Fire Services, State Fire Marshal, 7171 Bowling Drive, Suite 600, Sacramento, CA 95823, 916-427-4166 or from the Support Services Manager, Operations Coordination Center, P.O. Box 55157, Riverside, CA 92517, 714-782-4174.

Introduction

The Hazardous Materials organizational module is designed to provide an organizational structure that will provide necessary supervision and control for the essential functions required at virtually all Hazardous Materials incidents. This is based on the premise that controlling the tactical operations of companies and movement of personnel and equipment will provide a greater degree of safety and also reduce the probability of spreading contaminants. The primary functions will be directed by the Hazardous Materials Group Supervisor, and all resources that have a direct involvement with the hazardous material will be supervised by one of the functional leaders or the Hazardous Materials Group Supervisor.

The three functional positions of the Hazardous Materials Group (Entry Leader, Site Access Control Leader, and Decontamination Leader) require a high degree of control and close supervision. The Entry Leader supervises all companies and personnel operating in the Exclusion Zone. The Entry Leader has the responsibility to direct all tactics and control the positions and functions of all personnel in the Exclusion Zone. The Site Access Control Leader controls all movement of personnel and equipment between the control zones. The Site Access Control Leader has the responsibility for isolating the Exclusion and Contamination Reduction Zone and ensuring that citizens and personnel use proper access routes. The Decontamination Leader ensures all rescue victims, personnel, and equipment have been decontaminated before leaving the incident.

The Hazardous Materials Group Supervisor manages these three functional responsibilities which include all tactical operations carried out in the Exclusion Zone. All rescue operations, by definition, will come under the direction of the Hazardous Materials Group Supervisor. Evacuation and all other tactical objectives that are outside of the control zones are not the responsibility of the Hazardous Materials Group Supervisor. In addition to the three primary functions, the Group Supervisor will work with an Assistant Safety Officer, who is Hazardous Materials trained, and who must be present at the hazardous site. The Incident Safety Officer will have overall incident safety concerns, with the Assistant Safety Officer working directly with the Hazardous Materials Group Supervisor. The Group Supervisor may also supervise one or more Technical Specialists.

Tactical operations outside of the controlled zones, as well as many other hazardous materials related functions, will be managed by regular ICS positions. In most cases, the array of tactical objectives such as evacuation, isolation, medical, traffic control, etc., will be managed by Division/Group Supervisors. Other needs will be met by filling Command and General Staff positions.

Unified Command

It is assumed that all hazardous materials incidents will be managed under Unified Command principles because in virtually all cases fire, law enforcement, and public health will have some statutory functional responsibility for incident mitigation. Depending on incident factors, several other agencies will respond to a hazardous materials incident.

The Assisting Agencies section of this document lists some of the typical functional responsibilities of Law Enforcement and Health agencies. A matrix is provided showing typical responsibilities of agencies at the local, state, and federal levels.

Modular Development

A series of examples of modular development are included to illustrate one method of expanding the incident organization.

Initial Response Organization

Initial response resources are managed by the Incident Commander who will handle all Command and General Staff responsibilities. See Figure 1.

Reinforced Response Organization (3 to 15 fire and/or Law Enforcement units)

The Incident Commander has established a Hazardous Materials Group to manage all activities around the Control Zones and has assigned two Law Enforcement units to isolate the operational area. One Law Enforcement Officer has met with the Fire Incident Commander and together they have established Unified Command. The Incident Commanders have decided to establish a Planning Section to manage information. See Figure 2.

Multi-division Organization

The Incident Commanders have established most Command and General Staff positions and have established a combination of divisions and groups. See Figure 3.

Multi-branch Organization

The Incident Commanders have established all Command and General Staff positions and have established four branches. See Figure 4.

Hazardous Materials Position Descriptions and Functions

Hazardous Materials Group Supervisor

The Hazardous Materials Group Supervisor reports to the Operations Section Chief (or Hazardous Materials Branch Director if activated). The Hazardous Materials Group Supervisor is responsible for the implementation of the phases of the Incident Action Plan dealing with the Hazardous Materials Group operations. The Hazardous Materials Group Supervisor is responsible for the assignment of resources within the Hazardous Materials Group, reporting on the progress of control operations and the status of resources within the Group. The Hazardous Materials Group Supervisor directs the overall operations of the Hazardous Materials Group.

A. Check in and obtain briefing from the Operations Section Chief or Hazardous Materials Branch Director (if activated).
B. Ensure the development of Control Zones and Access Control Points and the placement of appropriate control lines.
C. Evaluate and recommend public protection action options to the Operations Chief or Branch Director (if activated).
D. Ensure that current weather data and future weather predictions are obtained.
E. Establish environmental monitoring of the hazard site for contaminants.
F. Ensure that a Site Safety Plan is developed and implemented.
G. Conduct safety meetings with the Hazardous Materials Group.
H. Participate, when requested, in the development of the Incident Action Plan.

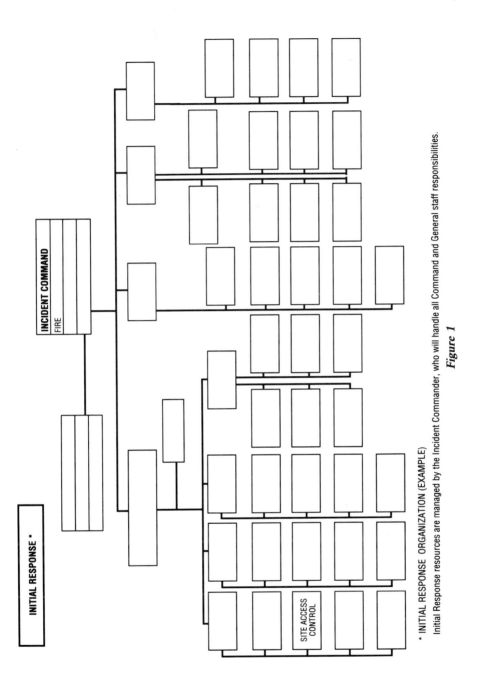

Figure 1

* INITIAL RESPONSE ORGANIZATION (EXAMPLE)

Initial Response resources are managed by the Incident Commander, who will handle all Command and General staff responsibilities.

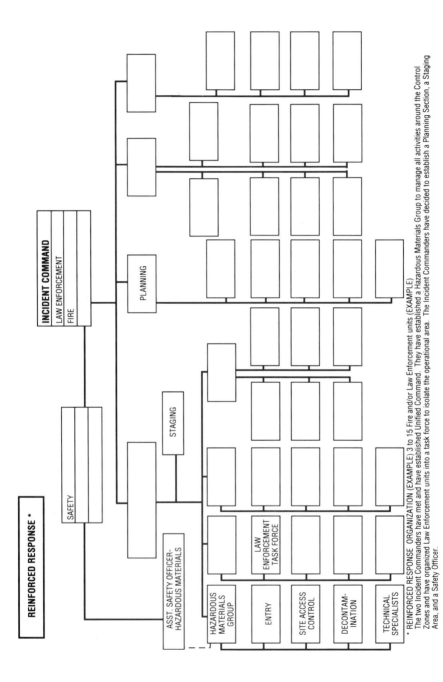

Figure 2

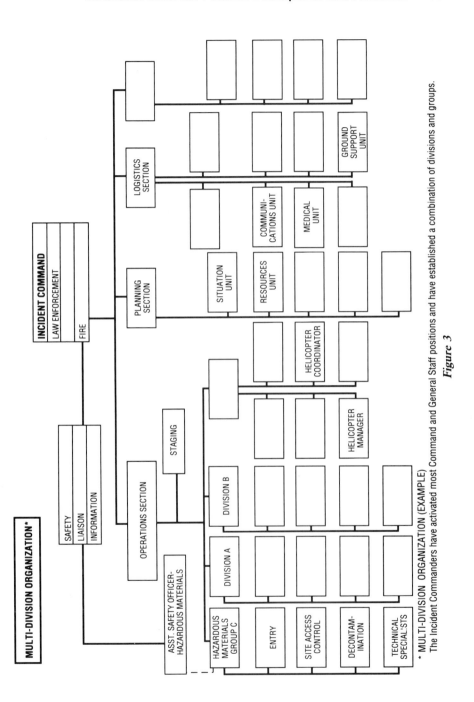

Figure 3

* MULTI-DIVISION ORGANIZATION (EXAMPLE)
The Incident Commanders have activated most Command and General Staff positions and have established a combination of divisions and groups.

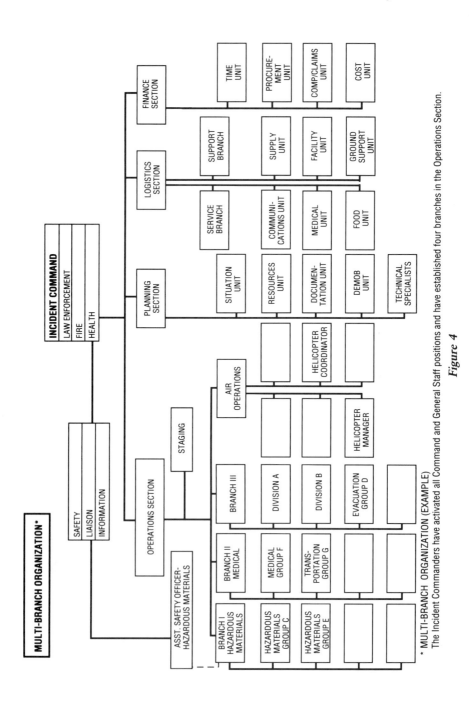

Figure 4

MULTI-BRANCH ORGANIZATION (EXAMPLE)

* The Incident Commanders have activated all Command and General Staff positions and have established four branches in the Operations Section.

I. Ensure that recommended safe operational procedures are followed.
J. Ensure that the proper Personal Protective Equipment is selected and used.
K. Ensure that the appropriate agencies are notified through the Incident Commander.
L. Maintain Unit Log (ICS 214).

Entry Leader

Reports to the Hazardous Materials Group Supervisor. The Entry Leader is responsible for the overall entry operations of assigned personnel within the Exclusion Zone.

A. Check in and obtain briefing from the Hazardous Materials Group Supervisor.
B. Supervise entry operations.
C. Recommend actions to mitigate the situation within the Exclusion Zone.
D. Carry out actions, as directed by the Hazardous Materials Group Supervisor, to mitigate the hazardous materials release or threatened release.
E. Maintain communications and coordinate operations with the Decontamination Leader.
F. Maintain communications and coordinate operations with the Site Access Control Leader.
G. Maintain communications and coordinate operations with Technical Specialist/Hazardous Materials Reference.
H. Maintain control of the movement of people and equipment within the Exclusion Zone, including contaminated victims.
I. Direct rescue operations, as needed, in the Exclusion Zone.
J. Maintain Unit Log (ICS 214).

Decontamination Leader

Reports to the Hazardous Materials Group Supervisor. The Decontamination Leader is responsible for the operations of the decontamination element, providing decontamination as required by the Incident Action Plan.

A. Check in and obtain briefing from the Hazardous Materials Group Supervisor.
B. Establish the Contamination Reduction Corridor(s).
C. Identify contaminated people and equipment.

D. Supervise the operations of the decontamination element in the process of decontaminating people and equipment.
E. Maintain control of movement of people and equipment within the Contamination Reduction Zone.
F. Maintain communications and coordinate operations with the Entry Leader.
G. Maintain communications and coordinate operations with the Site Access Control Leader.
H. Coordinate the transfer of contaminated patients requiring medical attention (after decontamination) to the Medical Group.
I. Coordinate handling, storage, and transfer of contaminants within the Contamination Reduction Zone.
J. Maintain Unit Log (ICS 214).

Site Access Control Leader

Reports to the Hazardous Materials Group Supervisor. Site Access Control Leader is responsible for the control of the movement of all people and equipment through appropriate access routes at the hazard site, and ensures that contaminants are controlled and records are maintained.

A. Check in and obtain briefing from the Hazardous Materials Group Supervisor.
B. Organize and supervise assigned personnel to control access to the hazard site.
C. Oversee the placement of the Exclusion Control Line and the Contamination Control Line.
D. Ensure that appropriate action is taken to prevent the spread of contamination.
E. Establish the Safe Refuge Area within the Contamination Reduction Zone. Appoint a Safe Refuge Area Manager (as needed).
F. Ensure that injured or exposed individuals are decontaminated prior to departure from the hazard site.
G. Track the movement of persons passing through the Contamination Control Line to ensure that long term observations are provided.
H. Coordinate with the Medical Group for proper separation and tracking of potentially contaminated individuals needing medical attention.
I. Maintain observations of any changes in climatic conditions or other circumstances external to the hazard site.
J. Maintain communications and coordinate operations with the Entry Leader.

K. Maintain communications and coordinate operations with the Decontamination Leader.

L. Maintain Unit Log (ICS 214).

Assistant Safety Officer — Hazardous Materials

Reports to the Incident Safety Officer as an Assistant Safety Officer and coordinates with the Hazardous Materials Group Supervisor (or Hazardous Materials Branch Director if activated). The Assistant Safety Officer - Hazardous Materials coordinates safety related activities directly relating to the Hazardous Materials Group operations as mandated by 29 CFR part 1910.120 and Subsection 5192, Title 8, CCR, and applicable State and local laws. This position advises the Hazardous Materials Group Supervisor (or Hazardous Materials Branch Director) on all aspects of health and safety and has the authority to stop or prevent unsafe acts. It is mandatory that an Assistant Safety Officer - Hazardous Materials be appointed at all hazardous materials incidents. In a multi-activity incident the Assistant Safety - Officer Hazardous Materials does not act as the Safety Officer for the overall incident.

A. Check in and obtain briefing from the Incident Safety Officer.

B. Obtain briefing from the Hazardous Materials Group Supervisor.

C. Participate in the preparation of and implement the Site Safety Plan.

D. Advise the Hazardous Materials Group Supervisor (or Hazardous Materials Branch Director) of deviations from the Site Safety Plan or any dangerous situations.

E. Has authority to alter, suspend, or terminate any activity that may be judged to be unsafe.

F. Ensure protection of the Hazardous Materials Group personnel from physical, environmental, and chemical hazards/exposures.

G. Ensure the provision of required emergency medical services for assigned personnel and coordinate with the Medical Unit Leader.

H. Ensure that medical related records for the Hazardous Materials Group personnel are maintained.

I. Maintain Unit Log (ICS 214).

Technical Specialist — Hazardous Materials Reference

Reports to the Hazardous Materials Group Supervisor (or Hazardous Materials Branch Director if activated). This position provides technical information and assistance to the Hazardous Materials Group using various reference sources such as computer data bases, technical journals, CHEMTREC, and phone contact with facility representatives. The Technical Specialist - Hazardous Materials Reference may provide product identification using hazardous categorization tests and/or any other means of identifying unknown materials.

A. Check in and obtain briefing from the Hazardous Materials Group Supervisor.
B. Obtain briefing from the Planning Section Chief.
C. Provide technical support to the Hazardous Materials Group Supervisor.
D. Maintain communications and coordinate operations with the Entry Leader.
E. Provide and interpret environmental monitoring information.
F. Provide analysis of hazardous material sample.
G. Determine personal protective equipment compatibility to hazardous material.
H. Provide technical information of the incident for documentation.
I. Provide technical information management with public and private agencies, ie: Poison Control Center, Tox Center, CHEMTREC, State Department of Food and Agriculture, National Response Team.
J. Assist Planning Section with projecting the potential environmental effects of the release.
K. Maintain Unit Log (ICS 214).

Assisting Agencies in Hazardous Materials Incident

Law Enforcement

The local law enforcement agency will respond to most Hazardous Materials incidents. Depending on incident factors, law enforcement may be a partner in Unified Command or may participate as an assisting agency. Some functional responsibilities that may be handled by law enforcement are:

A. Isolate the incident area.
B. Manage crowd control.
C. Manage traffic control.
D. Manage public protective action.

E. Provide scene management for on-highway incidents.

F. Manage criminal investigations.

Environmental Health Agencies

In most cases the local or state environmental health agency will be at the scene as a partner in Unified Command. Some functional responsibilities that may be handled by environmental health agencies are:

A. Determine the identity and nature of the Hazardous Materials.

B. Establish the criteria for cleanup and disposal of the Hazardous Materials.

C. Declare the site safe for re-entry by the public.

D. Provide the medical history of exposed individuals.

E. Monitor the environment.

F. Supervise the cleanup of the site.

G. Enforce various laws and acts.

H. Determine legal responsibility.

I. Provide technical advice.

J. Approve funding for the clean-up.

Glossary of Terms

29 CFR Part 1910.120: 29 of the Code of Federal Regulations, Part 1910.120 is the Hazardous Waste operations and Emergency Response reference document as required by SARA. This document covers employees involved in certain hazardous waste operations and any emergency response to incidents involving hazardous situations. Federal OSHA enforces this code.

Access Control Point: The point of entry and exit from the control zones. Regulates access to and from the work areas.

CHEMTREC: Chemical Transportation Emergency Center. A public service of the Chemical Manufacturers Association.

Compatibility: The matching of Personal Protective Equipment to the hazardous materials involved in order to provide the best protection for the worker.

STATE AGENCIES	Notification	Identification, Analysis; Technical Assistance	Coordination	Law Enforcement; Traffic Control	Rescue, Suppression, Containment	Cleanup and Disposal	Evacuation, Area Control	Emergency Medical	Public Health and Sanitation	Education and Public Information	Recovery	Natural Resources Protection and Damage Assessment	Conducts Criminal and Civil Investigation
Office of Emergency Services			X							X	X		
Highway Patrol	X	X	X	X		X	X				X		X
State Water Resources Control Board	X	X				X			X	X	X	X	X
Department of Fish and Game	X	X	X			X						X	X
Department of Conservation (Oil and Gas)	X	X			X	X							
State Lands Commission	X	X			X	X						X	
Department of Transportation	X					X	X				X		
Department of Health Services (DOHS)	X	X			X	X	X	X	X	X	X		X
Department of Food and Agriculture	X					X			X		X		
Department of Industrial Relations	X							X	X		X		X
Department of Water Resources	X				X				X			X	
Air Resources Board	X								X				
Department of Parks and Recreation						X	X					X	
Military Department	X			X	X		X	X					
Public Utilities Commission	X												
Attorney General											X	X	X
Department of General Services				X		X							
Emergency Medical Service Authority								X					
Department of Social Services			X								X		
Fire Services					X		X						

Responsibility Matrix State Agencies (Sample)

Responsibility Matrix (sample). Columns indicate responsibilities; "X" marks the assigned responsibility.

	Notification	Identification, Analysis, Technical Assistance	Coordination	Law Enforcement, Traffic Control	Rescue, Suppression, Containment	Cleanup and Disposal	Evacuation, Area Control	Emergency Medical	Public Health and Sanitation	Education and Public Information	Recovery	Conducts Criminal and Civil Investigation
LOCAL AGENCIES												
Fire Service	X	X		X	X		X		X			
Law Enforcement	X	X	X				X			X		X
County Department of Health Services (DHS)	X	X	X			X		X	X	X	X	X
Agriculture Commissioner	X			X	X			X	X	X		
Air Quality Management District	X							X	X			X
Public Works	X			X	X		X					
City Attorney											X	X
District Attorney											X	X
Hospitals	X							X				
FEDERAL AGENCIES												
Federal Emergency Management Agency (FEMA)			X					X		X	X	
United States Coast Guard	X	X	X	X	X	X						X
Environmental Protection Agency (EPA)	X	X	X			X						X
Others		X						X				
NON-GOVERNMENT												
Private Facility Owners	X	X	X		X	X	X			X	X	
Industry Co-Ops		X			X	X						
Private Hazardous Waste Services		X			X	X						
American National Red Cross							X			X		
Salvation Army							X			X		
Hospital, Ambulances	X							X				

Responsibility Matrix: Local Agencies, Federal Agencies, Non-Government (Sample)

Contamination Reduction Corridor (CRC):   That area within the Contamination Reduction zone where the actual decontamination is to take place. Exit from the Exclusion zone is through the Contamination Reduction Corridor (CRC). The CRC will become contaminated as people and equipment pass through to the decontamination stations.

Contamination Control Line (CCL):   The established line around the Contamination Reduction Zone that separates the Contamination Reduction Zone from the Support Zone.

Contamination Reduction Zone (CRZ):   That area between the Exclusion Zone and the Support Zone. This zone contains the Personnel Decontamination Station. This zone may require a lesser degree of personnel protection than the Exclusion Zone. This area separates the contaminated area from the clean area and acts as a buffer to reduce contamination of the clean area.

Control Zones: The geographical areas within the control lines set up at a hazardous materials incident. The three zones most commonly used are the Exclusion Zone, Contamination Reduction Zone, and Support Zone.

Decontamination (Decon):   That action required to physically remove or chemically change the contaminants from personnel and equipment.

Environmental: Atmospheric, Hydrologic and Geologic media (air, water and soil).

Exclusion Zone: That area immediately around the spill. That area where contamination does or could occur. The innermost of the three zones of a hazardous materials site. Special protection is required for all personnel while in this zone.

Evacuation: The removal of potentially endangered, but not yet exposed, persons from an area threatened by a hazardous materials incident. Entry into the evacuation area should not require special protective equipment.

Hazardous Categorization Test (Haz Cat):   A field analysis to determine the hazardous characteristics of an unknown material.

Hazardous Material: Any material which is explosive, flammable, poisonous, corrosive, reactive, or radioactive, or any combination, and requires special care in handling because of the hazards it poses to public health, safety, and/or the environment.

Hazardous Materials Incident: Uncontrolled, unlicensed release of hazardous materials during storage or use from a fixed facility or during transport outside a fixed facility that may impact the public health, safety and/or environment.

Mitigate: Any action employed to contain, reduce or eliminate the harmful effects of a spill or release of a hazardous substance.

Personal Protective Equipment (PPE):   That equipment and clothing required to shield or isolate personnel from the chemical, physical, and biologic hazards that may be encountered at a hazardous materials incident.

Refuge Area: An area identified within the Exclusion Zone, if needed, for the assemblage of contaminated individuals in order to reduce the risk of further contamination or injury. The Refuge Area may provide for gross decontamination and triage.

Rescue: The removal of victims from an area determined to be contaminated or otherwise hazardous. Rescue shall be performed by emergency personnel using appropriate personal protective equipment.

Safe Refuge Area (SRA):   An area within the Contamination Reduction Zone for the assemblage of individuals who are witnesses to the hazardous materials incident or who were on site at the time of the spill. This assemblage will provide for the separation of contaminated persons from non-contaminated persons.

Site: That area within the Contamination Reduction Control Line at a hazardous materials incident.

Site Safety Plan: An Emergency Response Plan describing the general safety procedures to be followed at an incident involving hazardous materials. This plan should be prepared in accordance with 29 CFR 1910.120 and the U.S. Environmental Protection Agency's "Standard Operating Safety Guides for Environmental Incidents" (1984).

Support Zone: The clean area outside of the Contamination Control Line. Equipment and personnel are not expected to become contaminated in this area. Special protective clothing is not required. This is the area where resources are assembled to support the hazardous materials operation.

Supplement 8: Guidelines for Decontamination of Fire Fighters and Their Equipment Following Hazardous Materials Incidents

The following supplement is a reprint of a document published by the Canadian Association of Fire Chiefs. It includes guidelines for decontaminating fire fighters and their equipment after exposure to hazardous materials and for planning decontamination procedures before an incident occurs. We are grateful for the opportunity to include this detailed, practical guide in the Hazardous Materials Response Handbook.

Note to Readers

The contents of this booklet are based on information and advice believed to be accurate and reliable. The Canadian Association of Fire Chiefs Inc., its officers, and members, jointly and severally, make no guarantee and assume no liability in connection with this booklet. Moreover, it should not be assumed that every acceptable procedure is included or that special circumstances may not warrant modified or additional procedures.

The user should be aware that changing technology or regulations may require a change in the recommended procedures contained herein. Appropriate steps should be taken that the information is current when used.

The suggested procedures should not be confused with any federal, provincial, state, municipal, or insurance requirements, or with national safety codes.

Introduction

The number of hazardous materials incidents to which the fire service is called increases year by year. At each of these incidents, there is a good chance that the responding fire fighters may become contaminated with the hazardous material. Frequently, however, the matter of decontamination is never thought of, or is only performed cursorily.

The Dangerous Goods Sub-Committee of the Canadian Association of Fire Chiefs, in 1986-87, undertook the preparation of a series of guidelines for decontamination, to be adopted by any fire department that wishes to do so, either as printed here or with local variations due to their own circumstances. Note that these are indeed *guidelines, not standards*.

The procedures listed are designed so they can be carried out by any department, rural or urban, volunteer or full-time, with a minimum of investment in special equipment.

Background of the Study and Rationale for Its Conclusions

In 1986 the Dangerous Goods Sub-Committee carried out an extensive study of fire services across the world to investigate the different approaches to decontamination of fire fighters following hazardous materials incidents.

Procedures were reviewed in detail from North America (Phoenix, San Francisco, Colorado, Metropolitan Toronto area) and England (Hampshire, Cambridgeshire, Greater Manchester, London, and the Home Office Guidelines). In addition, information was requested from Hong Kong, New Zealand, Australia, the People's Republic of China, France, Germany, Switzerland, Italy, Sweden, and the Netherlands. From these latter countries no formal replies were received, however, and indications are that decontamination procedures are either limited or absent. Japanese fire officials wrote back to indicate that they were studying various options but had not yet finalized any procedures.

Many magazine articles from the various periodicals published for the European and North American fire service market were reviewed. Furthermore, members consulted with chemical manufacturers, nuclear medicine physicists, hazardous waste disposal companies, industrial hygienists, toxicologists, and various jurisdictional agencies such as the Atomic Energy Control Board and provincial and federal Ministries of Health, Labour, and Environment.

All this research led to the conclusion that there were three basic philosophies in existence:

1. Wet and Dry Procedures;
2. Dispose or Retain Run-off;
3. Severity and Type of the Material.

The idea of a dry procedure made sense because of easier containment and no reaction with water. A single wet procedure, however, was not deemed to be sufficiently comprehensive; on the other hand some wet procedures called for making up solutions of a variety of chemicals, and these were deemed to be too complicated for use by every fire department.

The dispose or retain philosophy is usually considered a wet procedure; however, to base one's procedures solely on the concern about run-off appeared to be overly simplistic. Note, however, that concerns about run-off are addressed later in this document.

The third philosophy was examined in more detail. The methods of determining the severity and type of material were defined by various fire departments along the following lines:

1. By UN class
2. By effects on the environment
3. By chemical characteristics
4. By physiological effects on people
5. By broad groups.

The first alternative was deemed unsuitable because there are too many variables within a class (e.g., some flammable liquids are highly toxic, others are not). The second alternative was deemed to be of secondary importance to fire fighter safety (see the section on Environmental Considerations later in this document). It was found that alternatives three and four could in fact be used to arrive at alternative five, and this was the route taken to define the procedures in this document.

The procedures that were developed are therefore broken down as follows:

- Three general procedures for light, medium, and extreme hazards,
- Two specific procedures for substances that do not fit into the three general groups above (although they share many common factors), and
- One initial routine performed in some cases prior to the start of one of the other procedures.

The authors realize that if the procedures listed here are to be completed thoroughly, a number of decontamination operatives and a Decontamination Officer to oversee them are needed. Typically, this will require the services of at least one fire fighting company. Attention to detail and careful execution of all steps should lead to successful and safe completion of fire fighter decontamination.

Decontamination—General Observations

Six levels of decontamination are outlined. The incident commander will determine which level is applicable for the substance involved, using any reference sources that may state the applicable level. In the absence of such sources, advice should be sought from experts such as toxicologists, chemical company representatives, CANUTEC, CHEMTREC, etc.

The levels are:

A—for light hazards
B—for medium hazards
C—for extreme hazards
D—dry contamination for water-reactive and certain dry substances
E—for etiologic agents and certain dry pesticides and poisons
R—for radioactive materials.

Note that A-level decontamination, the most common, need only be done at the station. However, other levels need to be started at the incident scene as well as being continued on return to the station.

C-level decontamination, the most stringent level for the most toxic substances, may involve the destruction of all clothes worn.

In a few cases, scrubbing of clothes must be done while wearing SCBA as vapors released during cleaning may be harmful.

D-level decontamination is almost always followed by one of the other levels of decontamination, which will be dependent on the substance involved.

The procedures should be initiated if personnel are known or suspected to have been directly exposed to the chemical or its vapors, products of combustion, etc.

Officers should be aware of any cuts, wounds, lesions, or abrasions that their crews may have. If the apparatus is sent to an incident involving hazardous materials, such personnel should wherever possible exercise special care to avoid the chance of contamination through such wounds. Chemicals absorbed through the skin will be absorbed much faster if the skin is cut or abraded, thereby presenting a serious health hazard.

Adequate awareness is necessary to realize when decontamination will be required, so that early action can be taken to bring to the scene the equipment and manpower resources needed to set up and staff the decontamination area.

Decontamination Procedures

Level A for Light Hazards

On Return to Station

1. Wash down all protective clothing with a mild (1 to 2%) trisodium phosphate solution. Rinse with water.
2. Wash down SCBA cylinders and harnesses with a mild trisodium phosphate solution. Take care to wipe, not scrub, around regulator assembly. Rinse with clean water. If damage is suspected to any part of the unit, ensure it is sent for service.
3. Scrub hands and face with soap and water.

NOTE: Where the scrubbing of the protective clothing may release harmful vapors caught in the fibers, it may be necessary to wear breathing apparatus while washing down protective clothing. In these cases, monitor the atmosphere around the washing area. Release of vapors may indicate commercial cleaning is required.

Level B for Medium Hazards

At the Scene

1. Do not remove SCBA facepiece. Place helmet on back of neck.

2. Assistant to flush fire fighter downwards from head to toe with copious amounts of low pressure water. Include inside and outside of helmet, mask, harness, and inside of coat-wrists to the cuff.
3. Do not smoke, eat, drink, or touch face.

On Return to Station

4. Place apparatus temporarily out of service.
5. Remove all protective clothing and accessories. If possible, remove liner from helmet. Scrub all items, including the helmet liner, inside and out with a mild (1 to 2%) trisodium phosphate solution. Then flush copiously with water.

NOTE: Where the scrubbing of the protective clothing may release harmful vapors caught in the fibers, it may be necessary to wear breathing apparatus while washing down protective clothing. In these cases, monitor the atmosphere around the washing area. Release of vapors may indicate commercial cleaning is required.

6. Scrub all other protective gear such as gloves and breathing apparatus items likewise. Be sure to flush out gloves with water. If SCBA is stored in its case while returning from incident, scrub the case also.
7. Remove all clothing worn at the scene, including underwear, and place in garbage bag for laundering and/or dry cleaning (preferably the latter). Take all garbage bags with contaminated clothing to a place where they can be cleaned separately from other garments.
8. Shower, scrubbing all of the body with soap and water, with particular emphasis on areas around the mouth and nostrils and under fingernails. Shampoo hair and thoroughly clean mustache if you have one.
9. Do not smoke, drink, eat, touch face, or void until step #8 completed.
10. Put on clean clothes.
11. Do not put apparatus back in service until clean-up completed.

To Change SCBA Cylinders at the Scene

Flush empty cylinder and surrounding area of fire fighter's back with copious amounts of low pressure water. Also flush facepiece and breathing tube to prevent inhalation of harmful materials when regulator is disconnected.

Wear gauntlet-type rubber gloves, such as those used by linemen, when changing cylinders. Flush gloves after use before removing them.

Level C for Extreme Hazards

At the Scene

1. Do not remove SCBA facepiece. Place helmet on back of neck.
2. Assistant, wearing protective clothing and SCBA (plus disposable chemical suit wherever possible), to flush fire fighter downwards from head to toe with copious amounts of low pressure water. Include inside and outside of helmet, mask, harness, and inside of coat-wrists to the cuff.
3. Do not smoke, eat, drink, or touch face.
4. Put SCBA, used cylinders, and any equipment (including hoses and tarps) suspected or known to be contaminated in garbage bags. Seal bags and return them to the station. Where circumstances permit, remove and bag protective clothing also.

On Return to Station

5. Put bags returned from incident scene in exterior cordoned-off area away from public access. Place apparatus out of service.
6. Strip completely. Place all clothing (protective clothing and personal clothing) in plastic garbage bags. Place portable radios in a separate bag. Seal bags, place in exterior cordoned-off area.
7. Arrange for the supply of a number of steel drums. Upon their arrival, seal garbage bags with contaminated items into drums. Mark drums and place in exterior cordoned-off area, minimum 5-meter radius.
8. Arrange for the drums to be picked up and the contents analyzed. Some or all items may be destroyed; some may be able to be decontaminated and returned.
9. Shower, scrubbing all of the body with soap and water, with particular emphasis on areas around the mouth and nostrils and under fingernails. Shampoo hair. Thoroughly clean mustache if you have one.

Special Attention for Radioactive Incidents:

After showering, scan entire body with a radiation contamination monitor, paying special attention to hair, hands, and fingernails. Hold monitor approximately 3 cm from body. If any reading beyond normal background level is detected, the fire fighter should shower again, scrubbing with more soap than before.

10. Do not smoke, drink, eat, touch face, or void until step #9 is completed.
11. Put on clean clothes.

12. Report to hospital for medical examination. Inform physician which hazardous material was involved.

To Change SCBA Cylinders at the Scene

Flush empty cylinder and surrounding area of fire fighter's back with copious amounts of low pressure water. Also flush facepiece and breathing tube to prevent inhalation of harmful material when regulator is disconnected.

Wear gauntlet-type rubber gloves, such as those used by linemen, when changing cylinders. Flush gloves after use before removing them.

Place empty cylinder in black plastic garbage bag and seal for subsequent decontamination.

The person doing the flushing and cylinder-changing must wear protective clothing and SCBA, plus a disposable chemical suit if available.

Special Note

Where circumstances, local climate, and available resources permit, the performance of *all* steps at the scene (instead of performing steps 5-11 at the station) is preferable. The procedure is outlined as shown, however, in recognition of the fact that for many departments this will usually be impossible to achieve.

Level D for Water-Reactive Hazards

At the Scene

1. Set up a suitable vacuum cleaner with power supply. Provide a dry brush and a containment capture method for materials falling off the contaminated personnel. Assistants to don full protective clothing and SCBA, plus disposable chemical suits if available and appropriate.
2. If this is a radiation incident: The fire fighters suspected of being contaminated will be scanned carefully with a radiation monitor suitable for detecting surface contamination. All parts of their clothing and personal equipment will be scanned, including the soles of the boots. If no readings are found, the personnel that have been checked can leave the decontamination area.
3. If not a radiation incident, or if the fire fighter was found to be radioactively contaminated: Stand fire fighter in center of containment area, clean helmet and place on back of neck, then clean inside of helmet.

4. Commence cleaning from head downwards. Include all external areas. Slacken SCBA harness to allow cleaning behind straps and backplate. Likewise, loosen the hose-key belt and clean behind it.
5. When fire fighters have been fully vacuumed or brushed off, they will step out of the containment area. As they do so, their boots, including the soles, must be cleaned off so any contaminant will remain within the containment area.
6. Procedures will then continue as follows:

   • Radioactive incident—go to Level "R" routine
   • Etiological or dry pesticide incident—go to Level "E" routine
   • Other incidents—go to Level "B" routine (unless advice is received that Level "C" is more appropriate).

7. All used filters and collected waste are to be placed in a garbage bag, sealed and tagged, and disposed of in a manner acceptable to the agency having jurisdiction.

Level E for Etiologic Hazards

Special Equipment Required

A presentation spray can (such as used for pesticide spraying), biological neutralizing substance (such as bleach, commercial sterilizing agent, etc.), orange garbage bags, black garbage bags, sterilization bags as used by hospital laundries, and a box of surgical masks.

At the Scene

1. If using bleach, make up a 5% to 6% bleach solution in the spray can. Take note of the bleach concentrate percentage when calculating the make-up of the solution. Many brands as purchased in the store are already 6%.
   If using a commercial sterilizer, follow the manufacturer's directions.
2. Flush the fire fighter downwards from head to toe with low pressure *water*. SCBA facepiece can now be removed. Place helmets in black plastic garbage bag(s) and seal. Place surgical mask on fire fighter.
3. If using bleach, spray the fire fighter's boots (but not their bunker gear) and any tools, hoses, and other equipment used (except for portable radios) with the *bleach* solution in the spray can. Leave for 10 minutes, then flush with water.
   If using a commercial sterilizer, follow the manufacturer's instructions.

4. Remove SCBA. Place in black plastic garbage bag and seal. Remove fire fighter's protective clothing (except boots) and gloves. Place in orange plastic garbage bag and seal. Remove any portable radio worn. Place in black plastic garbage bag and seal. Discard surgical masks.
5. Do not smoke, eat, drink, or touch face.
6. Before leaving the scene, a fire fighter wearing SCBA should attempt to spray as much of the ground exposed to the material and the wash-down water as possible with bleach solution. Then flush the outside of the spray can with clean water.
7. Before leaving the scene, seal the orange garbage bags into the sterilization bags.

On Return to Station

8. Place apparatus temporarily out of service.
9. One fire fighter should dress in protective clothing and SCBA, and in an outside area perform the following tasks:

• Open the black plastic garbage bags, wipe all helmets, portable radios, SCBA sets, and used cylinders with a rag lightly dampened with a 6% bleach solution. After 10 minutes, wipe these items again with a rag dampened with clean water. If using a commercial sterilizer, follow the manufacturer's directions.
• Seal all used black garbage bags and rags into another bag and put out for normal garbage pick-up. If using bleach, empty the spray can and flush out to remove bleach residue.

10. Remove all clothing worn at the scene, including underwear, and place in garbage bag for laundering and/or dry cleaning (preferably the latter). Take all garbage bags with contaminated clothing to a place where they can be cleaned separately from other garments.
11. All personnel should shower, scrubbing all of the body with soap and water, with particular emphasis on areas around the mouth and nostrils and under fingernails. Shampoo hair and thoroughly clean mustache if you have one.
12. Do not smoke, eat, drink, touch face, or void until step #11 is completed.
13. Put on clean clothes. Place apparatus back in service when decontamination is completed.
14. Have cleaned firehose and SCBA checked by competent personnel before placing it back in service.

15. Arrange for the sterilization bags to be taken to a hospital laundry facility for cleaning and sterilization of the protective clothing, gloves, and any other garments sent in.

Reminder

Black garbage bags are to be used for items retained at the station. Orange bags are for items sent away for sterilization.

To Change SCBA Cylinders at the Scene

Flush empty cylinder and surrounding area of fire fighter's back with copious amounts of low pressure water. Also flush facepiece and breathing tube to prevent inhalation of harmful material when regulator is disconnected.

Wear gauntlet-type rubber gloves, such as those used by linemen, when changing cylinders. Flush gloves after use before removing them.

Place empty cylinder in black plastic garbage bag and seal for subsequent decontamination.

The person doing the flushing and cylinder-changing must wear protective clothing and SCBA.

Level R for Radioactive Hazards

At the Scene

1. *Preparation*
   A) Mark off a decontamination area with two parts.
   B) Make up a solution of detergent and water. Obtain scrub brushes.
   C) Set out a reserve air supply, preferably with a workline unit or otherwise with a spare SCBA.
   D) In the first part of the decontamination area, set up a runoff capturing method, either with wading pools or through the use of tarpaulins.
   E) If appropriate, a "walkway" of polyethylene sheeting (weighted down if necessary) can be placed from the exit from the incident scene to the decontamination area, to prevent possible contamination of the ground.

2. The decontamination crew will don SCBA and, where available, disposable chemical suits.

3. The fire fighters suspected of being contaminated will be scanned carefully with a radiation monitor suitable for detecting surface contamination. All parts of their protective clothing and personal equipment will

be scanned, including the soles of the boots. If no readings are found, the personnel that have been checked can leave the decontamination area.

4. Personnel found to be contaminated will be scrubbed down thoroughly with the detergent solution by the decontamination crew. This is followed by a flushing off using low pressure water. Efforts should be made to capture the runoff.

5. The fire fighters will then move to the second part of the decontamination area, where they will be scanned again with the radiation monitor. If any readings are found, they will return to the first part of the decontamination area and step 4 will be repeated.

6. When all personnel have been cleaned of contamination, the decontamination crew themselves will be hosed down. The matter of the captured runoff water will be discussed with environmental authorities and disposal arranged in a manner acceptable to them.

7. In the event fire fighters being decontaminated run out of breathing air, the reserve supply set out in step 1 will be passed to them. They should hold their breath while changing facepieces.

8. In the event that, despite repeated scrubbing, any fire fighters cannot be decontaminated, they will remove as much of their clothing as possible in the second part of the decontamination area, and don clean or spare clothing. The clothing that has been taken off will be sealed into garbage bags and returned to the station. This evolution must be executed in such a manner as not to contaminate the clean clothing.

9. Any equipment suspected or known to be contaminated will be sealed into garbage bags and returned to the station.

On Return to Station

Follow the Level "C" procedure steps 5 to 12 for those fire fighters who were found to be contaminated in step 3 above, and for any contaminated equipment.

To Change SCBA Cylinders at the Scene

Personnel emerging from the incident to have their breathing apparatus cylinder changed will be scanned with a radiation contamination monitor in a manner identical to step 3 above.

If no readings are found, the fire fighter can proceed to the SCBA cylinder change area and may then return to the incident with a fresh cylinder.

Personnel found to be contaminated may not return to the incident. They will be put through the full Level "R" decontamination procedure, and other fire fighters will be sent in to the incident to replace the fire fighters withdrawn.

Before the replacement fire fighters go in, they should attempt to obtain information as to where the other personnel might have received their contamination, in order to allow them to take the necessary caution when approaching that area.

NOTE: Steps 1 and 2 of the Level "R" procedure must be in place by the time the first fire fighter emerges from the incident. If circumstances permit, these preparations should be made before personnel even enter the incident area for the first time.

Decontamination—Specific Observations

Pre-incident Planning

Review the procedures and, if they are suitable for your location, assemble the equipment necessary into an easily transported container. Some departments, for instance, have all the special items needed for etiologic decontamination carried in a "Level E Decontamination Kit."

Many departments will have infrequent need to use these procedures. To prevent skill decay, and to prevent certain critical steps in the procedures being accidentally left out, it is suggested that a copy of the procedures be available at the scene and that regular training in the procedures take place. Executing these procedures accurately is not as easy as it would seem.

The time when you have twenty garbage bags with contaminated clothing sitting on your apparatus floor is not the time to start looking for a laundry that will clean them. Most commercial cleaning companies will not be interested in handling contaminated clothing.

Furthermore, it should be recognized that at some incidents the nature or extent of the contamination may be such that full decontamination is beyond the resources of the fire department (especially with Levels C, E, and R) and will require specialist treatment. With these three levels, consideration should be given to the destruction of all permeable items in case of serious exposure.

You should therefore make prior arrangements for the following:

* Obtaining steel drums at any time of the day or night. The drums must be clean and must have a removable lid—not just a bung and vent-hole.
* Analysis and expert decontamination of equipment and clothing contaminated by severely hazardous substances. This is needed for Level C and Level R, although different companies are likely to be needed for the two levels.

- Acceptable methods of disposal for items that cannot be cleaned, or that would be uneconomic to attempt to clean, for Level C, E, and R contaminants.
- The use of a hospital laundry service to perform Level E decontamination on protective clothing. This laundry should be approached for the loan of a number of sterilization bags, which are typically used in the hospitals to put dirty laundry in for shipment to the laundry service. Check that the hospital laundry service can take bunker coats—in some cases the buckles may bash the inside of their machines too much.
- Check the availability of replacement protective clothing and equipment that can be used while the original items are out being decontaminated under Levels C, E, or R.

You may want to establish a policy regarding personal items such as rings, wallets, watches, etc. Many of these, especially leather items, cannot be decontaminated and may have to be destroyed. Fire fighters should be aware of their department's policy with regard to recompense or replacement.

One further item of preplanning will always stand you in good stead; note the names and contact numbers of any local experts who could assist and advise you during the incident and its subsequent decontamination.

Plastic Bags

Notwithstanding the fact that throughout the preceeding procedures "black" garbage bags have been mentioned, there is a distinct advantage to using *clear*, plastic bags. These will permit the contents to be identified without opening them.

They should be at least 6 mil gauge and should be large enough that they can also be used as drum-liner bags.

Decontamination Area Layout

When choosing the location of the decontamination area, consider the following:

- Prevailing weather conditions (temperature, precipitation, etc.),
- Wind direction,
- Slope of the ground,
- Surface material and porosity (grass, gravel, asphalt, etc.),
- Availability of water,
- Availability of power and lighting,
- Proximity of the incident,

- Location of drains, sewers, and watercourses.

When setting up the area, provide the following features:

- Containment of wash-down water if that is necessary,
- Spare supply of breathing air (extra SCBA, extra cylinders, or workline units),
- A supply of industrial-strength garbage bags, double- or triple-bagged if necessary,
- Clearly marked boundaries, not just a rope lying on the ground,
- Clearly marked entry and exit points with the exit upwind, away from the incident and its contaminated area,
- A waiting location at the entry point where contaminated personnel can await their turn without spreading contamination further,
- Access to triage and other medical aid upon exit if necessary,
- Protection of personnel from adverse weather conditions,
- Security and control from the setting up of the area to final clean-up of the site.

Environmental Considerations

One fundamental concept forms the basis for these decontamination procedures: "The human being comes before the environment."

Notwithstanding the above, where containment of run-off is called for, genuine attempts must be made if only to avoid possible legal consequences. Examples of containment basins are:

- Children's wading pools,
- Portable tanks (as used in rural fire fighting),
- Tarps laid over a square formed by hard suction hose or small ground ladders,
- Diking with earth, sandbags, etc. covered with tarps.

Fire fighters stepping out of a containment basin should lift one foot, have it rinsed off so the water falls inside the basin, step out with that foot, and repeat for the other foot.

When the containment basin is full, it should be able to be siphoned or pumped off into drums or into a vacuum truck for controlled disposal in a manner acceptable to the authority having jurisdiction.

Any run-off that is not contained will eventually enter sewers and watercourses, or if it sinks into the ground will ultimately reach the water-table.

The Department of Mechanical and Fluid Engineering at Leeds (U.K.) University has determined that provided a chemical is diluted with water at the rate of approximately 2000:1, pollution of water-courses will be significantly reduced.

There is also a change in attitude coming with the environmental authorities, whereby they recognize that the small amount of chemical likely to be washed off contaminated fire fighters with adequate dilution will result in minimum damage to the environment, especially when compared to the results of the spill that generally led to the personnel contamination in the first place.

Any substances that enter sewers and water-courses should be reported to environmental authorities and to the sewage treatment plant likely to receive it. If necessary, advise water authorities downstream from the decontamination area of actual or potential pollution.

The most appropriate decontamination for materials that have a severe effect on the environment will usually be to use minimal amounts of water, with run-off containment. Other substances should be deluged off personnel with the 2000:1 factor as a minimum guideline.

Weather

If decontamination is done indoors because of bad weather, ensure that the drains go into a holding tank and not directly into the sewers.

If the hazardous material involved requires Level D decontamination, and it is raining or snowing, protect fire fighters from the precipitation until they have been processed.

Take care when using instruments in wet weather. Extreme cold may affect the operating effectiveness of instruments, especially delicate ones originally designed for use in laboratory environments.

Under extreme weather conditions (heat or cold), decontamination personnel must be rotated more frequently.

These decontamination procedures should be reviewed in light of your local climate and adapted if necessary where your area's weather conditions dictate.

Fluid Replacement

At hazardous materials incidents, especially when chemical suits are worn, serious dehydration can occur in fire fighters. Replacement of fluids should only be permitted if at least gross decontamination is performed—a washdown especially around the head and upper body.

The preferable method of consuming liquids is by means of drinking boxes with straws (the straw inserted by someone with uncontaminated hands), or by means of a squeeze bottle with an attached drinking tube as used by athletes.

The above should form part of comprehensive rehabilitation procedures which should be developed in consultation with your EMS providers.

Chemical Suit Decontamination

When a chemical suit is taken off its wearer, a suitably protected assistant should roll it in on itself in order to keep the outside of the suit from coming into contact with the wearer.

Because of the inherent smoothness and impermeability of chemical suits, it is usually only required that the on-scene washdown part of fire fighter decontamination is performed. Upon return to the station, instead of doing the steps listed in the appropriate procedure, fire fighters should wash and rinse the chemical suits and examine them carefully for damage caused at the incident. Zippers should be lubricated with their special lubricant.

Follow-up communication with the suit manufacturer as to the exposure, as well as follow-up from the exposing chemical's manufacturer, is useful in determining long-term effect of exposure to chemical-protective ensembles. Any questionable or unusual findings anywhere in the decontamination or testing process should be immediately referred to the manufacturers; the clothing should be placed out of service until it can be repaired or reevaluated. If a limited use (disposable) suit becomes contaminated, after gross decontamination it should be bagged and disposed of in a manner acceptable to the authority having jurisdiction.

Vacuum Cleaners for Level D

When selecting a vacuum cleaner, the following points should be taken into consideration:

• Can it operate off a generator, or is it unforgiving so far as voltage fluctuations are concerned?
• Will it operate safely in an area where it might get wet?
• How effective are the filters?
• Can the unit be safely cleaned out itself?
• Are replacement hoses easy to come by?

Although you won't want to operate your vacuum cleaner under water, it might accidentally get splashed so some basic water protection will be of benefit.

The degree of filtering achieved is important. Most wet/dry industrial type vacuums will achieve a reasonable effectiveness. Some specialized cleaners, equipped with HEPA (High Efficiency Particulate Air) filters will go down to 0.3 microns, but they are expensive. The small, cigarette-lighter plug powered car interior cleaners are not suitable as they filter very little, instead blowing most particulate they pick up back out through their exhaust ports.

Easy removal of contaminated filters will help, as will good access to the machine's insides for its own decontamination. You will usually find that it is impossible to guarantee the effectiveness of cleaning of the accordion-style hose, and you should consider replacing these if they become contaminated.

Remember not to operate a vacuum cleaner in a flammable or explosive atmosphere, unless yours is intrinsically safe.

Bleach

Bleach, as shown on the bottle's label, is corrosive. Do not spray bleach on fire fighters' skin—it hurts! It is also reactive—do not let it come in contact with fuels and solvents, as a heat-generating reaction (and possible fire) will result.

Do not spray bleach on protective clothing, as it deteriorates and discolors the garment. It also impairs its fire retardancy.

Regular bleach containers are often made of a hard plastic and are liable to crack or leak around the cap when carried on a vehicle. Consider transferring the bleach to a container such as a new, unused plastic gasoline can, which is far sturdier, but then be sure to label the gas can as containing bleach. Never put bleach in a metal container as it will react.

At room temperature, bleach shielded from sunlight will degrade about 1% per year, i.e., from 6% to 5% (faster in warmer temperatures, slower at colder temperatures). You should therefore replace the bleach at the appropriate time with a fresh supply, as its strength will have decreased with time.

It is recommended that, if possible, non-chlorinated bleach be used in case the bleach accidentally contacts the protective clothing.

Record Keeping

A member of the crew responsible for performing the decontamination should maintain written records of the following:

- Fire fighter's name, material involved, length of exposure,
- Level of decontamination performed,
- Any ill effects observed,
- Where fire fighter went, i.e.:

- returned to station
- sent to rest area
- removed to hospital
- reassigned to other duties at the scene
- etc.

At the station, entries should be made on the fire fighters' medical records of the incident date, material involved, and decontamination performed, where exposure is known or suspected. This will assist both in tracing future sickness through synergistic effects of chemicals in the body and with support of any later injury or sickness claims.

If appropriate, records should also be kept of the length of time each chemical suit was exposed, and what substance it was exposed to. This will permit the tracking of cumulative degradation of the suit material due to exposure to a variety of chemicals or due to repeated exposure to one particular substance.

Contamination of Vehicles

Any vehicle driven through a contaminated area must be washed down, including the undercarriage, chassis, and cab. Air filters on vehicle (and, where appropriate, generator) engines must be replaced. Porous items such as wooden hose beds, wooden equipment handles, seats, and cotton jacketed hose may be difficult to clean completely and may have to be discarded.

It is therefore better to take the "uphill and upwind" approach and keep vehicles at a suitable distance from incidents.

Precautionary Decontamination

There are occasions when an apparently normal alarm response turns into a hazardous materials incident. Frequently, most of the initial assignment crews will have gone into the incident area, and have the potential of being contaminated.

It is essential that all members so involved remove themselves from the area at once, call for decon capability, and stay together in one location. They must not wander around, climb on and off apparatus, and mix with other personnel since there is a potential for them to be contaminated.

Fire fighters so exposed should be given "gross decontamination" (a simple on-scene washdown) as a precautionary measure. Knowledgeable hazmat personnel such as the decontamination sector officer, in conjunction with the incident commander, should determine whether any further, more definitive decontamination is necessary.

Remember, the primary objective of decontamination must be to avoid contaminating anyone or anything beyond the hot zone. When in doubt about contamination, decon all affected personnel, equipment, and apparatus.

One Final Observation

The entire foregoing contents of this document have probably made you realize by now that it is much more desirable to handle hazardous materials incidents with chemical suits than with regular fire fighting turnouts. The cost of disposable suits is relatively cheap; even for a small department, throwing away a few hundred dollars' worth of disposable suits after one use will be cheaper than replacing fire fighting clothing or paying for commercial cleaning. In many jurisdictions, the fire service is permitted to recover the cost of destroyed equipment from the party responsible for the incident's occurrence, and the cost of disposable chemical suits can thus be recovered.

Always remember: if the emergency response crew is not equipped with gear suitable for entry into a hazardous or toxic atmosphere, then the option of "no go" should be considered the most appropriate tactic.

Speed—A Case for Exception?

Decontamination should emphasize thoroughness, not speed. Under noncritical conditions certain commonsense actions should be taken, such as decontaminating the fire fighter with the lowest air reserve first.

Speed is only important where a victim is involved and even then decontamination should be as thorough as is practicable.

Circumstances may dictate that emergency decontamination becomes necessary, examples of such situations being where a protective suit has become split or damaged, or when a fire fighter is injured. Emergency decontamination may also be applicable when contaminated civilians or other emergency workers (police, ambulance, etc.) are involved.

Emergency Decontamination Procedure

Paragraphs 1 to 6 below, although arranged in a basic chronological order, do not necessarily have to be undertaken in the exact sequence outlined. The officer-in-charge should act in the most expedient manner appropriate without worsening the situation.

The procedure outlined should be carried out as quickly as possible.

To protect the ambulance crew and hospital staff as well as the victim, every attempt must be made to perform at least this emergency procedure prior to transporting the victim to the hospital.

1. Remove the victim from the contaminated area into the decontamination zone and provide a supply of uncontaminated air or oxygen.
2. Remove fire helmet if worn and immediately wash with flooding quantities of water any exposed parts of the body that may have been contaminated.
3. If the victim is wearing SCBA, release the harness and remove the set leaving the face mask in position.
4. Remove all contaminated clothing (if necessary by cutting it off the victim) ensuring where practicable that the victim does not come into further contact with any contaminant. Maintain the washing of the victim while the clothing removal is taking place.
5. Remove the victim to a clean area. Render first aid as required, but do not apply mouth-to-mouth resuscitation. Send victim for medical treatment as soon as this emergency decontamination procedure has been completed.
6. Ensure hospital/ambulance personnel are informed of the contaminant involved.

Supplement 9: Packaging for the Transportation of Hazardous Materials

The emergency responder should understand thoroughly how hazardous materials are packaged. This supplement, a Technical Bulletin published in June 1989 by the Chemical Manufacturers Association, Association of American Railroads, and the Bureau of Explosives, provides an excellent summary of these packaging methods.

I: Overview

Purpose

The purpose of this Technical Bulletin is to assist emergency response personnel in identifying and describing packaging used to transport hazardous and non-hazardous materials.

This bulletin does not contain descriptions or illustrations of vessels used to transport materials by water (e.g., barges, tank ships, and bulk cargo vessels).

Illustrations throughout the bulletin are examples of the packaging being discussed — other designs may exist.

Definitions

For the purpose of this bulletin, the following definitions are presented:

Package

The word package means a packaging and its contents.

Packaging

Packaging (non-bulk or bulk) may be singular or plural and means anything that contains a material.

Non-Bulk Packaging

Non-bulk packaging is any packaging having a capacity meeting one of the following criteria:

1. Liquid — internal volume of 118.9 gallons (450 liters) or less;
2. Solid — capacity of 881.8 pounds (400 kilograms) or less; or
3. Compressed Gas — water capacity of 1000 pounds (453.6 kilograms) or less.

Non-bulk packaging may be single packaging (e.g., drum, carboy, cylinder) or combination packaging consisting of one or more inner packagings inside an outer packaging (e.g., glass bottles inside a fiberboard box). Non-bulk packaging may be palletized or placed in overpacks for transport in various transport vehicles, vessels, and freight containers. Examples of non-bulk packaging are bags, bottles, boxes, carboys, cylinders, drums, jerricans, and wooden barrels.

Bulk Packaging

Bulk packaging is any packaging, including transport vehicles, having a capacity greater than described under non-bulk packaging. Bulk packaging is further divided into two distinct types.

1. Bulk packaging that is placed on or in a transport vehicle or vessel for transportation using a crane, hoist, forklift, etc., for loading and unloading. Examples of this type bulk packaging are bulk bags and boxes, portable bins, portable tanks, intermodal portable tanks, and ton containers; and

2. Bulk packaging that is an integral part of the transport vehicle. Examples of this type bulk packaging are tank trucks, tank trailers, hopper trailers, tank cars, and hopper cars.

II: Non-Bulk Packaging

This Section addresses non-bulk packagings that may be a single packaging (e.g., drum, carboy, cylinder) or a combination packaging consisting of one or more inner packagings inside an outer packaging (e.g., glass bottles inside a fiberboard box).

The non-bulk packaging addressed includes:

- bags;
- bottles;
- boxes;
- multicell packaging;
- carboys;

- cylinders;
- drums;
- jerricans; and
- wooden barrels.

Bags

Bags are flexible packaging constructed of materials such as cloth, burlap, kraft paper, plastic, or a combination of these materials. They are mainly used for solid materials. They are generally enclosed on all sides except one, which forms an opening that may or may not be sealed after filling. Bags are closed by folding and gluing, heat sealing, tuck-in or self-closing sleeves,

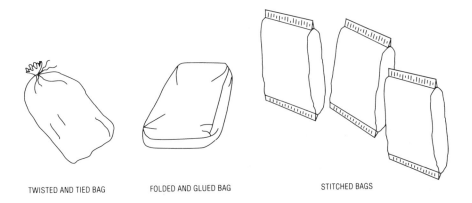

TWISTED AND TIED BAG   FOLDED AND GLUED BAG   STITCHED BAGS

stitching, crimping with metal, or twisting and tying. They typically contain up to 100 pounds of material and are usually shipped palletized. Examples of materials shipped in bags are cement, fertilizers, and pesticides.

Bottles

Bottles, sometimes referred to as jugs or jars, are used for liquids and solids. Most bottles are made of glass or plastic; however, metal and ceramic are sometimes used. Bottles are closed with threaded caps or stoppers. They range in capacity from ounces to 20 gallons or more (see carboy). Bottles are usually placed in an outside packaging (like fiberboard or wooden boxes) for transport. Examples of materials shipped in bottles are antifreeze, laboratory reagents, and certain corrosive liquids.

PROTECTED BOTTLE   PLASTIC BOTTLE   GLASS BOTTLE

Boxes

Boxes are rigid packaging having faces that completely enclose the contents. They are commonly used as the outside packaging for other non-bulk packages, including aerosol containers, bottles, and cans. Boxes may be made from fiberboard, wood, metal, plywood, reconstituted wood, plastic, or other suitable materials. Fiberboard boxes may contain up to 65 pounds of material, while wooden boxes may contain up to 550 pounds of material. Some boxes have inner packages that hold from one to nine gallons of material such as battery acid, laboratory reagents, and asphalt driveway crack sealers. Some boxes contain an inner package surrounded with absorbent and/or cushioning material, such as vermiculite.

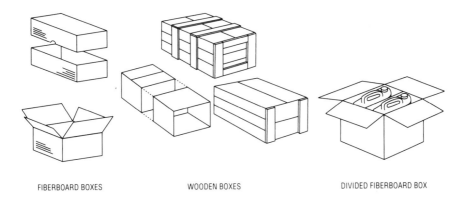

FIBERBOARD BOXES   WOODEN BOXES   DIVIDED FIBERBOARD BOX

Multicell Packaging

Multicell packaging consists of a formfitting, expanded polystyrene box encasing one or more bottles. The polystyrene box may be shipped with the two parts banded together, or packed in a box. When transporting certain Department of Transportation (DOT) hazardous materials, the maximum bottle capacity is 4 liters (just over 1 gallon); and up to six of these bottles may be placed in one multicell packaging. Examples of materials shipped in multicell packaging are specialty chemicals for the electronics industry, like hydrochloric and sulfuric acids, and various solvents.

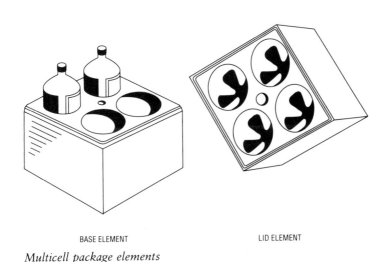

BASE ELEMENT   LID ELEMENT

Multicell package elements

Carboys

Carboys, used for liquids, are glass or plastic "bottles" that may be encased in an outer packaging such as expanded polystyrene boxes, wooden crates, or plywood drums. Carboys range in capacity to over 20 gallons. Examples of materials shipped in carboys are sulfuric acid, hydrochloric acid, ammonium hydroxide, and water.

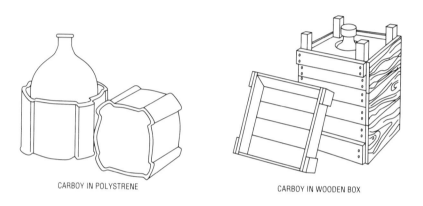

CARBOY IN POLYSTRENE   CARBOY IN WOODEN BOX

Cylinders

Cylinders normally contain liquefied, non-liquefied, dissolved gases, or mixtures thereof, but may also contain liquids or solids. Some cylinders will contain "gas only," while others contain aerosols (a mixture of compressed gas and other materials, like toiletries, spray paints, or whipping cream). Cylinders range in size from aerosol containers found in the home, such as spray deodorant, to the cryogenic (insulated) cylinders for liquid nitrogen that are approximately 24 inches in diameter and 5 feet high. Cylinders may be shipped either with or without an outer packaging, such as a box. Service pressures range from a few pounds per square inch to several thousand pounds per square inch.

All cylinders have a circular cross section with a valve or valve arrangement at one end of the cylinder. Some small cylinders have seals in place of the valve and are meant to be used with equipment having a valve arrangement. The majority of cylinders, other than those used for consumer commodities, have a pressure relief device (e.g., safety relief valve, rupture disc, or fusible plug that is designed to melt under fire conditions).

NOTE: Some cylinders, especially those containing oxygen and acetylene, may be interconnected (manifolded) and securely racked together with individual cylinder valves closed during transportation.

There are three basic types of cylinders:

- aerosol containers;
- uninsulated cylinders; and
- cryogenic (insulated) cylinders.

Aerosol Containers

Aerosol containers are small cylinders made of metal, glass, or plastic. Their ends may be flat or rounded, either inward or outward. Aerosol containers are transported in boxes. Examples of materials shipped in aerosol containers are cleaners, lubricants, paint, and toiletries.

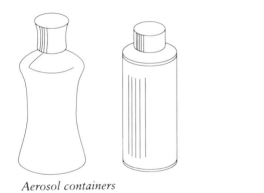

Aerosol containers   *Cryogenic cylinder*

Cryogenic (Insulated) Cylinders

Cryogenic (insulated) cylinders for cryogenic liquids consist of an insulated metal cylinder contained within an outer protective metal jacket. The area between the cylinder and the jacket is normally under vacuum. They are designed for a specific range of service pressures and temperatures. Cryogenic cylinders have a small protective ring at the top to protect the valves and a foot ring (a slight narrowing of the cylinder just above the bottom) that allows for handling with a special hand truck. These cylinders range in size up to 24 inches in diameter and 5 feet high. Examples of materials found in cryogenic cylinders are the cryogenic liquids argon, helium, nitrogen, and oxygen.

Uninsulated Cylinders

Uninsulated cylinders are typically made of steel although some, like self-contained breathing apparatus cylinders, are made of aluminum or fiberglass wrapped aluminum. They have rounded shoulders on top and screw-on caps or cylindrical rings to protect the valve. Uninsulated cylinders are typically up to 10 inches in diameter and 5 feet high. Examples of materials shipped in uninsulated cylinders are acetylene, gaseous nitrogen, liquefied petroleum gas, and oxygen.

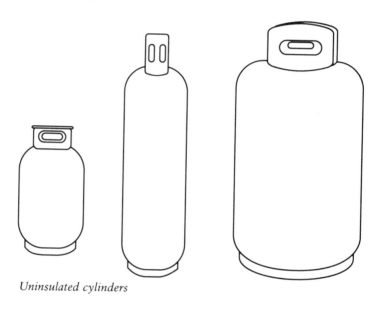

Uninsulated cylinders

Drums

Drums, used for liquids and solids, are usually cylindrical packagings made of metal, plastic, fiberboard, plywood, or other suitable materials. Drums are sometimes called buckets, cans, or pails. Typical drum capacities range up to 55 gallons. Metal and plastic drums can vary in size up to 23 inches in diameter and 34 inches high. Fiber (fibre) drums range from 8 inches in diameter and 4 inches high to 24 inches in diameter and 43 inches high. The most common salvage drums (or overpack drums), used to hold damaged or leaking non-bulk packaging for repacking or disposal, have a capacity of 85 gallons.

Drums have either removable or non-removable heads, referred to as "open head" and "tight or closed head" respectively. Removable heads are attached to a drum by a separate ring or built-in lugs. Some fiber drums have slip-on covers. Drums may have liners or linings depending on the contents.

The common metal, 55 gallon, closed head drum is approximately 23 inches in diameter and 34 inches high. The heads are joined to the body by folding the sheets together in a "chime." The chime is a metal ring around the top and bottom of the sidewall. The sides have two or more ridges, called "rolling hoops," for strength. A tight head drum usually contains two openings — one 2 inches in diameter and the other ¼ inch in diameter — closed with plugs called "bungs." A thin metal or plastic "tamper-proof seal," or "weather cap," is sometimes crimped over these "bungs."

Drums containing certain materials, such as hydrogen peroxide, have closures (bungs) in the head designed to vent pressure.

Virtually any liquid or solid material can be shipped in drums. Examples range from lubricating grease to acids and poisons.

5 GALLON DRUM
(PAIL, BUCKET, CAN)

METAL OPEN HEAD DRUM

TIGHT OR CLOSED HEAD METAL DRUMS

OPEN HEAD
PLASTIC DRUM

TIGHT OR CLOSED HEAD
PLASTIC DRUM

FIBER DRUMS

PLYWOO DRUM

Jerricans

Jerricans, used for liquids, are metal or plastic packagings of rectangular or polygonal cross section. The maximum capacity of a jerrican is approximately 15.8 gallons (60 liters). Examples of materials shipped in jerricans are antifreeze and other specialty products.

A jerrican

Wooden barrel or keg

Wooden Barrels

Wooden barrels, sometimes called "kegs," are used for liquids and solids. Made of natural wood, barrels have a circular cross section with convex walls consisting of staves and fitted with hoops. The barrel hoops are usually made of steel, iron, or suitable hardwood. The maximum capacity of a barrel is about 66 gallons (250 liters). An example of a material shipped in wooden barrels is distilled spirits.

III: Bulk Packaging that Is Placed on or in a Transport Vehicle or Vessel

This Section addresses bulk packaging that is placed on or in a transport vehicle or vessel using a crane, hoist, forklift, etc., for loading and unloading, including:

• bulk bags;
• bulk boxes;

- palletized non-bulk packages;
- portable tanks (intermodal tank containers and portable bins);
- ton containers; and
- protective overpacks for radioactive materials (also casks).

The type of bulk packaging addressed in this Section is often called an "intermediate bulk container" (IBC). Portable tanks over 118.9 gallons (450 liters), or less than 732.5 gallons (3,000 liters), and portable bins are often referred to as "rigid intermediate bulk containers" (RIBCs). Bulk bags and similar packaging are described as "flexible intermediate bulk containers" (FIBCs).

Bulk Bags

Bulk bags, used for solids, are preformed packaging made of flexible materials, e.g., woven polypropylene, and are available plain, coated, or with liners. Standard sizes range from 15 to 85 cubic feet with capacities varying from 500 to 5000 pounds. Bulk bags are transported in a variety of open and closed transport vehicles including rail box cars, intermodal containers, and box (van) trailers. Examples of materials shipped in bulk bags are fertilizers, pesticides, and water treatment chemicals.

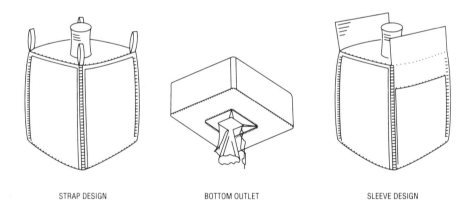

STRAP DESIGN   BOTTOM OUTLET   SLEEVE DESIGN

Bulk Boxes

Bulk boxes, used for solids, are rigid packagings usually made of multiwall fiberboard or plywood having closed faces that completely enclose the contents. Bulk boxes may contain up to 4000 pounds of materials and are designed to be shipped on a pallet. Bulk boxes are shipped in box (van) trailers and rail box cars. An example of a material shipped in bulk boxes is plastic pellets.

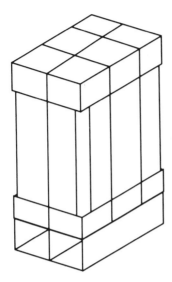

Bulk box

Palletized Non-bulk Packages

Non-bulk packages are often secured together on a pallet (typically 40 inches by 48 inches in size) for ease and safety of handling. This procedure, called "palletization," can be accomplished in several ways. The packages may be placed in a large corrugated box the size of the pallet; fastened together with plastic or steel banding; glued with a non-slip glue; held with a plastic film that is stretched or heat shrunk around the packages and pallet; or covered by a plastic shroud called a slip cover. Loaded pallets require the use of a forklift or other device for loading into and unloading out of the transport vehicle.

LOOSE BAGS ON PALLET   55 GALLON DRUMS ON PALLET   5 GALLON PAILS ON PALLET

Portable Tanks and Bins

Portable tanks and bins transport bulk gases, liquids, and solids. They are rigid bulk packaging equipped with skids, frames, or other mountings to facilitate handling by mechanical means. The tanks are made of metal or plastic with circular or rectangular cross sections with capacities from 118.9 gallons to 6300 gallons or more. Some portable tanks are insulated and some are equipped with steam and/or electric heating. They are transported in/on flat bed or van type trucks or trailers, box cars or flat cars, and vessels.

Definitions

Intermodal Tank Containers: Intermodal tank containers are portable tanks that are enclosed in a sturdy, metal supporting frame (either box type or beam type). This frame, built to international standards, is typically 8 by 8 by 20 feet, although there are some 40-foot frames. They may be transported by highway, rail, and water.

Portable Bins: Portable bins are portable tanks used to transport bulk solids. They are approximately 4 feet square and 6 feet high and may contain up to 7700 pounds. Most portable bins are loaded through the top and unloaded from the side or bottom. Dump type portable bins are shipped on flat bed trucks and trailers in agricultural areas and are used to transport ammonium nitrate fertilizer.

Types

Portable tanks can be described as one of three basic types:

* non-pressure;
* pressure; and
* specialized (including cryogenic portable tanks and tube modules).

Portable Tanks: Non-pressure		Pressure	Specialized
IMO-101	IMO Type 1	DOT Spec 51	Cryogenic
IM-102	IMO Type 2	IMO Type 5	IMO Type 7
DOT Spec 56	IMO Type 0		Tube module
DOT Spec 57			

Non-pressure Portable Tanks: Non-pressure portable tanks transport liquids and solids. These tanks may have rectangular, oval, or circular cross sections. They have capacities that generally do not exceed 6300 gallons (24,000 liters).

Although referred to as non-pressure, these tanks may contain internal pressures up to 100 psig. For example, the maximum allowable working pressure for IM 101 portable tanks ranges from 25.4 to 100 psig while the range for IM 102 is 14.5 to 25.4 psig. Examples of materials shipped in non-pressure portable tanks include food grade commodities, liquid fertilizers, resins, sodium cyanide, water treatment chemicals, and whiskey.

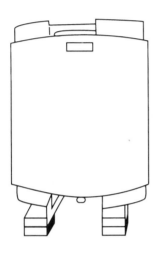 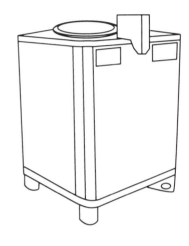

PORTABLE TANK   PORTABLE BIN

Non-pressure portable tank and bin

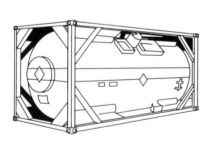 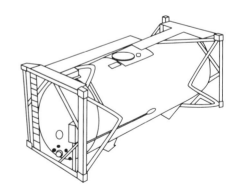

BOX TYPE FRAME   BEAM TYPE FRAME

Intermodal portable tanks

Pressure Portable Tanks: Pressure portable tanks transport liquefied compressed gases and liquids. The tanks have a circular cross section and may be as large as 6 feet in diameter and 20 feet long. Fittings for pressure intermodal tank containers are protected and found on the top, end, or bottom of the tank. Capacities range up to 5500 gallons with service pressures ranging from 100 to 500 psi. Examples of materials shipped in pressure portable tanks are anhydrous ammonia, bromine, liquefied petroleum gas (LPG), and sodium.

Pressure portable tanks include domestic intermodal tank containers built to DOT Spec. 51 requirements (and mounted within a frame) and international IMO Type 5 tanks.

 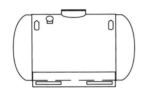 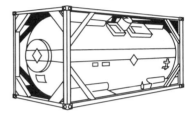

PRESSURE PORTABLE TANK   PRESSURE INTERMODAL TANK CONTAINER

Specialized Portable Tanks — Cryogenic: Cryogenic portable tanks transport cryogenic liquids and are approved for use by DOT special exemption. Containers built to international IMO Type 7 specifications are sometimes approved by DOT for cryogenic use. They consist of a tank-within-a-tank design with insulation between the inner and outer tanks. This space between the inner and outer tanks is normally maintained under vacuum. Examples of materials shipped in cryogenic liquid portable tanks are liquefied argon, ethylene, helium, nitrogen, and oxygen.

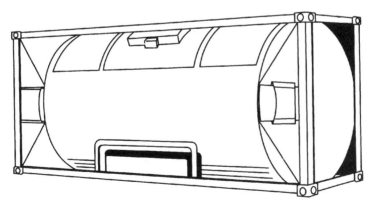

Cryogenic intermodal tank container

Specialized Portable Tanks — Tube Module: Tube modules, while not actually portable tanks, transport bulk gases. This rigid bulk packaging consists of several horizontal seamless steel cylinders, from 9 inches to 48 inches in diameter, permanently mounted inside an open frame with a box-like compartment at one end enclosing the valving. Service pressures range up to 2400 psi or more. Examples of materials shipped in tube modules include non-liquefied gases such as helium, nitrogen, and oxygen.

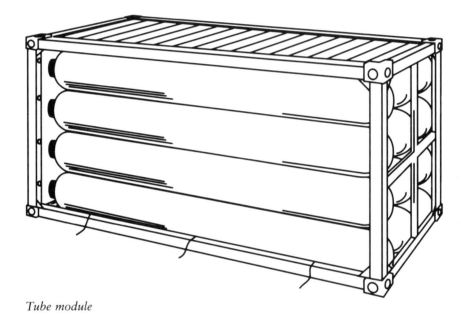

Tube module

Ton Containers

Ton containers, called multi-unit tank car tanks in Department of Transportation regulations, are rigid packaging that transport gases. They are cylindrical pressure tanks approximately 3 feet in diameter and 8 feet long with concave or convex heads. The name "ton container" comes from the packaging's capacity to transport one ton of chlorine. The ton container's valves are found at one end under a protective cap. Ton containers are transported on specially designed flat bed trailers and flat cars with racks and securing devices, and they are also transported in other motor vehicles and rail freight cars. Examples of materials shipped in ton containers are chlorine, phosgene, and sulfur dioxide.

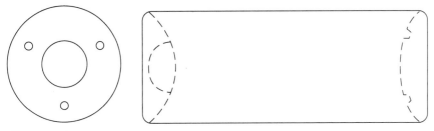

Ton container

Protective Overpacks and Casks for Radioactive Materials

Radioactive materials are found in all kinds of packaging. Some types of radioactive materials are found in protective overpacks and casks.

Protective Overpacks

Protective overpacks containing inner packages of radioactive materials are rigid packagings with either a cylindrical or box-like configuration.

Cylindrical overpacks are made of laminated or solid wood and may be covered with steel. These overpacks may have stiffener rings around the sides. They may weigh up to 6000 pounds when loaded.

The box-like overpack consists of two nested plywood boxes enclosed in a solid wooden box which is further reinforced with steel bars. These overpacks are painted with a paint that swells and forms a protective char when exposed to fire. They may weigh up to 3000 pounds when loaded.

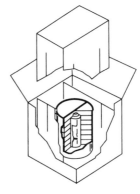

CYLINDRICAL   BOX-LIKE

Protective overpacks for radioactive materials

Casks

Casks transport certain radioactive materials. This rigid metal packaging ranges in size up to 10 feet in diameter and 50 feet long. Some casks have reinforcing rings or cooling fins.

Cask for radioactive materials

IV: Bulk Packaging that Is an Integral Part of a Transport Vehicle

This Section addresses a second type of bulk packaging that is an integral part of the transport vehicle. These transport vehicles include:

- motor vehicles (tank trucks, tank trailers, pneumatic hopper trailers); and
- rail freight equipment (covered hopper cars, tank cars, and gondolas).

Motor Vehicles

Bulk materials are transported by motor vehicle in cargo tanks and hoppers. Motor vehicles to which cargo tanks are permanently attached are called "tank trucks" and "tank trailers." Ores and some hazardous wastes, e.g., contaminated dirt, are transported in dump trucks and trailers.

Cargo Tanks (Tank Trucks and Tank Trailers)

Cargo tanks transport bulk liquefied and compressed gases, liquids, and molten materials. Cargo tanks are the tanks that are permanently mounted on a tank truck or tank trailer. However, a cargo tank may also be any bulk liquid or compressed gas packaging, not permanently attached to any motor vehicle, which by reason of its size, construction, or attachment to a motor vehicle can be loaded or unloaded without being removed from the motor vehicle.

Department of Transportation (DOT) cargo tanks are divided into three categories (other designs may exist):

* non-pressure;
* pressure; and
* specialized.

Cargo Tanks: Non-pressure		Pressure	Specialized
MC-303	MC-307	MC-330	Cryogenic
MC-304	MC-310	MC-331	MC-337
MC-305	MC-311		MC-338
MC-306	MC-312		Tub trailers

Non-pressure Cargo Tanks: Non-pressure cargo tanks transport liquids. They have either an oval or circular cross section. They are distinguished by a box-like structure and/or rollover (overturn) protection on top of the tank.

NOTE: Although referred to as non-pressure, these tanks may contain internal pressures up to 100 psig.

Non-pressure cargo tanks with an oval cross section typically transport materials that are lighter than water. This type cargo tank, generally made of aluminum, is designed for service pressures not to exceed 3 to 7 psig at 100°F. They have capacities up to 9000 gallons and may contain up to eight separate compartments. Examples of materials shipped in this type of non-pressure cargo tank are gasoline and solvents.

Non-pressure cargo tanks with a circular cross section typically transport materials that are heavier than water. Some of these tanks have external stiffener rings around the tank; however, these stiffener rings may be hidden by insulation and covered by an exterior jacket. Some insulated non-pressure cargo tanks do not appear circular in cross section, and, therefore, it is difficult to distinguish between cargo tanks transporting hazardous materials and those transporting milk. Non-pressure cargo tanks of this type are designed for pressures up to 25 psig with capacities up to 7000 gallons. This type cargo tank may have up to four separate compartments. Examples of materials shipped in this type non-pressure cargo tank are brine water, fertilizer solutions, and sulfuric acid.

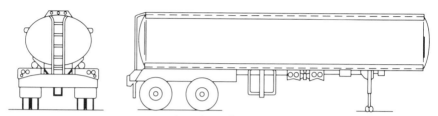

MC-306 non-pressure cargo tank with oval cross section

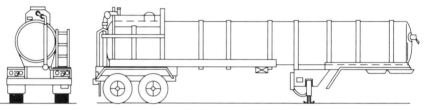

MC-312 non-pressure cargo tank with circular cross section

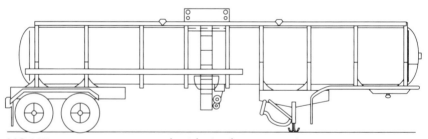

MC-307 non-pressure cargo tank with circular cross section

Some non-pressure cargo tanks have multiple compartments divided with bulkheads. The length of the box-like structure on top of the tank (for overturn or "rollover" protection) may help identify non-pressure cargo tanks with multiple compartments. If the box runs the length of the tank, the tank will generally have multiple compartments. If the box is only a few feet long, the tank will generally have only one compartment. Also, each compartment has a separate set of fill and delivery pipes and its own dome cover on top.

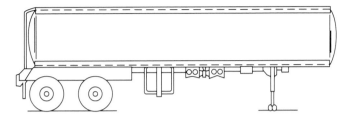

FULL LENGTH-MULTIPLE COMPARTMENTS

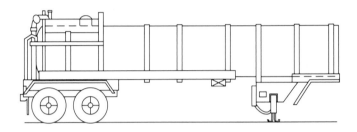

SHORT LENGTH-SINGLE COMPARTMENT

Overturn or "rollover" protection for single and multi-compartment cargo tanks

Pressure Cargo Tanks: Pressure cargo tanks transport liquefied and compressed gases. They have circular cross sections, approximately 8 feet in diameter, with hemispherical or ellipsoidal ends or heads. Some tank trucks have two 4-foot diameter tanks mounted side by side instead of the single larger diameter tank. Uninsulated pressure cargo tanks have the upper two-thirds painted white; others have a jacket of aluminum, stainless steel, or other bright, nontarnishing metal. Pressure cargo tanks are designed for service pressures from more than 100 psig to 500 psig. They have tank capacities up to 11,500 gallons. Examples of materials shipped in pressure cargo tanks are anhydrous ammonia and liquefied petroleum gas.

Pressure cargo tanks may have multiple compartments. The only distinguishing features of those tanks with multiple compartments are the separate temperature gauges for each compartment. Some tank trucks, built to transport liquefied petroleum gas, have two compartments. The smaller (front) compartment may be used as the fuel tank for the motor vehicle. In this case, there are no simple distinguishing features to indicate that these tanks have multiple compartments.

The smallest example of a pressure cargo tanks is the 500 to 1000 gallon nurse tank trailer of anhydrous ammonia found in agricultural areas.

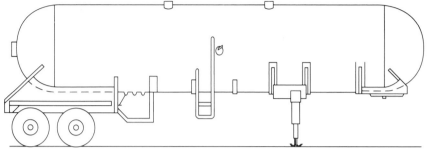

Pressure cargo tank trailer (MC-331)

Nurse tank trailer

Specialized Cargo Tanks — Cryogenic: Cryogenic cargo tanks transport cryogenic liquids. This type cargo tank is a tank within a tank. The space between the inner and outer tanks is filled with insulation and normally maintained under vacuum. The inner tank is circular in cross section and made of materials compatible with the product to be transported. They range in size from 500 to 14,000 gallons. The ends are dished. The valving is found in either a squared-off compartment on the back or mounted on the side just ahead of the trailer wheels. Design pressures range from 25.3 psig to 500 psig. Examples of materials shipped in cryogenic cargo tanks are liquefied argon, helium, hydrogen, and nitrogen.

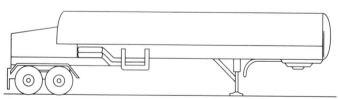

Cryogenic cargo tank

Specialized Cargo Tanks — Tube Trailer: Tube trailer is the common name for a semi-trailer that transports bulk non-liquefied compressed gases (although it is not a cargo tank). The tube trailer consists of a group of seamless steel cylinders, 9 to 48 inches in diameter, permanently mounted on a semi-trailer. The tube trailer may have as few as two large cylinders or more than twenty smaller cylinders. Cylinder service pressures range from 3000 to 5000 psi. Examples of materials shipped in tube trailers are helium, hydrogen, nitrogen, and oxygen. (All cylinders contain the same material.)

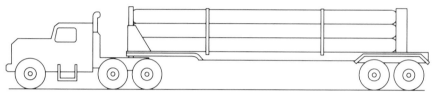

Tube trailer

Pneumatic Hopper Trailers: Covered hopper trailers that are pneumatically unloaded transport bulk solids. When viewed from the side, they have rounded sides and sloping ends. On the bottom are two or more cone shaped structures connected by a pipe approximately four inches in diameter. Covered hopper trailers that are pneumatically unloaded have capacities up to 1500 cubic feet. Examples of materials shipped in these covered hopper trailers are ammonium nitrate fertilizer, cement, and dry caustic soda.

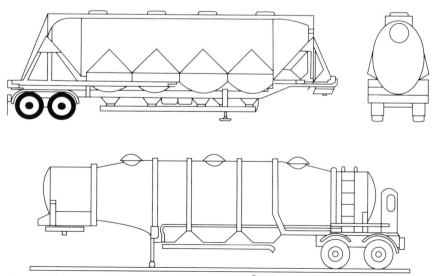

Covered hopper trailers

Rail Freight Cars

Bulk materials are generally transported in two types of rail equipment — covered hopper cars and tank cars. On occasion, gondola cars and box cars are used to transport bulk chemicals.

Covered Hopper Cars

Covered hopper cars transport bulk solids. They have flat or rounded sides and flat or angular ends, with two or more sloping sided bays on the bottom. A covered hopper with bays joined together with pipes usually has one compartment. A covered hopper car without pipes usually has as many compartments as the car has bays. Examples of materials shipped in covered hopper cars are adipic acid, ammonium nitrate fertilizer, and soda ash.

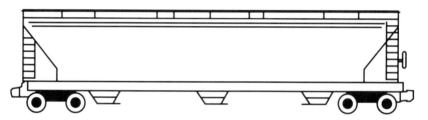

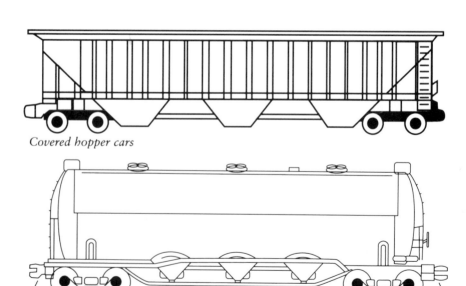

Covered hopper cars

Pneumatically unloaded covered hopper

Certain covered hopper cars are designed to pneumatically unload bulk granular solids. Some are built to tank car specifications and are unloaded using pressures of 15 psig or greater. Examples of materials shipped in pneumatically unloaded covered hopper cars are soda ash and polyvinyl chloride pellets.

Tank Cars

Tank cars transport bulk liquids, gases, and some solids. Typically, tank cars are built with two features in common — a circular cross section and rounded ends called heads. The heads of some tank cars appear flat when insulation or jacketed thermal protection is applied.

Definitions

Insulation: Is used to moderate the effect of outside temperature on the contents of the car. It can be applied to both pressure and non-pressure tank cars and is held in place by a metal jacket. Insulation is always found between the tanks in cryogenic tank cars.

Thermal Protection: Is used to keep the tank metal temperature at or below 800°F under various fire conditions. Thermal protection is used primarily on pressure tank cars transporting liquefied flammable gases, but may be found on other types of tank cars. Thermal protection can be either a sprayed-on material that swells on exposure to heat, or a blanket of mineral wool or various man-made ceramic fibers held in place by a metal jacket.

Types

For the purpose of this bulletin, tank cars are divided into three categories depending on their construction:

- non-pressure (with and without an expansion dome);
- pressure; and
- specialized (including cryogenic tank cars, high pressure tank cars, multiunit tank car tanks, pneumatically unloaded covered hopper cars, and wooden tank cars).

Non-pressure Tank Cars: Non-pressure tank cars transport hazardous and non-hazardous materials. Capacities range from 4000 to 45,000 gallons, and tank test pressures never exceed 100 psi.

NOTE: Although referred to as non-pressure, these tanks may contain pressures up to 100 psig.

Some non-pressure tank cars are insulated and/or thermally protected. Non-pressure tank cars are distinguished by either an expansion dome with

Tank car: Non-pressure		Pressure		Specialized
DOT-103	AAR-201	DOT-105	DOT-114	Cryogenic
DOT-104	AAR-203	DOT-109	DOT-120	DOT-113 AAR-204
DOT-111	AAR-206	DOT-112	AAR-120	High pressure —
DOT-115	AAR-211			DOT-107
				Wooden — AAR-208

Note 1:   Multi-unit tank car tanks classed as DOT-106 and DOT-110 are commonly called "ton containers." Due to their size and method of handling, ton containers are described in the section on bulk packaging that is placed on or in a transport vehicle or vessel for transportation.
Note 2:   Covered hopper cars that are unloaded pneumatically, classed as AAR-207, are described under Covered Hopper Cars due to their similarity in shape.
Note 3:   The specification marking, stencilled on the right side of the car while facing it, can be used to determine the tank car's class and construction.

visible fittings (on older cars) or the visible fittings without an expansion dome (on newer cars). In addition, some non-pressure tank cars have unloading valves on the bottom of the tank (bottom outlets).

Non-pressure tank car with expansion dome (inset)

Non-pressure tank car without expansion dome (inset)

Non-pressure tank cars with multiple compartments have one set of fittings for each compartment. Compartments may have different capacities and can transport different commodities at the same time. Examples of materials shipped in non-pressure tank cars are benzene, caustic soda, corn syrup, fruit juices, and whiskey.

A few non-pressure tank cars, typically those transporting nitric acid and ethylene oxide, have a protective housing similar to that found on pressure

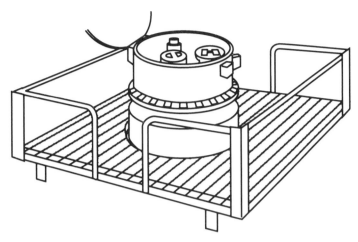

Nitric acid protective housing

tank cars. The housing for nitric acid cars has a flange in the center of the housing and is about twice as high as the housing for pressure tank cars.

Pressure Tank Cars: Pressure tank cars transport bulk flammable and nonflammable gases and poisons, as well as other hazardous materials. Their capacities range from 4000 up to 45,000 gallons, and tank test pressures range from more than 100 to 600 psi. Some pressure tank cars are insulated and/or thermally protected. Pressure tank cars are generally distinguished by the presence of a single protective housing on top that contains all valves and other fittings. The exceptions to this rule are pressure tank cars that have either a single safety relief valve or an auxiliary manway cover outside the protective housing. Examples of materials shipped in pressure tank cars are anhydrous ammonia, chlorine, and liquefied petroleum gas.

Specialized Tank Cars — Cryogenic: Cryogenic tank cars transport cryogenic liquids. The cryogenic tank car is a tank within a tank. The space

Pressure tank car with protective housing (inset)

between the inner and outer tanks is filled with insulation and normally maintained under a vacuum. Cryogenic tank cars are distinguished by the absence of top fittings, since the fittings are enclosed in cabinets either at ground level on both sides (between the trucks or wheels) or at one end of the car (just above the coupler). Examples of materials shipped in cryogenic tank cars are the cryogenic liquids argon, ethylene, hydrogen, and nitrogen.

Cryogenic tank car

Another cryogenic tank car is the box tank. The tank is built inside a 40-foot box car. The fittings for this car are located inside the doors on both sides.

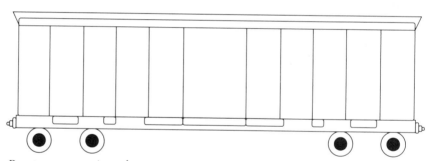

Box type cryogenic tank car

Specialized Tank Cars — High pressure: High pressure tank cars (tube cars) transport helium and hydrogen. Twenty-five to thirty seamless steel cylinders are mounted horizontally in a 40-foot frame with open sides. The enclosure at one end of the frame houses the fittings. The cylinders have high test pressures — from 3000 to 5000 psi. (All cylinders contain the same material.)

Specialized Tank Cars — Wooden tank cars: Wooden tank cars transport pickle liquor and vinegar. They are lined, coated, or treated wooden stave, metal hooped cylindrical tanks with flat ends. They have one manway and at least one fill and discharge opening on top of the tank and may have a bottom outlet.

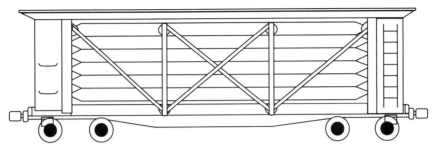

High pressure tank car (tube car)

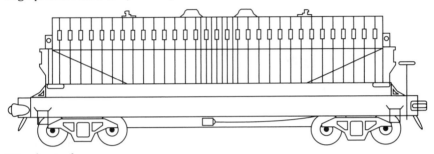

Wooden tank car

Gondolas

Gondolas transport bulk ores and other solid materials. The typical gondola has a solid floor fixed sides and ends, and may have a removable cover to provide weather protection. Gondola cars are not as tall as rail box, hopper, and tank cars. Examples of materials shipped in gondolas are contaminated dirt, flame sulfur, and iron ore.

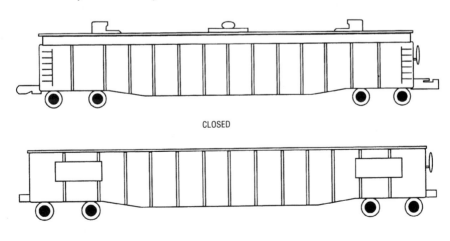

CLOSED

OPEN

Open and covered gondolas

V: Other Transport Vehicles and Freight Containers

This Section describes the most common types of motor vehicles, rail freight equipment, and freight containers used to transport non-bulk and bulk packages of the type that must be placed on or in a transport vehicle for transportation using a crane, hoist, forklift, etc. In addition to non-bulk and certain bulk packages, large items containing hazardous commodities, like rocket motors and large refrigeration machinery, are transported in or on various vehicles via all modes of transportation — highway, rail, water, and air.

Motor Vehicles (Highway)

Non-bulk and bulk (those that are placed on or in a transport vehicle) packages of materials are transported by various motor vehicles including flatbed trucks and trailers, dump trucks and trailers, enclosed van trucks and trailers, and intermodal containers on a chassis. Intermodal containers are transported by highway on a chassis and may look like a semitrailer. However, the intermodal container can be separated from the chassis.

Large items containing chemicals, like rocket motors and refrigeration machines, are transported on flatbed trailers, either without being packaged or enclosed only in a structure that provides weather protection.

Trucks

A truck is a powered motor vehicle with a non-detachable freight-carrying section. There are many truck body designs and almost every design can be used to transport non-bulk and some bulk packages. The freight-carrying section of the truck may be enclosed (such as a straight or panel truck) or have a deck with or without sides (flatbed and stakebed trucks).

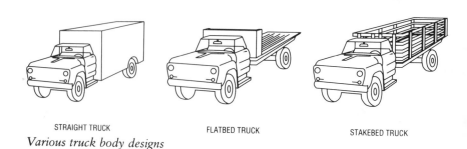

STRAIGHT TRUCK   FLATBED TRUCK   STAKEBED TRUCK

Various truck body designs

Semitrailers

A semitrailer is a vehicle without motor power, designed to be pulled by a truck or truck tractor. The semitrailer is equipped with one or more axles (usually located at the rear of the trailer) and constructed so that the front end and a substantial part of its weight rests on the truck tractor. Various designs exist, but the most common types used for transporting chemicals are the flatbed and box (van) semitrailer. Typically, trailers are approximately 8 feet wide, 12 feet high and range from 30 to 45 feet in length.

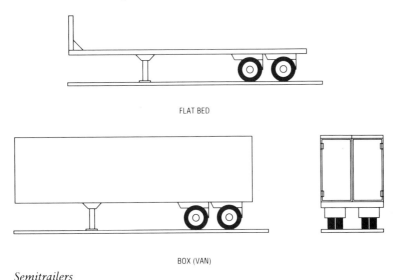

FLAT BED

BOX (VAN)

Semitrailers

Full Trailer

A full-trailer is a vehicle without motor power, designed to be pulled by a truck or truck tractor. The full-trailer is equipped with two or more axles and the full weight of the trailer rests upon its own axles. As with straight trucks and semitrailers described above, both open and closed designs exist.

Rail Freight Equipment

Non-bulk and certain bulk packages of chemicals are transported by rail on flat cars and in box cars. Van type trailers and intermodal containers (with and without chasis) are transported on flat cars. Large items containing chemicals, such as rocket motors and refrigeration machines, are transported on flat cars either without being packaged or enclosed only in a structure that provides weather protection.

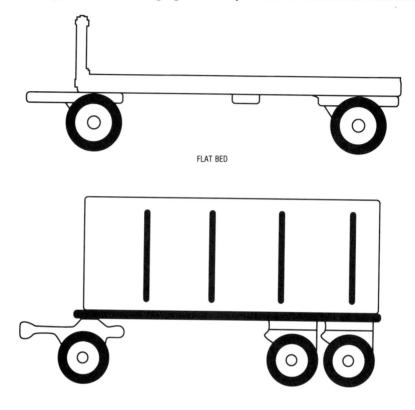

Full trailers

Box Cars

Box cars are enclosed rail freight cars usually with doors in the middle of both sides. Box cars are constructed primarily of steel, although some box cars have wooden interior linings. Box cars range in size up to 85 feet long with capacities up to 8000 cubic feet.

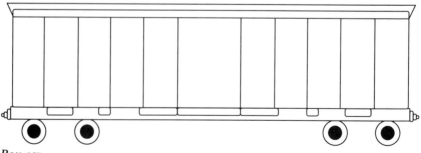

Box car

Gondola Cars

Gondolas may also be used to transport non-bulk packages. Some gondolas are equipped with covers to protect the contents.

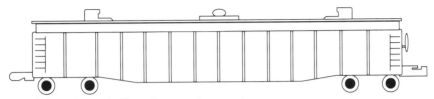

Gondola with non-bulk packages and protective cover

Flat Cars

Flat cars are rail freight cars that are primarily a flat surface riding upon a rail chassis. Various types of bulk packages — intermodal containers, van trailers, and items containing chemicals (e.g., refrigeration machines) are often shipped on flat cars. When items such as machinery are transported, they may be without outside packaging and enclosed in structures that provide weather protection.

TRAILER ON FLAT-CAR (TOFC)

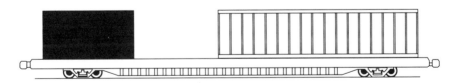

CONTAINER ON FLAT-CAR (COFC)

Freight Containers

Freight containers, often called a box, van, or simply container, transport materials in one or more modes of transportation without intermediate reloading.

Most freight containers are 8 foot by 8 foot by 20 foot or 8 foot by 8 foot by 40 foot and may be distinguished from semitrailer boxes by their corner castings, which allow them to be secured in stacks and lifted by cargo handling equipment. Freight containers are used primarily to unitize shipments of packages, but specialized designs are used to transport bulk solids.

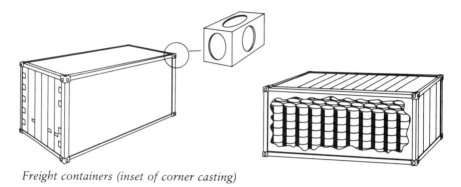

Freight containers (inset of corner casting)

Glossary

Bill of Lading — a document that describes the materials being shipped.

Bottle — see page 834.

Box — see pages 834–835.

Box Car — see page 862.

Bucket — see "Drum" pages 838–839.

Bulk Bag — see page 841.

Bulk Box — see pages 841–842.

Bulk Packaging — see pages 832–833.

Bung — a plug used to close a barrel or drum bung hole; see page 839.

Can — see "Drum" pages 838–839.

Carboy — see page 836.

Carrier — a person or company engaged in transporting materials.

Cargo Tank — See pages 848–853.

Cask — see page 848.

Chime — the connections between the side and ends of a metal drum or pail; see page 839.

Combination Package — a packaging consisting of one or more inner packagings and a non-bulk outer protective packaging. There are many different types of combination packagings.

Composite Packaging — packaging consisting of an inner receptacle, usually made of glass, ceramic, or plastic, and an outer protection (sheet metal, fiberboard, plywood, etc.) so constructed that the receptacle and outer protection form an integral packaging for transport purposes. Once assembled, it remains thereafter an integral single unit; it is filled, stored, shipped, and emptied as such.

Consignee — a person or company to which a material is being shipped.

Consist — a railroad document that lists the order of cars in a train.

COFC (Container-On-Flat-Car) — see page 863.

Container — see "Packaging" page 832.

Containership — cargo vessel designed and constructed to transport intermodal freight containers and tank containers (intermodal portable tanks).

Crate — an outer packaging of plastic, wooden, or metal construction with incomplete surfaces (open construction).

Cryogenic Cargo Tank — see page 852.

Cryogenic Cylinder — see page 837.

Cryogenic Liquid — a refrigerated liquefied gas having a boiling point colder than -130°F (-90°C) at one atmosphere, absolute.

Cryogenic Portable Tank — see page 845.

Cryogenic Tank Car — see page 858.

Cylinder — see pages 836–838.

Dangerous Cargo Manifest — a list of the hazardous materials carried as cargo on board a vessel; includes the location of the hazardous material on the vessel.

Drum — see pages 838–839.

FIBC (Flexible Intermediate Bulk Container) — see page 841.

Flat Car — see page 863.

Freight Container — see page 863.

Full Trailer — see page 861.

Gasket — a flexible material used to make a tight seal between two surfaces.

Gondola — see pages 859 and 863.

Hazardous Waste Manifest — a form required by the EPA and DOT for all modes of transportation when transporting hazardous wastes for treatment, storage, or disposal.

High Pressure Tank Car — see pages 858–859.

Hopper Car — see page 854.

Hopper Trailer — see page 853.

IBC (Intermediate Bulk Container) — see page 841.

Insulated Cylinder — see page 837.

Insulation — see page 855.

Intermodal Tank Container — see page 845.

Jar — see "Bottles" page 834.

Jerrican — see page 840.

Jug — see "Bottles" page 834.

Keg — see "Wooden Barrels" page 840.

Manifest — a shipping document that lists the commodities being transported on a vessel.

Multicell Packaging — see page 835.

Multi-unit Tank Car Tanks — see "ton containers" pages 846–847.

Non-Bulk Packaging — see page 832.

Non-Pressure Tank Cars — see pages 855–856.

NRC (Non-Reusable Container) — a container that is restricted by DOT for reuse in transporting certain regulated materials. Initials are also used to represent the National Response Center and the Nuclear Regulatory Commission.

Nurse Tank Trailer — see page 852.

Outside Container — outermost packaging used in transporting a material (other than a freight container).

Overpack — (1) a packaging used to contain one or more packages for convenience of handling and/or protection of the packages; (2) a term often used to describe the placement of damaged or leaking packages in a recovery drum (see "Drum" pages 838–839); (3) the outer packaging for radioactive materials (see page 848).

Package — see page 832.

Packaging — see page 832.

Pail — see "Drum" pages 838–839.

Palletization — see page 842.

Pneumatic Hopper Trailers — see page 854.

Portable Bin — see pages 843–845.

Portable Tank — see pages 843–848.

Pressure Tank Cars — see page 857.

Protective Overpack — see page 847.

RIBC (Rigid Intermediate Bulk Container) — see page 841.

Rail Freight Car — see page 854.

Receptacle — any packaging, other than a vessel or a barge, including a transport vehicle or freight container, in which hazardous materials are loaded with no intermediate forms of containment.

Salvage Drum — see "Drum" pages 838–839.

Semitrailer — see page 861.

Shipper — a person, company, or agency offering material for transportation.

Shipping Paper — a document that describes the shipper, contents of the shipment, and consignee (e.g., bill of lading, manifest, waybill).

STC (Single Trip Container) — container that, by DOT regulations, may not be refilled or reshipped with a DOT regulated material except under certain conditions.

Tank Car — see pages 855–859.

Tank Truck — see pages 848–853.

Thermal Protection — see page 855.

Trailer — a non-powered motor vehicle, designed for transporting freight, drawn by a truck or truck tractor.

TOFC (Trailer-On-Flat-Car) — see page 863.

Ton Container — see pages 846–847.

Transport Vehicle — cargo-carrying vehicle such as an automobile, van tractor, truck, semitrailer, tank car, or rail car used for the transportation of cargo by any mode. Each cargo-carrying body (trailer, rail car, etc.) is a separate transport vehicle.

Truck — a powered motor vehicle, designed for transporting freight, carrying its load on its own wheels.

Tube Car — see page 858.

Tube Trailer — see page 853.

Uninsulated Cylinders — see page 838.

Vessel — any watercraft used or capable of being used as a means of transportation by water.

Waybill — a railroad document describing a shipment for materials being transported by rail showing the shipper, consignee, routing, and weights used by the carrier for internal record and control, especially when the shipment is in transit.

Wooden Barrel — see page 840.

Wooden Tank Car — see pages 858–859.

Supplement 10: Chemical-Protective Clothing

Jeff Stull
Dan Gohlke
Steve Storment

This supplement consists of the transcripts of three presentations covering NFPA-compliant chemical-protective clothing originally presented on the Federal Emergency Management Agency's ENET satellite broadcast of June 24, 1992. The first transcript, presented by Jeff Stull, Chairman of the NFPA Subcommittee on Hazardous Chemical-Protective Clothing, addresses the NFPA standards for chemical-protective clothing. In the second, Product Specialist Dan Gohlke of W. L. Gore and Associates explains how to use the standards when purchasing NFPA-compliant chemical-protective clothing. The third transcript, presented by Deputy Chief Steve Storment of the Phoenix Fire Department, discusses NFPA-compliant clothing available at the time of the ENET broadcast. The papers appear here courtesy of the authors.

I. NFPA Standards on Chemical-Protective Clothing— Jeff Stull

We've heard why and how the NFPA standards were created. Our next subject covers exactly what is included in the standards. I will describe all three NFPA standards on chemical-protective clothing: NFPA 1991, *Standard on Vapor-Protective Suits for Hazardous Chemical Emergencies*; NFPA 1992, *Standard on Liquid Splash-Protective Suits for Hazardous Chemical Emergencies*; and NFPA 1993, *Standard on Support Function Protective Garments for Hazardous Chemical Operations*.

The NFPA standards are divided into six sections. These include Administration, which contains the scope, purpose, and definitions of the standard; Certification, which details manufacturer and third party organization responsibilities for product certification; Documentation Requirements, which dictates the information that the manufacturer must report; Design and Performance Requirements; Test Methods; and Referenced Publications.

There is also an appendix, which provides supplementary information and recommendations to end users of the standard.

The sections in Chapter 1 on scope, purpose, and definitions describe what products can be certified and their limitations.

NFPA 1991, for example, applies to vapor-protective suits used in hazardous chemical emergencies. The standard represents the highest level of protection and defines hazardous chemical emergencies as those activities within the hot zone.

Nevertheless, the suits covered in those standards are not to be used in fire fighting or in flammable or explosive atmospheres or against biological, radioactive, or cryogenic hazards. This is because the suit and suit materials are not tested for protection against these hazards.

NFPA 1992 defines performance criteria and test methods for suits also used in the hot zone, but limits these to suits intended to provide liquid splash protection only. These suits are not to be used in situations where vapor or gas exposure is expected.

Lastly, NFPA 1993, *Standard on Support Function Protective Garments for Hazardous Chemical Operations*, applies to garments worn by personnel outside the hot zone in support functions, such as decontamination or remedial cleanup. These are situations where the site is well characterized and the hazards are much less than those during a hazardous chemical emergency.

Chapter 2 in each standard addresses product certification. All new NFPA standards and those being revised require third-party certification. This means that an independent organization verifies that a product, a chemical-protective suit, meets all requirements of the standard.

Certification entails two principal parts: testing and quality assurance audits.

Testing is performed initially to determine product compliance, and then on a continued basis each year, to ensure that compliance with each performance requirement is maintained.

Since all products cannot be tested, quality assurance audits of the manufacturer's procedures and manufacturing methods are conducted. Visits to the manufacturer's facilities are performed to make sure that manufacturers follow stringent procedures maintaining a high level of quality in their fabrication of suits. These audits are conducted at least twice a year.

This process of certification helps to guarantee that protective clothing provides a minimum level of protection. Only organizations meeting certain requirements in the NFPA standard can certify protective clothing. Both the Safety Equipment Institute and Underwriters Laboratories are two such certification organizations.

Another important part of Chapter 2 is labelling.

Every certified suit has a label which shows its compliance with the respective NFPA standard. This label must provide certain information and warnings to the end user and will have the mark of the organization which certified it.

The label is your foremost means for determining whether you have a compliant suit or not. If it is not inside the suit, or if it does not bear a certification mark, it is not a compliant product.

Chapter 3 contains requirements for manufacturer documentation of the suits they produce or sell.

There are two parts—the Technical Data Package and the User Information.

The Technical Data Package includes a complete description of the protective suit, materials, and components, as well as all test data that show compliance of the suit to the NFPA standard.

Each NFPA standard requires that the manufacturer provide certain information to the user. This includes instructions for the use, maintenance, and storage, of the chemical-protective suit. Manufacturers must also provide recommendations for decontamination and retirement criteria for each suit.

This means that manufacturers must provide you with a complete package of information which details how the suit performed in laboratory tests as well as specific guidelines on how to use it.

The real meat of each standard is contained in Chapter 4. These are the design and performance requirements. Each suit model must meet every requirement. Failure in any one area means that the suit cannot be certified.

There are only a few design requirements in the standard. These pertain to areas where performance requirements were impractical. Performance requirements address several areas. These include requirements for the overall suit, the chemical resistance of the material, its durability, and its resistance to physical hazards. There also are performance requirements on the suit interfaces, seams and closures, and suit components.

I am going to cover the specific requirements in NFPA 1991, 1992, and 1993. For each performance area, I will explain to you which requirements apply and then how the testing is performed.

The first area is overall suit testing. Vapor-protective suits under NFPA 1991 are tested for gas-tight integrity, while all suits including both liquid splash-protective suits and support function garments are tested for liquid-tight integrity.

Pressure testing is performed to determine the integrity of the vapor-protective suits. This test involves inflating the suit to a given pressure and determining if the suit can maintain this pressure over a specified time interval. Most often this is accomplished by replacing an exhaust valve with a fitting that can be attached to a pressure gauge and compressed air source. Exhaust valves are not tested in this procedure and must be blocked off during the test. This test is performed on all suits.

The analog of the pressure test is the "shower" test for measuring the liquid-tight integrity of suits and garments. In this test, the suit is placed on a mannequin. The mannequin is dressed in a water-absorptive garment. The suited mannequin is then placed in a shower stall with nozzles directed towards the suit from several different directions. The suited mannequin is then sprayed for an hour, in four different orientations. Afterwards, the suit is removed from the mannequin and the inner garment inspected for signs of liquid penetration.

This testing is performed with water that has been treated with a non-foaming surfactant that allows easier liquid penetration and better simulates organic liquids, which have a low surface tension.

These tests are the only tests performed on the entire product. While they are not representative of exposures that may occur in the field, they are intended to measure how suits prevent penetration against both gases and liquids.

The key requirements in each standard are based on chemical resistance testing. Before we discuss these requirements, it is essential that you understand the different forms of chemical resistance.

There are three types of chemical resistance: degradation, penetration, and permeation. Each involves a different form of interaction with a protective clothing material.

Degradation occurs when the physical properties of a material change as the result of chemical contact. These changes may be evidenced by material discoloration, swelling, or deterioration. A material resisting degradation may still not act as a barrier to chemicals. It is for this reason that degradation resistance is not used in any of the standards.

Penetration is the bulk flow of liquid through a seam or closure, or pores and imperfections, in a material. Penetration also may occur as the result of material deterioration or degradation.

Penetration is an observable phenomenon. Penetration resistance testing is performed with a material specimen in a test cell. A liquid chemical is placed in contact with the material and the viewing surface of the material observed for visual penetration of liquid. The chemical is often dyed to enhance detection of penetration.

Permeation cannot be visually detected because it occurs on a molecular level. The chemical molecules on one side of the material absorb on the material's outer surface, diffuse through the material, and then desorb on the material's other side as a gas or vapor.

The actual test is conducted using a special test cell. A material specimen divides the cell into two chambers. In one chamber, the chemical is introduced. The other chamber is flushed with nitrogen, which is then analyzed for the presence of a permeating chemical.

The permeation resistance test gives two pieces of data: breakthrough time and permeation rate. Breakthrough time is measured from the time the chemical is introduced to the time it is first detected. Permeation rate is the amount of chemical that passes through a given area of material for a given unit of time. High permeation rates mean that larger amounts of chemical are passing through the material. Manufacturers must report both results, but suit performance is based only on breakthrough time.

In the NFPA standards on chemical-protective clothing, permeation testing is used for vapor-protective suits (NFPA 1991) and penetration testing is used for liquid splash-protective suits and support function protective garments in NFPA 1992 and 1993, respectively.

Chemical resistance requirements are applied to all primary materials in the protective suit. This includes the material used in the garment, visor of face shield, gloves, and boots or booties.

This represents a dramatic change over past practices. Until the NFPA standards were adopted, few manufacturers tested any material other than the garment material.

Two different lists of chemicals are applied for determining the chemical resistance of primary suit materials in each standard. NFPA 1991 employs a seventeen-chemical list, which has representative chemicals from different chemical classes, including 2 gases. NFPA 1992 and 1993 use a shorter list and are limited to liquids which are neither carcinogenic nor skin toxic, as defined by accepted references.

For all three standards, the chemical resistance requirement is based on a one-hour exposure, representing the worst case scenario.

For NFPA 1991, breakthrough time for the primary materials must be greater than one hour, for all seventeen chemicals. For both NFPA 1992 and 1993, the primary materials cannot show any penetration for one hour when tested against the eight-chemical battery.

Suit materials are also evaluated for the extremes in environments that may be experienced. These range from extreme cold to flame impingement.

Primary materials for NFPA 1991 and 1992 suits are tested for flame impingement resistance. Garment and glove materials are also subjected to a flexibility test at cold temperatures. Each test sets minimum performance requirements. For NFPA 1993, these tests are required for documentation purposes only.

The flame impingement test is the most misunderstood test in the standards. The reason we included this test was to eliminate use of suit materials that could increase wearer hazard in the event of flame contact. Passing materials or compliant suits do not offer protection against flame contact. I must emphasize this point, especially since some compliant suit configurations appear to connote this performance.

The test as designed is intended to prevent users from becoming "walking torches," under the worst of circumstances. The test answers two basic questions: (1) Does the material easily ignite?, and (2) If so, does the material continue to burn once ignited?

In this test, a folded edge of the material is contacted with the flame for a three-second period. The flame is withdrawn, and a determination is made whether the material ignited. If the material ignites during this first exposure, it fails. If there is no ignition, the material is then contacted for an additional twelve seconds and an observation made on sample ignition. If the material ignites, the time the material continues to burn and the length that the sample burns is measured. NFPA 1991 and 1992 require that the material cannot burn any longer than ten seconds, or at a distance representing half of the specimen.

In the cold flexing tests, material specimens are placed on a special test fixture at a temperature of $-25°C$ or $-13°F$. The stiffness of the material is measured and cannot exceed a specified value.

Material durability is assessed through testing for chemical resistance following abrasion and flexing. These requirements apply only to NFPA 1991 and 1992 suits, which are used in the hot zone.

The abrasion test is conducted using a device which has a coarse sandpaper as an abradant. This abradant is rubbed against the material for one hundred cycles. Smaller material specimens are cut from the abraded material and then subjected to either permeation or penetration tests.

Flex durability is accomplished with a Gelbo flex tester. A material sample is placed on the apparatus and then flexed in a compressive and twisting fashion over several cycles. As in abrasion testing, smaller specimens are taken and then tested for chemical resistance.

The strength and physical hazard resistance of material is assessed through a variety of test techniques. These are applied to materials of suits in each standard. Two additional tests are used in NFPA 1993 to overcome performance limitations that were noted in field studies for lightweight materials.

Material burst strength is measured to simulate how the suit material prevents rupture from protruding objects, such as the wearer breathing apparatus.

A puncture propagation tear test simulates how well materials resist snagging, as from a protruding nail. The tensile strength of materials is also measured as a means of assessing overall strength of the garment.

Likewise, tear strength test results are an indication of how well materials resist tearing once initially torn.

Equally important for the protection offered by the suit are the performance of the suit interfaces, which include seams and closures. The suit is only as good as its weakest link, and this often is the seams or closures.

Like the suit materials, both the seam and closure are evaluated for chemical resistance and strength. Both items must perform as well as the other primary materials in NFPA 1991 and 1992 compliant suits. There are no seam or closure requirements in NFPA 1993 for support function protective garments.

The last area of requirements pertain to the functional performance of suit components. Visors, for example, must provide sufficient clarity as not to inhibit the wearer's vision, particularly since a breathing apparatus mask will be worn in conjunction with the suit.

Exhaust valves, used in vapor-protective suits, perform several functions and also undergo two tests for their performance. Exhaust valves function to release air from inside the suit that comes from the user's breathing apparatus. When the pressure inside the suit is the same as the outside environment, the valve remains in a closed position. As air accumulates within the suit, the pressure inside the suit becomes positive. This positive pressure forces the valve to open and release air. The pressure at which the valve opens is known as the cracking pressure.

The exhaust valve's cracking pressure can affect the comfort in wearing the suit and the ease of movement. Small cracking pressures help lift the weight of the suit off the wearer. Cracking pressures that are too high will make movement difficult. NFPA 1991 requires that exhaust valve cracking pressures not exceed three inches water gauge pressure.

While the function of the exhaust valve is to release air, it also represents a significant area where contamination can enter the suit. Inside the suit is a volume of air between the wearer and the suit wall. If the wearer goes through rapid movements such as bending down, he or she compresses this volume, causing a negative pressure as compared to the outside environment. This negative pressure can draw in the contaminated outside environment, allowing chemical to contact the wearer's skin. For this reason, NFPA requires that all exhaust valves be tested for inward leakage resistance. This test involves the same procedures and criteria used to qualify exhaust valves used in respirators for limiting inward leakage. NFPA 1991 also requires a protective cover over the exhaust valves.

This concludes my description of the performance requirements in NFPA 1991, 1992, and 1993. These tests taken together represent a very rigorous set of requirements and present a difficult challenge for product certification. They demand a high level of suit quality and a comprehensive approach for defining product performance. The results are suits that consistently offer a minimum level of protection.

II. Using the NFPA Standards to Select Chemical-Protective Clothing—Dan Gohlke

First of all, I would like to make the point that these are voluntary standards. They have been crafted by a balanced group of users, manufacturers, and third-party interests to provide a significant advancement in the quality, reliability, and capability of protective clothing to meet the needs of the first responder to a haz-mat incident. But they will remain only interesting curiosities unless they become the basis for purchasing decisions. It is you, the user, that give these standards importance.

The advantages of a comprehensive set of minimum performance requirements, common ground for testing and reporting, documentation provided in a technical data package, and third-party testing and certification are significant added values. You get these values only when garments compliant with one of these NFPA standards are purchased. These features are significant values because they give the user confidence he or she is buying an all-around good product. In the past there have been abuses of the confidence between buyer and seller. Products have been sold into the market where the various components of the suit did not have the same protective capabilities so, for example, a user might buy a suit based on the garment material's characteristics but get an inferior nonequivalent visor in it.

Of, when comparing information from two different vendors, a buyer would have to look at data on different chemicals, tested in different ways, with different sensitivities, forcing the buyer to look at inconsistent information and possibly to draw incorrect conclusions.

Often the buyer would have to look at an incomplete set of information. Very commonly, only permeation data on the garment material would be available; nothing on the physical properties of the garment material, let alone on the other components or integrity of the suit.

It has even happened that misinformation, mistakes, and misunderstandings have occurred between buyer and seller.

Buying against one of the NFPA standards goes a long way to eliminating these kinds of problems.

The minimum performance requirements apply to the garment, gloves, boots, and visor, so you have at least a common minimum capability in all the major components.

The tests are well defined and reporting is consistently done, so it's easy to compare data. This particular area has been where the technical community has had to work hardest to achieve comparability from lab to lab. And still it may not be perfect, but it's far better than it was before.

The Technical Data Package that comes with every suit is an extremely valuable element of these standards because, for the first time, it gives you all the information about a suit; not just whether it passes or fails, but the actual data that characterizes the suit. It will also contain relevant information about care, maintenance, retirement, testing, repair, donning, doffing, replacement parts, etc.; in other words, a basic owner's manual for the suit, not unlike an owner's car manual.

The certification features provide an independent intermediate third-party to vouch for the quality and reliability of the product and information given by the seller.

The most familiar way of describing protective clothing has been the EPA design level classifications. These are the Level A, Level B, Level C, and Level D categories. Choosing protective clothing has usually been done by combining an EPA design level classification with an evaluation of the garment material's permeation resistance. This is the approach used, for example, in the "Guidelines for the Selection of Chemical-Protective Clothing," which is widely circulated by ACGIH. This practice has a number of shortcomings and often results in inconsistent selection of chemical-protective clothing.

One of the shortcomings is that EPA only provides a design specification instead of a performance specification. So, for example, you might have an EPA Level A fully encapsulating suit but have no idea whether it meets your protective need or not. Generally, the EPA standards would conceptually match the NFPA standards and performance expectations as follows:

Performance Required	NFPA Standard	EPA Standard
Vapor protection	NFPA 1991	Level A (gas-tight)
Liquid splash protection	NFPA 1992/NFPA 1993	Levels B and C

The problem is, you have no assurance that the EPA classification will deliver the expected performance, while, with the NFPA standards, you do.

Another problem is that permeation data is used to evaluate material performance regardless of the protective need. You remember, as has already been discussed, that permeation is the transfer of a chemical through a barrier on a molecular level. This is consistent in concept with vapor protection, and so NFPA 1991, which provides for vapor-protection suits, uses the permeation test to evaluate materials.

On the other hand, the liquid splash-protective suits use the penetration test. Penetration is the bulk flow of a liquid through porous material's seams, closures, pinholes, or other imperfections in a protective clothing material. This is consistent in concept with liquid splash protection. In tabular form, it would be like this:

Performance Required	NFPA Standard	EPA Standard
Vapor protection	Permeation	Not required
Liquid splash protection	Penetration	Not required

The NFPA standards associate permeation testing with vapor protection, and they associate penetration testing with liquid splash protection. The EPA classification gives no guidelines in this area at all. EPA allows inconsistencies to arise such as using permeation data to characterize the barrier performance of aprons, lab coats, and coveralls, which obviously are not intended for vapor protection at all by their design. The NFPA standards will hopefully bring a clearer understanding of what the garment is expected to do and why it is being worn.

These standards then form a more practical basis for the selection of chemical-protective clothing. And the critical decisions you must make to select among them is much simpler.

As can be seen from Diagram 1, there are two basic issues that separate performance into the three standards — outside the hot zone vs. inside the hot zone and liquid splash protection vs. vapor protection.

Areas outside the hot zone, for the purposes of these standards, are characterized by an environment where there is no vapor threat, there is no flammable threat, and the risk of exposure is low. These characteristics are consistent with most practical definitions of a hot zone. NFPA 1993 describes support function garments for use outside the hot zone. These garments would cover activities like decon, bottle changing, dressing, emergency medical care, training, and most industrial plant operations. Because these areas contain no vapor threat, these garments are only liquid splash-protective suits and are evaluated with the shower integrity test and chemical penetration testing. Because these areas contain no flammable threat, there is no flame resistant requirement for these garments. These garments may burn. Because the risk of exposure is low in these areas, the physical strength requirements are half what they are for garments used in the hot zone, there is no chemical penetration testing on the seams, and there is no flex or abrasion preconditioning for the chemical penetration testing.

Diagram 1

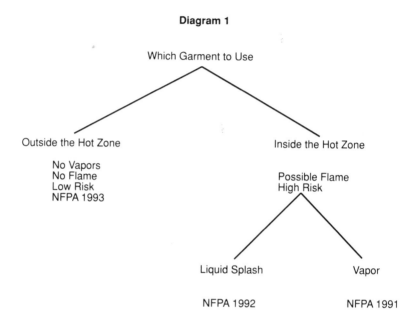

Which Garment to Use

Outside the Hot Zone

No Vapors
No Flame
Low Risk
NFPA 1993

Inside the Hot Zone

Possible Flame
High Risk

Liquid Splash

Vapor

NFPA 1992

NFPA 1991

Once you are planning to go inside the hot zone, you are entering an area where the risk of exposure is higher and a flame hazard might exist. Therefore, the strength requirements are higher, all seams are tested for chemical resistance, and the chemical testing is done after preconditioning steps of flexing and abrasion. The garment materials must also pass a flame test.

One point that needs to be made clear here is that this flame test is a modified vertical flame test. It is modified in a way that makes it a less severe test than a normal vertical flame test such as might be used on turnout ensembles. The intent of this test is to be sure that the chemical suit does not behave like a torch when exposed to a flame. The chemical suit should not contribute to your burn injury. This requirement should not be interpreted to mean that these NFPA 1991 or NFPA 1992 suits require flash fire protection. There is no flash fire exposure test in these standards currently, and so no claim should be made about flash fire protection for these garments unless it has been otherwise demonstrated. Many NFPA 1991 or 1992 chemical suits are sold with flash fire protective overcovers for the purpose of passing this flame test or perhaps the abrasion preconditioning to chemical testing. For these reasons, the suits must be worn together at all times, whether a flash fire hazard exists or not.

But please do not be misled into thinking that these standards require flash fire protection or flash fire overcovers, because they do not. Many compliant suits are available that do not incorporate a flash fire overcover.

Above and beyond this point, a second point is that there is no standard test technique currently available for evaluating the effectiveness of flash fire overcovers, so buyer beware. Some of these items may not be providing the performance you are expecting.

In the hot zone, one final differentiation can be made between vapor protection and liquid splash protection. NFPA 1991 provides vapor protection inside the hot zone. NFPA 1992 provides liquid splash protection in the hot zone. This differentiation is an important one that allows some flexibility in price, design, and productivity when it can be made. The decision to use a vapor-tight suit or a liquid-splash suit is one that requires some judgment. All liquids give off vapors, so any liquid splash exposure inherently involves some vapor exposure. The decision that must be made is, "Does the vapor exposure represent a hazard?"

The NFPA standards have given a criterion that helps to make this determination. If a chemical is documented to be a known or suspected carcinogen in Sax or the NIOSH pocket guide for the CHRIS list, or if it has a skin notation in the TLV Book, then a vapor-tight suit (NFPA 1991) should be used. Otherwise, a liquid-splash suit can be used.

In practice, this means that most monitoring and containment operations can use NFPA 1992 suits as well as most vehicular accidents, unless they involve certain chemical spills. This would easily capture more than half and perhaps as much as 80 to 90% of your responses.

III. NFPA-Compliant Chemical-Protective Clothing — Steve Storment

As users of chemical-protective clothing on hazardous materials response, we are faced with decisions on what clothing we should wear and when we should wear it. Before the NFPA standards were established in 1990, there just wasn't any guidance available except our own personal experience, some of it bad.

Today, fire departments can choose from several chemical-protective suits and garments, which have met the tough requirements required to be certified to each NFPA standard. For NFPA 1991, seven suits have been certified. Five suits have been certified to NFPA 1992, and five garments have received certification to NFPA 1993.

There still are many products out there that are not certified. This may be because the manufacturers have not tested their products, but it is more likely that their products cannot pass the NFPA requirements. A suit or garment must meet *all* of the requirements. If it fails in any one area, it fails entirely.

I will discuss each of the chemical-protective suits and garments that have been certified to meet NFPA standards.

But first, it is important that you know how to recognize a certified product. The quickest way is to look for the label. This label must be clearly visible and permanently attached inside the suit, so that you can see it when you are putting it on.

The label shows that the suit is compliant with the standard and gives you instructions for using the suit. It lists the chemicals to which the suit has been tested; identifies the manufacturers, the date the suit was manufactured, and the construction materials; and it gives appropriate warnings.

NOTE: The suits discussed in this section are those that met the appropriate NFPA standard at the time of the ENET broadcast, June 24,1992. Since that time, there may be additional complying CPC ensembles that have been certified.

In addition, the label will have the mark of the certification organization that certified the product. If the mark is not on the label, the suit is not certified. It is just that simple.

To find out what products are certified, you can contact the certification organizations directly. There are currently only two of them—the Safety Equipment Institute or "SEI," and Underwriters Laboratory. Nearly all products have been certified by SEI.

First, let's look at several suits that have been certified to NFPA 1991, the standard on vapor-protective suits, what many of you may be calling Level "A" suits. As I said earlier, there are seven different products. That means there are many choices you can make to achieve the same minimum protection. The manufacturers themselves have chosen different materials and ways to configure their products. These suits all have certain features in common. They completely encapsulate the wearer and their breathing apparatus. They have combinations of gloves to protect the hands, and they have booties made from the garment material.

They differ in the materials used in their construction and in the way they are designed—particularly, the type of visor used, the location of the closure and exhaust valves, and their general fit. Manufacturers also offer different options for their respective suits.

For NFPA 1991 suits, the first difference you will encounter is that some suits use single layer garment materials, while others use two layers. Both types of suits are certified. It is just that manufacturers have chosen different ways to meet NFPA 1991 requirements.

The first product, Responder, from Lifeguard, is an example of a two-layer product. It has an inner suit made of responder material, which provides broad chemical resistance. The overcover, which is either an aluminized PBI/Kevlar or fiberglass, must be worn over the inner suit to have a compliant product.

If you do not wear the overcover, you are not wearing an NFPA-compliant suit. The overcover allows the entire product to meet *all* NFPA 1991 requirements, particularly flame and abrasion resistance.

This suit has a flexible Teflon FEP visor on both the inner suit and the overcover; it also has booties made from the garment material. An outer aluminized fabric boot is provided to be worn over the bootie or over a bootie/rubber boot combination. Lifeguard offers the responder suit with several different glove system options. This particular suit is outfitted with a north Silvershield inner glove, a middle north glove, and an outer Golden Needles Kevlar knit glove.

Other glove systems for the suit employ combinations of safety 4 4H gloves, North Viton gloves, and Guardian flame resistant neoprene gloves.

The NFPA 1991 responder suit is equipped with two exhaust valves located at the top of the hood and on the back. The overcover also has a separate exhaust valve. The exhaust valves are protected from chemical splashes by inverted pockets.

This suit is designed with a front closure system that crosses the body diagonally. It also is available with a rear closure system. This closure, like all NFPA 1991 suit closures, is made to be pressure sealing or leak free. This particular closure uses steel alloy teeth and a PVC tape material. The outercover uses a conventional zipper with a cover flap. This suit is also available with a number of pass-throughs, which are installed on the mid-torso region of the suit. The pass-throughs allow connection of supply air hoses, with the breathing apparatus or internally worn cooling devices. The responder suit is also provided with an internal strap system for proper sizing.

The next suit is similarly designed. It is a MSA Hazmat suit which uses an inner chemical resistant garment with an aluminized PBI/Kevlar undercover. The suit has a single Teflon FEP visor, and the overcover is attached to the inner suit by a series of snaps around the edge of the visor. The inner suit uses a PVC/metal pressure sealing closure, and the overcover has a conventional zipper with a Velcro flap. A four-glove system is used that consists of a cotton glove, an inner north Viton glove, with an outer flame resistant north Silvershield glove, and an outer Golden Needles Kevlar knit glove. This model is only available with a front closure system.

As with the previous suit, the gloves are attached together onto the sleeve by an inner elastic band and hard glove ring combination. The glove can be removed for replacement, but we recommend that you follow the manufacturer's recommendations for any suit repairs.

Aluminized PBI/Kevlar overboots are provided to be worn over garment material, booties, or a second outer boot for physical protection.

The suit has three exhaust valves at the top back of the suit that are used to exhaust breathing air from your SCBA. These exhaust valves are protected from liquid splashes by inverted pockets. The pockets are designed so that the valves can be removed and inspected.

Another two-layer suit is the Chemrel Max from Chemron. The suit is constructed of Chemrel Max, a proprietary plastic laminate, with the overcover made from an aluminized fiberglass fabric. The particular suit is manufactured so that the overcover cannot be removed or the inner suit worn separately. The two exhaust valves for the suit penetrate both layers and are permanently installed at the top and back of the suit. The visor is comprised of a Teflon FEP film over a rigid PVC lens material. Snaps are used to keep the overcover in place. The glove system consists of an inner north Silvershield glove and an outer Guardian flame resistant neoprene glove. Booties are made of the Chemrel Max material and are integral to the suit. Aluminized fiberglass outer boots are provided with the suit.

Lakeland, formerly Fyrepel, has certified two different products. Both are two-layer system suits. The first suit is interceptor. This suit uses an inner white plastic laminate material worn with an aluminized PBI/Kevlar overcover. Snaps are included around the visor for attaching the overcover. Reflective trim is included on the portions of the overcover. The inner suit uses a PVC/metal pressure-sealing zipper mounted on the back. Velcro strips are used to secure overcover opening. The suit has a single Teflon FEP visor and a three-layer glove system including an inner north Silvershield glove, an intermediate north butyl glove, and an outer aluminized PBI/Kevlar glove. The aluminized gloves are secured on the overcover by Velcro strips. A single exhaust valve is provided on the back lower torso of the suit. This suit has been certified with a number of pass-throughs including those from Scott, ISI, and MSA. It is designed with back closure only. Lakeland's second suit, Forcefield, uses a material that is much different from what we have discussed thus far. The inner material in this suit is a Teflon-coated Nomex called Forcefield. The overcover is identical to the one worn with the interceptor suit. It uses the same visor and glove system as the interceptor. The fitting has a plastic plug that must be removed prior to use.

The first single layer suit we will examine is the Challenge 6000 from Chemfab Corporation. This suit is constructed of a proprietary Teflon/Fiberglass woven material laminate, which by itself meets all NFPA 1991 chemical and flame resistance requirements. The suit is designed with a Teflon FEP visor and a north Silvershield and Guardian neoprene glove system. The Challenge 6000 is designed with a back closure system that uses a neoprene/brass pressure sealing zipper. A single exhaust valve is installed on the back upper hand to vent respirator air. The suit also has reinforced knees and elbows. Like the other suits, it uses booties.

Since booties alone may not provide all the necessary *physical* protection to the foot, outer boots are typically worn over the booties. These outer boots are the Hazmax boots from Bata Shoe. They have been tested to applicable NFPA 1991 requirements and meet or exceed chemical resistance and flame resistance minimum performance levels. The Challenge 6000 suit is designed with splash guards, which are pulled over the top of the boot to prevent liquid from collecting inside the boot. Similarly, splash guards are provided on the sleeve cuffs if outer gloves are worn.

Challenge 6000 has been certified for a number of respirators and one cooling system pass-through.

The last NFPA 1991 suit we will discuss is the Trelleborg HPS. This suit involves the only elastomeric-based garment material in a single layer construction. The HPS material is a Viton/Butyl rubber laminate with a plastic layer on the inside. It uses a rigid plastic laminate visor and a two glove system composed of an inner plastic laminate glove with an outer vitric rubber glove. The outer rubber glove is attached by an elasticized band. The suit has three exhaust valves, two located on the hood, the third on the side. Both the protective pockets and the hoods are removable. The suit has an internal air distribution system for cooling, which requires the appropriate respirator and air line attachments.

Many of these suits offer the option of supplied air. There are advantages and disadvantages to using supplied air. The principal advantage is an unlimited supply of air. The disadvantages are that they reduce responders mobility because air supply hoses are limited to 300 feet and can become entangled. They can also increase the risk of chemical exposure.

Some combination self-contained breathing apparatus allow operation from both an air bottle and a supplied air source. These systems may afford flexibility for some hazardous material responses.

The suits you use must have a pass-through that is specific to the type of SCBA or supplied air system you are using.

There are a number of other suits on the market. Just because they are designed to be fully encapsulating and to pass pressure testing does not mean that they meet the requirements of NFPA 1991.

Suits constructed of PVC, Butyl rubber, and neoprene do not have the broad based chemical resistance required by NFPA 1991, nor do they resist ignition when contacted by flame. Only the suits discussed now meet NFPA 1991.

Like NFPA 1991, manufacturers have used different approaches for designing products to meet NFPA 1992 requirements. Although they do not have to, NFPA 1992 suits generally completely cover the wearer and their breathing apparatus.

Most suits have garment "booties," but gloves may be attached or loose depending on the manufacturer. Two of these suits are similar to their NFPA 1991 counterparts.

The first suit is the NFPA 1992 Responder from Lifeguard. Again, this is a two-layer suit with inner responder material and an overcover of either aluminized PBI/Kevlar or fiberglass.

Aluminized fabric booties are worn over inner material booties. The chief differences between this suit and the NFPA 1991 version are that the seams are only taped on one side and single gloves, the Guardian neoprene gloves, are used. This suit uses a different closure, which is not pressure sealing but is liquid penetration resistant. It does not have any exhaust valves but instead uses a protected opening for venting exhaust air. The Responder is available in a number of configurations and offers an internal strapping system, which allows form fitting wearing.

The second suit is an NFPA 1992 Interceptor suit. This also is a two-layer suit system with inner interceptor material, an aluminized PBI/Kevlar overcover. It is also similar to its NFPA 1991 counterpart but differs in its seam construction and zipper type. The suit uses a Teflon FEP visor but has a Kevlar knit outer glove over the inner Butyl glove.

The Pacesetter from Lion Apparel uses a woven fabric laminate, which includes a W.L. Gore Goretex liner for preventing liquid penetration. The material is also breathable, that is, it allows air and moisture permeation. The visor is a FEP film over a PVC flexible lens, and it uses Guardian neoprene gloves and Seruus rubber Chembreauer or Bata Shoe Hazmax boots. Its two-piece construction includes a lower coverall with an upper hooded protective top with visor that also has sleeves. The suit comes with a hard hat and spare glove rings.

The Chemfab Challenge 5000 is constructed of a Teflon-coated, nonwoven Nomex material. It also has a two-piece design. The lower coverall has suspenders while the upper top is hooded with sleeves and a visor. The visor is made of flexible Teflon FEP film, while the suit uses Guardian neoprene gloves. The suit is constructed with booties, and outer boots may be worn for additional physical protection to the feet.

The newest suit to be certified is the Stasafe CPE suit from Standard Safety. It is constructed of a single layer, flame-retardant, chlorinated, polyethylene or CPE over a woven nylon fabric. It also uses a PVC visor that is .040 inch thick. This is an encapsulating suit similar in design as those used for NFPA 1991. It uses a pressure-sealing zipper and is gas-tight. It comes with three separate exhaust valves installed in the back of the suit, but it does not meet some of the other requirements for NFPA 1991. The suit is designed for use with the Guardian neoprene glove and Bata Hazmax boot. The suit sleeves use Standard Safety's patented "seal-tight" glove cuff system. Instead of garment material booties, the suit has neoprene booties that are worn inside the Bata Hazmax boot. The suit is available with two different respirator pass-throughs.

The last series of certified products that we will look at are support function garments that meet NFPA 1993. As the name implies, this standard applies only to the garment and not to the face shield, gloves, or boots unless these items are an integral part of the garment.

Support function garments are used outside the "hot" zone. The NFPA intends that these garments be used in support roles like decontamination, or remedial cleanup, once the significant hazards have been removed or mitigated.

This first NFPA 1993 garment is called Comfortgard III from Scott-Durafab. This garment uses Goregard, a breathable lightweight material that incorporates a Goretex film. This garment is available in two designs. The first design has two pieces—a hooded coverall which is entered through the neck. A drawstring is used to close the garment around the wearer's neck. To cover openings in this region, the second piece, a hooded top with sleeves, is worn. This top includes an elasticized opening to fit around the respirator mask.

The second design is of one-piece construction with a Teflon FEP visor. This garment is entered through an opening across the shoulders, much the same way as a diving dry suit. Flaps of material with Velcro strips are used to seal the suit—so there is no zipper. Both designs have elasticized wrists with splash covers, for both gloves and outer boots.

A second NFPA 1993 garment, designed without a closure, is Fortress from Abanda. It is constructed of a disposable, Saran-coated polypropylene. It is of one-piece construction. The suit is designed as a "sack" suit with a drawstring. It has a Teflon FEP visor on a hood that is attached to the suit by an adhesive tape. This garment has soft "booties" made from the garment material and elasticized sleeve cuffs to accept gloves.

Scott-Durafab's Barricade also uses a hooded design with booties and elasticized splash covers for gloves and outer boots. In this case, a film-coated material, Barricade from Dupont, is used as the principal barrier material. Instead of the shoulder entry system, this garment has a front system closure that uses a PVC metal pressure-sealing zipper. It has an open port for an exhaust valve.

Front Line is another NFPA 1993 hooded garment with a visor, soft booties, and elasticized openings for gloves. The Front Line material is a proprietary plastic film over a nonwoven fabric. It is manufactured by Kappler, a sister company of Lifeguard. The suit uses a back closure system with an inner conventional zipper and an outer two-track closure. It also has elasticized wrist with booties and splash covers for gloves and outer boots.

The last NFPA 1993 garment we will discuss at is CPC 2000 from Lion Apparel. CPC 2000 is constructed of a material that uses a woven fabric and Goretex composite. This garment is constructed like a conventional coverall with a front liquid-resistant zipper and flap with Velcro strips. There are also Velcro closures around the wrists and ankles. The garment also uses a bonnet-like hood that protects the head from liquid splashes. There is also a more extensive hooded top with sleeves and a Teflon FEP visor that cover the upper torso.

This completes the discussion of all the NFPA-compliant chemical-protective suits and garments. Though different, these products are all designed (and have been tested) to provide a minimum level of protection during hazardous material response.

Which suit you should purchase is a decision you have to make. If your organization handles hazardous materials responses, your team should be equipped with suits or garments meeting each of the three standards.

When you choose a suit or garment, you should take into consideration other equipment in your inventory, like SCBAs, as well as the types of responses your group handles and the resources available for their purchase. One of the best ways to decide which protective clothing to buy is to inspect the clothing first hand and "try it out" if possible. Only then can you get an appreciation of other factors not covered in the standard, such as sizing, comfort, and ease of donning.

We have discussed features of each suit and their qualities in meeting the respective NFPA standard. We have not intended to show one suit to have advantages other another. If you are interested in prices, please contact the manufacturer or your safety equipment distributor. In selecting suits, we recommend that you compare only NFPA-compliant suits.

Supplement 11: California Hazardous Materials Medical Management Protocols

This supplement, excerpted from the hazardous materials medical management protocols of the California Emergency Medical Services Authority, provides excellent guidelines for developing effective protocols for the management of medical problems that may result from hazardous materials incidents. These particular excerpts address prehospital chemical contamination, chemical contamination in the emergency department, and basic decontamination protocols. Thanks to Dr. Gus Koehler, Hazardous Materials Project Manager for the California EMS Authority, for his assistance.

Introduction to Protocols

Health care providers who care for injured persons exposed to hazardous materials must know how to evaluate and manage a contaminated victim's medical problems while protecting themselves and others from potential hazardous exposure (secondary contamination). The following treatment protocols provide succinct, step-by-step information on how to manage medical problems arising from the most common kinds of hazardous materials ("hazmat") episodes.

These protocols are designed for use by "EMS hazmat entry team members," paramedics or other rescue health workers in the field, and hospital emergency department physicians and nurses. The protocols are intended as *guidelines*. They may require modification depending on the resources of a particular hospital or the needs of a particular patient. It is essential for the safety of health care personnel and patients that hospitals and emergency medical services agencies have a written plan for management of the contaminated victim, and that their personnel are trained to follow it. In all incidents, health care providers should immediately contact their base hospital, if prehospital care provider, or Regional Poison Control Center for advice on managing victims of hazardous materials exposure.

Controversies abound in the evolving field of environmental toxicology. For this reason, the protocols are sometimes vague or ambiguous. For example, no consensus exists on what specific protective gear, if any, is appropriate for emergency departments and prehospital medical care providers because most authorities agree that it is unacceptable to provide sophisticated protective gear to persons who have not been previously properly fitted and trained in its use.

It is imperative that proper decontamination has been initiated by the hazmat team or other trained responders in the hot zone/decontamination area. Rescuers who are trained to use self-contained breathing apparatus, to select the appropriate chemical protective suits, and know how to function in them are the only ones who should assist with decontamination or enter the hot zone.

Federal OSHA has established new hazardous materials training requirements for responders who are called to a spill. The requirements are identified in OSHA 29 CFR 1910.120. California OSHA is developing similar regulations (Section 5192, Title 8, California Code of Regulations). These regulations will be no less stringent than that currently required by federal OSHA. California OSHA enforces these standards, not the Authority.

The Federal 29 CFR 1910 final rule applies to EMS. If an employer expects to respond to a hazardous materials incident, then he must train his employees about the hazards involved and the role that they will be expected to play. The rule states that: "Training shall be based on the duties and function to be performed by each responder of an emergency response organization" (p. 9329).

The NIOSH EPA hazardous waste site operations document is helpful in defining the EMT's role. It categorizes "medical support" as involving "off-site personnel" (p. 3-3). Ambulance personnel "provide emergency treatment procedures appropriate to the hazards on site." Decontamination is carried out by others under the direction of the Decontamination Station Officer(s) (p. 3-4). In California, the Authority has taken the position that the person doing the decontamination would probably be a fire fighter (hopefully training as an EMT-I) who is responsible for decontaminating personnel as well as victims. Given most EMS personnel's daily medical duties, training, and responsibilities at a hazmat spill site (particularly if they are not part of the fire service) they should be trained at least at the "First Responder Awareness Level."

EMT-IIs and paramedics should be trained at the "First Responder Operations Level" if they are expected to select and don protective equipment, conduct rescues, decontaminate victims or response personnel. In any case, these two EMS classifications should take a course on the medical management of hazmat victims that is based on these protocols or their equivalent. Fire service personnel, who may also be EMTs, will be trained at a higher level because of their fire service duties and functions. All of this must be consistent with the hazmat role that the local EMS agency has defined for EMS personnel. Again, it is the employer's responsibility to see that their personnel are properly trained to meet these requirements.

"First responders at the awareness level are individuals who are likely to witness or discover a hazardous substance release and who have been trained to initiate an emergency response by notifying the proper authorities of the release. They would take no further action beyond notifying the authorities of the release" (p. 9329). Given their typical daily responsibilities and training, EMTs clearly are not responsible for making a rescue wearing protective gear, for controlling and containing the release, stopping its spread, or for decontamination of protective equipment. However, they can provide medical care to a fully decontaminated victim. The federal rule does not specify how many hours of training are necessary for this level. We have evaluated the rule and recommend *no less than four hours of training* as being sufficient to meet the OSHA requirements for this category of responder. Additional medical training to manage hazardous materials victims would probably be necessary.

"First responders at the operations level are individuals who respond to releases or potential releases of hazardous substances as part of the initial response to the site for the purpose of protecting nearby persons, property or the environment from the effect of the release. They are trained to respond in a defensive fashion without actually trying to stop the release from a safe distance, keep it from spreading, and prevent exposures" (p. 9329). *Eight hours of training is required* for this level by federal OSHA. Again, additional medical training to decontaminate and manage victims would probably be necessary. A four to six hour medical management course would fill this need.

The Authority's interpretation of Federal 29 CFR 1910.120 and of the draft California OSHA regulations is that EMTs should not be required to wear special protective gear, SCBA, or respirators unless they have been trained to use them and their responsibilities at the scene require it. In the vast majority of cases, they should be trained to recognize a hazardous materials incident and be able to initiate a response. Decontaminated victims should be brought to them for medical care so that untrained and unprotected EMTs are not put at risk.

In all cases, employers should consider SARA III and all OSHA requirements for training their employees (Federal OSHA, FR 54: 9294-9336; Cal OSHA, General Industrial Safety Orders, Section 5192, Draft 8, December 13, 1990; and, Code of California Regulations, Sections 3203, 3220, 5141, 5144, 5155, 5192, and 5194). According to California OSHA staff, many of these requirements apply to hospital emergency departments too. Your local California OSHA office should be contacted if you have any questions. Again, the Authority does not enforce state or federal OSHA regulations.

A hazardous materials Basic First Responder Course approved by California Specialized Training Institute (805-549-3535) and EMS hazardous material medical management and planning courses offered by the University of California Davis' Hazardous Substances Program (1-800-752-0881), provide a good introduction to how a hazmat response is organized, what to do if a prehospital health care provider is first on the scene, management of medical care, and planning a response.

These protocols do *not* address accidents involving *radioactive materials*. Radioactive materials incidents require unique strategies, monitoring equipment, and specialized consultants. Well-established protocols already exist for their management (call Oak Ridge National Laboratories — 615-576-3131 — for information).

The Concept of Secondary Contamination

An essential question to ask is, "What is the risk of *secondary contamination* (to rescuing personnel, transport vehicles, hospital emergency departments) from this chemical?" It is traditionally axiomatic in hazardous materials emergency management that chemicals should be considered both highly toxic and highly contaminating to personnel, vehicles, and the environment. However, a great many chemicals are very highly toxic *only* in the high concentrations found in the immediate exposure area (hot zone) but pose *little or no risk* to persons outside the hot zone. Small amounts of some chemicals may produce relatively little acute toxicity, but because they are suspected of causing cancer or other chronic disease they are considered to create a risk of secondary contamination.

Tables 1 and 2 list selected examples of hazardous substances which carry a high vs. a low risk for *secondary contamination*. The lists are meant to be illustrative, not exhaustive. Note that highly toxic chemicals may be found in *either* list. The Regional Poison Control Center or base hospital can assist you in determining the potential for secondary contamination of other hazardous materials.

Substances with Serious Potential for Secondary Contamination

Unless the victim has been properly decontaminated, substances like those listed in Table 1 may persist in significant amounts on the victim's clothing, skin, hair, or personal belongings, and may jeopardize health care workers or other attendants. Recommended protective gear should be worn (Table 5 or Table 7). Reducing the potential for chemical exposure from any form of mouth-to-mouth resuscitation, including use of pocket one-way valve mouth-to-mouth resuscitation devices should be carefully considered when the victim has been exposed to one of the listed gases. If resuscitation efforts are necessary, a bag valve mask with reservoir device or manually triggered oxygen powered breathing device, should be applied to the patient. Contact with even lightly contaminated skin or clothing should be minimized prior to decontamination. *Proper decontamination by adequately protected personnel must be carried out before the victim is treated by prehospital or emergency department personnel.*

Table 1   Substances with a High Risk for Secondary Contamination

Examples:
- Acids, alkali, and corrosives (if concentrated)
- Asbestos (large amounts, crumbling)
- Cyanide salts and related compounds (e.g., nitriles) and hydrogen cyanide gas
- Hydrofluoric acid solutions
- Nitrogen-containing and other oxidizers which may produce methemoglobinemia (aniline, aryl amines, aromatic nitro-compounds, chlorates, etc.)
- Pesticides
- PCBs (polychlorinated biphenyls)
- Phenol and phenolic compounds
- Many other oily or adherent toxic dusts and liquids

Substances with Little Risk for Secondary Contamination

Many of the substances listed in Table 2 are highly toxic. However, even if they persist in the victim's clothing, skin, hair, or personal belongings after removal from hot zone, they are not likely to jeopardize health care workers or rescuers and are not likely to secondarily contaminate vehicles or the emergency department. *On-scene decontamination, if indicated, is desirable (especially clothing removal and victim wash) but not essential.*

Table 2   Substances with a Low Risk for Secondary Contamination

Examples:
- Most gases and vapors unless they condense in significant amounts on the clothing, skin, or hair
- Weak acids, weak alkali and weak corrosives in low concentrations (excluding hydrofluoric acid)
- Weak acid or weak alkali vapors (unless clothing soaked and excluding hydrofluoric acid vapor)
- Arsine gas
- Carbon monoxide gas
- Gasoline, kerosene, and related hydrocarbons
- Phosphine gas
- Smoke/combustion products (excluding chemical fires)
- Small quantities of common hydrocarbon solvents (e.g., toluene, xylene, paint thinner, ketones, chlorinated degreasers)

Basic Decontamination Protocol

In a properly functioning hazardous materials response, victims will be decontaminated in the decontamination corridor (Diagram 1) by properly suited hazmat team members. This will include removal of wet or exposed clothing, flushing affected skin and hair with water, and soap or shampoo wash if needed (i.e., for oily or adherent substances). The following basic decontamination protocol should be followed for all contaminated victims.

Table 3   Basic Decontamination Protocol

1. Determine the need for decontamination by consulting the appropriate protocol and calling your Regional Poison Control Center.
2. For advice on selection of specific protective clothing, you may also contact CHEMTREC at (800) 424-9300 or the AAR Bureau of Explosives at (202) 835-9500. If the proper protective equipment is not available, or prehospital or hospital staff have not been trained to use it, call for assistance from the local, usually fire department, hazmat team.
3. Evaluate ABCs, stabilize spine (if trauma suspected), establish patent airway and breathing, if indicated. Move victim away from contact with hazardous material to a clean area. Rescuers in level "A" (fully encapsulated suit with self-contained breathing apparatus) equipment may not be physically able to do anything more than drag victims onto a back board and then drag them out of the Hot Zone. If not breathing, and if physically possible to quickly accomplish, give oxygen using bag valve mask with reservoir device or manually triggered oxygen powered breathing device.
4. If ambulatory, victims should be directed to leave the hot zone, others assist with evacuation, and decontaminate themselves following the directions below under the direction of the decontamination supervisor.
5. If clothing has been contaminated, strip the victim and double-bag clothing, then flush the entire body with plain water for 2-5 minutes. Clothing contaminated with dust should be removed dry with care taken to minimize any dust becoming airborne. If circumstances, time, and practice allow, a dust mask or respirator should be placed over the victim's nose or mouth. Dust should be brushed off of the face prior to fitting the mast or respirator.
6. Flush exposed eyes and other body surfaces with copious plain water for 2-5 minutes. Eye irrigation should continue for at least 10-15 minutes, preferably with saline.
7. If contaminant is oily or greasy, soap and/or shampoo may be used followed by additional water flushing.
8. Clean under nails with scrub brush or plastic nail cleaner.

Prehospital Care

A. Field Response

Because chemicals are used extensively in our society, the potential for hazardous materials accidents exists almost everywhere. Hazardous materials incidents range from relatively confined site-specific events to rapidly expanding accidents that endanger a sizable community. Regardless of its size, an incident's successful management requires pre-planning and interagency coordination.

Managing the victims of a hazardous materials incident necessitates the coordination of many resources and agencies. Roles of various agencies vary to some extent according to the county's hazardous material area plan. Generally, fire fighters and law enforcement officers are the first to arrive on scene and may obtain important information about the chemicals involved. They will designate an Incident Commander to manage incident operations at the scene. Special Hazardous Material (Hazmat) Units (either Health Department or Fire Department) may be available to provide additional guidance in identifying and managing the hazardous materials and to perform decontamination of equipment, environment, victims, and personnel. Emergency Medical Services (ambulance) personnel transport the victims who have already been decontaminated (if necessary) and manage their medical problems en route to the hospital. In the event of a disaster, the county Office of Emergency Services and the local EMS agency will become involved in resource coordination. Finally, the local hospital emergency department will receive and care for the victims.

The emergency medical service prehospital providers responding to a hazardous materials incident have five goals:

Table 4   Five Goals of Prehospital Provider

- To protect themselves and other prehospital responders from any significant toxic exposure;
- To obtain accurate information on the identity and health effects of the hazardous materials and the appropriate prehospital evaluation and medical care for victims;
- To minimize continued exposure of the victim and secondary contamination of health care personnel by ensuring that proper decontamination (if necessary) has been completed prior to transport to a hospital emergency department;
- To provide appropriate prehospital emergency medical care consistent with their certification; and
- To prevent unnecessary contamination of their transport vehicle or equipment.

B. Hazard Information about Specific Chemicals

Every effort should be made to obtain accurate information about the health hazards of the toxic materials involved in the incident, the potential for secondary contamination, and the level of decontamination required, if any. Information may be obtained from the Incident Command Safety Officer, the base hospital, or the Regional Poison Control Center.

C. Prehospital Provider Protection

Prehospital health care providers who are not members of the hazardous materials team and properly outfitted with protective gear should not enter the contaminated area (hot zone and decontamination corridor, as shown in Diagrams 1 and 2) but instead must wait at the perimeter for decontaminated victims to be brought to them. It is assumed that members of the hazardous materials team working in the hot zone and decontamination area are trained and capable of providing initial airway and spine stabilization and basic decontamination. Rescuers wearing level "A" (fully encapsulated suit with self-contained breathing apparatus) equipment will probably experience several factors that will limit their ability to provide emergency care in the hot zone such as: Vision impairment, reduction in dexterity (lifting, disentangling, etc.), limited air support, and heat stress. Other factors such as the number of rescuers allowed into the hot zone will also limit what care can be given.

The table on page 902, "EMS Vehicle Equipment for Hazardous Materials Incidents," identifies how an ambulance should be outfitted to respond to a hazmat incident.

D. Prehospital Decontamination

Unprotected EMS responders must advise on and observe the decontamination procedures from a distance to ensure that they are properly carried out. They should practice with the local hazmat team to become familiar with the steps involved. **If there is any doubt about the potential for secondary contamination, decontaminate the victim.** A contaminated appendage can be washed without wetting the whole body if that is the only part contaminated. Clothing covering the rest of the body and exposed skin should be carefully checked for contamination.

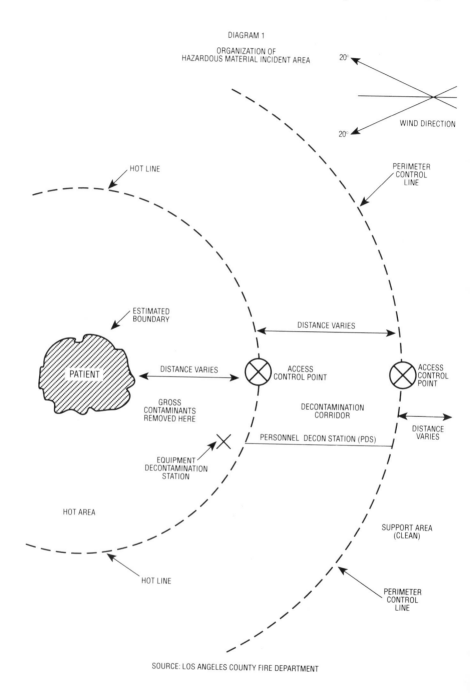

DIAGRAM 1

ORGANIZATION OF
HAZARDOUS MATERIAL INCIDENT AREA

SOURCE: LOS ANGELES COUNTY FIRE DEPARTMENT

DIAGRAM 2

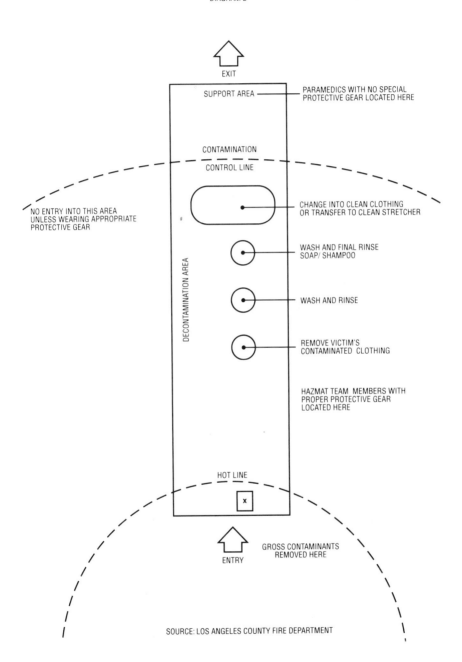

EXIT

SUPPORT AREA ——— PARAMEDICS WITH NO SPECIAL
PROTECTIVE GEAR LOCATED HERE

CONTAMINATION

CONTROL LINE

NO ENTRY INTO THIS AREA
UNLESS WEARING APPROPRIATE
PROTECTIVE GEAR

CHANGE INTO CLEAN CLOTHING
OR TRANSFER TO CLEAN STRETCHER

WASH AND FINAL RINSE
SOAP/ SHAMPOO

DECONTAMINATION AREA

WASH AND RINSE

REMOVE VICTIM'S
CONTAMINATED CLOTHING

HAZMAT TEAM MEMBERS WITH
PROPER PROTECTIVE GEAR
LOCATED HERE

HOT LINE

x

ENTRY

GROSS CONTAMINANTS
REMOVED HERE

SOURCE: LOS ANGELES COUNTY FIRE DEPARTMENT

HAZARDOUS MATERIALS RESPONSE HANDBOOK

Table 5   EMS Vehicle Equipment for Hazardous Materials Incidents*

- Binoculars to assess scene from a safe distance.
- Plastic (10-12 mil, preferably clear) trash bags (3 or 4 mil) to isolate and dispose of contaminated articles and toxic vomitus. Plastic sheeting to cover floor of ambulance in the rare case where a contaminated victim must be transported, or if the victim might vomit ingested toxic material.
- A large supply of oxygen to treat breathing problems caused by exposure to hazardous materials (more than is usually carried).
- A large wash basin, bucket, or plastic waste basket which can be lined with a trash bag to collect contaminated eye wash water or vomitus.
- Disposable plastic-coated blankets (or "chucks") to soak up and isolate liquids from a decontaminated patient. Use these for absorbing toxic vomitus.
- Disposable gowns and slippers for patients who must remove contaminated clothes at the scene and for EMS personnel (long sleeve gowns) to cover outer clothes.
- Disposable surgical or examination gloves.
- Surgical or other paper masks.
- Waterproof disposable shoe covers.
- Splash goggles or face shields to protect EMS personnel from splashes while they work on the patient.
- Inexpensive stethoscopes, blood pressure cuffs and other gear which can be discarded if contaminated.
- Isotonic saline and IV tubing for eye irrigation.
- A bag valve mask (BVM) or similar device in lieu of mouth-to-mouth respiration. (Pocket masks are NOT acceptable.)
- Liquid soap for washing off oily contaminants.
- Epsom salts for soaking hydrofluoric acid burns.
- Shears or sharp knife for removing clothing from victim.
- Copy of the current "D.O.T. Emergency Response Guidebook," a copy of these protocols, and other appropriate medical management protocols.

*Source: Based on a list prepared by the Contra Costa/Solano County Joint Emergency Medical Services Hazardous Materials Response Program. Additional equipment is necessary for handling radiation contamination. See: "Emergency Department Radiation Accident Protocol." Leonard RB, Ricks RC. Annals of Emergency Medicine, 9:9:462-70, 1980. Also, see *Medical Management of Radiation Accidents.* Mettler FA, Kelsey CA, Ricks RC. CRC Press, Florida, 1989.

If victims are already properly decontaminated before they are brought to health care providers at the perimeter of the hot zone/decon area, they will pose very little, if any, risk to the prehospital health provider or their vehicle. Thus, health care providers will not generally need to use any specialized protective gear, even for substances considered as potential secondary contaminants.

In many cases (e.g., corrosive materials in the eye; oily pesticide skin exposure), prehospital health care personnel may need to repeat or continue decontamination procedures (e.g., eye irrigation; soap/water skin wash) after receiving the victim at the perimeter. Although specialized protective gear should not be necessary, it is prudent for providers to don the protective gear listed in the Table 5. (Some of these items are often carried as a "communicable disease" kit.) All leather items, wool or other highly absorbent materials that cannot be decontaminated should be removed prior to providing care.

No provider should put on a respirator or other specialized gear unless that worker has been previously fitted and trained in its use.

If the transport vehicle is inadvertently contaminated, advice from the local environmental health department, hazardous materials team, or local hazardous materials spill cleanup companies should be sought on how to determine the level and location of the contamination and on how to clean it up. Advice should also be sought on how to preserve evidence for law enforcement, and dispose of or clean contaminated clothing and personal items.

E. Prehospital Triage

Victims with obvious significant illness or injury will need rapid transport and treatment after initial stabilization and basic decontamination is carried out. In virtually all cases, patients with serious trauma or medical illness can be quickly stripped and flushed with water prior to delivery to prehospital health providers outside the hot zone. This is true even in cold or inclement weather. If this cannot be performed because of acute life-threatening conditions or other circumstances, then the vehicle must be protected and those providing care during transport and driving the vehicle must be properly fitted and trained with the appropriate level of specialized protective gear. However, every effort should be made to decontaminate the victim at the scene if the means to do so are available. In those jurisdictions where a prehospital provider might be placed in such a situation without assistance from a properly trained hazmat specialist, advance arrangements for additional training and protective equipment should be made.

Consult the specific protocols for recommended prehospital care of exposed victims. *Note that some of the management protocols may exceed the EMT-II or paramedic scope of practice in a local area. Refer to your local EMS agency medical director for guidance.*

Victims with few or minimal symptoms are not necessarily safe from progression of illness. Many toxic substances have delayed onset effects, which may appear several hours later, after the victim has returned home. If the toxic substance is known, obtain consultation from the Regional Poison Control Center to determine if delayed effects might be seen and for guidance on triage of asymptomatic or mildly symptomatic exposure victims. Any persons suspected of being exposed should be seen and evaluated by emergency department staff.

F. Decontamination of Prehospital Personnel

Prehospital workers will not normally need personal decontamination. In those rare circumstances where they have been in the hot zone or have attended to a victim who was not properly decontaminated, they should consider themselves to be potentially contaminated. Consult the lists above or knowledgeable sources to determine the risk of secondary contamination, since in many, if not most, cases no personal decontamination will be necessary. Information can be obtained from the Incident Command Safety Officer at the scene, the base station hospital, or the Regional Poison Control Center. **If in doubt, decontaminate.**

G. Victim and Response Personnel Follow-up

The names, addresses, and telephone numbers of all personnel and victims who have been or may have been exposed at a hazmat scene should be recorded for future notification if it is subsequently determined that medical evaluation or treatment is required.

Emergency Department Care

In managing a victim who has been exposed to a hazardous material and who may be contaminated or who is not known to have been adequately decontaminated before arrival at the hospital, the emergency department staff has five goals:

Table 6   Emergency Department Goals

- To protect hospital staff members from any significant toxic exposure;
- To minimize any additional exposure of the victim to the toxic substance (e.g., in the event that the victim's clothing is soaked);
- To evaluate quickly whether the victim is in immediate danger of dying and needs immediate endrotracheal intubation, CPR, or other emergency procedures;
- To quickly determine the toxic identity and effects of the hazardous materials and to provide specific treatment if indicated; and
- To prevent hospital contamination and to protect passers-by from any significant toxic exposure.

A. Preplanning and Need to Determine Risk for Secondary Contamination

An important part of any chemical disaster pre-planning is to survey the area surrounding the hospital to determine which types of hazardous materials are used by local industries. It is noteworthy that the JCAH Accreditation Manual for Hospitals, 1986, calls for hospitals to participate in community planning whenever feasible (Section 3.1.1.1). The emergency department administrator should become familiar with the county Hazardous Material Area Plan, which identifies procedures to be used to coordinate the management of hazardous materials and to establish roles and responsibilities for government agency actions in response to a hazardous material incident. The name of the agency responsible for preparing and maintaining the Hazardous Material Area Plans in your region can be obtained from the Office of Emergency Services, Hazardous Materials Division (916-427-4287).

In order to obtain more detailed information on specific chemicals used by nearby industries, some emergency departments obtain copies of the Material Safety Data Sheets (MSDSs) from local industries and keep them on file. According to federal and state legislation, employers must provide the information contained on an MSDS to health care providers who need the information to care for an affected patient.

The information must be provided without regard to "trade secrets" in an emergency. However, it is not generally practical to keep large numbers of MSDSs in an indexed, usable filing system which can be relied upon in an emergency. MSDSs contain basic chemical, reactivity, and toxicology data, but are usually very limited in medical treatment information and are of variable technical quality. In addition, they rarely provide information regarding the potential for secondary contamination or recommendations for decontamination. The Regional Poison Control Centers are the best sources of acute health effects information on hazardous materials.

To locate information about the risk for secondary contamination of health care personnel, other patients in the department, and the hospital facility, call the Regional Poison Control Center.

B. The Contaminated Victim

In the ideal situation, victims will already be properly decontaminated before they are brought to the emergency department, and they will pose very little, if any risk to the hospital health provider or the facility.

However, a written protocol must be prepared for those situations where a victim, heavily contaminated with a highly toxic chemical, arrives at the emergency department (e.g., a walk-in). If a victim contaminated by a substance with serious potential for secondary contamination has already entered the emergency department, separate zones should be set up by the emergency room charge person:

1. The contamination area,
2. A designated decontamination area (preferably outside), and
3. A clean zone.

The contaminated area should be marked and isolated. Personnel must not be allowed to indiscriminately enter or leave these zones unless checked for contamination.

The best course of action for most facilities is to call the fire department hazmat team (if there is one) to come to the emergency department and set up a decontamination area *outside* the ambulance entrance. A practical alternative is to provide simple but effective decontamination *outside* the ambulance entrance using an inflatable "kiddie" pool, or shower, and soap (Green soap®, New Dawn®, or any mild dishwashing detergent). The victim can often remove his/her own clothing and wash off the material. Provide plastic bags for double-bagging contaminated clothing and, if available, a tent or curtain for victim privacy. Victims who are not ambulatory can be decontaminated by appropriately protected and trained hospital staff on a protected gurney in the same area. Establishing the decontamination area *outside* of the emergency department is important because of the potential risk of secondary contamination by inhalation of toxic vapors or dusts.

If the hospital emergency department is located in a highly industrialized area and can expect to receive contaminated victims, consideration should be given to training staff to use self-contained breathing apparatus (SCBA) and other appropriate protective equipment. The local hazmat team or an industrial hygienist should be consulted about training requirements, equipment, frequency of training, and other relevant safety details. The county health department should be contacted to determine how to dispose of the contaminated water.

C. Protective Clothing for Hospital Staff

If proper decontamination has been carried out prior to transport, no specialized protective gear should be required for hospital staff. Disposable surgical gowns, aprons, gloves and shoe coverings may be appropriate (Table 7). In the vast majority of circumstances, the equipment in the following table will adequately protect emergency department staff as they remove soaked clothing, wash the victim's skin/hair with soap/shampoo, or perform eye irrigation. With very concentrated acids or caustics or with substantial amounts of oily or lipid-soluble liquids (e.g., pesticides), disposable Tyvek or Saranex coveralls and unmilled nitrile gloves will probably offer sufficient protection until the victim can be decontaminated. Advice on appropriate suits and gloves can be obtained from the local hazardous materials team, or the Regional Poison Control Center. Hospital staff should remove all leather items, wool clothing and other materials that cannot be easily decontaminated. Consideration should be given to obtaining disposable medical equipment. Personnel without adequate personal protective equipment should not be in close proximity to victims who are grossly contaminated or being decontaminated.

If simple outdoor decontamination is not possible, arrangements should be made in advance with a qualified industrial hygienist to obtain special protective respiratory equipment and to provide training in its proper use.

No provider should be asked to put on a respirator or other specialized gear unless that worker has been fitted and trained in its use.

D. Decontaminating the Victim

If decontamination is required, a thorough wash-down of the victim's skin for a few minutes with plenty of soap and water is generally adequate (see p. 897 for basic decontamination protocol). However, for chemical

Table 7   Suggested Equipment List for Management of Hazardous
Materials Contamination, Part I

The following emergency supplies should be stored in an area near the emergency department rear entrance and checked periodically (e.g., quarterly):

- Written procedures for handline chemically contaminated victims.
- Protective clothing for staff:

For most circumstances: Disposable gowns, surgical masks, plain latex gloves (enough for double gloving), shoe covers, splash goggles (at least two pair), aprons, caps. At least some of the gowns, aprons and shoe covers should be impervious to water.

For heavy chemical or corrosive contamination: At least 2 Tyvek or Saranex suits and 2 pair unmilled nitrile gloves (be sure to check with Poison Control or hazmat team to see if they are compatible with the particular hazardous substance involved).

Note: Respiratory protective gear is not generally available and, in addition, should not be used unless it is properly maintained and staff have been properly fitted and trained in its selection and use. Therefore, if inhalation exposure is a risk, decontamination should be done *outside*.

- Decontamination supplies:

Inflatable "kiddie" pool (large) with foot operated air pump (or other means of collecting decontamination water), large plastic tarp to place under pool forming an outer containment area, 55-gallon plastic trash cans to hold water, mild dishwashing detergent or soap in squeeze bottle, sponges, absorbent pads for washing, nail brush, tent or curtain for privacy, exterior wall water outlet/ shower nozzle hooked up to lukewarm (or cold) water supply.

Metal gurney or morgue table for non-ambulatory patients.

Alternatively, if a dedicated decontamination room is provided, plans must be made for separate exhaust ventilation, adequate ventilation (at least 6 changes/ hour), plastic sheeting to cover floor, 2-inch tape to secure plastic, means of containing contaminated water, and respiratory protective gear for staff who may be in an enclosed space with volatile hazardous materials. (Note, any employee required to use a respirator must be medically cleared, fitted, and trained.)

Plastic bags for double-bagging contaminated clothing (preferably clear).

Diking or absorbent material: Dikes can be made by taping the edges of a large plastic tarp or sheets of plastic draped over a ladder turned on its side or rope strung horizontally. Absorbent materials such as kitty litter, pillows, diapers, or other similar material may be useful to absorb spills.

Saline and IV tubing for eye irrigation set-up.

Note that special "decontamination solutions" and neutralizing agents are not recommended except in specific rare circumstances (e.g., hydrofluoric acid). Water (and perhaps soap) are the recommended means of decontaminating victims. Extra care needs to be given to victims contaminated with water-reactive substances: Consult your poison control center.

- Other Supplies:

Wall suction with disposable collection bag to hook up to gastric tube to remove and isolate toxic vomitus.

Table 7, continued

Extra medical supplies or equipment which could be taken out of service temporarily if contaminated (including crash cart with ambu bags, defibrillator, EKG monitoring equipment, IV stands, etc.).

Inexpensive medical equipment which could be disposed of if contaminated (including stethoscope, blood pressure cuff, etc.).

Tape and rope for marking off perimeters.

Plastic sheeting (4 mil) for covering floor or covering entrance to and floor of decontamination area for materials with high potential for secondary contamination.

2-inch tape for securing plastic.

Cotton-tipped applicators and stoppered glass containers for swabs of hazardous materials for laboratory analysis, or evidence for later prosecution of the party responsible for the hazmat spill.

- Special medical treatment supplies:
  See specific treatment protocols.

*Source: Based on a list prepared by the Contra Costa/Solano County Joint Emergency Medical Services Hazardous Materials Response Program. Additional equipment is necessary for handling radiation contamination. See: "Emergency Department Radiation Accident Protocol." Leonard RB, Ricks RC. Annals of Emergency Medicine, 9:9:462-70, 1980. Also, see *Medical Management of Radiation Accidents*. Mettler FA, Kelsey CA, Ricks RC. CRC Press, Florida, 1989.

contamination of an open wound, gentle scrubbing or irrigation of the wound for 5-10 minutes or longer is advisable, using lukewarm water. With eye exposures, irrigation of the eyes with sterile saline should be carried out for at least 15-30 minutes. Check conjunctival sac pH if exposure was to an acid or alkaline material. Contaminated facial and nose hair and ear canals should be gently irrigated with normal saline, using frequent suction. **If there is any doubt about contamination, decontaminate the victim.** A contaminated appendage can be washed without wetting the whole body if that is the only part contaminated. Clothing covering the rest of the body and exposed skin should be carefully checked for contamination. Following decontamination, specific medical management of the victim can be addressed.

Gastric lavage should be performed if ingestion is suspected. Use wall suction and an isolated collection bag to avoid exposure to liquid or vapors of toxic vomitus. Administer activated charcoal after lavage is completed.

E. Medical Management of the Victim

In a life-threatening emergency, a decision to delay patient care because of concerns about contamination and possible exposure of hospital staff will require considerable clinical judgment. Delay may be necessary with certain extremely hazardous substances present in significant quantities on or near the victim. The Regional Poison Control Center can provide emergency assistance in making these decisions. In reality, a delay in starting treatment because of such concerns will only rarely be required. Attention to the basic ABCs of life support (airway, breathing, and circulation) should be given if it does not pose a significant risk to the care giver.

The most important step for the Emergency physician is to get information about *what* hazardous substances are involved and what estimated *dose* the victim received. This information will often have been obtained by the EMS personnel, fire, police, or hazmat team responding to the episode. If this information does not accompany the victim to the Emergency Department, the hospital staff can direct the EMS, fire, or police who accompany the patient to obtain the information from the Incident Commander at the scene.

F. Decontamination of Hospital Staff and Cleanup

Health care workers who attend to victims who have not been previously decontaminated should consider themselves to be potentially contaminated. Contact the Regional Poison Control Center to determine the risk of secondary contamination. In many (if not most) cases, no personal decontamination will be necessary. However, *if in doubt, decontaminate.*

Procedures for post-emergency cleanup including disposal of contaminated wastes should be addressed by written protocol. Hazardous waste must be disposed of properly, and not mixed with non-hazardous trash. Advice on disposal of hazardous waste can often be obtained from the Health Department or the hazmat team.

In dealing with the problem of contaminated corpses, the important objectives are to limit the spread of contamination within the hospital and to protect transport personnel and personnel in the coroner's office or other pathology staff. A corpse with known or suspected significant contamination can be easily decontaminated in the emergency department, particularly if a contamination zone has already been established and other decontamination activities are being carried out. Depending on the nature of the contaminant, the clothing can be removed and double-bagged, and the body washed. Be careful to save samples or swabs of the material, if not already identified, as legal evidence. These can be saved in a sealed, clean test tube or specimen container.

The contaminated corpse should be double body-bagged. The body bag should have a prominent label indicating that the corpse is contaminated and the nature of the contaminant. Emergency department staff should record the telephone number of the coroner or pathologist on the label for more information about the nature of the contamination. All deaths resulting from toxic exposure are coroner's cases and the hospital staff should notify the coroner's office.

G. Security

The emergency department hazardous materials incident protocol should indicate that hospital security or engineering staff will be notified to help with isolating and managing a potential contamination problem.

A protected and trained security person assigned to the decontamination zone can assist in handing equipment and supplies into the contaminated zone, and completing the double-bagging of contaminated clothing or other articles before handing them out to the clean zone. Security or engineering personnel can also assist in preventing the spread of contaminated puddles of water on the floor (by the use of dikes, for example), in securing the ventilation system if necessary so that contaminated air does not circulate to the rest of the building, and in setting up an outdoor decontamination station.

H. After the Incident

Hospitals are subject to two major reporting requirements with regard to hazardous materials victims:

* Occupational Illness or Injury: Illnesses or injuries occurring in the course of employment must be reported by the treating physician in a "Physician's First Report of Occupational Illness or Injury." New versions of this form were released in 1989 and are available by calling (415) 557-1924.

    In the event of a death occurring in the course of employment, employers are responsible for calling the Occupational Safety and Health Administration (OSHA). Cal OSHA is responsible for all employers except federal agencies.
* Pesticide Poisoning: The State of California mandates that illnesses due to pesticide poisoning, *even if not occupational in origin* must be reported by telephone within 24 hours to the local health officer for the area in which the poisoning occurred. A follow-up written report submitted on a "Physician's First Report of Occupational Illness or Injury" or a comparable form must be submitted within one week of the telephone report.

Following the completion of the response to the incident, a critique should be conducted with all of the staff involved. Only by a thorough review of the events can mistakes be corrected and procedures modified for improving the management of future incidents.

Editor's Note: The following are just two examples of medical management protocols contained in the California guidelines.

Hazardous Materials Medical Management Protocols: Unknown Material

Forms

This section assumes that a victim has been exposed to a hazardous material which cannot be identified in the form of a gas or vapor, liquid, or solid/dust.

Background

Every attempt should be made to identify the substance involved using placards, shipping papers, or other means. However, if such identification is impossible, responders should make worst case assumptions about the material. Rescuers should assume that the material may be:

a. Poisonous by inhalation, ingestion, and cutaneous absorption;
b. Corrosive (either acidic or alkaline);
c. Lipid soluble, and therefore able to penetrate certain types of protective clothing and protective gear, and able to be absorbed through intact skin;
d. Oily and persistent on skin and clothing, and therefore difficult to decontaminate; and,
e. Reactive and likely to give rise to irritant or poisonous gases on contact with water or heat.

Potential for Secondary Contamination

Victims contaminated with an unknown liquid or solid/dust material should be assumed to carry a risk of causing secondary contamination. If the victim's only exposure was to small amounts of gas or vapor, the risk of secondary contamination to health care personnel away from the scene is probably very small. Theoretically, small amounts of gas might be trapped in a victim's clothing. In such a situation once the clothing had been removed and double-bagged, the risk to rescuers would be minimal. However, if the exposure involved an aerosol which might condense on a victim's skin or clothing, there would be a potential for secondary contamination until decontamination had been carried out. For exposures involving direct contact with an unknown liquid or solid material or dust, rescuers should assume that the victim poses a risk of secondary contamination until decontaminated. When in doubt, decontaminate the victim (see "Basic Decontamination Protocol," p. 897).

Patient Management in the Hot Zone/Decon Area

1. In general, until the possibility of fire, explosion, or serious reactivity has been ruled out, rescuers will not enter the Hot Zone. Once entry appears to be feasible, rescuers should don fully encapsulated protective clothing and gloves capable of withstanding both corrosives and hydrocarbon solvents, and self-contained breathing apparatus.
2. Quickly evaluate and support ABCs. Stabilize the spine (if trauma suspected), establish airway and breathing, and consider high flow supplemental oxygen by bag valve mask with reservoir, if possible and practical.
3. Flush the victim with water spray, and remove and double-bag clothing and flush skin for 1-2 minutes. If the victim complains of eye irritation, have the victim remove contact lenses if able to do so. Irrigate exposed eyes if symptomatic.

Prehospital Management after Initial Decontamination

1. If victim is NOT decontaminated and responder is properly trained, don protective equipment (self-contained breathing apparatus) capable of withstanding brief exposure to both corrosives and hydrocarbon solvents. Activate basic decontamination protocol (see "Basic Decontamination Protocol," p. 897). If these requirements cannot be met, request assistance from the local hazmat team or your Regional Poison Control Center. In addition, wash oily contaminated areas, including skin or hair, with soap and/or shampoo.

2. Re-evaluate airway, intubating the trachea if victim is unconscious or has developed severe respiratory distress. Continue to provide high-flow oxygen by mask. Attach cardiac monitor.
3. Support BP if needed, with IV crystalloid solutions. Treat bradycardia with atropine or other modality appropriate to the patient's clinical status.
4. Consider aerosolized bronchodilators if significant wheezing is present.
5. Continue to flush affected skin and eyes with copious water or saline. Remove contact lenses and irrigate eyes with saline via plain IV tubing for at least 10-15 minutes or until symptoms of pain or irritation have resolved.
6. Even if significant ingestion is suspected, do not induce vomiting. Instead, if the victim is conscious and able to protect the airway, immediately dilute with 1 glass of water and give activated charcoal 60-100 grams if available. *Do NOT give activated charcoal if a corrosive is suspected.*
7. Continue to irrigate injured eyes or exposed areas of skin for at least 15 to 20 minutes if the victim continues to complain of discomfort.
8. Treat seizures with diazepam (Valium):
   5-10 mg IV for an adult; and
   1-2 mg IV for children.

Management in the Hospital

1. If victim is NOT decontaminated and responder is properly trained, don protective equipment capable of withstanding brief exposure to both corrosives and hydrocarbon solvents, and self-contained breathing apparatus. Activate basic decontamination protocol (See "Basic Decontamination Protocol," p. 897). If these requirements cannot be met, request assistance from the local hazmat team or your Regional Poison Control Center. In addition, wash oily contaminated areas with soap and/or shampoo.
2. Evaluate and support ABCs (airway, breathing, and circulation).
3. Obtain arterial blood gases, chest X-ray, and electrocardiogram in seriously symptomatic patients. Administer high-flow oxygen if the victim has respiratory distress or altered mental status. Aerosolized bronchodilators will probably be helpful, and are seldom contraindicated, for most cases of bronchoconstriction due to hazmat exposures. Monitor cardiac rhythm.
4. Diagnostic Considerations — For a hazmat victim with cardiorespiratory collapse, consider the diagnosis of generalized cellular poisoning (cyanide, azide, sulfide, for example). Always consider the diagnosis of carbon mon-

oxide poisoning. In the appropriate setting, consider the diagnosis of exposure to anti-cholinesterase pesticides and the possible use of atropine in a patient with bradycardia, wheezing, seizures, and/or other signs of cholinergic stimulation. Consider the diagnosis of methemoglobinemia for patients with cardiopulmonary distress. If there is a concomitant likelihood of cyanide exposure, as for example through smoke inhalation, nitrites should generally not be used to treat methemoglobinemia. If the hazmat victim appears to have suffered substantial skin exposure to a corrosive liquid, be aware that patients with significant hydrofluoric acid exposure, as manifested by painful burns or dramatic respiratory injury, may require prophylactic IV calcium as well as specialized treatment for exposed skin areas.

5. Treat seizures with diazepam (Valium):
   5-10 mg IV for an adult; and
   1-2 mg IV for children.

6. Appropriate lab studies, in addition to other routine and indicated studies such as electrolytes and glucose and anion and osmolar gap, might include carboxyhemoglobin, methemoglobin, calcium, plasma and RBC cholinesterase levels, liver function studies, methanol level, and serum lactate.

7. Treat skin and eye exposures with copious irrigation, for at least 15 to 20 minutes. If eye irritation persists, perform a fluorescein and slit-lamp examination to rule out corneal injury.

8. In patients who present with initial symptoms of respiratory irritation or distress, be alert for the development of delayed onset pulmonary edema, up to 24 hours after the exposure.

9. In cases of significant ingestion, treat as for other types of toxic ingestion with gastric lavage and/or administration of activated charcoal (unless ingestion of a corrosive is suspected); do not induce vomiting. Consider saving a sample of gastric contents for possible subsequent lab analysis, but isolate them in a closed container as soon as possible. Be aware that gastric washing may contain volatile material that could potentially expose hospital personnel to noxious vapors. If corrosive injury to the victim's esophagus is suspected, consider consultation with gastroenterologist or surgeon for possible endoscopy.

For Advice On Clinical Management, Call Your Regional Poison Control Center At: (   ) -_____ - _____

Acids and Acid Mists (Not Including Hydrofluoric Acid)

Forms

Gas, liquid (variable concentrations), mixtures with water, and aerosolized dusts.

Background

Acids act as direct irritants and corrosive agents to skin and moist mucous membranes. Severe burns may result. Generally, these substances have very good warning properties; even fairly low airborne concentrations of acid mists, or vapors produce rapid onset of eye, nose and throat irritation. Inhalation of higher concentrations can produce cough, stridor, wheezing, chemical pneumonia or non-cardiogenic pulmonary edema. Occasionally, pulmonary edema may be delayed for several hours, especially with low-solubility gases such as nitrogen oxides (given off by nitric acid). Ingestion of acids can result in severe injury to the airway, esophagus and stomach.

Potential for Secondary Contamination

Small amounts of acid mists can be trapped in clothing after an overwhelming exposure but are not usually sufficient to create a hazard for health care personnel away from the scene. However, clothing which has become soaked with concentrated acids may be corrosive to rescuers. Once the victim has been stripped and flushed with water, there is no significant risk of secondary contamination. Decontamination is not necessary for victims with inhalation exposure only.

Patient Management in the Hot Zone/Decon Area

1. Rescuers should don agent-specific protective clothing and gloves, and self-contained breathing apparatus particularly if mists or vapors are present.
   Ambulatory patients should be instructed to remove themselves from the hot zone and to decontaminate themselves under the direction of the decontamination supervisor.

2. Quickly evaluate and support ABCs. Stabilize the spine (if trauma suspected), establish airway and breathing, and consider high flow supplemental oxygen by bag valve mask with reservoir, if possible and practical.
3. Flush exposed skin with water spray, and if clothing has been soaked by acid or acid spray, remove and double-bag clothing and flush skin for 1-2 minutes. Remove contact lenses and irrigate exposed eyes if symptomatic.

Prehospital Management after Initial Decontamination

1. If victim is **NOT** decontaminated and responder is properly trained, don appropriate agent-specific protective equipment and self-contained breathing apparatus. Activate basic decontamination protocol (see "Basic Decontamination Protocol," p. 897). If these requirements cannot be met, request assistance from the local hazmat team or your Regional Poison Control Center.

   If victim is decontaminated, don appropriate protective equipment consistent with risk of secondary contamination ("The Concept of Secondary Contamination," p. 895, and Table 5, p. 902).
2. Evaluate and support ABCs (airway, breathing and circulation). Re-evaluate airway, intubating the trachea if victim has developed severe respiratory distress. Provide high-flow oxygen by mask. Attach cardiac monitor.
3. Aerosolized bronchodilators (e.g., metaproterenol) may be helpful for victims with wheezing.
4. Continue to flush affected skin and eyes with copious water or saline. Remove contact lenses and irrigate eyes with saline via plain IV tubing, for at least 10-15 minutes or until symptoms of pain or irritation have resolved.
5. Victims with minimal or quickly resolving symptoms probably do not require immediate evaluation in the emergency department. However, remember that with certain acids and low-solubility gases (e.g., fuming nitric acid forming nitrogen oxides) pulmonary edema may occur after a delay of 12-24 hours.
6. **Ingestion: DO NOT induce vomiting.** Immediately dilute with 1 glass of water or milk.

Management in the Hospital

1. If victim is **NOT** decontaminated and responder is properly trained, don appropriate agent-specific protective equipment and self-contained

breathing apparatus. Activate basic decontamination protocol (see "Basic Decontamination Protocol," p. 897). If these requirements cannot be met, request assistance from the local hazmat team or your Regional Poison Control Center.

If victim is decontaminated, don appropriate protective equipment consistent with risk of secondary contamination ("The Concept of Secondary Contamination," p. 895, and Table 7, pp. 908-909).

Remove and double-bag clothing if not already done. Wash exposed skin copiously with water. Decontamination is probably not needed for acid exposures unless the victim's skin or clothing has been soaked with acid liquid.

2. Evaluate ABCs (airway, breathing, and circulation). Watch for signs of airway closure and laryngeal edema, such as hoarseness, stridor, or retractions.

3. Administer oxygen by mask. Bronchodilators may be helpful for wheezing. Intubate if patient manifests severe respiratory distress from pulmonary edema or upper airway swelling. Obtain arterial blood and chest X-ray if respiratory distress is present. If respiratory distress is present or if exposed to low-solubility gases such as nitrogen oxides, admit and observe 24 to 48 hours for possible delayed onset of pulmonary edema. Severe upper airway edema may necessitate endotracheal intubation or cricothyrotomy.

4. If the patient complains of eye irritation, check for the presence of contact lenses and remove, then irrigate eyes copiously with saline via plain IV tubing for at least 10-15 minutes, or until symptoms of pain or irritation have resolved. Consider fluorescein or slit-lamp examination to rule out corneal injury.

5. If a significant ingestion occurred, consider endoscopy to evaluate injury to the esophagus and stomach.

6. Advise patient that full recovery is generally the rule, but cases of chronic airway disease have been reported following severe exposures. Advise and arrange for follow-up in case victim begins to experience respiratory distress. After exposure to oxides of nitrogen, sudden severe relapse may occur two to three weeks later.

For Advice on Clinical Management, Call Your Regional Poison Control Center At: (   ) -_____ - _____

Supplement 12: Protection of the Health Care System

Jonathan Borak, MD
Michael Callan
William Abbott

This supplement is an excerpt from the book Hazardous Materials Exposure: Emergency Response and Patient Care by Jonathan Borak, MD, Michael Callan, and William Abbott. It provides an excellent discussion of the hazards of contamination, which extend from the scene of an incident to the emergency facility to which victims may be transported. Our thanks to Brady/Prentice, a division of Simon & Schuster, for allowing us to reprint this material.

Introduction

Goal

On completion of this chapter the student will have an understanding of the importance of protecting the "downstream" health care system and the procedures to be followed to guard against contamination.

Objectives

Specifically the student will be able to:

• Name the people and things in the health care system requiring protection, and discuss the importance of protection for each.

• Discuss the ways in which EMS personnel can be protected from contamination by hazardous materials.

• Describe the procedures for protecting EMS equipment and vehicles from contamination by hazardous materials.

• Describe the procedures for protecting the emergency department and its staff from contamination by hazardous materials.

• Discuss the kinds of protective equipment and supplies that are needed for ambulance-based EMS personnel to protect themselves and their ambulances.

Overview

This chapter describes some procedures and preplanning issues that can help protect the health care system and its personnel from contamination during hazardous materials incidents. Emergency responders involved at an incident should be constantly aware that risk of exposure and injury exists. Moreover, it should be understood that rescuers and incident victims are not the only ones at risk. Contamination can be transmitted "downstream" from the site of the original incident. As a result, contamination can spread to involve other rescuers, rescue equipment and vehicles, and even the hospital and its emergency department to which victims are transported.

To protect themselves, rescuers should take care to avoid unnecessary exposures and to use appropriate personal protective equipment when entering contaminated places. Protecting the downstream health care system requires that rescuers carefully perform decontamination and follow other incident safety procedures. In addition, a need exists for preplanning so that personnel, ambulances and hospital facilities can be adequately protected from contamination.

Goal

The goal of protecting the downstream health care system is to prevent contamination of health care workers and the facilities in which they work. By avoiding contamination, it is possible to assure that all facilities and personnel will continue to function and provide care to the community's sick and injured. A real danger exists that personnel, ambulances, hospitals, or emergency departments may be unable to function and provide care if they are not protected from contamination.

Protection should be provided to all personnel, equipment, and facilities that are involved in caring for victims of hazardous materials exposure. Those who might become contaminated through incidental contact with victims, personnel, or equipment should also be protected. The following partial list of people and things make up the emergency health care system and require some protection:

1. EMS personnel
2. EMS vehicles and equipment
3. Emergency department personnel
4. Emergency department patients
5. Emergency department facility
6. Hospital

When EMS or emergency department personnel become contaminated by hazardous chemicals, they risk personal illness. In addition, a possibility exists that they will pass the contamination along to their patients, colleagues, and others. When EMS equipment and vehicles become contaminated, a risk exists that the equipment will be out of service until decontamination has been completed. Sometimes, equipment or vehicles cannot be decontaminated and must be destroyed. Contamination of the emergency department or other areas of the hospital may require that those areas be closed to patients until decontamination has been completed.

A contaminated health care system threatens the health and well-being of health care workers. Moreover, that health care system can lose its ability to provide needed services. Such an event can cause much greater hardship than was caused by the hazardous material incident that created the emergency in the first place.

Protection of EMS Personnel

EMS personnel, like other emergency responders who work at the actual site of a hazardous material incident, are at risk of contamination. The risk can be minimized by limiting the number of personnel who have access to the contaminated area, assuring that the appropriate protective equipment is used, and enforcing careful decontamination of all personnel and equipment. Some specific guidelines and procedures to provide such protection are described subsequently.

Restriction of Access to the Contaminated Area

Only the fewest possible rescue personnel should be allowed to enter zones of high contamination. EMS personnel should not be allowed to enter contaminated areas for reasons other than victim assessment or rescue.

Limitation of Number of Involved Personnel

A natural tendency exists for crowds to form at the scenes of accidents and disasters. Even among rescuers, attraction and fascination with the disaster can lead to unnecessary crowding. Unlike most other types of accidents, however, hazardous materials can affect and harm observers and bystanders. For this reason, the number of EMS personnel involved at an incident should be limited to the fewest actually needed. Surplus personnel should remain at a staging site distant from the area of contamination.

Protect Pregnant Staff

A great deal is not known about the effects that most acute chemical exposures have on fetuses. Enough suspected harm exists, however, to argue that pregnant women should not be allowed to enter an area of chemical contamination. The purpose of this guideline is to protect the unborn child from potential toxic harm.

Decontamination of All Exposed Personnel

Any emergency responder who becomes exposed to harmful chemicals must undergo decontamination. In the excitement of responding to an industrial accident, EMS personnel must not neglect their own decontamination.

Appropriate Use of Personal Protective Equipment

EMS personnel must not enter heavily contaminated areas without proper personal protective equipment. The choice of personal protective equipment should be based on identification of the contaminating chemicals and determination of their environmental concentrations.

Personnel who do not actually enter the incident hot zone, but who have physical contact with potentially contaminated victims, should also be protected from becoming contaminated. Those rescuers should at least consider use of disposable chemical protective clothing, and latex or vinyl gloves. A list of protective equipment and supplies that should be carried in an ambulance or rescue vehicle is presented later.

Ventilation of Ambulance during Transport

A danger always exists that residual contamination remains on a victim, rescuers, or rescue equipment. When contaminated victims or rescuers are in the unventilated patient compartment of an ambulance, dangerous concentrations of chemicals can develop. Exposure of victims and EMS personnel can occur and EMS responders may suffer exposure injury. In these cases, the ambulance should be ventilated while the victim is being transported to hospital. Ventilation should be maintained whenever the ambulance crew is uncertain about the adequacy of decontamination.

Maintenance of Personal Exposure Log

Emergency responders involved in hazardous materials incidents should keep a diary that records all exposures and contaminations that they have suffered. The diary should include names of chemicals causing exposure, dates of the exposures, and any symptoms that developed. In addition, responders should undergo regular, periodic medical examinations and routine laboratory testing. The information contained in the exposure log should be made available to the responder's personal physician at the time of medical examinations.

Medical surveillance involves periodic medical examinations of workers who are regularly exposed to high levels of hazardous chemicals. Medical surveillance must also be provided to workers who develop symptoms as a result of toxic exposures. Many employers are required by federal regulations to provide medical surveillance examinations to their employees who work with hazardous materials. Unfortunately, most emergency responders and EMS personnel do not fit into the groups of workers for whom periodic examinations are required.

Response personnel who anticipate exposure to hazardous materials should have a complete physical examination before beginning emergency response work ("preplacement examination"), an annual physical examination, an examination at the termination of response work ("exit examination") and an examination following each documented hazardous material exposure. The purpose of these examinations is to recognize intoxication promptly when it occurs and to assure that workers receive appropriate medical care when needed. A written report for each examination should be included in the personal exposure log.

Protection of EMS Equipment and Vehicles

Victims of hazardous materials exposure who are transported in an ambulance or other EMS vehicle can contaminate that vehicle and its equipment. This is especially likely when victims with severe injuries and need for rapid transport to a health care facility are transported after only superficial decontamination. Once contaminated, an ambulance may not be usable until decontamination has been performed. The vehicle may be out of service for days or longer.

This potential problem can be avoided by a small effort and proper planning. A simple, seven-step approach to protecting an ambulance from contamination follows.

1. *Identify and set aside needed equipment:* The ambulance equipment that will likely be needed for patient care should be removed from the vehicle and set aside. This should include equipment and supplies that are normally stored in drawers and lockers in the patient compartment. This equipment will be returned to the ambulance shortly.

2. *Remove unnecessary equipment:* Transporting vehicles often carry equipment that will not be needed during a hazardous materials incident such as stair chairs and extra stretchers. This equipment should be removed and stored for later use (Figure 1).

3. *Tape closed all ambulance compartments:* Once all necessary equipment has been removed from the ambulance drawers and lockers, those compartments should be sealed closed with duct tape. If the vehicle is equipped with a storage locker under the bench seat, the doors of that locker should also be taped shut. The tape should be applied so that the compartments and storage areas are sealed closed from the patient compartment. Figure 2 illustrates the sealing of storage compartments.

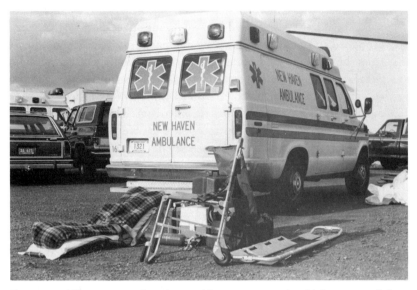

Figure 1 *All nonessential and portable equipment should be removed from the transporting vehicle.*

Figure 2 *The cabinet doors are being sealed with heavy-duty plastic sheets and heavy duct tape.*

Figure 3 *Sheets of heavy-duty plastic are draped along the ambulance walls and are secured with heavy duct tape.*

4. *Encapsulate the patient compartment:* The inside of the patient compartment should be encapsulated by heavy gauge plastic or Tyvek® sheets that are taped to the four sides of the compartment. Sheets should be taped from the ceiling to the floor on each wall and along the front of the patient compartment. A sheet of plastic or a fire department salvage tarpaulin should be placed on the compartment floor with the edges rolled to create a catchment basin for water or other contaminants. If possible, a plastic sheet should be taped to the ceiling as well. Figure 3 illustrates an ambulance being prepared in this way to transport hazardous materials exposure victims.

   The radio, main oxygen supply, and suction equipment will now have been sealed behind plastic sheets and will not be available to the patient compartment. Portable equipment should be used for treatment at the scene and during transport.

5. *Return all needed equipment to ambulance:* The equipment that will needed to care for victims and that was removed earlier from the ambulance should now be returned to the patient compartment. This equipment is placed within the plastic covered area in which patients will be treated.

6. *Use disposable and portable equipment:* Whenever possible, use portable or disposable equipment for treating hazardous materials victims. This suggestion will make decontamination of equipment as easy as possible and will avoid loss of a major vehicle if a component piece of equipment becomes contaminated.

7. *Monitor contamination:* After victims have been transported to the emergency department, it is recommended that the vehicle and crew return to the hazardous material incident for decontamination. If the ambulance will not be needed for additional exposure victims, the plastic and tape should be removed and disposed of in an area designated for contaminated wastes and materials. The vehicle should be monitored and decontaminated if necessary (Figure 4). The crew should go through decontamination to the extent required by their level of contamination and exposure.

Protective Equipment and Supplies

A minimum amount of protective equipment and supplies are needed so that ambulance-based EMS personnel can protect themselves and their ambu-

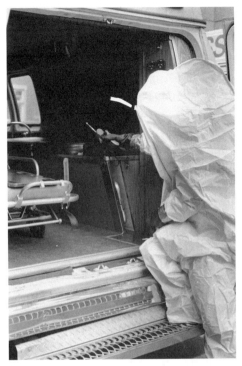

Figure 4 The transporting vehicle is monitored
for contaminants.

lances. The equipment listed subsequently is appropriate for ambulance crews
who remain outside of heavily contaminated areas. It is also appropriate for
rescue personnel who care for victims who have received incomplete decontamination. Rescuers who are exposed to higher levels of contamination
require higher levels of personal protection.

1. *Gloves* should be worn by all rescuers who have physical contact with
   exposure victims. Generally, use of double layers of vinyl or latex surgical gloves provides adequate protection when victims have undergone
   decontamination. A large supply of vinyl or latex examining gloves should
   be available in the ambulance.

   If rescuers are likely to care for heavily contaminated victims, then higher
   levels of glove protection should be considered. Use gloves that are made
   of materials effective against the chemicals to be encountered. Rescuers
   should refer to compatibility charts when selecting the glove to use at a

hazardous materials incident. Such compatibility charts are usually provided by glove manufacturers. Compatibility charts are also available in reference books such as *Guidelines for the Selection of Chemical Protective Clothing*.

When special protective gloves are worn by rescuers, vinyl or latex gloves should also be worn as an inner lining. These higher-protection gloves tend to be thick and clumsy. They can interfere with hand dexterity and sensitivity of touch. Wearing them will restrict the ability of rescuers to perform delicate EMS procedures. They may also limit the ability of a rescuer to perform a secondary survey.

Specific types of commonly used chemical protective gloves are described subsequently. They are listed from least expensive to most expensive. Also presented are general recommendations about the classes of chemicals for which each provides good or bad protection. Because many exceptions to these recommendations exist, rescuers should verify the appropriateness of each type of glove before it is used.

- *Natural rubber* gloves are flexible and inexpensive. They offer only limited protection, however. They are useful for protection against alcohols and dilute acids and bases. They offer poor protection against most organic chemicals and solvents.
- *Nitrile rubber* gloves provide good general purpose protection. These gloves are relatively flexible and inexpensive. They are good protection against alcohols, oils and fuels, alkali, amines, and phenols. They are poor protection against aromatic and halogenated hydrocarbons, amides, ketones, and esters.
- *Neoprene rubber* gloves offer good protection against strong alkali, dilute acids, alcohols, phenols, fuels and oils, and aliphatic hydrocarbons. They are not protective against aromatic and halogenated hydrocarbons, ketones, and concentrated acids. They are several times more expensive than nitrile rubber gloves.
- *Butyl rubber* gloves are effective protection against alkali and many organic chemicals, but they are not protective against aliphatic, aromatic, or halogenated hydrocarbons or gasoline. They are nearly twice as expensive as neoprene rubber gloves.
- *Viton* gloves provide good protection against organic solvents (such as aliphatic, aromatic and halogenated hydrocarbons, and acids). They do not provide protection against ketones, esters, aldehydes and amines. They are more than ten times more costly than nitrile and neoprene rubber gloves.

2. *Overboots* of latex or neoprene should be worn on the feet to protect rescue personnel from contamination caused by water run-off and other spilled liquids.

3. *Face shields or safety goggles* should be worn to prevent splash exposure of a rescuer's eyes, nose, or mouth. Shields and goggles also protect against contamination by dusts. Equipment of this type does not protect against inhalation of toxic gases.

4. *Chemical resistant jumpsuits* can protect rescue personnel from exposures resulting from liquids and dusts. Some suits can also protect EMS personnel from exposure to blood and other body fluids. In general, laminated suits offer more resistance than nonlaminates. Examples of suitable protective fabrics include Tyvek® (The DuPont Company, Wilmington, Del.) coated with either polyethylene or Saranex® (The Dow Chemical Company, Midland, Mich.).

   Protective suits that are disposable are usually least expensive. Suits should have hoods. Keep available either a selection of sizes or only extra-large sizes to be certain that suits will fit all personnel.

5. *Sheets* of plastic of Tyvek® in sufficient size should be available so that the inside of the ambulance can be encapsulated.

6. *Duct tape* should be available.

7. *Stretchers* should be used that do not absorb hazardous materials. For example, metal or plastic stretchers are adequate. Stretchers with cloth, wood, or leather parts will absorb chemicals and pose a risk of contamination to future victims and rescuers.

8. *Reference books* should be available to EMS personnel who must research the health effects of specific hazardous materials exposures.

Protection of the Emergency Department and Its Staff

It is important that the emergency department and hospitals to which exposure victims are transported not become contaminated. If contamination did occur, it might be necessary for the entire hospital or its emergency department to be closed until decontamination had been completed. The emergency department staff and any patients who happen to be in the emergency department must also be protected.

The following guidelines and procedures describe some ways to avoid contamination of the emergency department facility and its staff when hazardous materials exposure victims are transported to hospital.

1. *Perform triage outdoors:* When possible, triage and assessment of victims should occur outside of the hospital and its emergency department. The fewer the number of contaminated patients who enter the emergency department and the more slowly they enter, the more likely that contamination can be controlled. It is possible that some triaged patients will not require admission to the emergency department and that others can receive further decontamination before emergency department entry [Figures 5(a) and 5(b)].

2. *Use a separate emergency department entrance:* Potentially contaminated victims should enter the emergency department through a separate hazardous materials entrance that leads them to an area where decontamination and assessment can be performed (Figure 6). In this way, it is less likely that other patients and public areas will be contaminated.

3. *Use a designated treatment room:* Victims should be fully decontaminated before they are allowed to enter the open emergency department. If contaminated patients need emergency care before complete decontamination, they should be evaluated and treated in designated areas that are not simultaneously used by other patients (Figure 7).

   If possible, these rooms should be closed, self-contained spaces to limit the spread of contaminants further. The room should be clearly marked so that hospital staff are warned of contamination dangers. Exposure victims should not be allowed to make contact with other patients who happen to be in the emergency department at the same time.

4. *Restrict access to treatment room:* Only personnel directly involved in the care of exposure victims should be allowed to enter the treatment area. Visitors and guests must not be permitted to visit these patients until all contamination risks have been addressed.

5. *Limit staff contact with exposure victims:* The number of emergency department staff who have contact with exposure victims should be limited to the minimum number actually required. Those staff members should not have responsibility to other uncontaminated patients who happen to be in the emergency department at the same time. Pregnant staff members should not be assigned to care for victims of hazardous chemical exposures.

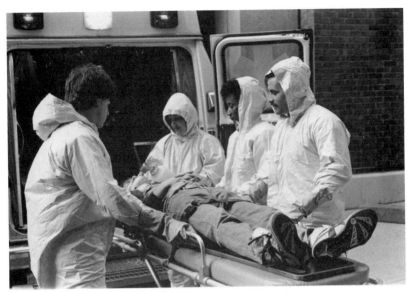

Figure 5(a) *The rescuers and hospital personnel, dressed in jumpsuits to avoid possible contamination, lower the victim from the ambulance.*

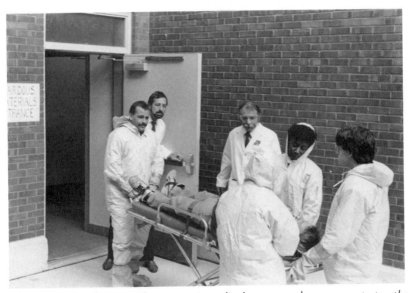

Figure 5(b) *Emergency department medical personnel come out to the transporting vehicle to greet the rescuers and guide the rescue team through the hazardous materials emergency entrance.*

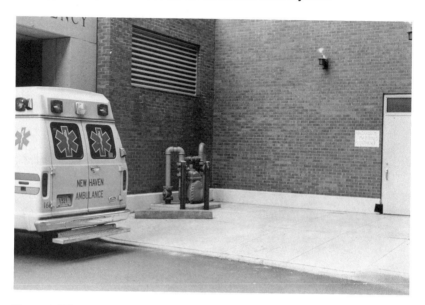

Figure 6 Whenever possible, a separate entrance to the emergency department should be used when transporting victims of hazardous emergencies.

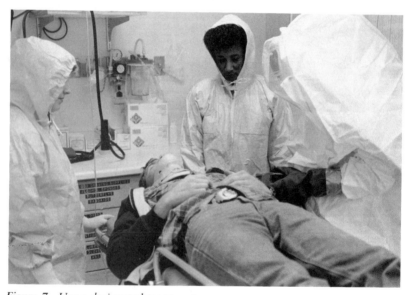

Figure 7 Use a designated treatment room.

6. *Consider use of personal protective equipment:* Even in the emergency department, the use of personal protective equipment can serve to protect staff from contamination. Disposable Level B jumpsuits provide relatively inexpensive protection of staff members performing triage and decontamination of patients. Double layers of vinyl or latex gloves can also be used for additional staff protection.

7. *Protect emergency department from contaminants:* The floor over which exposure victims and rescuers travel should be covered with heavy gauge plastic sheets so that contaminants do not make contact with the emergency department floor. Disposal bags and barrels should be available to contain victims' clothing and other contaminated materials.

8. *Maintain a personal exposure log:* All emergency department staff who have direct contact with hazardous materials or hazarous materials victims should keep a diary in which they record all exposures or contaminations that they have suffered. The diary should include the names of chemicals that caused exposure, the date of the exposure, and any symptoms that develop. In addition, responders should undergo regular, periodic medical examinations and routine laboratory testing. The information contained in the exposure log should be made available to the responder's personal physician at the time of each medical examination.

   Any staff member who has suffered a documented hazardous materials exposure should have a medical examination directed to the possible effects of that exposure. A written report of the examination findings should be included in the personal log.

9. *Monitor cleanup:* After exposure victims have been treated and either discharged or transferred, the emergency department treatment rooms should be monitored to assure that no residual contamination is present before those rooms are used for other patients.

10. *Isolate drainage and ventilation:* When possible, exposure victims should be washed and treated in hospital areas that have water drains leading to isolated holding tanks and ventilation systems that are independent of those used by the rest of the hospital. Unfortunately, these concerns have only rarely been addressed in the design and construction of hospitals. As a result, the wash water in many hospitals is drained into community sewer systems, and a risk often exists that airborne contaminants will be carried throughout the hospital via ventilation systems.

    One way to contain contaminated wash water effectively is by use of a decontamination stretcher with hosing that drains the water into a self-contained collection system or into a reservoir or 55-gallon drum [Figures 8(a) and 8(b)]. A morgue table can be readily turned into a decontamination stretcher for this purpose.

Summary

Emergency responders involved with hazardous materials incidents must help to protect the downstream health care system from contamination. Their goal should be to assure that the entire health care system continues to function properly despite the presence of hazardous chemicals. To achieve that goal, it is necessary to restrict the spread of contamination. Protection must be provided to EMS personnel, EMS vehicles and ambulances, emergency department staff and patients, and the physical facility of the emergency department and hospital. The protocols and guidelines presented earlier provide methods for protecting the various components of the health care system.

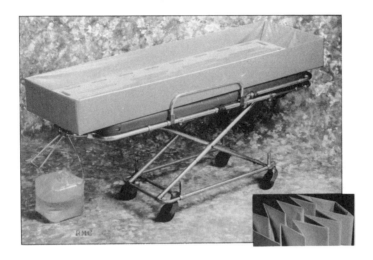

Figures 8(a) and (b)   *Two examples of commercial decontamination stretchers with self-contained collection systems. (Courtesy of Radiation Management Consultants, Philadelphia, Pa.)*

References

Bronstein, Alvin C., and Phillip L. Currance, *Emergency Care for Hazardous Materials Exposure*. St. Louis: The C. V. Mosby Company, 1988.

Carlson, Gene P., ed., *HazMat Response Team Leak and Spill Guide*. Stillwater, Okla.: Fire Protection Publications, 1984.

Guidelines for the Selection of Chemical Protective Clothing. Cambridge, Mass.: Arthur D. Little, 1987.

Hazardous Materials Response for First Responders. Washington, D.C.: U.S. Environmental Protection Agency, 1987.

Proceedings of "The Medical Management of HazMat Incidents," presented by Michael S. Hildebrand, American Petroleum Institute, and Prince George Fire Department's Hazardous Materials Response Team; sponsored by the EMS degree program, George Washington University, Washington, D.C., June 1988.

Noll, Gregory G., Michael S. Hildebrand, and James G. Yvorra, *Hazardous Materials: Managing the Incident*. Stillwater, Okla.: Fire Protection Publications, 1988.

Supplement 13: Medical Surveillance Guidelines

This supplement provides examples of medical surveillance procedures for hazardous materials response personnel that are intended to help local agencies formulate their own policies. Our thanks to Dr. Mary Jo McMullen, a member of the Hazardous Materials Response Personnel Committee and medical advisor to the Summit County, Ohio, Hazardous Materials Response Team, and to Peter McMahon, Chairman of the Hazardous Materials Response Personnel Committee, for providing information from the Erie County, New York, Hazardous Materials Response Team.

Summit County SOPs

Definition of Medical Surveillance

Worker medical surveillance is the ongoing, systematic evaluation of employees at risk of suffering adverse effects of workplace exposure for the purpose of achieving early recognition and, ideally, prevention of these effects.

While medical surveillance is often interpreted as including only standardized, usually legally mandated periodic examinations looking for evidence of disease, a sound state-of-the-art medical surveillance program should possess other elements designed to protect employee health, including:

A. Preplacement Examinations

The purpose of these exams is to assess fitness-for-duty and to provide baseline data for subsequent examinations. Fitness-for-duty must be assessed in the context of the physical requirements of the job.

B. Periodic Examinations

The purpose of these examinations is:

1. To detect clinical or pre-clinical signs of adverse health effects which might signal developmentof workplace-related adverse health effects;
2. To detect problems not caused by work that might affect an employee's ability to safely perform his/her job duties.

C. Exposure-Related Examinations

The purpose of these examinations is to detect any changes from baseline that might have occurred as a result of a specific workplace exposure, or to evaluate whether a medical problem experienced by an employee might be related to a workplace exposure.

D. Return-to-Work Examinations

This is a special type of fitness-for-duty examination, conducted to insure that employees returning to work after a work-related or non-work-related illness or injury are capable of safely performing their job duties without harm to themselves or others.

Comments

Examination types C and D above are often performed by employees' personal physicians. In a well-designed medical surveillance program, it is important for a physician familiar with the physical requirements and chemical and physical hazards of the job to perform or review these examinations so that worker health is better safeguarded and appropriate information is included in the workplace medical record.

E. Cross-Sectional and Longitudinal Analyses of Examination Results

The purpose of these analyses is to detect trends in symptoms, signs and/or laboratory results that may indicate a workplace health hazard. Analysis of exposure records and worker absentee records may also be useful for these analyses.

OSHA Requirements for Haz Mat Team Medical Surveillance

OSHA Regulation 1910.120 regarding "hazardous waste operations and emergency response" requires medical surveillance for Haz Mat teams. Other relevant OSHA regulations include 1910.20 (access to employee exposure and medical records) and 1910.134 (respiratory protection).

The essential elements of the OSHA-mandated surveillance applicable to Haz Mat teams are as follows:

A. Frequency of Examinations

1. Prior to assignment.
2. Annually. May be performed less frequently if deemed appropriate by physician overseeing program, but at least every two years.
3. On termination of employment or termination as member of Haz Mat team, if no exam performed within prior six months.
4. As soon as possible upon notification by employee of overexposure (even if asymptomatic) or development of signs or symptoms indicating possible overexposure. Follow-up examinations as determined by physician.

B. Examination Contents

1. To be determined by "attending physician."
2. To include fitness-for-duty and ability to wear PPE (personal protective equipment).

C. Information Provided to Physician

1. Copy of relevant OSHA standards.
2. Description of employee's duties, exposures, required PPE, previous examination results.

D. Written Record by M.D. to Employer Will Include:

1. Physician's opinion concerning "any medical conditions which would place the employee at increased risk of material impairment of the employee's health" from work on Haz Mat team and use of PPE;
2. Any recommended work limitations;
3. A statement that the employee has been informed by the physician of the results of the medical examination and any medical conditions which require further examination or treatment.
4. The material given to the employer "shall not reveal specific findings or diagnoses unrelated to occupational exposures."

E. Rights of Employees

1. All employees and their "designated representatives" have the right to obtain copies of their medical records.
2. OSHA requires that these examinations be "made available" to covered employees; OSHA does not make these examinations mandatory for employees.

F. Record Keeping

1. The records must be retained for the duration of employment plus 30 years.

On-Scene Medical Surveillance

The following procedures are intended for on-scene medical surveillance of emergency personnel using totally encapsulating suits.

The *"pre-suit" exam* is intended to determine whether or not someone is physically fit, at that point in time, for suit entry. Therefore, a pre-suit exam *should be completed on those individuals who are going into encapsulating suits.* This would include entry team and decon as appropriate.

The *"post-suit" exam* is intended to determine whether the individual has suffered any immediate effects from a chemical exposure or from the heat stress of the suit. It is intended to serve as a tool for medically monitoring those team members who have had any chemical exposures. It also serves as part of the data base for the biannual exams. Therefore, a *post-suit exam should be conducted on anyone who has been in an encapsulating suit and anyone who has* been in the hot zone. Additionally, the Haz Mat control officer may request any other personnel to receive a post-suit exam at his discretion.

For example, if all the work at a given scene is done in structural fire fighting protective clothing, no one would need pre-suit exams, but everyone in the hot zone would need post-suit exams. It is also important to remember that copies of these forms, filled out completely, need to be received by the medical advisor as quickly as possible after an incident, preferably within twenty-four hours.

To recap: A pre-suit exam is necessary for anyone going into an encapsulating suit. A post-suit exam is necessary for anyone coming out of the hot zone.

A. Pre-Suit Exam

1. History: new medical problems within the past two weeks?
   all medications including over-the-counter (OTC) drugs
   medication allergies
   alcohol in the past 24 hours? past 2 hours?
2. PE: vital signs — temp, pulse, resp rate, BP, and body weight
   skin lesions?
   lung sounds?
   rhythm strip (optional)
   brief mental status exam
3. Recommended pre-suit oral hydration: 8-16 oz water, Exceed, or diluted (2:1) Gatorade

B. Criteria to Deny Entry

1. History: new onset heart/lung problems, hypertension, diabetes
   recent (within 72 hours) nausea/vomiting, diarrhea, fever, heat exhaustion
   new prescription medication (check with M.D.), OTCs — cold meds, decongestants, antihistamines
   heavy alcohol intake in past 24 hours, any alcohol intake in past 2 hours
2. PE: T > 99.5°F oral or 100.5°F core, resp rate >24, BP > 105 diastolic
   P > 70% of max; max heart rate (220-age) x .7 (see table p. 942)
   skin — open sores, large areas of rashes, sunburn
   lungs — wheezing or congested lung sounds
   hard contact lenses
   as determined by the safety officer or Haz Mat control officer

C. Post-Suit Exam

1. History: any symptoms of chemical exposure, heat stress or cardiovascular collapse

Age-Predicted Heart Rates		
Age	70%	85%
20–25	140	170
25–30	136	165
30–35	132	160
35–40	128	153
40–45	125	149
45–50	122	145

2. PE: vital signs immediately and at 5–10 minutes (BP, P, RR, T, Wt, orthostatics as indicated)
   skin
   lungs
   brief mental status exam
   rhythm strip if pulse is irregular

Treatment Protocol for Haz Mat Team Members

1. All personnel — rest time equals suit time plus oral rehydration (Exceed, Gatorade, pop, coffee, not water). Give all personnel signs and symptoms to watch for.
2. If at 10 minutes a team member is not within 10% of baseline, do orthostatic vital signs.
3. IF > 3% body weight loss (4½ # in a 150 # person), or positive orthostatics (pulse increase by 20 or systolic BP decreases by 20 at 2 minutes standing)
   or > 85% max pulse at 10 minutes
   or T > 101°F oral (102°F core)
   or nausea, altered mental status or any other symptoms

   THEN IV fluid hydration RL or NS @ rate (usually w/o) to get P < 100, systolic BP > 110
   oxygen 4–6 liters/minute NC, may increase as needed
      use Alvin Bronstein and Phillip Currance, *Emergency Care for Hazardous Materials Exposure,* or appropriate field guide, or med control for treatment of specific symptoms/types of exposure
   all such personnel are to be transported to the hospital
   complete a standard squad run report, pink copy to EMS C/O

PERSONNEL INCIDENT LOG

To be completed on all suited team members, any personnel who become symptomatic, and at request of I/C or Haz Mat C/O. To be forwarded to medical advisor within 24 hours of incident for review. Think about local fire department's initial responders.

Name _____   Date _____

Dept _____   SSN _____

Location _____   Assignment _____

Incident C/O _____   EMS C/O _____

Materials involved _____

_____

Level of protection _____   Exposure time _____

Decontamination type _____

Type of exposure (circle): inhalation, ingestion, skin absorption, suit only

Pre-Suit Exam:

Hx: Ok_____   Not ok_____   Explain _____

          Meds _____

          Allergies _____

PE: Time_____   P _____   BP_____   RR _____   T _____   Wt _____

          Skin lesions? _____

          Lung sounds? _____

          Mental status exam? _____

Post Exposure Exam

Hx: any complaints? _____

PE: Time_____   P _____   BP_____   RR _____   T _____   Wt _____

PE: Time_____   P _____   BP_____   RR _____   T _____   Orthos _____

          Skin lesions? _____

          Lung sounds? _____

          Mental status exam? _____

          ☐ Ok to return to work   ☐   Not ok   Explain _____

          _____

          ☐ Required medical attention, see run sheet attached

Signed _____

Protocol for Medical Follow-up of Hazardous Materials Incidents

The medical advisor is to be notified by the incident commander as soon as possible when Haz Mat personnel are transported from the scene to the hospital. This will be on check sheet. This will facilitate coordination of immediate and follow-up care.

All personnel incident logs are to be forwarded to the medical advisor immediately after the Haz Mat incident and will be filed with their physical examinations. Follow-up will be recommended as necessary. Cost of follow-up evaluations and occupational medicine specialist consultation will be paid for by the employer (per SARA). It may be possible to bill this to party responsible for incident through emergency management.

In cases where multiple personnel were exposed (over five), the zone coordinators for the involved zones will be expected to work with the medical advisor to facilitate follow-up.

It is incumbent upon individual members of the Haz Mat team to report to the medical advisor *any* illness (medical signs or symptoms) occurring within 72 hours of a hazardous materials exposure and any illness (signs or symptoms) occurring at any time which the member or his physician feels may be related to a hazardous materials exposure.

Erie County On-Scene Medical Surveillance Forms

The following are additional forms provided for reference by the Erie County Hazardous Materials Response Team.

MEDICAL OFFICER REPORT

Incident # _____   Date _____   Location_____   Time _____

Medical Officer _____   Tech_____

Tech_____   Tech_____

Medical Officer General Responsibilities

Initial When Completed

1. Locate Medical Sector adjacent and to the rear of Decon Exit. _____

2. Refer to Reference Material for signs and symptoms of over-exposure. _____

3. Evaluate, release or restrict team personnel prior to suit-up per checklist. _____

4. Coordinate with Command for transport of injured persons and notify hospital(s). _____

5. Maintain Patient Log Sheet. _____

6. Evaluate teams after Decon per checklist. _____

7. Copies of all medical records and forms to IC. _____

8. Clean-up of Medical Sector. _____

Pre-Entry Instructions

1. Prior to donning appropriate protective clothing, all entry-level personnel will undergo a physical assessment and hydration check.

2. Entry will be denied to any personnel with:

Temperature	100° Orally
Respirations	>24
Pulse	>110
Blood Pressure	170/88

3. Hydration—Each entry-level member will be given 16 oz. of water or a 4:1 electrolyte solution.

Post-Entry Instructions

1. After Decon, Entry personnel will undergo an assessment and hydration check.

2. If temperature is >100°F, rest in a cool environment.

3. If temperature is >102°F, rest in cool environment and apply tepid towels to the neck, groin and underarms.

4. If respirations, pulse or blood pressure exceed parameters in the pre-entry section, or 10% of the pre-exposed value, personnel should be kept at rest, and reassessed every 5 minutes until return to normal and/or acceptable amounts.

5. If there is no sign of heat disorders or dehydration, administer cool fluids by mouth in 8 oz. increments until saturated.

6. If dehydrated, administer O_2—10 Liters via non-rebreather.

7. Further assessment is necessary if indicated by toxicity of substance and exposure.

Attending Medical Person Signature _____

Medical Officer Signature _____

PERSONNEL MEDICAL EVALUATION

Incident # _____   Date _____   Location _____   Time _____

Team Member _____   Department _____

Type of Protective Clothing to Be Worn: _____

_____

Pre-Entry Evaluation   **Post-Entry Evaluation**

Skin Color & Temperature _____   Skin Color & Temperature

Respirations   _____   Respirations   _____

Pulse   _____   Pulse   _____

Blood Pressure   _____   Blood Pressure   _____

Time In   _____   Time Out   _____

Entry Will Be Denied to Any Person with:

Temperature	100° Orally
Respirations	>24
Pulse	>110
Blood Pressure	170/88

Hydration: Each entry-level member will be given 16 oz. of water or a 4:1 electrolyte solution.

Entry and Back-up Team personnel are to read the following:

While in protective clothing, you are to monitor your own physical condition. Any changes in gait, speech, or behavior will require immediate withdrawal and decontamination, and assessment. Any feelings of chest pain, dizziness, shortness of breath, headache, or weakness also require immediate withdrawal and decontamination.

Working time in the suits varies with ambient temperature and humidity. Time in SCBA or in protective clothing must be monitored and must have at least 15 min. reserve time or tag line access for decontamination.

Circle the temperature which applies:

Temperature	**Time**
73°F to 99°F	15 - 20 min
100°F and greater	10 min

Attending Medical Person Signature _____

Medical Officer Signature _____

HEALTH EXPOSURE REPORT

Name _____   Unit Number _____   Assignment_____

Incident Number   □□□□□□   Date Exposed   □□□□□□   Date
Filed   □□□□□□

Level of Treatment □   Safety Equip. Used □

1-On Scene   1-Firefighting Turnout/SCBA

2-Hospital   2-Protective Gloves

3-Special Center   3-Protective Gloves, Goggles, Mask and Disposable Suit

4-Doctor's Office   4-Encapsulated Suit Type _____

5-Other _____   5-Other _____

6-None at this time   6-None

Substances

Substance[s] Exposed to   □□   □□   □□   □□

Duration of Exposure   □□ |□ □   □□ |□ □   □□ |□ □   □□ |□ □
                        hours minutes hours minutes hours minutes hours minutes

01-Acetic Acid	13-Chlorine	25-Muratic Acid
02-Acetone	14-Hepatitis B	26-Nitric Acid
03-Acetylene	15-HIV [AIDS]	27-PCBs
04-Alcohol [Ethyl]	16-Hydrochloric Acid	28-Polyvinyl Chloride
05-Alkaline	17-Hydrocyanic Acid	29-Rabies
06-Anhydrous Ammonia	18-Hydrogen Chloride	30-Radioactive Materials
07-Asbestos	19-Hydrogen Sulfide	31-Sulphur [Sulfur] Dioxide
08-Asphalt	20-Insecticide	32-Tuberculosis
09-Battery Fluid Alkaline	21-Ketone	33-Lye
10-Calcium Hypochlorite	22-Liquefied Petroleum Gas	34-Meningoccal Meningitis
11-Carbon Dioxide	23-Sulfuric Acid	35-Haz-Mat Not Identified
12-Carbon Monoxide	24-Mononucleosis	36-Fire Gases Not Identified
		37-Other _____

Method of Exposure _____

_____

Extent of Injuries at This Time _____

_____

Others Exposed _____

_____

HAZARDOUS MATERIALS RESPONSE HANDBOOK

PATIENT LOG

Incident #_____   Date_____   Location _____

Triage Officer _____   Page _____   of _____

_____

Name/Tag Number   Age   Hospital   Chief Complaint

_____
_____
_____
_____
_____
_____
_____
_____
_____
_____
_____
_____
_____
_____
_____
_____
_____
_____
_____
_____
_____
_____
_____
_____
_____
_____

Supplement 14: EMS Sector Standard Operating Procedures

This supplement contains standard operating procedures for a hazardous materials incident EMS sector developed by the Montgomery County, Maryland, Department of Fire and Rescue Services along with sample medical protocols developed by the State of Maryland. Thanks to Deputy Chief Mary Beth Michos of the Montgomery County Department of Fire and Rescue Services.

Montgomery County Standard Operating Procedures

I. General Information

On all hazardous materials incidents requiring levels A or B protective clothing or specialized protective clothing, an EMS sector officer (designated Haz Mat EMS) shall be assigned by the Haz Mat sector officer.

Where possible, the EMS sector officer should be trained to a minimum of the EMT-P level.

At the discretion of the Haz Mat sector officer, an EMS sector officer may be assigned on other incidents where conditions may subject personnel to severe stress.

Haz Mat EMS shall assume responsibility for the medical care of all Haz Mat response personnel and advise the EMS control officer (FRC ICS) on the care of other emergency services personnel and civilian victims.

II. Haz Mat EMS Functions

The Haz Mat EMS sector shall be conducted in accordance with the following:

1. The health and emergency care of all Haz Mat personnel shall be the prime mission of the Haz Mat EMS sector.
2. At least one ALS unit shall remain committed to provide advanced care whenever Haz Mat personnel are engaged in entry operations.
3. The Haz Mat EMS officer shall assign ALS personnel to perform pre-entry and post-entry medical evaluations of entry team personnel in accordance with *Section V* and *Section VI* of this procedure.
4. The Haz Mat EMS officer shall have the authority to deny entry to any Haz Mat personnel for medical reasons, after evaluation. The Haz Mat EMS officer shall also retain the authority to order any Haz Mat personnel to undergo medical evaluation and transport to a hospital for further evaluation and/or treatment.
5. The Haz Mat EMS officer will coordinate with the research officer on the acute, delayed, and chronic effects of exposure and treatment for each hazardous material. The Haz Mat EMS officer shall ensure that such information is provided to the EMS control officer and any receiving hospitals.
6. The Haz Mat EMS officer shall ensure that a complete and accurate log is maintained of all EMS functions including pre- and post-entry evaluations, treatment of personnel, and communications with the EMS Control Officer.
7. The Haz Mat EMS officer shall complete the Hazardous Materials EMS Checklist, as appropriate, on the following pages.

III. Haz Mat EMS Resources

☐ Hazardous Materials EMS Checklist
☐ ALS unit immediately available

☐ BLS equipment, monitor and drug kit from Haz Mat 7

☐ Haz Mat medical evaluation forms

☐ HIRT Internal Communications Capability

☐ Note pad and pencil

IV. Incident Operations Structure

The Haz Mat EMS officers shall be responsible for maintaining the following lines of communications with other sectors. This directive shall not preclude direct communications with other sector officers where necessary.

```
HAZ MAT SECTOR OFFICER <  - - - - - - - - - - - - - - - - - - - - - -
                         |                                          ¦
                         |                                          ¦
                         |                                          ¦
                         |------------------------------------------|
                         |                                          |
                         |                                          |
OPERATIONS OFFICER < - - - - - - - - - - - - - - - - - - - - - >SAFETY
                                                                    |
                                                                    |
                                                                    |
                                                                  EMS
```

Incident operations structure

V. Personnel Medical Evaluations

Medical evaluations of physical condition are to be conducted on all personnel prior to suiting up and engaging in activities while wearing chemical protective clothing.

Assessment of the following physical signs shall be conducted and recorded for a pre-entry baseline. The pre-entry baseline should be compared to the medical data carried on board Haz Mat 7. Any significant deviations may, in the opinion of the Haz Mat EMS officer, preclude an individual from operating as part of an entry team.

An assessment of the same physical signs shall be conducted and recorded for all personnel immediately upon leaving decon. Any significant changes from the pre-entry baseline may, in the opinion of the Haz Mat EMS officer, preclude an individual from participating in another entry.

If, in the opinion of the Haz Mat EMS officer, personnel require further evaluation at a medical facility, personnel shall be transported without delay and shall not deny transport.

VI. Physical Signs

1. **Blood Pressure:** Entry shall be denied to personnel with a blood pressure exceeding 150 systolic or 100 diastolic where there is 20 or more points deviation from their normal resting pressure.
2. **Pulse:** Entry shall be denied to personnel with a pulse greater than 110 or irregular without prior history.
3. **Respirations:** Entry shall be denied to personnel with a respiratory rate greater than 24.
4. **Temperature:** Entry shall be denied to personnel with an oral temperature greater than 99.2°.
5. **EKG:** All personnel shall have a minimum 10 second rhythm strip obtained.
6. **Weight:** Re-entry shall be denied personnel who show a loss of more than 2% pre-hydration weight.

Prior to entry, all personnel are to hydrate with *AT LEAST 1 PINT* of Exceed or water. Weight should be taken prior to hydration to prevent false fluid loss readings.

EMS SECTOR OFFICER CHECKLIST

Nature of Incident: _____

_____

Location: _____

EMS Sector Officer: _____

_____

☐ EMS Sector Officer Identified by EMS Vest

☐ EMS SECTOR PERSONNEL / ASSIGNMENTS

1. _____ / _____

2. _____ / _____

3. _____ / _____

4. _____ / _____

COMMAND

Haz Mat Sector Officer: _____

Haz Mat Safety Officer: _____

Haz Mat Operations Officer: _____

Decon Officer: _____

EMS Control Officer: _____

Command Post Location: _____

SITE SET-UP

☐ *TREATMENT AND TRIAGE SITE SET-UP (patients)*

Location (describe area, should be close to decon): _____

_____

☐ *MEDICAL EVALUATION SITE SET-UP (pre- and post-entry evaluations)*

Location (describe area, should be close to haz mat units):

_____

☐ *COMMAND, ENTRY, SAFETY, AND DECON OFFICERS NOTIFIED OF LOCATION*

☐ *MEDICAL EVALUATION AND TREATMENT SUPPLIES AND EQUIPMENT AVAILABLE*

☐ *TRANSPORT VEHICLES AVAILABLE* (one must be an ALS unit)

    ☐ Unit Number _____ ☐ ALS [ ]

    ☐ Unit Number _____ ☐ ALS [ ]   ☐ BLS [ ]

    ☐ Unit Number _____ ☐ ALS [ ]   ☐ BLS [ ]

    ☐ Unit Number _____ ☐ ALS [ ]   ☐ BLS [ ]

    Other (bus, helicopter, etc.)

    _____

    _____

☐ *PROTECTIVE CLOTHING FOR EMS PERSONNEL DETERMINED*

Level of Protection: _____ Type of Clothing: _____

☐ *PRIMARY RECEIVING HOSPITAL IDENTIFIED:* _____

(Check with EMS Control Officer.)

☐ *RECEIVING HOSPITAL NOTIFIED OF DECONTAMINATION PROCEDURES*

(Consult Decon Officer for recommendations.)

CHEMICAL INFORMATION

☐ *NAME(S) OF CHEMICALS INVOLVED (obtained from research)*

1. _____
2. _____
3. _____
4. _____
5. _____
6. _____

☐ *SIGNS / SYMPTOMS OF EXPOSURE AND ONSET*

Chem   ☐ ACUTE   ☐ DELAYED. . . . .Time _____

1.		
2.		
3.		
4.		
5.		
6.		

☐ *ADDITIONAL CHEMICALS LISTED IN NOTES SECTION*

MEDICAL TREATMENT

☐ *EXPOSURE TREATMENT*   ☐ Protocol   ☐ By Case

☐ Physician Contacted   Who _____   Time _____

1. _____

2. _____

3. _____

4. _____

☐ *ANTIDOTES*

1. _____   3. _____

2. _____   4. _____

CONTRAINDICATIONS

1. _____   3. _____

2. _____   4. _____

☐ *FACILITY CONTACTED FOR TREATMENT / ANTIDOTE INFORMATION*

☐ Poison Control Center   ☐ Phone Number: _____

☐ Other:

Facility: _____   Phone Number: _____

Contact Person: _____

Facility: _____   Phone Number: _____

Contact Person: _____

☐ *AVAILABILITY OF DRUGS / ANTIDOTES ESTABLISHED*

Locations: _____

_____

_____

ENTRY TEAM SAFETY

☐ *ENTRY TEAM BRIEFED ON EFFECTS OF CHEMICAL(S)*

☐ *SAFETY OFFICER BRIEFED ON EFFECTS OF CHEMICAL(S)*

☐ *E.M.S. PERSONNEL BRIEFED:*
    ☐ on effects of chemical(s)
    ☐ treatment procedures
    ☐ medical monitoring procedures review procedures sheet

☐ *PRE-ENTRY PHYSICALS CONDUCTED (see attached sheet)*

☐ *PROTECTIVE CLOTHING FOR E.M.S. DETERMINED*

☐ *LEVEL OF PROTECTION* ☐ A ☐ B ☐ C ☐ D

☐ *POST-ENTRY PHYSICALS CONDUCTED (see attached sheet)*

☐ *POST-TRANSPORT DECON REQUIRED FOR:*

    ☐ Rescuers   ☐ Vehicles   ☐ Equipment

☐ *RECEIVING HOSPITAL NOTIFIED OF DECON REQUIREMENTS AND PROCEDURES*

    ☐ Personnel   ☐ isolated area required?
    ☐ Equipment   ☐ isolated area required?

NOTES:

(include significant observations such as time in suit, inappropriate behaviors, intuitive feelings, etc.)

MEDICAL EVALUATION FORM—Entry Team No. ____

PRE-ENTRY EVALUATION

NAME	B/P	PULSE	RESP.	TEMP.	WEIGHT	EKG

POST-ENTRY EVALUATION

NAME	B/P	PULSE	RESP.	TEMP.	WEIGHT	EKG

5 MINUTES POST-ENTRY

NAME	B/P	PULSE	RESP.	TEMP.	WEIGHT	EKG

Maryland Medical Protocols for Hazardous Materials Exposure

1. Introduction

1.1 This protocol assumes that the ambulance is the first and only unit to arrive on the scene. Should there already be other units on the scene, the Incident Commander's instructions should be strictly adhered to in conjunction with this protocol.

2. En Route to and Approaching the Incident Scene

2.1 If hazardous material involved in the incident is known while responding to the scene, begin to research the hazardous material using appropriate reference material [e.g., Manufacturer Safety Data Sheets (M.S.D.S.), D.O.T. *Emergency Response Guidebook*, pre-incident plans]. Become familiar with the following:

2.1.1 Potential health hazards.

2.1.2 Proper level of personal protection equipment indicated by the hazardous material.

2.1.3 Other potential hazards.

2.1.4 "Safe distance" (the distance from the incident that is considered to be free from hazards).

2.2 If at all possible, approach the incident from uphill and upwind.

2.3 While nearing the scene, be observant for environmental clues (e.g., the lean of the trees may indicate wind direction; unusual odors or vapor clouds may indicate a hazardous condition).

2.4 Begin to don the proper level of protective clothing and equipment if available and trained in its use.

3. Arrival at the Incident Scene

3.1 Position the ambulance vehicle outside the "hot" zone at a safe distance.

3.2 *Immediately* establish a "hot" zone and deny access by anyone into that area. Upon arrival of additional units, stage as necessary and establish "warm" and "cold" zones as appropriate.

3.3 Evaluate the magnitude of the incident and gather as much specific information as possible on the hazardous material involved without endangering personnel.

3.4 Call for appropriate assistance.

3.4.1 Coordinate closely with other responding units and/or agencies. Confirm hazardous material involved, advise best route of travel, etc.

3.4.2 Advise potential receiving hospitals of hazardous material involved and possible number of patients involved.

3.4.3 Contact appropriate poison control center to receive detailed health implications of hazardous material involved and product-specific treatment protocols. The following may be required when calling the designated poison control center:

- The chemical name of the hazardous material
- Length of exposure
- State (i.e., gas, solid, or liquid) of hazardous material
- Route of introduction

3.5 Complete the donning of personal protective clothing and equipment if available and trained in its use.

4. Gain Access to the Patient(s)

4.1 Ambulatory patients (persons able to remove themselves from the "hot" zone).

Assume that anyone egressing the "hot" zone is contaminated. They should be treated as such until properly assessed and decontaminated.

4.1.1 Move these patients to, and contain them in, a controlled area at the perimeter of the "hot" zone.

4.1.2 Do not make physical contact with these persons until the proper level of personal protective clothing and equipment has been donned.

4.1.3 Move these personnel to the decontamination area in an organized fashion.

4.2 Non-ambulatory patients.

4.2.1 Attempt to remove these patients from the "hot" zone if the proper level of personal protection and personnel trained in their use are available.
4.2.2 Treatment in the "hot" zone should be limited to gross airway management, cervical spine immobilization, and control of obvious hemorrhage. No invasive procedures should be performed, as this would provide a direct route of introduction of the hazardous material into the patient.
4.2.3 Move patient to the decontaminated area.

5. Decontamination Procedures

5.1 Remove gross contaminants.
5.2 Remove all the contaminated clothing. Articles that remain on the patient and cannot be removed should be isolated from the environment.
5.3 Further decontamination should be completed based upon the patient's condition, environmental conditions, and resources available.
5.3.1 Take care not to introduce contaminants into open wounds.
5.3.2 Contain all runoff from decontamination procedures for proper disposal.
5.4 Isolate patient from the environment to prevent the spread of any remaining contaminants.
5.5 Transfer patient to a "clean, protected" crew for transport if resources are available.

6. Assessment of Patient(s)

6.1 Complete primary and secondary surveys as conditions allow. Bear in mind the product-specific information received from the designated poison control center.
6.2 In multiple patient situations, begin proper triage procedures.

7. Treatment Procedures

7.1 Treat presenting signs and symptoms as appropriate and when conditions allow.

7.2 Administer orders of "on-line" medical direction.

7.2.1 IV therapy should be administered only with physician direction.

7.2.2 Invasive procedures should be performed only in fully decontaminated areas where conditions permit. These procedures may create a direct route for introduction of the hazardous material into the patient.

7.2.3 Reassess the patient frequently, as many hazardous materials have latent physiological effects.

8. Transport to Hospital

8.1 Recontact receiving hospital.

8.1.1 Update on treatment provided and any other information received from appropriate poison control center.

8.1.2 Obtain specific instructions regarding entering the hospital.

8.2 Transport patient.

8.2.1 Land transport — Protect vehicle and equipment from contaminants.

8.2.2 Air transport — Is inappropriate for contaminated patients.

9. Transferring Responsibility for Patient to Hospital Personnel

9.1 Await direction from hospital personnel before entering hospital.

9.2 Assist hospital personnel with patient decontamination and treatment as requested.

9.3 Arrange for personal decontamination.

10. Decontamination Procedures

Arrange for decontamination of the following in accordance with information received from expert resources.

10.1 Personnel.

10.2 Emergency care equipment.

10.3 Vehicles.

11. Medical Follow-up of Personnel

All public safety personnel who came into close contact with the hazardous material should receive an appropriate medical examination, post incident, based upon information from the designated poison control center. This should be completed within 48 hours of the incident and compared with the findings of any recent, preincident examination. Personnel who routinely respond to hazardous materials emergencies should have periodic preincident examinations. Personnel should be advised of possible latent symptoms at the time of their exam.

Maryland Medical Toxic Gas (Smoke Inhalation) Field Protocol

1. Introduction

1.1 A specialty referral program for carbon monoxide/cyanide inhalation has been developed by the MIEMSS Department of Hyperbaric Medicine.

Exposure to these gases may be related to a fire scene and smoke inhalation or inhalation of exhaust fumes or some paint solvents.

It is not feasible to list the absolute triage for every possible toxic gas inhalation patient, nor to define on a statewide basis which patients should go directly from the field to the department of hyperbaric medicine at MIEMSS or to go first to a closer facility. Time, distance, weather, and proximity to MIEMSS are all factors that must be considered in making an individual patient decision based upon the patient assessment and mechanism of exposure.

The following guidelines are intended to aid in the decision making, with the understanding that appropriate consultation should be obtained if there is any question regarding appropriate referral, treatment, and/or transport. Consultation is particularly important if the nature of the toxic gas is known, e.g., carbon monoxide (CO), general smoke inhalation, cyanide gas, hydrogen sulfide (H_2S), or methylene chloride (CH_2Cl_2-dichloromethane), because exposure to these substances can be treated with hyperbaric oxygen therapy.

2. Candidates for Hyperbaric Medicine Referral

2.1 Toxic gas inhalation only.
2.1.1 Unconscious patients.
2.1.2 Incoherent, does not follow verbal commands.
2.1.3 Combative, does not follow verbal commands.

NOTE: All combative patients are to be transported via ambulance.

2.2 Burn patients with toxic gas inhalation.

2.2.1 Adult:

a. Patient 50 years of age or less who is unconscious, combative, and/or incoherent and who does not follow verbal commands with:

   1. 20% BSA burns or less (second and third degree)
   2. Above 20% BSA burns (second and third degree) should be referred to the appropriate burn center.

b. Patients over 50 years of age with burns (second and third degree) over greater than 10 percent BSA should also be referred to the appropriate burn center.

2.2.2 Pediatric:

a. Patient unconscious, combative, and/or incoherent, who does not follow verbal commands with:

   1. 10% BSA burns or less (second and third degree) if 10 years of age or younger
   2. 20% BSA burns or less (second and third degree) if over 10 years of age.

b. All other patients should be referred to the appropriate facility after consultation with the appropriate pediatric trauma center and the appropriate burn center.

3. Clinical Data and Referral

3.1 Prior to obtaining consultation, have as much of the following data as possible collected:

3.1.1 Age, race, and sex of the patient.

3.1.2 Vital signs: Pulse, BP, respirations, and breath sounds.

3.1.3 Nature of the injury: toxic gas inhalation or toxic gas inhalation with burns and/or other associated injuries.

3.1.4 Specifics to the exposure.

a. Type of fire, namely the combustible involved, e.g., PVC, acrylics, styrofoams, urethanes, wood, petroleum products

b. In cases other than fire-related, the type of gas involved

c. Type of occupancy (e.g., dwelling, vehicle)

d. Duration of exposure, when possible
e. Level of consciousness (utilize Glasgow coma scale) with clinical signs and symptoms
f. Prehospital care rendered.

NOTE: Once as much of the above information as possible has been obtained, radio contact with MIEMSS systems communication (SYSCOM) or other appropriate facility should be made and the information passed on. A copy of the ambulance report form should be transported with the patient if at all possible.

4. Advanced Life Support (ALS) Unit Responses

4.1 An ALS unit should be dispatched to all suspected toxic gas inhalation incidents and immediately requested when not dispatched initially.
4.2 Upon arrival of the ALS unit, blood samples are to be collected via closed vacutainer technique prior to starting the IV.
4.3 For conscious patients, informed consent must be obtained prior to drawing blood samples. In the case of unconscious patients, blood samples may be obtained at the time the IV infusion is started.
4.4 The blood sample will be placed in one 5-ml purple-top tube to be transported with the patient. The tube containing the blood sample should also contain the patient's name along with the date and time the sample was obtained.
4.5 SYSCOM will be notified whenever any unconscious toxic gas inhalation patient is taken to a facility other than the MIEMSS Department of Hyperbaric Medicine.

NOTE: Some concern has been expressed regarding the type of tube in which to place the blood sample. The main concern of blood analyzing facilities is that the blood remain uncoagulated. Therefore, if purple-top tubes are not immediately available, any tube (e.g., gray tops) containing an anticoagulant would suffice.

5. Treatment

5.1 Toxic gas inhalation only.
5.1.1 Remove patient from toxic environment.

5.1.2 Obtain vital signs every 5 minutes.

5.1.3 Administer oxygen, 50-100%.

5.1.4 Initiate cardiac monitoring.

5.1.5 IV.

a. For symptomatic or unconscious adult patients, D5W KVO (20-30 ml/hr)

b. For pediatric patients, if an IV is indicated, the infusion should be lactated Ringer's solution titrated to a systolic BP of 80 mmHg

c. For adult patients with burns or associated trauma, the IV should be lactated Ringer's solution titrated to a systolic BP of 100 mmHg

d. For pediatric patients with burns, the IV solution shall be lactated Ringer's solution KVO unless medical direction specifies otherwise.

5.1.6 MAST as indicated.

NOTE: Patients with respiratory burns should not have an esophageal airway inserted.

5.1.7 Due to the extreme toxicity and cumulative effects of certain by-products of combustion and/or incomplete combustion, any person suspected of toxic gas exposure who refuses treatment shall be advised of the possible consequences and advised against returning to the toxic environment. The necessary signature on the ambulance report form shall be obtained.

Index

Markings, 246, 248
Portable, see Intermodal tank containers
Underground storage, 530
Vent and flaring systems, 275
Target organ effects, 213-214, 267-268
Technical information, 593-594, see also
  Chemicals; Data bases; Hazards analysis;
  Material safety data sheets (MSDS)
  Emergency response plans, 789
  Medical, 391-392, 443-446, 910, 929,
    959
  Pesticides, 502
  Protective equipment, personal, 346-347,
    348-349
  Reference manuals, 258
  Response options, 346-348
  At scene, 899
  Transportation, 558-559
Technical information centers, 258-259, see
  also Poison control centers
Technical specialists, 442
  Logistics, 783
  Operations, 780
  Planning, 781
  References/information, 259, 801-802
Telephone emergency rosters, 574-577, 701,
  see also Hotlines
Telephone service, 445
Temperature
  Critical, 263-264
  Heat stress monitoring, 495-499, 941-
    942, 952
  Of product, 262
Termination procedures, 785-786
  Definition, 133, 381-382
  By EMS/HM Level I Responder, 389,
    411-412
  By EMS/HM Level II Responder, 417,
    432-433
  By incident commander, 324, 786, 787-
    789
  Reports, see Critiques
Tests
  Emergency response plans, 591-592
  Hazardous categorization, 806
  Hazards of chemicals, 516-518
  Splash-protective suits, 872-876, 879
  Vapor-protective suits, 872-876
Thermal protection, 51-53, 855

Thermal stress, see Heat stress monitoring
  (HSM)
Threshold limit value ceiling (TLV-C), 278
**Threshold limit value short-term exposure
  limit (TLV-STEL),** 278
**Threshold limit value time-weighted average
  (TLV-TWA),** 278, 420
Threshold planning quantity (TPQ), 604-
  605
**Title III SARA (Emergency Planning and
  Community Right-to-Know Act of 1986)**
  Chemical lists, 617
  Community right-to-know requirements,
    606-609, 611, 617
  Definition, 626
  Emergency notification, 602, 605-606,
    617
  Emergency planning, 527-529, 567, 603-
    605, 611
  Enforcement, 612
  Extremely hazardous substances, 617
  Implementation, 602-617
  Information flow requirements, 615-616
  Key dates, 614
  Material safety data sheets (MSDS), 506
  Preemption, 611
  Review of systems, 613
  Toxic chemical release reporting, 602,
    609-611, 617
  Trade secrets, 605, 611-612
  Training, 601, 612-613, 894
Toll-free telephone numbers, see Hotlines
Toluene
  Hazardous materials incidents, 504
  MSDS data, 508-514
  Protective clothing/equipment for, 489,
    492-493
Ton containers, 846-847
**Totally encapsulating chemical-protective
  suit (TECP),** see Vapor-protective cloth-
  ing
Tote bins, 342
Toxic chemicals
  Definition, 11, 120, 212-213, 376
  EPA chemical emissions inventory, 609-
    611
  Monitoring equipment, 345
  Routes of exposure to, 394
  SARA Title III chemical release reporting,
    602, 609-611, 617
Toxic gases, 964-967